Courtroom Use and Misuse of Mathematics, Physics and Finance

Courtroom Use and Misuse of Mathematics, Physics and Finance

Cases, Lessons and Materials

Ashley Saunders Lipson
Professor of Law
University of LaVerne College of Law

Carolina Academic Press
Durham, North Carolina

Library of Congress Cataloging-in-Publication Data

Lipson, Ashley S.
[Mathematics, physics and finance for the legal profession]
Courtroom use and misuse of mathematics, physics and finance : cases, lessons and materials / Ashley Saunders Lipson.
pages cm
Includes bibliographical references and index.
ISBN 978-1-61163-376-4 (alk. paper)
1. Evidence, Expert--United States. 2. Forensic statistics--United States. 3. Forensic accounting--United States. 4. Forensic sciences--United States. 5. Physics--Popular works. 6. Mathematics--Popular works. 7. Lawyers--United States--Handbooks, manuals, etc. I. Title.

KF8961.L57 2013
347.73'6--dc23

2013009773

Carolina Academic Press
700 Kent Street
Durham, North Carolina 27701
Telephone (919) 489-7486
Fax (919) 493-5668
www.cap-press.com

Printed in the United States of America

Dedicated to:
Rene Descartes, Sir Isaac Newton, Albert Einstein, and Cupcake

Contents

Preface

The subject matter can be deceiving, not with respect to content, which is quite clear from the title, but rather with respect to difficulty and tediousness. Moreover, the book is designed to take an important trio of crucial topics and spoon feed them as comfortably as possible to the people who need them the most but like them the least—law students. Many jokingly state that they became law students because they couldn't handle numbers in college. A rude awakening, however, confronts them in practice, where numbers become the primary focus of what they do.

As the first pedagogical book combining the subjects of Mathematics, Physics, Finance and Law, this treatise is built on the notion that lawyers are ill prepared to face the world dominated by numbers and the multitudes who know how to distort and misrepresent them. Thieves such as Bernie Madoff, who managed to deceive the entire legal staff of the Securities and Exchange Commission for decades, were not merely flukes.

Using a similarly named course ("*Mathematics, Physics and Numbers for Lawyers*"), I experimented with the cases and subjects in order to best determine how to capture and maintain the students' interest, while at the same time building confidence in skills that are otherwise foreign to law school curricula. Nearly every chapter of the book contains some material that lawyers must possess; some of the information is indispensable. Like it or not, we all live in a scientific world, a digital world—one that is ruled by numbers, equations, formulas and statistics.

The chapter headings and subtitles present some of the most complex and baffling subjects that confront today's scientists. Students are urged not to be afraid. The book provides the softest approaches possible.

Cases have been edited, some severely due to size constraints. Where it was to possible to do so seamlessly, I have omitted annoying ellipses, asterisks and "*citations omitted*" messages.

Acknowledgments

Thanks must be directed to Kenneth Settle, Eric Kelly, Deborah Zexter, Dean Emeritus Randall Rubin, Kevin Marshall, Josh Effron and Rea Belk, in addition to my lovely wife Anna and the students at the University of La Verne School of Law for their comments and assistance.

I also want to thank Linda Lacy, Chris Harrow, Tim Colton and the other outstanding employees of Carolina Academic Press for their continued assistance and valuable advice.

Courtroom Use and Misuse of Mathematics, Physics and Finance

Chapter 1

Introducing Numbers to the Legal Profession

"Uncovering the divine truth and beauty of Mathematics would be easier if numbers didn't get in the way."

—Anonymous

As current and former students of the law, many of us if not most, were mathematically abused children. Our earliest confrontations came from teachers assigned to cover several different elementary school subjects, without regard to preference. And it's safe to say that for many, arithmetic and physics were not among their favorites. As a result, they hammered us with the same maltreating number exercises that their predecessors inflicted upon them; abusive conduct, after all, is handed down.

These teachers may have been intelligent, skilled educators, but few if any were real mathematicians, physicists or scientists who devoted their lives to such matters. Their teaching credentials carried no such requirements. In fact, many probably inherited a fear and loathing for the subjects from their own teachers. The exciting mysteries and true value of the curricula, therefore, would have been lost upon them because it was lost upon those who taught them; and sadly, the loss would continue to fall upon successive generations.

Suffice it to say that our early instruction did not come from scientists who loved their subject and were thus able to convey the elegant truths, limitless power, and beauty of mathematics, not to mention the breathtaking mysteries of physics, astronomy and the many other disciplines that flow from them. This drought of high quality, professional instruction, extending all the way from elementary school through high school, no doubt contributed to the declining status of our students as compared to those of other industrialized countries.[1] We have definitely been heading in a negative direction.[2]

---

1. In math, our students lag others in 23 out of 30 industrialized countries. Glod, *U.S. Teens Trail Peers Around World on Math-Science Test*, The Washington Post, Wednesday, December 5, 2007. See also, Lee, *Math, Science, Reading Scores Show U.S. Schools Slipping Behind*, PBS Newshour, December 10, 2010 and Tomsho, *U.S. Math Scores Hit a Wall*, The Wall Street Journal, October 15, 2009. Currently, checking the Internet shows little change.

2. The following is not a college examination, nor a high school test. It was an exam characteristically given to 8th Graders in Salina, Kansas in 1895. Even if Problems 1 through 9 don't embarrass would-be lawyers, Problem 10 is certain to do so.

(1) Name and define the Fundamental Rules of Arithmetic.

(2) A wagon box is 2 ft. deep, 10 ft. long, and 3 ft. wide. How many bushels of wheat will it hold?

(3) A wheat load weighs 3,942 lbs.; what is it worth at 50cts/bushel, deducting 1,050 lbs. for tariffs?

(4) District No 33 has a valuation of $35,000. What is the necessary levy to carry on a school seven months at $50 per month, and have $104 for incidentals?

We never truly encountered professionals in their fields until college, but by then it was too late for most. Victims of abuse rarely seek more of the same; so for many of us, scientific subjects were avoided in college altogether. Those with number skills became accountants and investment bankers. The more traumatized victims among us opted for other non-mathematical endeavors.[3] At the graduate level, many sought the deceptive safe harbor of Law, believing the dangerous myth that lawyers could avoid mathematics, physics and numbers altogether. Nothing could be further from the truth.

Practicing law is at worst, a war, and at best, a game. Either way, it involves strategic thinking, planning, gamesmanship, and geometric logic. It is evident to every experienced practitioner that numbers play an indispensable role. As a lawyer, if you do not thoroughly understand *the numbers* and how they are used and misused, you are in no better position than the new poker player at the table who is unable to identify the patsy; and, as everyone knows, when you can't identify the patsy, you're most likely *it*.

There are countless ways that skillful lawyers use numbers, charts, statistics and equations to gain an upper hand during legitimate attorney interaction, without regard to actionable misconduct or unethical behavior. Sadly, however, like the victim of a perfect scam, the unknowledgeable attorney seldom knows that he or she has been taken advantage of. But despite our deficiencies, there is hope.

There is both bad and good news for those who thought that a career in law meant leaving mathematics, physics and accounting far behind. The bad news is that such a thought is clearly a myth; the good news is that anyone who survived the LSAT (Law School Admissions Test) should be able to master the necessary basics painlessly. In particular, because you achieved a high enough LSAT score to enter an ABA accredited law school, you are probably not as stupid as you think, even with respect to advanced math and physics. For, even though pure mathematics has been eliminated from the LSAT over the years, the substituted logic games assure that you have the requisite brain power.[4]

---

(5) Find the cost of 6,720 lbs. coal at $6.00 per ton.
(6) Find the interest of $512.60 for 8 months and 18 days at 7 percent.
(7) What is the cost of 40 boards 12 inches wide and 16 feet long at $20 per meter?
(8) Find the bank discount on $300 for 90 days (no grace) at 10 percent.
(9) What is the cost of a square farm at $15 per acre, the distance of which is 640 rods?
(10) Write a Bank Check, a Promissory Note, and a Receipt.

3. There was a time when Mathematics, and some basic science courses, which might have included Chemistry, Biology and Physics, were required for a college degree. Things have changed. Here is a list of actual courses that recently appeared on university transcripts that accompanied law school applications; unlike the parenthetical comments, the course titles are actual and unedited:

(1) *Mother Sun and Her Neighbors* (Kiss goodbye to father "Astronomy").
(2) *Reading the Newspaper*
(3) *How to Use YouTube*
(4) *Musical Political Protest*
(5) *Human Sexuality*
(6) *Wine Making*
(7) *Brain Repair* (Prerequisite—Metal Shop?)
(8) *Feelings and Emotions*
(9) *Self-Love*
(10) *Introduction to Marriage*
(11) *How to Take Notes* (Prerequisite must be—Possession of a good memory)
(12) *The Beatles*

4. The The American Bar Association (ABA) does not impose upon its member schools mandatory LSAT minimum scores. It does, however, prescribe standards for admission which normally include LSAT scores as a major admissions component.

Despite an un-admitted friendship, Law relies heavily upon the mathematical sciences at many levels, practical, logical, theoretical and ethical. The sciences, however, require no such reciprocity from Law. Where laws are permitted to develop haphazardly through stare decisis and political maneuvering, science utilizes rigorous definitions, axioms and postulates. Logic tends to count and there are no minority or dissenting opinions once the truth has been established. Law, on the other hand, provides a pretense of logical structure. Sometimes, the results are brilliant; but as we know, foolishness is also commonplace.

In *Civil Procedure I*, you might recall how confusing law professors were able to make an ultra-simple topological concept such as the ability of each of 50 sets (states) to exercise personal jurisdiction over the elements (persons) of other sets. Such a modest concept should have been subject to uniform legislative action initiated by congressional competence. Instead of Congress, however, the flag bearer for defining jurisdiction was a crooked, twice-bigamist, wife-deserting pedophile lawyer by the name of John Mitchell (who cheated on the wives he neglected to desert), assisted by a racist publisher named Sylvester Pennoyer and a judge who loathed them both.[5] No math or logic spoken there.

Did you ever wonder why our haphazard system of territorial jurisdiction never squarely addressed the single most significant factor in the confusing mix—i.e., the distance that a defendant would be required to travel to respond to a suit? Did the legal thinkers of the time truly believe that traveling from New York to California in the 1800s burdened due process to the same extent as going from Rhode Island to Connecticut? Instead of learning a simple, well written principle to define the exercise of territorial jurisdiction, we spent much of a law school term attempting to reconcile a stream of disordered and often unreconcilable decisions. Compare the volumes of judicially created confusion to the physicist's less-than-ten symbols for accurately describing the movement of every particle in the physically known Universe.[6]

Our not-so-civil procedure laws are by no means isolated examples of judicial fiat. In general, while jurists and law professors delight in taking simple matters and making them unnecessarily complex, scientists, on the other hand, take things that are very complex and make them simple. Having survived *Civil Procedure*, you probably already suspect this.

When the worlds of mathematics and law collide, as they inevitably do, the results are sometimes baffling and even entertaining. Other times, the consequences can appear to be incomprehensible. In any event, if there is a common point to be demonstrated, it is this—Those lawyers who lack at least a basic appreciation and understanding for the numerical sciences are often at a distinct disadvantage.

In this text, we scratch many surfaces, perhaps more deeply than one might assume. Three main goals come to mind: First, learning to recognize and guard against the most common deceptive tactics accompanying the misuse of mathematics, statistics, science and accounting. Second, gaining greater confidence in our own ability to solve those basic problems that we are most likely to confront the legal professional, and lastly, attaining a greater appreciation mathematics-based sciences.

---

5. *Pennoyer v. Neff*, 95 U.S. 714 (1877) formed the cornerstone of interstate jurisdiction when attorney John Mitchell improperly attached Neff's land and sold it to Sylvester Pennoyer (whom Judge Deady referred to as "Slypester the Annoyer"). Sylvester would later become the governor of Oregon. Many years later *International Shoe v. Washington*, 326 U.S. 310 (1945) and its progeny took up the banner of confusion and carried it into the present.

6. Schrodinger's Equation (a fundamental tenet of quantum mechanics) will be very briefly addressed later in this text.

§ 1.10 Numbers for the Lay Public

Everything in the Universe is expressible in numbers. Not just time, distance, force and other concepts associated with science, but everything. That includes color, light, and smell. Even the thoughts and emotions that run through our brains are expressible in terms of electrons, axons and neurons, which in turn are reducible to atoms and subatomic particles which, in turn are predicated solely upon the *number* of electrons, neutrons, protons and quarks that comprise them. For us lawyers, numbers are no less significant.

In contrast to lay persons, few practitioners actually fail to recognize the importance of numbers and equations. Most rationalize a lack of knowledge with a false belief that experts and accountants can pick up the slack. They cannot. What we don't know *can* hurt. Lost opportunities, or worse, falling victim to the schemes of those who *do* have a command of the subject matter is quite common. Many topics throughout this text are devoted to recognizing those schemes.

§ 1.20 Numbers for the Judiciary

It is not possible to list the circumstances where numbers have played a role in determining the outcome of legal disputes, whether civil or criminal. Even when we are not consciously aware of it, numbers and physics often play an important role in the courtroom. There are many arithmetical and mathematical principles that we accept and employ without discussion or consideration. But just as an opposing witness begins his testimony by answering benign foundational questions before insidiously moving on to speculation and hearsay, so too, are our most trusted truths (math and physics) often "twisted by knaves to make a trap for fools."[7] The lawyer without the skills required to recognize a slight of hand in the shell game of numbers is no less detrimental to his client than the trial attorney who lacks evidence skills. With that in mind, there are two basic gatekeeping functions to regulate the manner in which science and its accompanying numbers are presented to the trier of fact:

§ *1.21 Judicial Notice*

From an admissibility perspective, the laws of mathematics and physics seem to occupy a special place in the courtroom. Subject to judicial notice, the judge may accept those laws without requiring expert testimony. Judicial notice is a substitute for evidence, not a supplement. Rule 201(b) of the Federal Rules of Evidence provide that:

> The court may judicially notice a fact that is not subject to reasonable dispute because it: (1) is generally known within the court's territorial jurisdiction; or (2) can be determined from sources whose accuracy cannot reasonably be questioned.

Not all scientific principles enjoy a *judicial notice* status. There was once a relatively clear dividing line between established scientific doctrine (such as physics and mathematics, sub-

---

7. Apologies to Rudyard Kipling (1865–1936), from the poem "*If*."

jects once thought to be unassailable) and other, newer quasi-scientific spin-offs. Today, however, more matters are subject to closer scrutiny, even our most trusted laws. Perhaps, this increased doubt constitutes a positive influence on the relationship between law and science. It is interesting to note that for many years, the population generally referred to Einstein's *theory* of relativity, which had long since been proven to be a correct *law* as opposed to a mere theory. On the other hand, we still refer to "Newton's *Laws* of Motion," which, without Einstein's modifications, are clearly *incorrect*. Close inspection, therefore, can never hurt.

Despite its shortcomings, the word "science" carries enormous weight with a jury and therefore warrants close scrutiny for anything pretending to deserve such a label. Scrutiny often comes in the shape of three formidable landmarks (considered *infra*): *Frye v. United States*, 293 F. 1013 (D.C. Cir., 1923), *Daubert v. Dow*, 509 U.S. 579 (1993), and *Kumho Tire Company, Ltd. v. Carmichael*, 526 U.S. 137 (1999), in addition to the widely accepted Rule 702 of the Federal Rules of Evidence and its state counterparts. In the face of these hurdles, however, courts are still likely to continue to place mathematics and physics on a pedestal. So, despite the new wave of suspicion, even the leader of the Argus-eyed trio acknowledged:[8]

> Indeed, theories that are so firmly established as to have attained the status of scientific law, such as the laws of thermodynamics, properly are subject to judicial notice under Federal Rule of Evidence 201 (pertaining to Judicial Notice).

Courts must take judicial notice of the elementary laws of physics. *Elwood v. City of New York*, 450 F.Supp. 846 (D.C.N.Y., 1978); *Belcher v. Buesking*, 175 Ind.App. 322, 371 N.E.2d 417 (1978); *Surman v. Ohio & Pennsylvania Oil & Gasoline Co.*, 116 Ohio App. 453, 183 N.E.2d 386 (1962); *Prince v. Bennett*, 322 S.W.2d 886 (Mo., 1959); *Pendley v. Union Bankers Ins. Co.*, 99 Ga.App. 189, 107 S.E.2d 910 (1959); *Alldread v. Mills*, 211 Ark. 99, 199 S.W.2d 571 (1947); see also Am.Jur Evidence, § 98. Appellate courts, upon review of a judgment, will not hesitate to take judicial notice of unquestioned laws of mathematics, and where the trial court has failed to recognize such laws and has entered a judgment supported by some evidence which is so far at variance with undisputed physical facts as to demonstrate that the evidence is false and verdict unjust, that judgment will be set aside. *Winterberg v. Thomas*, 126 Colo. 60, 246 P.2d 1058 (1952).

Courts will most likely continue to take judicial notice of the laws of the so-called hard sciences. *Boeing Company v. U.S.*, 75 Fed.Cl. 34 (10th Cir., 2007); *Commonwealth v. Bakke*, 46 Va.App. 508, 620 S.E.2d 107 (2005); *State v. Sturgill*, 1991 WL 15990 (Ohio App., 1991). Parties should, however, always be provided with an opportunity to challenge judicial notice, even with respect to the laws of physics and mathematics. *Beasley v. U.S. Welding Service, Inc.*, 129 Fed.Appx. 901, (5th Cir., La., 2005); *Counihan v. Azadeh*, 1992 WL 47975 (S.D.N.Y., 1992).

Lawyers must have some threshold knowledge of the scientific principles that they intend to articulate when attempting to impress the court. See *Schmitt v. Mertel*, 208 Ill.App.3d 741, 567 N.E.2d 594 (1991) chiding: "The plaintiff invokes Newtonian laws of physics, without specifying which laws, to establish a 'fact' upon which her claim depends. We take judicial notice that there are three Newton's Laws of Motion."

8. *Daubert v. Dow*, 509 U.S. 579 (1993), Fn. 11.

Alldread v. Mills

211 Ark. 99, 199 S.W.2d 571 (1947)

McFADDIN, Justice.

This appeal stems from a motor vehicle collision. Appellee was plaintiff in the trial court, and was awarded a verdict and judgment for $4,000. From an unavailing motion for new trial, the appellant (defendant below) brings this appeal, urging here for reversal the six assignments herein listed and discussed. For convenience, we refer to the parties as they were styled in the trial court.

First Assignment: 'The verdict and judgment are contrary to clearly established and uncontroverted physical facts in the case.'

Shortly after dark on Saturday night, October 28, 1944, the plaintiff, George Mills, accompanied by his wife and two other persons, was driving his Ford sedan northerly towards Monticello, and the defendant's servant (Monroe Franklin) was driving defendant's 1½ ton Ford truck southerly from Moticello. The two vehicles collided on State Highway No. 81, and plaintiff received injuries and damages, the nature and extent of which will be mentioned in the discussion of Assignment No. 6. At the place where the collision occurred the road is practically straight, and a heavy black line indicates the center of the road. The plaintiff was attempting to drive on the east side, and the defendant on the west side. The hard-surface highway is 18 feet and 4 inches wide.

Plaintiff's witnesses testified that the plaintiff was on the east side, and that the defendant's truck crossed the center line and caused the collision. The defendant's witnesses testified that the defendant's truck was on the west side, and that the plaintiff's car crossed the center line and caused the collision. The cars did not hit head-on; but the left front of each vehicle received the impact, and each vehicle continued some distance before coming to a stop.

Plaintiff's car had the left front wheel and fender damaged, and the left front door crushed. From the collision point, the car angled to the west (left) side of the road before stopping in the ditch about 50 yards from the collision. The defendant's truck had the left front wheel torn from the axle, and knocked 'back up under' the truck, with the front or running part of the wheel turned out. Also, there was damage to the body of the truck on the left side just back of the cab door. The truck traveled about 50 yards southwest from the collision. The back end skidded to the right, and the truck stopped with the rear end in the ditch on the west side of the highway. As the truck went down the highway, the rubber from the front wheels left skid marks on the pavement, which skid marks remained visible for several weeks.

It is conceded by the defendant that there is sufficient evidence to sustain the verdict of the jury if the physical facts be ignored, but it is most earnestly contended that three physical facts establish the defendant's case and necessitate reversal of the jury's verdict. Learned counsel for defendant invokes the rule that a verdict will be set aside if it be against incontrovertible physical facts. [Citations omitted]. To these cases might well be added the following authorities: 3 Am.Juris. 451; 4 C.J. 861; 5 C.J.S., Appeal and Error, § 1649, p. 631 and note p. 640; 46 C.J. 183; and the annotation in 21 A.L.R. 141 on 'Evidence contrary to scientific principles or laws of nature.' Mr. Justice Hart, in St. Louis S. W. Ry. Co. v. Ellenwood, 123 Ark. 428, 185 S.W. 770, stated the rule in this language:

> Appellate courts take notice of the unquestioned laws of nature, of mathematics, of mechanics, and of physics. So, where there are undisputed facts shown in the evidence, and by applying to them the well-known laws of nature, or math-

> ematics, and the like it is demonstrated beyond controversy that the verdict is based, upon what is untrue and what cannot be true, this court will declare as a matter of law that the testimony is not legally sufficient to warrant the verdict.

With this rule in mind, we proceed to examine the physical facts which defendant says (a) are incontrovertible; and (b) prove beyond controversy that the collision must have occurred on the west side of the center line of the highway. These physical facts are:

(1) The skid marks on the pavement;

(2) the condition of each vehicle after the collision; and

(3) the final point of rest of each vehicle.

(1) The skid marks. As previously stated, definite skid marks were made on the pavement by the front wheels of the defendant's truck; and these marks began about ten inches west of the center line, and continued with the movement of the truck. Defendant argues that the collision happened where the skid marks began; and that, since the skid marks began on the west side of the center line, therefore, the defendant's driver was not on the east or wrong side of the road. But several witnesses testified that there was dirt (caused by the collision) on the east side of the center line; and that the truck tire skidded in the dirt for some distance before starting to make the rubber skid marks on the pavement. Also, witnesses testified that the rubber tires of the truck did not make the skid marks until the front wheels became locked by being driven west across the center line. As typical of the plaintiff's evidence on this point, we quote a portion of the testimony of Dempsey Polk, city marshall of Monticello:

Q. From the signs made from this on your investigation, tell the jury whether or not Mr. Mills' car was across the center line when the collision occurred, or whether or not the truck was on Mr. Mills' side when it occurred.

A. The track made over here where this track started in the dirt, knocked off this car, was within six inches of the edge here and gradally went on across.

Q. Then the truck according to your testimony is bound to have hit this car on the east side of the center line of the highway?

A. That's right.

It is thus made to appear that the plaintiff's witnesses claimed (a) that the skid marks began only after the wheels left the dirt, which had fallen from each vehicle at the point of collision; and (b) that the beginning of the rubber skid marks did not indicate the point of collision. If this evidence be true—and that was for the jury—the skid marks did not establish incontrovertible evidence as to the point of collision. The fixing of that point was still left to the evidence of witnesses. The jury took the plaintiff's version, and we cannot say that the jury's verdict was wrong as a matter of law.

(2) The condition of each vehicle after the collision. We have previously detailed the damage to each vehicle. Defendant submits that the marks of impact and the resulting physical conditions of the vehicles prove that the collision occurred west of the center line. But we cannot agree. The same marks of impact, and the same resulting physical conditions could have occurred, regardless of which side of the center line was the point of the collision.

(3) The final point of rest of each vehicle. Since the car and the truck both stopped in the ditch on the west side of the road, defendant argues that the car, being the lighter vehicle, must have been headed west before the collision, and merely 'sideswiped' the truck. This argument is reinforced by the fact that the left front headlight of the truck was not burning before the collision, and the plaintiff thought—until a moment before the collision—the it was the right front headlight of the truck that was not burning.

But we cannot say that the course traveled by the vehicles from the collision to the final point of rest furnishes 'incontrovertible physical facts' that the collision occurred on the west side of the center line. The Missouri Court of Appeals, in Lang v. Missouri Pac. R. Co., 115 Mo.App. 489, 91 S.W. 1012, 1013, uses this language, which expresses our views here:

> 'So frequently do unlooked for results attend the meeting of interacting forces that courts, in such cases, should not indulge in arbitrary deductions from physical law and fact, except when they appear to be so clear and irrefutable that no room is left for the entertainment, by reasonable minds, of any other.'

Regarding the defendant's first assignment, about incontrovertible physical facts in this case, it is not for us to substitute our conclusions for those of the jury, unless the physical facts demonstrate beyond a doubt that the verdict was erroneous. We cannot so declare in this case. We take the view, so well expressed by the Missouri Court of Appeals in Stokes v. Metropolitan St. R. Co., 173 Mo.App. 676, 160 S.W. 46, 47:

> We are asked to reverse the judgment on the ground that plaintiff's version of her injury is so contrary to physical law and so incredible that it should not be accorded any probative value. We would not be justified in treating as substantial that which has no substance, in committing the solecism of holding in effect that testimony might be true which common experience and common knowledge of physical laws would reject as palpably false; and should we find the evidence of plaintiff 'is so contrary to the daily experience of common life, so at war with the conceded physical fact', as to be beyond reasonable belief, we would not hesitate to brush it aside as wholly valueless.
>
> On the other hand, it is our duty, as an appellate tribunal, to exercise great caution in applying the tests of common sense and common knowledge of physical law to a given state of facts. The testimony of unimpeached witnesses should not be lightly waved aside as impossible or incredible. Common experience and observation teach us that strange and astonishing things sometimes happen in the physical world, and it would not do to give to dogmatic and undemonstrated conclusions, respecting natural laws, precedence over the testimony of apparently credible witnesses.

We adopt as our own the words of the Missouri Supreme Court in Dempsey v. Horton, 337 Mo. 379, 84 S.W.2d 621, 624:

> Only in those extraordinary cases where deductions from physical laws and facts so clearly and irrefutably disprove the testimony of witnesses that reasonable minds may not entertain any other conclusion, are courts justified in ruling sworn testimony inherently unbelievable and impossible.

[Discussions pertaining to the 2nd through the 6th assignments of error are omitted.]

On the whole case we find no error, and the judgment is affirmed.

Hunter v. New York, Ontario & W. Ry. Co.

New York Court of Appeals, 116 N.Y. 615 (1889)

In an action to recover damages for injuries alleged to have been caused by defendant's negligence to plaintiff, a brakeman in its employ, plaintiff's evidence was to the effect that he was instructed to ride on top of the cars at places where it might be necessary to apply the brakes, and that on approaching such a point he went on top of a box car and sat down where the brake was and remembers nothing more. It appeared that a tunnel is located at that particular spot on defendant's road, which for two hundred feet from its west end, at which plaintiff's train entered, is twenty feet high; that then a brick arch commences and continues for eighty-five feet, reducing the height of the tunnel to fifteen feet nine inches, measured from the rail, and the space between the top of the car and the bottom of the arch was four feet and seven inches. Plaintiff was found beside the track, about three hundred feet from the east end of the tunnel, with a gash in his forehead, which, if received from having come in contract with the arch while sitting as he testified, would have called for a height of body of four feet eight inches, or a full height of over eight feet.

No evidence was given as to plaintiff's height. The court charged the jury that if plaintiff was sitting down it was for them to say whether his head would reach as high as the bottom of the arch. A motion to dismiss the complaint, made on the ground that the facts proved were insufficient to constitute a cause of action, was denied. Held (BRADLEY and VANN, JJ., dissenting), error; that the court will take judicial notice of the height of the human body and that it would be impossible for a man, unless of a height unprecedented, sitting as plaintiff testified, to strike his head against the arch.

Courts may take judicial notice of facts which are a part of the general knowledge of the country and which are generally known and have been duly authenticated in repositories of facts open to all, and especially so of facts of official, scientific or historical character, and within this rule the court may take judicial notice of the size and height of the human body.

While all reasonable intendments should be indulged in to support a judgment, the court will not assume the existence of a fact which is contrary to the ordinary course of nature, solely because the party raising the question did not give evidence in regard thereto upon the trial.

APPEAL from judgment of the General Term of the Supreme Court in the second judicial department, entered upon an order made December 14, 1886, which affirmed a judgment in favor of plaintiff, entered upon a verdict.

This action was brought to recover damages for injuries received by plaintiff while in defendant's employ as a brakeman. The facts are sufficiently stated in the opinion.

William Vanamee for appellant. There was no proof which warranted the submission of the question of defendant's negligence to the jury, and the motion for a nonsuit should have been granted (most citations omitted). The physician having been called in to treat and amputate the plaintiff's leg, and having acted solely in that capacity, could testify only as to such information as is necessary to enable him to act in that capacity.

The motion for a nonsuit should have been granted because the plaintiff wholly failed to show that he was free from contributory negligence. The burden is upon the plaintiff to establish affirmatively the absence of contributory negligence. Plaintiff's failure to descend in the first case, or the act of ascending in the second case, constitutes contributory negligence.

Railroad employees are presumed to understand the nature and hazards of the employment when they engage in the service, and they assume all the ordinary risks and obvious perils incident thereto. Such risks are presumably within the employee's contract of service. The tunnel arch was a necessary and proper structure, and had been constructed upwards of seven years. The duty of the master is fulfilled when he provides safe structures, such as a prudent person would furnish for his own use, and gives the warnings and signals designated by law. It is the duty of a master to provide his servants with suitable and safe structures, machinery, track and road bed, other materials and appliances necessary for the proposed work.

Plaintiff had the right to assume that defendant had exercised due care in that respect and that he would not be unduly exposed. The defendant was negligent in not informing plaintiff of the arching built in the tunnel. This the plaintiff could not know of by going through the tunnel except by some such experience as he had, at the risk of his life. The servant, by accepting his employment, assumes the usual risks and perils incident to the use of the machinery and property of the defendant, so far as the risks are apparent. The negligence of the conductor or other employees of the defendant in not informing plaintiff of the arching that had been built in this tunnel will not, as defendant insisted at the trial and before the General Term, exonerate the defendant from liability. (Cone v. D. L. & W. R. R. Co., 81 N. Y. 207; Wilds v. H. R. R. R. Co., 23 How. 492; Booth v. B. & A. R. R. Co., 73 N. Y. 18.) The court committed no error in excluding the evidence of the physician. The charge of the trial court was eminently fair to defendant, and defendant has been justly defeated on every point. The jury saw the plaintiff, saw how tall he was, saw all the witnesses and heard their evidence, and where inferences are to be drawn by a jury from facts and circumstances, submitted for their consideration, the appellate court will, with great reluctance, if ever, interfere with their determination. (Naugatuck Co. v. Babcock, 22 Hun, 481; Rounds v. D. L. & W. R. R. Co., 64 N. Y. 138; Cothan v. Collins, 29 How. 155; Head v. Smith, 44 id. 476; Dollenr v. Leitz, 9 Daly, 17.)

BROWN, J.

Assuming that the plaintiff was struck upon the head by the brick arch within the tunnel, and that he was, as a result of that blow, thrown from the cars and injured, I think there was ample evidence for the jury to determine that the defendant was guilty of neglect producing the accident, and that the plaintiff was free from carelessness contributing to it. The jury were warranted in finding that the only notice that the plaintiff had of the existence of the arch was that received from the tell-tale. This was located about two hundred feet west of the west entrance of the tunnel. It served as a warning of the approach to the tunnel, but it gave no notice of the obstruction within the tunnel. A person receiving its warning and noticing the height of the tunnel might naturally suppose that the height at the entrance would be maintained throughout its length, and if the height was at any point reduced that notice of that fact would be given. Relying, therefore, upon what would be apparent to his observation, he was exposed to a danger of which he had no notice or information.

The defendant evidently perceived the weakness of this part of its case and gave evidence of actual notice of the existence of the arch to plaintiff; but upon that question the verdict of the jury is in accord with the plaintiff's testimony. The plaintiff also testified that he was instructed to ride upon the top of the cars at places where it might be necessary to apply the brakes. He was approaching such a point on the road when the accident happened, and it was a fact for the jury to determine whether, in going from the top of the car at the time he did, he was guilty of any negligence. Certainly, if he had a right to assume, as we think he had, in the absence of notice as to the existence of the arch, that there was room in the tunnel to maintain a standing position, he would not necessarily be careless in occupying such a position on the car near his post of duty. The difficult question in the case is to reconcile the plaintiff's theory of the accident with the evidence.

It appears that the tunnel at the west entrance was twenty feet high. Two hundred feet from the entrance the brick arch began and continued for a distance of eighty-five feet. It reduced the height of the tunnel to fifteen feet nine inches, measured from the rail. The plaintiff testified that he left the engine and went on the top of a box car and sat down. His exact words are: 'I was sitting down on the box-car, on the head box-car, where the brake was; I went out on the box-car and sat down, and that is the last I remember.'

This car was identified and proven by a witness called by plaintiff to have been eleven feet two inches high from the rail to the foot-board. that the men walk on in the centre of the roof. There was, therefore, a space of four feet and seven inches between the top of the car where plaintiff was sitting and the bottom of the arch. There was a cut or gash on plaintiff's forehead, which it is claimed he received by coming in contact with the arch, so that his head, to have received a blow at that point, must have been at least four feet eight inches above the foot-board on the top of the car.

The plaintiff's counsel in his brief has endeavored to show the space between the car and the arch to have been much less, but the measurements appear, without contradiction, to be those I have stated. The theory of the plaintiff's case was that he was rendered unconscious by a blow received on his head from coming in contact with the arch, and that he was carried by the train to a point about three hundred feet east of the east end of the arch, where he fell from the car and was run over and his foot crushed. That he might have been carried to the point where he was found is not improbable, but that he could have received a blow on the head from the arch while sitting upon the top of the car would appear to be physically impossible.

There was no evidence given on the trial as to the plaintiff's size or height, and the argument is now made that as the jury saw him, and could, therefore, judge of his size, it must be assumed that it was not impossible for his head to have reached as high as the arch; and the learned judge who presided at the trial appears to have submitted this question to the jury saying: 'If the plaintiff was sitting down, it is for you to say whether his head would reach to that height.'

The verdict of the jury rests upon an affirmative answer to this question, and we are now called upon to say whether we will accept that finding and sustain the judgment, or whether we will take judicial notice of the height of the human body and the measurements of its separate parts, and so taking notice of those facts reverse a judgment that is based upon a finding clearly contrary to the laws of nature.

No exception was taken to the charge, but by an exception to the denial of the motion to dismiss the complaint made at the close of the plaintiff's case, and renewed at the close of the testimony, on the ground that the facts proven were insufficient to constitute a cause of action, the question is properly before this court.

Courts are not bound to take judicial notice of matters of fact. Whether they will do so or not depends on the nature of the subject, the issue involved and the apparent justice of the case.

The rule that permits a court to do so is of practical value in the law of appeal, where the evidence is clearly insufficient to support the judgment. In such case judicial notice may be taken of facts which are a part of the general knowledge of the country, and which are generally known and have been duly authenticated in repositories of facts open to all, and especially so of facts of official, scientific or historical character.

Thus it has been held that courts will take judicial notice of matters of public history, such as the existence of the late civil war and the particular acts which led to it. (Swinnerton v. Columbian Ins. Co., 37 N. Y. 174; Woods v. Wilder, 43 id. 164.) Of the general course of business in a community, including the universal practice of the banks. (Nat. Bk. v. Hall, 83 N. Y. 338; Yerkes v. Nat. Bk., 69 id. 382, 387.) That books of general record giving descriptions and standing of all ships, known as 'American Lloyds,' 'The Green Book' and 'The Record Book,' are referred to by business men for the purpose of ascertaining the condition, capacity and value of ships. (Slocovich v. Orient Mutual Ins. Co., 108 N. Y. 56, 63.) Of the value of 'pounds' in our money, and in rendering judgment convert them into dollars. (Johnston v. Hedden, 2 Johns. Cas. 274.) Of the expectation of human life. (Johnson v. Hudson R. R. Co., 6 Duer, 634.) Of the course of seasons and husbandry, and the general course of agriculture, and that a crop at a certain date would not have matured. (Ross v. Boswell, 60 Ind. 235; Floyd v. Ricks, 14 Ark. 286.) Of the time of the rising and setting of the sun and moon. (Case v. Perew, 46 Hun, 57.)

And generally of those things which happen according to the ordinary course of nature; the course of time and the movements of the heavenly bodies, the coincidence of the days of the week with the days of the month; ordinary public fasts and festivals and legal weights and measures. (1 Greenl. on Ev. chap. 11, § 5). And to ascertain such well-known facts recourse is had to such documents and references as are worthy of confidence.

Within this rule the court may take notice of the size and height of the human frame, and, doing so, we know that the plaintiff's head could not have reached to a height sufficient to come in contact with the arch.

We know that the average height of man is less than six feet. That the average length of the body from the lower end of the spine to the top of the head is less than thirty-six inches. That the measurement varies but little in adults, and that the chief difference in the height of men is in the length of their lower limbs. To assume, therefore, as we must, in order to sustain the judgment, that the top of the plaintiff's head, when in a sitting position, was four feet seven inches above the board on which he was sitting, is to assume him to have been not only far above the average height of man, but of a height beyond that of which we have any authentic record.

It has not been claimed by the respondent that the plaintiff was a man of extraordinary height, and if he was, I think a fact so rare in the course of nature should be made apparent in some way on the record.

It can be asserted, I think, without contradiction, that a man whose forehead would be four feet seven inches above a seat upon which he was sitting would have a frame at least nine feet high. History affords no authenticated instance of men attaining such height. Buffon, in his Natural History, records instances of men attaining extraordinary height, but modern writers do not accept his statements. Pliny tells of an Arabian nine feet high, but the story is not authenticated.

In the article upon 'Giants' in the Encyclopedia Britannica, it is stated that the tallest man whose stature has been authentically reported was Frederick the Great's Scotch giant, who was eight feet three inches tall. In the College of Surgeons in London there is a skeleton of an Irishman, who was named Charles Bierne, which measures eight feet high. Such heights are of rare occurrence, and the height of nine feet has probably never been attained by man.

Suppose the proof had shown that, upon approaching the entrance to the tunnel the plaintiff was standing up and his body had been found between the entrance and the west end of the arch, would it be assumed that his head had struck the roof of the tunnel, which would have been eight feet ten inches above the top of the car. In other words, would the court, to sustain the judgment, assume him to have been over eight feet and ten inches in height. Yet that assumption would call for no greater exercise of the imagination than to suppose his head to have reached the bottom of the arch when he was in a sitting posture. To assume either fact requires us to believe that the plaintiff was nearly, if not fully, nine feet in height.

I think, therefore, the court may take judicial notice of the fact that a man could not strike his head against an obstruction four feet and seven inches above the place on which he was sitting, and that being so, the negligence of the defendant was not established.

In no other way or manner is it suggested that the defendant was negligent, except in the maintenance of the arch and the failure to warn the plaintiff of its existence. Unless the plaintiff's injuries can, by a preponderance of evidence, be attributed to those causes, his case must fail. Unless the proof shows that he struck his head against the arch, the judgment can only be sustained on pure speculation. There are numerous ways in which the accident might have happened, but none other have any support in the testimony, and if the case is left in such a condition that it is just as possible the injury came from one cause as another, the judgment must be reversed. (Taylor v. City of Yonkers, 105 N. Y. 202.)

It has been said that an appellate court will not take judicial notice of facts not proven on the trial for the purpose of reversing a judgment. While all reasonable intendments should be indulged in to support a judgment, the court is not called upon to assume the existence of a fact which is contrary to the ordinary course of nature, solely because the party raising the question did not give oral testimony upon it at the trial.

In a case like this, in which it is well known that it can be submitted to a jury with generally but one result, the judgment should not be upheld when it is apparent that the verdict is not supported by the evidence.

Here the finding, which must exist to support the judgment, is so contrary to our general knowledge, and so far outside of common occurrence, that it may, in the absence of further proof, be regarded as contrary to nature, and hence untrue, and substantial justice will be done by reversing the judgment and granting a new trial.

Upon such trial, if the plaintiff was a giant in stature, or if, as claimed by the learned counsel for respondent, the space above the car was less than I have stated, such facts may be made clear.

The judgment should be reversed, and a new trial granted, costs to abide event.

FOLLETT, Ch. J., POTTER and PARKER, JJ., concur;

BRADLEY and VANN, JJ., dissent for the reason that the ground upon which Judge BROWN founds his conclusion was not specifically raised at the trial, and it does not necessarily appear that it might not have been obviated if it had been so raised there; and that, in such case, judicial notice of a fact, upon which no evidence was given or point made on the trial, should not be taken for the purpose of reversing a judgment;

Judgment reversed.

Notes and Problems

1. Can you distinguish *Hunter* from *Alldread?*

2. The *Hunter* case was decided in 1889, when statistically, people were much shorter. Would the New York Court of Appeals render the same decision today?

3. In *Alldread,* what is your reaction to the statement by the court: "It is conceded by the defendant that there is sufficient evidence to sustain the verdict of the jury *if the physical facts be ignored* ..."(Emphasis supplied)?

4. Sections 9(1) and 9(2) of The Uniform Evidence Code contained the language "propositions of generalized knowledge," which was not carried over into the federal rules. For purposes of judicial notice, how would you distinguish between those principles of mathematics and physics that are *generalized knowledge* and those that are accurate but not *generalized?* Principles of quantum mechanics are recognized as "accurate," though not widely understood.

5. Although judicial notice is a substitute for evidence (not a supplement), it should *never* be assumed that the acceptance of any given fact is *automatic.* Counsel should always be prepared to provide documentary information (dictionaries, encyclopedias, recognized scientific literature, etc.) to the court in support of a judicial notice request. Often an attorney seeking judicial notice must provide information to the court, even for very basic propositions. See *Carley v. Wheeled Coach*, 991 F.2d 1117 (3rd Cir., Virgin Islands,1993), a products liability action by a medic who was injured when the ambulance in which she was riding rolled over; the district court was held to have erred when it took judicial notice of the simple fact that motor vehicles with high centers of gravity have a greater tendency to roll over. The appellate court curiously reasoned that most people know little, if anything about how high centers of gravity cause vehicular accidents; therefore, the fact was not "readily ascertainable," as required by Rule 201(b) of the Federal Rules of Evidence.

6. Not every physical movement is explainable; that fact in and of itself should, perhaps, be subject to judicial notice, particularly when there is a violent impact caused by a moving body. For the "*Freakish Effects Doctrine,*" see *Spiro v. St. Louis Transit Co.*, 102 Mo.App. 250, 76 N.W. 684 (1903). The *Stokes* case (cited in *Alldread*) noted: "Common experience and observation teach us that strange and astonishing things sometimes happen in the physical world ..." As a known and accepted tenet of quantum physics, subatomic particles frequently disappear and then suddenly appear somewhere else almost instantaneously (ergo, the term "quantum leap"). This means that the probability of a set of particles leaping together is greater than zero. Therefore, it's possible though not probable that you might suddenly find yourself in Las Vegas in the next split-second. When quantum physics is involved, nothing is certain. (Even that which *stays* in Las Vegas may not *stay* in Las Vegas.)

7. For a southern perspective on the subject, see *St. Louis Southwestern Ry. Co. V. Ellenwood*, 123 Ark. 428, 185 S.W. 768 (1916): "Appellate courts take notice of the unquestioned laws of nature, of mathematics, of mechanics, and of physics. So, where there are undisputed facts shown in the evidence, and by applying to them the well-known laws of nature, of mathematics, and the like it is demonstrated beyond controversy that the verdict is based upon what is untrue and what cannot be true, this court will declare as a matter of law that the testimony is not legally sufficient to warrant the verdict."

8. In addition to the basic laws of Newtonian Physics, appellate and trial courts have taken judicial notice of some interesting facts including these: "A banana tree holds water," "electricity is dangerous," "the stock market fluctuates," "waves will fill in a hole that is dug on the beach," and "not everything that a cow eats turns into milk."

Schenk v. Lane

Court of Appeals of Tennessee, 1999 WL 58601 (1999)

FARMER, J.

Defendant Raymond D. Lane appeals a jury verdict awarding $297,000.00 to Plaintiff Pamela L. Schenk for injuries sustained as a result of an automobile accident occurring between Lane and Schenk. For the reasons set forth below, we affirm in all respects.

Factual and Procedural History

On September 21, 1988, Schenk and Lane were involved in an automobile accident on Mt. Juliet Road in Wilson County, Tennessee occurring between 6:30 and 7:00 p.m. Where the collision occurred, Mt. Juliet Road is a five lane highway with two northbound lanes, two southbound lanes, and one center turning lane. Schenk was driving her Chevrolet Cavalier north on Mt. Juliet Road. Lane was driving his Chevrolet pickup truck east on Interstate 40 and had just taken the Mt. Juliet Road Exit. Lane stopped at the stop sign at the end of the exit ramp and then proceeded to make a left turn across the two northbound lanes of Mt. Juliet Road. After noticing Lane's vehicle, Schenk slammed on her brakes and changed to the outer northbound lane. Schenk was able to avoid Lane's truck but collided with the eighteen foot trailer that Lane was pulling behind the truck. The accident was witnessed by Jimmy Chafin, who was traveling northbound on Mt. Juliet Road immediately behind Schenk and John Steele, who was exiting I-40 at Mt. Juliet Road behind Lane. Schenk testified that, at the time of the impact, Lane's truck and trailer were both positioned across her lane of traffic. Lane and Steele, however, both testified that Lane's truck was in the center turning lane and that his trailer was positioned across the inside northbound lane at the time of the collision.

Following the accident, Schenk was taken to the emergency room where she was treated for a "busted" mouth, a chipped tooth, scratches on her chin and forehead, and a "big bump" where Schenk collided with her rear view mirror. She was x-rayed, given muscle relaxants, and released that same evening. For approximately two weeks following the accident, Schenk was sore "from head to toe." Beyond this two week period, Schenk experienced persistent pain and soreness in her hips that eventually proceeded down to her legs. In early 1989, Schenk consulted Dr. Mark Houston regarding this continued pain. After conducting tests, Dr. Houston advised Schenk that there was nothing wrong with her and that she needed to relax. Approximately two months later, Schenk sought a second opinion from Dr. Stein, who took x-rays and a CT scan of Schenk and then informed her that she had a ruptured disk in her lower back. Dr. Stein suggested that Schenk undergo back surgery but Schenk decided against this option. Instead, Schenk underwent a limited amount of physical therapy and took mild muscle relaxants prescribed by Dr. Stein. In late 1992, Schenk began experiencing pain in her lower back and through her hips. She returned to Dr. Spengler, who ordered more physical therapy and gave Schenk a steroid injection to decrease inflammation in these areas. Schenk was also examined by Dr. Vaughan Allen, who referred Schenk to Dr. Leon Ensalada, a pain medicine specialist. Dr. Ensalada performed several procedures on Schenk, including two more steroid injections, three nerve blocks, and a facet rhizotomy. According to Dr. Ensalada, when Schenk reaches maximum medical improvement, she will still retain a permanent partial impairment of the whole person. Dr. Spengler testified that, under the American Medical Association Guidelines, Schenk would receive a permanent partial impairment rating of approximately eight to ten percent. Both Dr. Ensalada and Dr. Spengler agreed that Schenk's injuries may require further medical treatment but were unable to estimate the costs of such treatment.

Schenk filed a complaint against Lane seeking damages for personal injuries sustained as a result of the accident. At the conclusion of the trial, the jury found that Schenk was responsible for one percent of the negligence causing the accident, that Lane was responsible for ninety-nine percent of the negligence causing the accident, and that Schenk had sustained a total of $300,000.00 in damages as a result of the accident. Consistent with this finding, the trial court granted a judgment in favor of Schenk in the amount of $297,000.00. Lane appeals the findings of the jury as well as the trial court's denial of his motion for new trial.

Issues
(Issues involving medical testimony are omitted)

[Appellant argued that] the Court improperly did not allow the appellant to argue the extrapolation of the distance and time which would have elapsed between when the appellee first saw the appellant and when the impact occurred....

Closing Arguments

During closing arguments, counsel for Lane attempted to argue that Schenk had sufficient reaction time to avoid Lane's vehicle. An exchange took place between counsel and the trial judge as follows:

Counsel for Lane:	Miss Schenk testified she was on this end of the bridge. There's 250 feet. And I had her draw the bridge above it. And she acknowledged all that, the bridge between her and Mr. Lane. Now, Miss Schenk testified she was traveling 45 miles an hour. That extrapolates out to 66 feet a second. In other words, in a second of time, her car would have gone 66 feet.
Counsel for Schenk:	I'm going to object to that, Judge. There's no basis for that in the record. Nobody has testified to that, and I don't think—
The Court:	I sustain the objection.
Counsel for Lane:	Your Honor, if I may, I submit that that's just math.
The Court:	What's that?
Counsel for Lane:	You divide 5,280 feet a second ...
The Court:	Ladies and gentlemen, that may or may not be the case, but we can't argue anything that's not in front of you at this point in time. All we can argue is matters that, in fact, evolved as testimony. So go ahead.

Lane now contends that the trial judge erred in sustaining the objection. Specifically, Lane argues that the calculation of Schenk's reaction time is a matter of "simple mathematics" of which the trial judge could have taken judicial notice.

We agree that it is not improper to argue the mathematical formula of "Rate × Time = Distance." The testimony was varied concerning both Schenk's rate of speed and the distance between the location where she first noticed Lane's vehicle and the location of the collision. Schenk testified that she thought that the speed limit on Mt. Juliet Road was between thirty-five and forty-five miles per hour and that she was traveling within the speed limit. When questioned about her rate of speed immediately preceding the collision, however, Schenk responded that she had no idea how fast she was going. Chafin, who was traveling northbound on Mt. Juliet Road directly behind Schenk, testified that both he and Schenk were "moseying" along "at a reasonable rate of speed." Additionally,

Schenk testified that she first noticed Lane's vehicle as she approached the bridge on Mt. Juliet Road. Lane testified that, subsequent to the accident, he measured the distance between the southern end of the bridge and the end of the exit ramp, the point from which he began to turn left onto Mt. Juliet Road. When asked about this distance, Lane stated that the length of the bridge is "approximately two hundred—350 foot" and that the end of the exit ramp is "20-something-foot past the bridge itself."

When asked by counsel for Schenk if this distance was 200 feet, Lane nodded affirmatively and then stated that it was about 250 feet. Lane later testified that, on the morning of trial, he drove from the south end of the bridge to the end of the exit ramp and the odometer on his car indicated that this distance was exactly one-tenth of a mile.

It is within the province of the jury to determine factual differences. After reviewing the record as a whole, we have determined that, even if the trial court was in error in sustaining the objection, we cannot say it more probably than not affected the judgment nor that it resulted in prejudice to the judicial process.

Conclusion

In light of the foregoing, we conclude that the trial court did not err in denying Lane's motion for a new trial. For the reasons set forth above, the judgment is affirmed.

Notes and Problems

1. Was the Courts' reasoning sound? Should Counsel for Lane have requested that the court merely take judicial notice of the method of calculation?

2. Would a judicial notice request have overcome the failure by counsel to present evidence or testimony about the calculation during the course of the trial?

3. Is 45 miles per hour in fact 66 feet per second?

Searcy v. Sellers

Court of Civil Appeals of Texas, 470 S.W.2d 103 (1971)

ELLIS, Chief Justice.

This is a case involving the application of the doctrine of discovered peril in a rear-end automobile collision case. The case was tried before a jury which found that the defendant-appellant was guilty of primary negligence in certain particulars, but also found that plaintiff-appellee was guilty of contributory negligence. Additionally, the jury found the appellant guilty of discovered peril negligence, and on such basis the trial court entered judgment against appellant in the sum of $20,052.00, the amount of damages found by the jury. From such judgment the appellant has duly perfected his appeal. Neither party is challenging the jury's findings with respect to either the primary negligence of appellant or the contributory negligence of appellee. The disputed question for determination in this appeal is whether under the facts of this case the doctrine of discovered peril is applicable. Reversed and rendered.

Among other matters, the appellant is asserting 'no evidence' points. Basically, it is appellant's position that there is no evidence to support the jury's findings that (1) appellant discovered and realized the appellee's perilous position in time to have that last clear chance accorded him by law to avoid the collision; and (2) appellant committed discovery peril negligence in the time and space afforded him to act after his discovery of the appellee's position of peril. These fundamental 'no evidence' issues have been duly joined by counter-points set out in appellee's brief.

The rule is well established that in considering 'no evidence' points, the evidence is to be viewed in the most favorable light in support of the jury's findings of fact, considering only the evidence and reasonable inferences which support the findings and rejecting the evidence and reasonable inferences which are contrary to the findings. In Re King's Estate, 150 Tex. 662, 244 S.W.2d 660 (1951). Also, we recognize that in evaluating 'no evidence' points in this case, it is not permissible to give any weight to the jury's other findings, and to determine whether after disregarding the jury's findings on primary and contributory negligence, there is evidence in the record of probative force to support the jury's findings on the discovered peril issues. Enloe v. Barfield, 422 S.W.2d 905 (Tex.Sup.1967). The ultimate issue to be determined is whether the facts and circumstances as disclosed by the record in this case, together with all reasonable inferences therefrom, constitute some evidence of probative force to support the jury's finding to the effect that the appellant actually did discover and realize that the appellee was in a position of peril in time to have avoided the collision by the use of all means at his command consistent with the safety of himself and his vehicle.

This particular legal confrontation on the vigorously contested issues involving the doctrine of discovered peril had its inception in a suit brought by plaintiff-appellee, Parlie Sellers, against defendant-appellant, Eugene Ross Searcy, seeking recovery for personal injuries she sustained as a result of the collision of the pickup vehicle driven by appellant with the rear portion of her automobile. The setting for the occurrence giving rise to this controversy is disclosed by the undisputed evidence hereinafter set out. On May 13, 1968, appellee, Mrs. Sellers, accompanied by her passenger, Mrs. Jerri Lester, was driving her automobile in a southerly direction in the right or west traffic lane of U.S. Highway 87, a four lane north-south highway, divided by a median with paved crossovers between the east and west traffic lanes at various intervals.

Also, prior to the collision, appellant, Mr. Searcy had been driving his pickup southerly on the same right or west traffic lane of Highway 87 and following some distance behind the appellee's automobile. The collision occurred approximately seven miles north of Plainview, Texas. A third party, R. J. Moses, who had been traveling in a northerly direction on U.S. Highway 87, shortly before the collision, had negotiated a left turn on a paved crossover preparatory to returning to his house located to the west of the two southbound traffic lanes. Just prior to the collision, R. J. Moses stopped in the crossover facing westward waiting for the two southbound vehicles of appellant and appellee to pass before proceeding across the southbound lanes of the highway into the driveway leading to his house on the opposite side. It is further undisputed that at the scene of the collision there were only the three vehicles in sight, the appellee's automobile with the appellee as driver accompanied by her passenger, Mrs. Jerri Lester, the appellant's pickup with the appellant as the sole occupant, and the R. J. Moses automobile in which he was the sole occupant. These four named persons were the sole witnesses to the events occurring prior to and at the time of the collision and each appeared and testified during the trial of the case. It is also undisputed that the collision occurred in daylight and clear weather, and there were no obstruction of view to prevent appellant from observing the appellee's vehicle ahead of him.

We deem it appropriate to review that portion of the evidence from the witnesses, together with the attendant conditions and circumstances, pertinent to the vital question as to the existence or nonexistence of the requisite evidence to establish liability against the appellant for discovered peril negligence. Preliminarily, it is here noted that recovery under the doctrine of discovered peril is confined within the limits of rather exacting guidelines and presents a comparatively narrow issue. R. T. Herrin Petroleum Transport Co. v. Proctor, 161 Tex. 222, 338 S.W.2d 422 (1960). Further, the jury's findings on primary and contributory negligence are immaterial in passing upon the issue of discovered peril negligence. Parks v. Airline Motor Coaches, Inc., 145 Tex. 44, 193 S.W.2d 967 (1946); Enloe v. Barfield, supra. Additionally, when the defendant has been found guilty of primary negligence and the plaintiff guilty of contributory negligence proximately causing the occurrence in question, discovered peril negligence is based upon a new duty which arises only after discovery and realization of the perilous position of the other party. R. T. Herrin Petroleum Transport Co., et al., supra; Texas and Pacific Railway Company v. Meeks, 338 S.W.2d 169 (Tex.Civ.App.—Eastland 1960, writ ref'd n.r.e.).

Another significant principle is that discovered peril negligence involves the existence of a last clear chance to avoid the injury after discovery and realization of another's perilous position. In this connection, in the case of Schuhmacher Co. v. Posey, 147 Tex. 392, 215 S.W.2d 880, 882 (1942), the court stated: "The doctrine of discovered peril or last clear chance means that the last clear chance must be a clear one that implies thought, appreciation, mental direction and lapse of sufficient time to act effectually upon the impulse to save another from injury." As has been so frequently said, this doctrine is a humanitarian one, and evidence convicting a person of the rather grave fault of neglecting to act to avoid injuring another whom he has discovered in a position of peril must not be imputed or presumed.

Other principles and standards governing the applicability of the discovered peril doctrine as established by the case law of this state will be set out following a review of certain testimony of witnesses and circumstances considered significant in this case.

The appellee, Parlie Sellers, testified that she was driving at a speed of about 55 to 60 miles per hour as she approached the area in which the collision occurred. Prior to her arrival to a point even with the crossover above mentioned, she observed the Moses automobile make a left turn from the northerly lane of traffic into the crossover with its signal lights blinking, and coming to a complete stop inside the crossover. Also, by looking into her rearview mirror, she had previously observed the appellant's pickup following in the same traffic lane behind her. She was not aware of the distance that the pickup was behind her at the time she observed the Moses vehicle turn into the crossover. She further testified that upon seeing the Moses vehicle make its turn into the crossover she raised her foot from the accelerator so she could slow down and have a chance to stop if the Moses vehicle pulled in front of her. By taking her foot off the accelerator she slowed down her vehicle, and she testified that at no time prior to the collision did she touch the brake pedal, give any signal to the vehicle behind her, apply her brakes or stop her vehicle. She further testified that it was 'a short period of time' after she raised her foot from the accelerator until her vehicle was hit from the rear by the appellant's vehicle.

Mrs. Jerri Lester, appellee's passenger, generally corroborated the testimony of Mrs. Sellers regarding speed, lack of application of brakes or stopping, noted no significant change of speed of appellee's automobile prior to the collision and further testified that Mrs. Sellers 'had no reason to stop.' Mr. R. J. Moses who was stopped in the crossover facing westward, toward the southbound lanes of traffic observed the collision as it occurred in front of him. He stated that when he first noticed the approaching vehicles, the appellant's pickup was approximately one hundred yards behind the appellee's automobile and both vehicles were traveling at an estimated speed of forty-five miles per hour. He further testified that the appellee's automobile came to a complete stop in the west lane of the southbound traffic lanes and that appellant's pickup ran into the stopped vehicle. He was unable to state with certainty as to whether the stopping by appellee was gradual or sudden. He also stated that he was only watching the vehicles and not the persons, and that the appellant's pickup came heading on toward appellee's vehicle after she was in a stopped position. He further stated that there was a barrow ditch along and beyond the paved shoulder on the west side of the southbound traffic lanes and that there was a dip in the barrow ditch area where his driveway crossed the ditch. He stated also that the time between the appellee's slowing and stopping was 'a very short time' and that it 'all happened pretty fast.'

The appellant, Mr. Searcy, testified that prior to the collision he was traveling in the west lane of the southbound traffic lanes at a speed of about fifty miles per hour. He stated that he initially observed the appellee's automobile ahead of him at a distance of possibly a mile, and, at a distance of from one-quarter to one-half mile he noticed that he was gaining on her and would possibly be required to pass her, but he didn't keep his 'eyes glued on her all the time.' He also gave testimony to the effect that he subsequently noticed that she 'stopped fast' or suddenly when she was about 80 to 100 feet in front of him and that this particular instant was the first time that he was aware she was not continuing to move and had stopped. Also, he said that he didn't see her slow down or make any manifestation by brake-lights, signals, or otherwise, to warn him that she was going to stop her vehicle. He stated, also, that following his discovery and realization that she had stopped in front of him, he threw on his brakes and tried to turn to the right leaving forty-three feet of skid marks prior to colliding with the rear portion of appellee's automobile.

Further, he stated that the impact resulting from the collision caused his vehicle to go off the pavement to the right and knocked appellee's vehicle forward off the pavement to the left. Appellant also testified that upon discovery that appellee had stopped, he was nearer to the right than left side of appellee's vehicle, and that he turned to the right rather than to the left because it appeared to him that the right turn was the better of the two alternatives when turning with his wheels locked by his brakes. In this connection he stated that his turning to the left would have required the sharper turn as it would be necessary to attempt to go between the left side of appellee's automobile and the front of the Moses vehicle stopped in the crossover just to the left of appellee's stopped automobile under circumstances indicating considerable likelihood of colliding with the Moses automobile.

The elements of discovered peril involve the exposed condition of the plaintiff (appellee), an actual discovery by the defendant (appellant) in time to avert the injury by the use of all means at his command commensurate with his own safety, and failure to use such means. Turner v. Texas Co., 138 Tex. 380, 159 S.W.2d 112 (1942). As applied in Texas, an actual discovery of the perilous position of the injured party is essential in establishing discovered peril negligence and it is not enough that the defendant should have or could have by reasonable care realized or had knowledge of the perilous position of the plaintiff, or was negligent in not discovering, it, but it must be shown that he actually possessed such knowledge. Turner v. Texas Co., supra; Keith v. Ledbetter, 431 S.W.2d 433 (Tex.Civ.App.—Amarillo 1968, writ ref'd n.r.e.). In addition to such proof of discovery and realization of appellee's perilous position, it must be shown that appellee was pursuing and probably would pursue a course of action likely to result in an injury. Ford v. Panhandle & Santa Fe Ry. Co., 151 Tex. 538, 252 S.W.2d 561 (1952); Safeway Stores, Inc. v. White, 162 Tex. 473, 348 S.W.2d 162 (1961). Also, the Ford case pointed out that the standard of proof required to support the jury's findings on the elements of discovered peril is proof of such facts and circumstances, as taken together with all reasonable inferences therefrom, constitute some evidence of probative force of their existence.

The burden of proof rests upon the appellee to establish by a preponderance of the evidence the elements of discovered peril. Gentry v. Southern Pacific Company, 457 S.W.2d 889 (Tex.Sup.1970) and cases cited therein. It cannot be presumed that appellant was guilty of violating the new duty imposed upon him. Parks v. Airline Motor Coaches, supra. The fact of timely realization of the perilous position of the injured person may be established by circumstantial evidence, but in the absence of direct evidence, appellee is required to prove facts and circumstances from which it can be fairly inferred that appellant became aware of the peril. Safeway Stores, Inc. v. White, supra; Creech v. Thompson, 156 Tex. 561, 297 S.W.2d 817 (1957). A jury is not privileged to indulge in surmise or speculation, for there must be a factual basis for the inferences drawn. Heavy Haulers, Inc. v. Precise, 348 S.W.2d 653 (Tex.Civ.App.—Texarkana 1961, writ ref'd n.r.e.).

A review of the testimony given by the various witnesses, without considering at this time the testimony of the appellant, supplies no evidence that Searcy actually discovered the appellee's position of peril before the collision in time to avoid it. Neither appellee, Mrs. Sellers, Mrs. Lester, her passenger, nor the third party witness, Mr. Moses, testified as to any words spoken or overt acts by the appellant, or concerning any fact or circumstance that would indicate that he, the appellant, actually discovered or realized that appellee had stopped in front of him, or had changed her previous course of action in any manner, in time for him to take any evasive action.

The testimony of Roy Jordan, Mrs. Bill Warren and Correne Tanner in no way served to establish timely discovery and realization of appellee's perilous position prior to the collision. In effect, such testimony to the effect that he didn't see her automobile until he hit her tends to support exoneration of the appellant from discovered peril negligence. Safeway Stores v. White, supra; Turner v. Texas Co., supra.

An analysis of the testimony of the appellant Searcy fails to disclose any evidence that he actually discovered and realized that appellee had entered a position of peril in time for him to avoid the collision. It is recognized that Searcy is a party to the suit; however, the court pointed out in Texas & N.O.R. Co. v. Grace, 144 Tex. 71, 188 S.W.2d 378 (1945) that the fact that testimony of an interested witness is not accepted in his favor does not operate to convert it into evidence against him. In the case of R. T. Herrin Petroleum Transport Co. v. Proctor, supra, 338 S.W.2d p. 427, the court stated: "[T]estimony from an interested witness, while it need not be accepted as true by a jury, is not evidence that the exact opposite of what the witness said is true. There must be some circumstance supporting the conclusion that a situation opposite to that depicted by the witness actually existed."

A consideration of the various circumstances does not operate to disprove, but rather to corroborate the testimony of the appellant. We find no circumstances from which it might be inferred that appellant had accomplished actual discovery and realization of plaintiff's perilous position in time to avert the injury. Although he may have been generally aware of her automobile when he was within one-half mile of her and realized there might come a time when he should pass her, there is no evidence of the existence of a condition of peril at that time, and there is no evidence that she warned him in any manner that she was changing her previous course of action, and when he actually discovered her perilous position he was within 80 to 100 feet of her. Thus, the 'actual discovery' evidence, is supplied solely by appellant's testimony.

Contradictory inferences were required to come from other sources and no such contradictory evidence is in the record. Texas & N.O.R. Ry. Co. v. Grace, supra; R. T. Herrin Petroleum Co. v. Proctor, supra. When testimony furnished by the appellant is accepted to supply the essential element of actual discovery, it is proper to receive and accept his testimony to the effect that at the time of such discovery of appellee's perilous position, he was traveling at the rate of 50 miles per hour at a distance of 80 to 100 feet from his pickup to appellee's stopped vehicle in his path. At the instant of such discovery, appellant applied his brakes, skidded 43 feet while at the same time attempting to turn to the right before colliding with appellee's automobile and these matters are not contradicted by other evidence. See Missouri-Kansas-Texas Railroad Company of Texas v. McFerrin, 156 Tex. 69, 291 S.W.2d 931, 944 (1956); Welch v. Ada Oil Company, 302 S.W.2d 175, 179 (Tex.Civ.App.—Texarkana 1957, writ ref'd n.r.e.). In this connection, it is pointed out that courts will take judicial notice of natural forces as well as the primary laws of physics, mechanics, and mathematics. Levlon v. Dallas Ry. & Terminal Co., 117 S.W.2d 876 (Tex.Civ.App.—Dallas 1938, writ ref'd).

Also, it is a matter of common knowledge that skidding of an automobile results in loss of control by the driver. Dallas Ry. & Terminal Co. v. Garrison, 45 S.W.2d 183 (Tex.Com.App., 1932). Further, it is generally recognized that three-fourths of a second is required for a normal driver to react to danger. Murphy v. Whitehurst, 300 S.W.2d 758 (Tex.Civ.App.—San Antonio 1957, writ ref'd n.r.e.).

As a physical fact, if the appellant's automobile were traveling 50 miles per hour (73 1/3 feet per second), and the discovery of appellee's perilous situation were accomplished when he was 100 feet behind her, mathematical computation establishes that had he not applied his brakes the 100 feet would have been covered in 1.36 seconds, and that his vehicle would have traveled 55 feet before he could have reacted to apply the brakes or turn the steering wheel, leaving only 45 feet of space between the pickup and appellee's automobile.

These circumstances combined with the appellant's testimony that he left 43 feet of skid marks after applying his brakes reasonably support the inference that the time and space were too short after actual discovery for him to have the recognized last clear chance. Rather, it appears obvious that under all of the reasonable inferences from the circumstances and conditions, the period of time between discovery and reaching the vehicle ahead of him was too short to allow anything more than instinctive and impulsive reactions to the immediate danger which are insufficient to constitute discovered peril negligence. Keith v. Ledbetter, supra; Waldeck v. Watts, 326 S.W.2d 913.

The appellee has pointed out that the appellant made the statement to the effect that he saw appellee 'stop fast' or 'suddenly stopped' and that Mr. Moses, the third party witness, made the statement that he (Moses) saw appellee 'stopped' when appellant was about 100 yards to her rear. However, Moses did not testify that he and appellant saw appellee 'stopped' at the same time. We cannot agree that testimony of this character from the two parties constituted probative support for an inference that appellant actually saw or discovered appellee's automobile in a stopped position when he was about 100 yards behind her. Further, consideration should be given to the differences in position, motion, line of vision and perspective of the two parties.

The fact that there was no obstruction of view in the vicinity of the collision and appellant was in a position to observe appellee's vehicle for a mile or so before the collision are not within themselves circumstances from which timely actual discovery and realization can be inferred. As was pointed out in Turner v. Texas Co., supra, the fact that the driver of the automobile had the other vehicle under observation for the time it took to drive a mile or more is no evidence tending to prove when the truck driver's position became perilous, or when the automobile driver discovered or realized it was perilous.

The present applicable law in Texas in this regard is summarized in the following language set out in Heavy Haulers, Inc. v. Precise, supra: "No case has come to the attention of the court in which the operator's realization of the injured party's peril, a subjective condition of the operator's mind, has been inferred simply because the injured party was visible at all times to the operator of the moving vehicle inflicting injury."

Thus, a full consideration of all pertinent testimony of the witnesses and the circumstances supports the conclusion that no witness or circumstance contradicts the testimony of the appellant. The appellee has cited various cases, but the great majority of such cases involve an admission or statement by the operator of timely discovery and the existence of circumstances from which it could be reasonably inferred that after such discovery the operator failed to take appropriate evasive action.

The case of Ford v. Panhandle & Santa Fe Ry. Co., supra, has been strongly urged by appellee in support of her position. It is our opinion that the instant case is distinguishable on the facts. In the Ford case, actual discovery of the peril occurred at a time when the movements of the automobile which was stricken were, under the evidence, reasonably predictable, and the train was far enough back that it could have been reasonably inferred that the operator failed to apply his brakes when he had time to do so, and could thereby have reduced the speed of the train sufficiently to have averted the collision. A review of the pertinent cases indicates clearly that the holding of the Ford case has generally been rejected where the peril is brought on by the injured party changing from one course of conduct to another, or by committing an act which the party charged is not required to anticipate from what he actually saw on the occasion in question.

We do not agree that Ford is applicable in the present case. Also, the appellant has cited the recent case of Hinojosa v. Stephens, 461 S.W.2d 232 (Tex.Civ.App., 1970), involving the application of the doctrine of discovered peril in a rear-end automobile collision case. It is our opinion that the facts in that case are also distinguishable from those in the instant case. Under the evidence in the Hinojosa case, when the defendant put on his bright lights, he saw 'two people together in a car parked there' one block to one and one-half blocks away. Also, the evidence showed that the defendant was traveling not more than an estimated 37 miles per hour. The defendant applied his brakes but took no further evasive action. Also, the court's opinion points out that the defendant realized that if he continued in the lane he was in at the speed he was going that there would be a collision. In the instant case we have no comparable evidence or inferences of timely discovery and realization of the peril or of the driver's failure to use the means available to avoid the collision.

A review of the record indicates that all of the evidence available in this case on the discovered peril issues was fully developed. After considering all the testimony and circumstances and the reasonable inferences therefrom, we hold there is no evidence as required showing that appellee's evidence measures up to the quantum of proof essential to support the jury's findings on the discovered peril issues. Since both appellant and appellee have been found guilty of negligence proximately causing the collision in question and such findings are in no way challenged, in view of our holding herein on the question of discovered peril, we deem it unnecessary to pass upon the other points presented.

The judgment of the trial court is reversed and judgment is herein rendered that plaintiff-appellee take nothing.

Notes and Problems

1. In *Searcy*, the court cavalierly stated that it was a matter of "common knowledge" that the skidding of an automobile results in a loss of control by the driver. Is this necessarily true? See *infra*, § 7.14 (There Shall Be friction).

2. In *Winterberg v. Thomas*, 126 Colo. 60, 246 P.1058 (1952), *infra*, § 7.13, the appellate court used its own *Judicial Knowledge* to calculate speeds and distances to determine who ran a red light. How, if at all, does this differ from judicial notice?

3. Some jurisdictions pay lip service to an "assured clear distance" standard, requiring that drivers never travel at a speed that would prevent safe stopping under all circumstances. Is this practical?

4. For quick conversions of miles per hour and feet per second, see *infra* § 2.24.

§ 1.22 Expert Evidence

Many basic laws of physics and mathematics fall within the scope of judicial notice. But not all numbers enjoy such status. Statistics, for example, is a "science" that has been working its way into courtrooms at any increasing pace. But at an equally increasing pace, it has abused and misused its quasi-scientific status, as we shall see in Chapter 5. Like statistics, many forms of scientific evidence, particularly novel ones, face some formidable hurdles.

In 1923 and for many years thereafter, the landmark *Frye v. United States*, 293 F. 1013 (D.C. Cir., 1923) set the tone for new or questionable scientific evidence by creating a very strict admissibility standard known as the "general acceptance" test. The tough standard required that before scientific evidence could be admitted, it must first be demonstrated that the principles upon which it was based, were generally accepted by the relevant scientific community. The states and the federal circuits were split over whether or not a different standard should apply. In 1975, the promulgation of Rule 702 of the Federal Rules of Evidence addressed but failed to resolve the issue. Nor did its Committee Notes offer any resolution of the dispute. The Rule, as originally drafted, required only that expert witnesses possess some knowledge, skill, experience, training or education in their particular area of expertise.

For the next several years, the circuits were split as to whether or not Rule 702 overruled the *Frye* standard in the federal courts. Some held that the failure of the rules drafters to mention or specifically overrule *Frye* implied an intention to preserve the strict "general acceptance" standard. Others claimed that the new federal rule had subsumed the issue and thereby overruled *Frye*. The matter wasn't settled until 1993, when *Daubert v. Merrell Dow Pharmaceuticals, Inc.*, 509 U.S. 579, 113 S.Ct. 2786 (1993) expressly overruled *Frye* by creating a multi-factor test for the admissibility of new scientific evidence, which interestingly included "general acceptance" as one its factors. Today, in the federal courts and in the courts of many states, *Daubert* and its progeny basically rule the admissibility of scientific evidence. California, which adheres to a general acceptance standard, is an exception.

Daubert v. Merrell Dow Pharmaceuticals, Inc.

United States Supreme Court, 509 U.S. 579, 113 S.Ct. 2786 (1993)

Justice BLACKMUN delivered the opinion of the Court.

In this case we are called upon to determine the standard for admitting expert scientific testimony in a federal trial.

I.

Petitioners Jason Daubert and Eric Schuller are minor children born with serious birth defects. They and their parents sued respondent in California state court, alleging that the birth defects had been caused by the mothers' ingestion of Bendectin, a prescription antinausea drug marketed by respondent. Respondent removed the suits to federal court on diversity grounds.

After extensive discovery, respondent moved for summary judgment, contending that Bendectin does not cause birth defects in humans and that petitioners would be unable to come forward with any admissible evidence that it does. In support of its motion, respondent submitted an affidavit of Steven H. Lamm, physician and epidemiologist, who is a well-credentialed expert on the risks from exposure to various chemical substances. Doctor Lamm stated that he had reviewed all the literature on Bendectin and human birth defects—more than 30 published studies involving over 130,000 patients. No study had found Bendectin to be a human teratogen (i.e., a substance capable of causing malformations in fetuses). On the basis of this review, Doctor Lamm concluded that maternal use of Bendectin during the first trimester of pregnancy has not been shown to be a risk factor for human birth defects.

Petitioners did not (and do not) contest this characterization of the published record regarding Bendectin. Instead, they responded to respondent's motion with the testimony of eight experts of their own, each of whom also possessed impressive credentials. These experts had concluded that Bendectin can cause birth defects. Their conclusions were based upon "in vitro" (test tube) and "in vivo" (live) animal studies that found a link between Bendectin and malformations; pharmacological studies of the chemical structure of Bendectin that purported to show similarities between the structure of the drug and that of other substances known to cause birth defects; and the "reanalysis" of previously published epidemiological (human statistical) studies.

The District Court granted respondent's motion for summary judgment. The court stated that scientific evidence is admissible only if the principle upon which it is based is "sufficiently established to have general acceptance in the field to which it belongs." 727 F.Supp. 570, 572 (S.D.Cal.1989), quoting United States v. Kilgus, 571 F.2d 508, 510 (CA9 1978). The court concluded that petitioners' evidence did not meet this standard. Given the vast body of epidemiological data concerning Bendectin, the court held, expert opinion which is not based on epidemiological evidence is not admissible to establish causation. 727 F.Supp., at 575. Thus, the animal-cell studies, live-animal studies, and chemical-structure analyses on which petitioners had relied could not raise by themselves a reasonably disputable jury issue regarding causation. Ibid. Petitioners' epidemiological analyses, based as they were on recalculations of data in previously published studies that had found no causal link between the drug and birth defects, were ruled to be inadmissible because they had not been published or subjected to peer review.

The United States Court of Appeals for the Ninth Circuit affirmed. 951 F.2d 1128 (1991). Citing Frye v. United States, 54 App.D.C. 46, 47, 293 F. 1013, 1014 (1923), the court stated that expert opinion based on a scientific technique is inadmissible unless the technique is "generally accepted" as reliable in the relevant scientific community. The court declared that expert opinion based on a methodology that diverges "significantly from the procedures accepted by recognized authorities in the field ... cannot be shown to be generally accepted as a reliable technique."

The court emphasized that other Courts of Appeals considering the risks of Bendectin had refused to admit reanalyses of epidemiological studies that had been neither published nor subjected to peer review. Those courts had found unpublished reanalyses "particularly problematic in light of the massive weight of the original published studies supporting [respondent's] position, all of which had undergone full scrutiny from the scientific community." Contending that reanalysis is generally accepted by the scientific community only when it is subjected to verification and scrutiny by others in the field, the Court of Appeals rejected petitioners' reanalyses as "unpublished, not subjected to the normal peer review process and generated solely for use in litigation." The court concluded that petitioners' evidence provided an insufficient foundation to allow admission of expert testimony that Bendectin caused their injuries and, accordingly, that petitioners could not satisfy their burden of proving causation at trial.

We granted certiorari, 506 U.S. 914, 113 S.Ct. 320, 121 L.Ed.2d 240 (1992), in light of sharp divisions among the courts regarding the proper standard for the admission of expert testimony ...

II.

A.

In the 70 years since its formulation in the Frye case, the "general acceptance" test has been the dominant standard for determining the admissibility of novel scientific evidence at trial. See E. Green & C. Nesson, Problems, Cases, and Materials on Evidence 649 (1983). Although under increasing attack of late, the rule continues to be followed by a majority of courts, including the Ninth Circuit.

The Frye test has its origin in a short and citation-free 1923 decision concerning the admissibility of evidence derived from a systolic blood pressure deception test, a crude precursor to the polygraph machine. In what has become a famous (perhaps infamous) passage, the then Court of Appeals for the District of Columbia described the device and its operation and declared:

> "Just when a scientific principle or discovery crosses the line between the experimental and demonstrable stages is difficult to define. Somewhere in this twilight zone the evidential force of the principle must be recognized, and while courts will go a long way in admitting expert testimony deduced from a well-recognized scientific principle or discovery, the thing from which the deduction is made must be sufficiently established to have gained general acceptance in the particular field in which it belongs." 54 App.D.C., at 47, 293 F., at 1014 (emphasis added).

Because the deception test had "not yet gained such standing and scientific recognition among physiological and psychological authorities as would justify the courts in admitting expert testimony deduced from the discovery, development, and experiments thus far made," evidence of its results was ruled inadmissible.

The merits of the Frye test have been much debated, and scholarship on its proper scope and application is legion. Petitioners' primary attack, however, is not on the content but on the continuing authority of the rule. They contend that the Frye test was superseded by the adoption of the Federal Rules of Evidence. We agree.

Indeed, the debates over Frye are such a well-established part of the academic landscape that a distinct term—"Frye-ologist"—has been advanced to describe those who take part. See Behringer, Introduction, Proposals for a Model Rule on the Admissibility of Scientific Evidence, 26 Jurimetrics J. 237, 239 (1986), quoting Lacey, Scientific Evidence, 24 Jurimetrics J. 254, 264 (1984).

We interpret the legislatively enacted Federal Rules of Evidence as we would any statute. Beech Aircraft Corp. v. Rainey, 488 U.S. 153, 163, 109 S.Ct. 439, 446, 102 L.Ed.2d 445 (1988). Rule 402 provides the baseline:

> "All relevant evidence is admissible, except as otherwise provided by the Constitution of the United States, by Act of Congress, by these rules, or by other rules prescribed by the Supreme Court pursuant to statutory authority. Evidence which is not relevant is not admissible."
>
> "Relevant evidence" is defined as that which has "any tendency to make the existence of any fact that is of consequence to the determination of the action more probable or less probable than it would be without the evidence." Rule 401. The Rule's basic standard of relevance thus is a liberal one.

Frye, of course, predated the Rules by half a century. In United States v. Abel, 469 U.S. 45, 105 S.Ct. 465, 83 L.Ed.2d 450 (1984), we considered the pertinence of background common law in interpreting the Rules of Evidence. We noted that the Rules occupy the field, id., at 49, 105 S.Ct., at 467, but, quoting Professor Cleary, the Reporter, explained that the common law nevertheless could serve as an aid to their application:

> "In principle, under the Federal Rules no common law of evidence remains. 'All relevant evidence is admissible, except as otherwise provided....' In reality, of course, the body of common law knowledge continues to exist, though in the somewhat altered form of a source of guidance in the exercise of delegated powers."

We found the common-law precept at issue in the Abel case entirely consistent with Rule 402's general requirement of admissibility, and considered it unlikely that the drafters had intended to change the rule. Id., at 50–51, 105 S.Ct., at 468–469. In Bourjaily v. United States, 483 U.S. 171, 107 S.Ct. 2775, 97 L.Ed.2d 144 (1987), on the other hand, the Court was unable to find a particular common-law doctrine in the Rules, and so held it superseded.

Here there is a specific Rule that speaks to the contested issue. Rule 702, governing expert testimony, provides:

> "If scientific, technical, or other specialized knowledge will assist the trier of fact to understand the evidence or to determine a fact in issue, a witness qualified as an expert by knowledge, skill, experience, training, or education, may testify thereto in the form of an opinion or otherwise."

Nothing in the text of this Rule establishes "general acceptance" as an absolute prerequisite to admissibility. Nor does respondent present any clear indication that Rule 702 or the Rules as a whole were intended to incorporate a "general acceptance" standard. The drafting history makes no mention of Frye, and a rigid "general acceptance" requirement would be at odds with the "liberal thrust" of the Federal Rules and their "general approach of relaxing the traditional barriers to 'opinion' testimony." Beech Aircraft Corp. v. Rainey, 488 U.S., at 169, 109 S.Ct., at 450 (citing Rules 701 to 705). See also Weinstein, Rule 702 of the Federal Rules of Evidence is Sound; It Should Not Be Amended, 138 F.R.D. 631 (1991) ("The Rules were designed to depend primarily upon lawyer-adversaries and sensible triers of fact to evaluate conflicts"). Given the Rules' permissive backdrop and their inclusion of a specific rule on expert testimony that does not mention "general acceptance," the assertion that the Rules somehow assimilated Frye is unconvincing. Frye made "general acceptance" the exclusive test for admitting expert scientific testimony. That austere standard, absent from, and incompatible with, the Federal Rules of Evidence, should not be applied in federal trials.

B.

That the Frye test was displaced by the Rules of Evidence does not mean, however, that the Rules themselves place no limits on the admissibility of purportedly scientific evidence. Nor is the trial judge disabled from screening such evidence. To the contrary, under the Rules the trial judge must ensure that any and all scientific testimony or evidence admitted is not only relevant, but reliable.

The primary locus of this obligation is Rule 702, which clearly contemplates some degree of regulation of the subjects and theories about which an expert may testify. "If scientific, technical, or other specialized knowledge will assist the trier of fact to understand the evidence or to determine a fact in issue" an expert "may testify thereto" (emphasis added). The subject of an expert's testimony must be "scientific ... knowledge." The adjective "scientific" implies a grounding in the methods and procedures of science. Similarly, the word "knowledge" connotes more than subjective belief or unsupported speculation. The term "applies to any body of known facts or to any body of ideas inferred from such facts or accepted as truths on good grounds." *Webster's Third New International Dictionary* 1252 (1986). Of course, it would be unreasonable to conclude that the subject of scientific testimony must be "known" to a certainty; arguably, there are no certainties in science. See, e.g., Brief for Nicolaas Bloembergen et al. as Amici Curiae 9 ("Indeed, scientists do not assert that they know what is immutably 'true'—they are committed to searching for new, temporary, theories to explain, as best they can, phenomena"); Brief for American Association for the Advancement of Science et al. as Amici Curiae 7–8 ("Science is not an encyclopedic body of knowledge about the universe. Instead, it represents a process for proposing and refining theoretical explanations about the world that are subject to further testing and refinement" (emphasis in original)).

But, in order to qualify as "scientific knowledge," an inference or assertion must be derived by the scientific method. Proposed testimony must be supported by appropriate validation—i.e., "good grounds," based on what is known. In short, the requirement that an expert's testimony pertain to "scientific knowledge" establishes a standard of evidentiary reliability.

Rule 702 further requires that the evidence or testimony "assist the trier of fact to understand the evidence or to determine a fact in issue." This condition goes primarily to relevance. "Expert testimony which does not relate to any issue in the case is not relevant and, ergo, non-helpful." 3 Weinstein & Berger ¶ 702[02], p. 702–18. See also United States v. Downing, 753 F.2d 1224, 1242 (CA3 1985) ("An additional consideration under Rule 702—and another aspect of relevancy—is whether expert testimony proffered in the case is sufficiently tied to the facts of the case that it will aid the jury in resolving a factual dispute"). The consideration has been aptly described by Judge Becker as one of "fit." "Fit" is not always obvious, and scientific validity for one purpose is not necessarily scientific validity for other, unrelated purposes. See Starrs, Frye v. United States Restructured and Revitalized: A Proposal to Amend Federal Evidence Rule 702, 26 Jurimetrics J. 249, 258 (1986). The study of the phases of the moon, for example, may provide valid scientific "knowledge" about whether a certain night was dark, and if darkness is a fact in issue, the knowledge will assist the trier of fact. However (absent creditable grounds supporting such a link), evidence that the moon was full on a certain night will not assist the trier of fact in determining whether an individual was unusually likely to have behaved irrationally on that night. Rule 702's "helpfulness" standard requires a valid scientific connection to the pertinent inquiry as a precondition to admissibility.

That these requirements are embodied in Rule 702 is not surprising. Unlike an ordinary witness, see Rule 701, an expert is permitted wide latitude to offer opinions, including those that are not based on firsthand knowledge or observation. See Rules 702 and 703. Presumably, this relaxation of the usual requirement of firsthand knowledge—a rule which represents "a most pervasive manifestation of the common law insistence upon the most reliable sources of information," Advisory Committee's Notes on Fed.Rule Evid. 602, 28 U.S.C.App., p. 755 (citation omitted)—is premised on an assumption that the expert's opinion will have a reliable basis in the knowledge and experience of his discipline.

C.

Faced with a proffer of expert scientific testimony, then, the trial judge must determine at the outset, pursuant to Rule 104(a), whether the expert is proposing to testify to (1) scientific knowledge that (2) will assist the trier of fact to understand or determine a fact in issue. This entails a preliminary assessment of whether the reasoning or methodology underlying the testimony is scientifically valid and of whether that reasoning or methodology properly can be applied to the facts in issue. We are confident that federal judges possess the capacity to undertake this review. Many factors will bear on the inquiry, and we do not presume to set out a definitive checklist or test. But some general observations are appropriate.

Ordinarily, a key question to be answered in determining whether a theory or technique is scientific knowledge that will assist the trier of fact will be whether it can be (and has been) tested. "Scientific methodology today is based on generating hypotheses and testing them to see if they can be falsified; indeed, this methodology is what distinguishes science from other fields of human inquiry." Green 645. See also C. Hempel, Philosophy of Natural Science 49 (1966) ("[T]he statements constituting a scientific explanation must be capable of empirical test"); K. Popper, Conjectures and Refutations: The Growth of Scientific Knowledge 37 (5th ed. 1989) ("[T]he criterion of the scientific status of a theory is its falsifiability, or refutability, or testability").

Another pertinent consideration is whether the theory or technique has been subjected to peer review and publication. Publication (which is but one element of peer review) is not a sine qua non of admissibility; it does not necessarily correlate with reliability, see S. Jasanoff, The Fifth Branch: Science Advisors as Policymakers 61–76 (1990), and in some instances well-grounded but innovative theories will not have been published, see Horrobin, The Philosophical Basis of Peer Review and the Suppression of Innovation, 263 JAMA 1438 (1990). Some propositions, moreover, are too particular, too new, or of too limited interest to be published. But submission to the scrutiny of the scientific community is a component of "good science," in part because it increases the likelihood that substantive flaws in methodology will be detected. See J. Ziman, Reliable Knowledge: An Exploration of the Grounds for Belief in Science 130–133 (1978); Relman & Angell, How Good Is Peer Review?, 321 New Eng.J.Med. 827 (1989). The fact of publication (or lack thereof) in a peer reviewed journal thus will be a relevant, though not dispositive, consideration in assessing the scientific validity of a particular technique or methodology on which an opinion is premised.

Additionally, in the case of a particular scientific technique, the court ordinarily should consider the known or potential rate of error, see, e.g., United States v. Smith, 869 F.2d 348, 353–354 (CA7 1989) (surveying studies of the error rate of spectrographic voice identification technique), and the existence and maintenance of standards controlling the technique's operation, see United States v. Williams, 583 F.2d 1194, 1198 (CA2 1978) (noting professional organization's standard governing spectrographic analysis), cert. denied, 439 U.S. 1117, 99 S.Ct. 1025, 59 L.Ed.2d 77 (1979).

Finally, "general acceptance" can yet have a bearing on the inquiry. A "reliability assessment does not require, although it does permit, explicit identification of a relevant scientific community and an express determination of a particular degree of acceptance within that community." United States v. Downing, 753 F.2d, at 1238. See also 3 Weinstein & Berger ¶ 702[03], pp. 702-41 to 702-42. Widespread acceptance can be an important factor in ruling particular evidence admissible, and "a known technique which has been able to attract only minimal support within the community," Downing, 753 F.2d, at 1238, may properly be viewed with skepticism.

The inquiry envisioned by Rule 702 is, we emphasize, a flexible one. Its overarching subject is the scientific validity and thus the evidentiary relevance and reliability—of the principles that underlie a proposed submission. The focus, of course, must be solely on principles and methodology, not on the conclusions that they generate.

Throughout, a judge assessing a proffer of expert scientific testimony under Rule 702 should also be mindful of other applicable rules. Rule 703 provides that expert opinions based on otherwise inadmissible hearsay are to be admitted only if the facts or data are "of a type reasonably relied upon by experts in the particular field in forming opinions or inferences upon the subject." Rule 706 allows the court at its discretion to procure the assistance of an expert of its own choosing. Finally, Rule 403 permits the exclusion of relevant evidence "if its probative value is substantially outweighed by the danger of unfair prejudice, confusion of the issues, or misleading the jury...." Judge Weinstein has explained: "Expert evidence can be both powerful and quite misleading because of the difficulty in evaluating it. Because of this risk, the judge in weighing possible prejudice against probative force under Rule 403 of the present rules exercises more control over experts than over lay witnesses." Weinstein, 138 F.R.D., at 632.

III.

We conclude by briefly addressing what appear to be two underlying concerns of the parties and amici in this case. Respondent expresses apprehension that abandonment of "general acceptance" as the exclusive requirement for admission will result in a "free-for-all" in which befuddled juries are confounded by absurd and irrational pseudoscientific assertions. In this regard respondent seems to us to be overly pessimistic about the capabilities of the jury and of the adversary system generally. Vigorous cross-examination, presentation of contrary evidence, and careful instruction on the burden of proof are the traditional and appropriate means of attacking shaky but admissible evidence. See Rock v. Arkansas, 483 U.S. 44, 61, 107 S.Ct. 2704, 2714, 97 L.Ed.2d 37 (1987). Additionally, in the event the trial court concludes that the scintilla of evidence presented supporting a position is insufficient to allow a reasonable juror to conclude that the position more likely than not is true, the court remains free to direct a judgment, Fed.Rule Civ.Proc. 50(a), and likewise to grant summary judgment, Fed.Rule Civ.Proc. 56. Cf., e.g., Turpin v. Merrell Dow Pharmaceuticals, Inc., 959 F.2d 1349 (CA6) (holding that scientific evidence that provided foundation for expert testimony, viewed in the light most favorable to plaintiffs, was not sufficient to allow a jury to find it more probable than not that defendant caused plaintiff's injury), cert. denied, 506 U.S. 826, 113 S.Ct. 84, 121 L.Ed.2d 47 (1992); Brock v. Merrell Dow Pharmaceuticals, Inc., 874 F.2d 307 (CA5 1989) (reversing judgment entered on jury verdict for plaintiffs because evidence regarding causation was insufficient), modified, 884 F.2d 166 (CA5 1989), cert. denied, 494 U.S. 1046,

110 S.Ct. 1511, 108 L.Ed.2d 646 (1990); Green 680–681. These conventional devices, rather than wholesale exclusion under an uncompromising "general acceptance" test, are the appropriate safeguards where the basis of scientific testimony meets the standards of Rule 702.

Petitioners and, to a greater extent, their amici exhibit a different concern. They suggest that recognition of a screening role for the judge that allows for the exclusion of "invalid" evidence will sanction a stifling and repressive scientific orthodoxy and will be inimical to the search for truth. See, e.g., Brief for Ronald Bayer et al. as Amici Curiae. It is true that open debate is an essential part of both legal and scientific analyses. Yet there are important differences between the quest for truth in the courtroom and the quest for truth in the laboratory. Scientific conclusions are subject to perpetual revision. Law, on the other hand, must resolve disputes finally and quickly. The scientific project is advanced by broad and wide-ranging consideration of a multitude of hypotheses, for those that are incorrect will eventually be shown to be so, and that in itself is an advance. Conjectures that are probably wrong are of little use, however, in the project of reaching a quick, final, and binding legal judgment—often of great consequence—about a particular set of events in the past.

We recognize that, in practice, a gatekeeping role for the judge, no matter how flexible, inevitably on occasion will prevent the jury from learning of authentic insights and innovations. That, nevertheless, is the balance that is struck by Rules of Evidence designed not for the exhaustive search for cosmic understanding but for the particularized resolution of legal disputes.

IV.

To summarize: "General acceptance" is not a necessary precondition to the admissibility of scientific evidence under the Federal Rules of Evidence, but the Rules of Evidence—especially Rule 702—do assign to the trial judge the task of ensuring that an expert's testimony both rests on a reliable foundation and is relevant to the task at hand. Pertinent evidence based on scientifically valid principles will satisfy those demands. The inquiries of the District Court and the Court of Appeals focused almost exclusively on "general acceptance," as gauged by publication and the decisions of other courts. Accordingly, the judgment of the Court of Appeals is vacated, and the case is remanded for further proceedings consistent with this opinion.

It is so ordered.

Chief Justice REHNQUIST, with whom Justice STEVENS joins, concurring in part and dissenting in part.

The petition for certiorari in this case presents two questions: first, whether the rule of Frye v. United States, 54 App.D.C. 46, 293 F. 1013 (1923), remains good law after the enactment of the Federal Rules of Evidence; and second, if Frye remains valid, whether it requires expert scientific testimony to have been subjected to a peer review process in order to be admissible. The Court concludes, correctly in my view, that the Frye rule did not survive the enactment of the Federal Rules of Evidence, and I therefore join Parts I and II-A of its opinion. The second question presented in the petition for certiorari necessarily is mooted by this holding, but the Court nonetheless proceeds to construe Rules 702 and 703 very much in the abstract, and then offers some "general observations." Ante, at 2796.

"General observations" by this Court customarily carry great weight with lower federal courts, but the ones offered here suffer from the flaw common to most such observations—they are not applied to deciding whether particular testimony was or was not admissible, and therefore they tend to be not only general, but vague and abstract.

This is particularly unfortunate in a case such as this, where the ultimate legal question depends on an appreciation of one or more bodies of knowledge not judicially noticeable, and subject to different interpretations in the briefs of the parties and their amici. Twenty-two amicus briefs have been filed in the case, and indeed the Court's opinion contains no fewer than 37 citations to amicus briefs and other secondary sources.

The various briefs filed in this case are markedly different from typical briefs, in that large parts of them do not deal with decided cases or statutory language—the sort of material we customarily interpret. Instead, they deal with definitions of scientific knowledge, scientific method, scientific validity, and peer review—in short, matters far afield from the expertise of judges. This is not to say that such materials are not useful or even necessary in deciding how Rule 703 should be applied; but it is to say that the unusual subject matter should cause us to proceed with great caution in deciding more than we have to, because our reach can so easily exceed our grasp.

But even if it were desirable to make "general observations" not necessary to decide the questions presented, I cannot subscribe to some of the observations made by the Court. In Part II-B, the Court concludes that reliability and relevancy are the touchstones of the admissibility of expert testimony. Ante, at 2794–95. Federal Rule of Evidence 402 provides, as the Court points out, that "[e]vidence which is not relevant is not admissible." But there is no similar reference in the Rule to "reliability." The Court constructs its argument by parsing the language "[i]f scientific, technical, or other specialized knowledge will assist the trier of fact to understand the evidence or to determine a fact in issue, ... an expert ... may testify thereto...." Fed.Rule Evid. 702. It stresses that the subject of the expert's testimony must be "scientific ... knowledge," and points out that "scientific" "implies a grounding in the methods and procedures of science" and that the word "knowledge" "connotes more than subjective belief or unsupported speculation." Ante, at 2794–95. From this it concludes that "scientific knowledge" must be "derived by the scientific method." Ante, at 2795. Proposed testimony, we are told, must be supported by "appropriate validation." Ante, at 2795. Indeed, in footnote 9, the Court decides that "[i]n a case involving scientific evidence, evidentiary reliability will be based upon scientific validity." Ante, at 2795, n. 9 (emphasis in original).

Questions arise simply from reading this part of the Court's opinion, and countless more questions will surely arise when hundreds of district judges try to apply its teaching to particular offers of expert testimony. Does all of this dicta apply to an expert seeking to testify on the basis of "technical or other specialized knowledge"—the other types of expert knowledge to which Rule 702 applies—or are the "general observations" limited only to "scientific knowledge"? What is the difference between scientific knowledge and technical knowledge; does Rule 702 actually contemplate that the phrase "scientific, technical, or other specialized knowledge" be broken down into numerous subspecies of expertise, or did its authors simply pick general descriptive language covering the sort of expert testimony which courts have customarily received? The Court speaks of its confidence that federal judges can make a "preliminary assessment of whether the reasoning or methodology underlying the testimony is scientifically valid and of whether that reasoning or methodology properly can be applied to the facts in issue." The Court then states that a "key question" to be answered in deciding whether something is "scientific knowledge" "will be whether it can be (and has been) tested." Ante, at 2796. Following this sentence are three quotations from treatises, which not only speak of empirical testing, but one of which states that the "criterion of the scientific status of a theory is its falsifiability, or refutability, or testability."

I defer to no one in my confidence in federal judges; but I am at a loss to know what is meant when it is said that the scientific status of a theory depends on its "falsifiability," and I suspect some of them will be, too.

I do not doubt that Rule 702 confides to the judge some gatekeeping responsibility in deciding questions of the admissibility of proffered expert testimony. But I do not think it imposes on them either the obligation or the authority to become amateur scientists in order to perform that role. I think the Court would be far better advised in this case to decide only the questions presented, and to leave the further development of this important area of the law to future cases.

Notes and Problems

1. *Daubert* is now considered the authoritative landmark with respect to the admissibility of scientific evidence. Along with its child, *Kumho Tire Company, Ltd. v. Carmichael*, 526 U.S. 137 (1999), et al., *Daubert* and its progeny gave us the following list of non-exclusive factors to determine whether or not a given expert's scientific evidence should be admitted:

> 1. Whether the expert's theory or technique can be, or has been tested;
>
> 2. Whether the expert's theory or technique has been subjected to peer review;
>
> 3. The known or potential rate of error;
>
> 4. The existence and maintenance of standards controlling the technique's operation; and,
>
> 5. General acceptance in the scientific community.

Following *Daubert* and its ever-expanding wake, the Federal Rules of Evidence were amended several times in an effort to keep pace with changing technology and the substantial number of *Daubert's* grandchildren. Even though some states provide their own brand of gate-keeping for scientific evidence, Rule 702 (in its present restyled form) seems to deliver the most widely accepted model:

> A witness who is qualified as an expert by knowledge, skill, experience, training, or education may testify in the form of an opinion or otherwise if:
>
> (a) the expert's scientific, technical, or other specialized knowledge will help the trier of fact to understand the evidence or to determine a fact in issue;
>
> (b) the testimony is based on sufficient facts or data;
>
> (c) the testimony is the product of reliable principles and methods; and
>
> (d) the expert has reliably applied the principles and methods to the facts of the case.

2. Madam Trixie is recognized as the best astrologist on the planet by the NAA (National Association of Astrologists). All concede that her knowledge, training, experience and skills exceed all others. Defense counsel for Teddy Tarver a.k.a. Teddy the Termite wants her to testify as an expert at his upcoming bank robbery trial in U.S. District Court. Trixie intends to state that she created a "chart" for Teddy and studied it very carefully, after following all of the recognized rules governing the stars as they relate to the date and time of Teddy's birth. She also intends to testify that because of the manner in which Teddy's rising moon tilts 45 degrees above his Uranus moon, it is virtually impossible that Teddy's

personality would permit him to commit robbery. The prosecutor catches wind of this defense and makes a motion in limine to suppress the testimony. Based on Rule 702 of the Federal Rules of Evidence, how would you rule on the motion? Based on *Daubert*, how would you rule?

3. Over the years, several methodologies have passed *Daubert* muster, including fingerprinting, voiceprinting and ballistics. The ones that do not, such as graphology (using handwriting to determine personality traits) and Tarot card readings, are referred to as "junk science" and deemed inadmissible in all courts of law. Still other "scientific" devices remain questionable, such as the penile plethysmograph (PPG) test, an unspeakable contraption designed to quantitatively measure and predict a male's level of perversion. See *Commissioner v. Sousa*, 4 Mass.App.Ct. 1128, 910 N.E.2d 418 (2009).

Not every form of scientific evidence requires a presentation with respect to *Daubert* compliance. Those that rely solely upon accepted principles of physics, (even precepts involving accident reconstruction) may not require separate *Daubert*-like hearings in order to achieve admissibility. In *State v. Furbush*, 131 Conn.App. 733, 27 A.3d 497 (2011), involving a prosecution for second degree manslaughter with a motor vehicle, methods used by the state's experts to reconstruct a vehicular collision did not involve innovative scientific techniques, and thus the trial court was not required to hold a hearing on the admissibility of the experts' testimony under the *Daubert* standard.

4. *Kumho Tire Company, Ltd. v. Carmichael, supra*, involved an accident that occurred when the right rear tire of a minivan blew out. One of the passengers in the vehicle died; others were severely injured. The success of the Carmichael lawsuit hinged on the testimony of a tire-failure expert who relied on various features of tire technology that the manufacturer did not dispute, as well as background facts about the particular tire on the subject minivan. The expert concluded that a defect in the tire caused the accident; these conclusions, by contrast, were based on certain observations about the tire that Kumho Tire had vigorously disputed. Kumho also disagreed with certain aspects of the tire expert's methodology, and requested that the federal district court exclude it under Rule 702 of the Federal Rules of Evidence. Taking its marching orders from *Daubert* (which had sanctified a gatekeeping role for trial judges considering expert testimony), the district court determined that the tire expert's methods were not scientifically valid, and therefore excluded his testimony and evidence. Carmichael appealed to the Eleventh Circuit, which reversed the district court's ruling. It had opined that *Daubert* was expressly limited only to scientific expert testimony and did not apply to skill" or "experience-based observation." An appeal was taken to the Supreme Court. The High Court noted that the line between "scientific" and "technical" knowledge is not always clear, stating:

> "Pure scientific theory itself may depend for its development upon observation and properly engineered machinery. And conceptual efforts to distinguish the two are unlikely to produce clear legal lines capable of application in particular cases."
>
> The Supreme Court concluded that the district court correctly refused to admit the tire expert's testimony. The district court was required to determine whether or not the tire expert's methods could reliably determine what had caused the tire explosion on Carmichael's van. The expert, who had been employed by Michelin for over 10 years, possessed the required experience. Nor was the visual inspection of the tires an issue. A problem, however, arose when the expert testified that his inspection of the tire led to the conclusion that a

defect caused the tire to explode because he could not see evidence of any other cause. The district court acted within its discretion to exclude the evidence proffered by the tire expert in light of these concerns. Accordingly, the Supreme Court reversed the Eleventh Circuit's decision overruling the district court.

§ *1.23 Legal Arithmetic*

For a light, softball introduction to the chapters that follow, we might take a brief detour to consider some arithmetic controversies as the litigants determine whether or not a plaintiff's claim meets the minimum jurisdictional amount for a given court. Among the first math problems encountered in law school are those pertaining to subject matter jurisdiction in federal district courts. The familiar *$75,000* minimum for federal diversity jurisdiction often required aggregation (i.e., addition). The *Pendley* case is presented as a reminder that as you wander fearfully into physics, statistics and quantum mechanics, there are those practicing lawyers who can't even add.

Pendley v. Union Bankers Ins. Co.

99 Ga.App. 189, 107 S.E.2d 910 (1959)

QUILLIAN, Judge.

While it is true that this court has no jurisdiction of an appeal by direct bill of exceptions from a judgment in the Civil Court of Fulton County in a case wherein the amount involved is less than $300 (Citations omitted), this court will not dismiss the writ of error in a case where it appears that, although the plaintiff prays recovery of the sum of $293.50 for hospital expenses and medicines under the terms of a policy of insurance upon which the suit was brought, the allegations of the petition reveal that in computing the total of the enumerated sums alleged to be due the plaintiff by the defendant insurance company, counsel for the plaintiff has committed a simple error in addition and that the true total sought to be recovered under the terms of the policy is $302.50.

The facts alleged in a petition and not the prayer are determinative of the relief sought (Kinzy v. Waddell, 203 Ga. 689, 47 S.E.2d 872), and this court is required to know judicially, as a matter of common knowledge of the science of mathematics that the sum of '$75 for doctor's bills, $20 for operating room, $10 for X-rays, medicines $7.50, penicillin $10,' is $122.50 and not '$113.50' and that the sum of $122.50 and '$180 for hospital bills' is $302.50 and not '$293.50.' The true amount for which suit is brought being in excess of $300, this court has jurisdiction of the writ of error.

In view of the ruling made in the foregoing note, it is unnecessary to determine whether or not the 25 percent penalty alleged to be due the plaintiff under the provisions of Code, § 56-706 constitutes a part of the 'amount involved' within the meaning of the act governing appellate procedure in the Civil Court of Fulton County (Ga.L.1933, p. 290 et seq.). The defendant's request that the case of General Assurance Corp. v. Roberts, 92 Ga.App. 834(1), 90 S.E.2d 70, be reviewed and overruled insofar as it is therein ruled that such penalty does constitute a legitimate part of such 'amount involved,' is denied and the motion to dismiss the writ of error is denied.

A general demurrer, like an oral motion to dismiss a petition for failure to set forth a cause of action, may be filed at any time (Miami Butterine Co. v. Frankel, 190 Ga. 88, 93, 8 S.E.2d 398, and cit.), and it is not error for the trial court, after duly and properly opening a default, to permit the defendant to file a general demurrer to the petition, nor is it error to overrule the plaintiff's motion to dismiss such demurrer as contravening the provisions of Code, § 81-403, which requires all dilatory pleas to be filed on or before the appearance day.

'The time when an insurance policy shall become effective is an essential element of the contract, and the parties may fix a future date upon which it shall become effective. Clark, Rosser & Co. v. Brand, 62 Ga. 23(1); Metropolitan Life Insurance Co. v. Thompson, 20 Ga.App. 706(7), 93 S.E. 299; Electric City Lumber Co. v. New York Underwriters' Insurance Co., 43 Ga.App. 355(1), 158 S.E. 620; Matlock v. Hollis, 153 Kan. 227, 109 P.2d 119, 132 A.L.R. 1316; 6 Couch on Insurance, § 1328; 29 Am.Jur. 225, § 219; 32 C.J. 1164, § 277. A policy bearing a given date, and purporting to insure for the future only, cannot be made the basis of an action to recover for a loss occurring on a prior date. Fowler v. Preferred Accident Insurance Co., 100 Ga. 330(3), 28 S.E. 398.' Boswell v. Gulf Life Insurance Co., 197 Ga. 269, 272, 29 S.E.2d 71, 73.

Where, under an application of the foregoing principles of law announced in the preceding note, it appears in an action on a policy of insurance that on September 25, 1957, the plaintiff made application for a policy of insurance, paid the sum of $19 and was given a receipt 'for registration fee (if required) and three months premium,' and it is stipulated in the receipt that 'no liability is created or assumed by the company except for refund of this payment, until the policy applied for has been issued,' and by the terms of the policy it is not to become effective until October 15, 1957, such policy does not insure against injury sustained by the plaintiff on October 10, 1957, five days prior to the effective date of the policy. The trial court did not, consequently, err in sustaining the general demurrer to the petition.

Judgment affirmed.

Notes and Problems

1. Simple arithmetic can be tricky where aggregation takes the place of addition. Satisfaction of the minimum-amount requirement for federal diversity jurisdiction (per 28 U.S.C. § 1331) often depends on whether or not the claims of a party can be added together in order to exceed the required $75,000.00. Would the requirement be satisfied if a single plaintiff, for example, sues a single defendant wherein the complaint has two separate related causes of action seeking $40,000 for each claim? What if the two claims were unrelated?

2. Consider another 28 U.S.C. § 1331 problem involving multiple parties instead of claims. Would the requirement be satisfied if a single plaintiff sued several defendants seeking $40,000 from each of them? Can two plaintiffs each having claims in the sum of $40,000 against a single defendant "gang up" up on him?

3. What is the minimum amount for your state's court of general jurisdiction? What is the maximum amount for its small claims court?

§ 1.30 Numbers for the Legislature

Some might be surprised to learn the degree of involvement and respect that our founding fathers had for the subjects of mathematics and physics. George Washington, John Adams, James Madison and Thomas Jefferson, to name a very few, were all quite knowledgeable in at least one of those disciplines. Jefferson in particular was fascinated by Euclidean geometry.

In 1760 at age 17, he attended William and Mary College (the second oldest college in the country next to Harvard), where he then studied mathematics and philosophy. Jefferson once wrote the following: "The faculties of the mind, like the members of the body, are strengthened and improved by exercise. Mathematical reasoning and deductions are, therefore, a fine preparation for investigating the abstruse speculations of the law."[9]

With other founding fathers sharing a similar respect for numbers, it is no wonder that the U.S. Constitution contains many. In addition to its most basic concept, "majority rule," the document requires a 2/3 majority for the senate to ratify a treaty, a vote of 60/100 senators to prevent a filibuster and 538 presidential electors (As opposed to actual numbers, the Constitution instead provides a formula for determining the number of presidential electors). The numbers no doubt flow, in part, from the mathematical predilections of the drafters.[10]

Often called "The Father of the Constitution," James Madison, the fourth president of the United States, was the document's principal author and was responsible for writing more than one third of the federalist papers. As one trained in mathematics, Madison frequently analogized its teachings and principles and developed crucial concepts such as pluralism, along with checks-and-balances.

§ *1.31 Statutory Gymnastics*

Unfortunately, words by persons less skilled than our founding fathers must often be used to describe and legislate the mathematical operations required for governmental statutes, taxing authorities, regulations and standards. Unlike equations, which are usually precise, employing the English language to describe in words those matters that translate into numbers, is not nearly as clear. In terms of the competition, numbers will always be less ambiguous than words. For that reason, regulations containing numerical examples might best accompany statutes that require mathematical operations.

Some situations, particularly those involving tax statutes, are so poorly constructed or drafted that even regulations can cause problems. Of the many examples strewn throughout the Internal Revenue Code, the following regulation apparently garnered an award;

9. Letter from Thomas Jefferson to Bernard Moore (1765).

10. Michael I. Meyerson, *Mathematics and the Legal Imagination: A Response to Paul. Edelman*, 19 University of Baltimore School of Law, Constitutional Commentary 447 (Summer 2002).

the prize, known as the: "*Most Incomprehensible Government Regulation Award,*" was awarded to the Internal Revenue Service for temporary Treasury Regulation 469-1T (f)(2)(i)(C).[11]

> *Significant participation passive activities.* If the taxpayer's passive activity gross income from significant participation passive activities (within the meaning of § 1.469—2T(f)(2)(ii)) for the taxable year (determined without regard to § 1.469—2T(f)(2) through (4)) exceeds the taxpayer's passive activity deductions from such activities for the taxable year, such activities shall be treated, solely for purposes of applying this paragraph (f)(2)(i) for the taxable year, as a single activity that does not have a loss for such taxable year.

If truth be told, for tax practitioners, the above award winner is far less ambiguous than many provisions of the Internal Revenue Code. Ambiguity does, however, carry purpose, the most important involves keeping from the average taxpayer the truth about loopholes for special interests.

A blunder by United States Supreme Court resulted in one of the most famous leveraging schemes in tax history. This "*gift*" to tax lawyers, tax dodgers and inventive accountants consisted of a single infamous footnote in the case of *Crane v. Commissioner*, 331 U.S. 1 (1947) which implied that wealthy individuals could borrow money to invest in worthless property and thereby generate a depreciation deduction. The loans were non-recourse and, therefore, would never be repaid. To fully appreciate *Crane* an understanding of some elementary arithmetic tax rules concerning the purchase and sale of capital assets is required.[12]

Regina v. Ojibway

8 Crim. L.Q. 137 (1965)

BLUE, J. August, 1965

This is an appeal by the Crown by way of a stated case from a decision of the magistrate acquitting the accused of a charge under the Small Birds Act, R.S.O., 1960, c. 724, s. 2. The facts are not in dispute. Fred Ojibway, an Indian, was riding his pony through Queen's Park on January 2, 1965. Being impoverished, and having been forced to pledge his saddle, he substituted a downy pillow in lieu of the said saddle. On this particular day the accused's misfortune was further heightened by the circumstance of his pony breaking its foreleg. In accord with Indian custom, the accused then shot the pony to relieve it of its awkwardness. The accused was then charged with having breached the Small Birds Act, s. 2 of which states: "2. Anyone maiming, injuring or killing small birds is guilty of an offence and subject to a fine not in excess of two hundred dollars." The learned magistrate acquitted the accused holding, in fact, that he had killed his horse and not a small bird. With respect, I cannot agree.

In light of the definition section my course is quite clear. Section 1 defines "bird" as "a two legged animal covered with feathers." There can be no doubt that this case is covered by this section.

Counsel for the accused made several ingenious arguments to which, in fairness, I must address myself. He submitted that the evidence of the expert clearly concluded that the animal in question was a pony and not a bird, but this is not the issue. We are not inter-

---

11. McAllister, *Giving IRS the Prize for Gobbledygook; Accountants Baffled By Tax Regulation.* The Washington Post, April 13, 1990.

12. See *infra*, § 9.52.

ested in whether the animal in question is a bird or not in fact, but whether it is one in law. Statutory interpretation has forced many a horse to eat birdseed for the rest of its life.

Counsel also contended that the neighing noise emitted by the animal could not possibly be produced by a bird. With respect, the sounds emitted by an animal are irrelevant to its nature, for a bird is no less a bird because it is silent.

Counsel for the accused also argued that since there was evidence to show accused had ridden the animal, this pointed to the fact that it could not be a bird but was actually a pony. Obviously, this avoids the issue. The issue is not whether the animal was ridden or not, but whether it was shot or not, for to ride a pony or a bird is of no offence at all. I believe counsel now sees his mistake.

Counsel contends that the iron shoes found on the animal decisively disqualify it from being a bird. I must inform counsel, however, that how an animal dresses is of no consequence to this court. Counsel relied on the decision in Re Chicadee, where he contends that in similar circumstances the accused was aquitted. However, this is a horse of a different colour. A close reading of that case indicates that the animal in question there was not a small bird, but, in fact, a midget of a much larger species. Therefore, that case is inapplicable to our facts.

Counsel finally submits that the word "small" in the title Small Birds Act refers not to "Birds" but to "Act", making it The Small Act relating to Birds. With respect, counsel did not do his homework very well, for the Large Birds Act, R.S.O. 1960, c. 725 is just as small. If pressed, I need only refer to the Small Loans Act, R.S.O. 1960, c. 727 which is twice as large as the Large Birds Act.

It remains then to state my reason for judgment which, simply, is as follows: Different things may take on the same meaning for different purposes. For the purpose of the Small Birds Act, all two-legged, feather-covered animals are birds. This, of course, does not imply that only two-legged animals qualify, for the legislative intent is to make two legs merely the minimum requirement. The statute therefore contemplated multi-legged animals with feathers as well. Counsel submits that having regard to the purpose of the statute only small animals "naturally covered" with feathers could have been contemplated. However, had this been the intention of the legislature, I am certain that the phrase "naturally covered" would have been expressly inserted just as the word "Long" was inserted in the Longshoreman's Act.

Therefore, a horse with feathers on its back must be deemed for the purposes of this Act to be a bird, and a fortiori, a pony with feathers on its back is a small bird.

Counsel posed the following rhetorical question: If the pillow had been removed prior to the shooting, would the animal still be a bird? To this let me answer rhetorically: Is a bird any less of a bird without its feathers?

Appeal allowed.

Notes and Problems

1. *Regina v. Ojibway* demonstrated the absurdity to which literal statutory interpretation can lead. The case provides an example of "*textualism*," one of the three general methods of statutory interpretation. A favorite of constitutional strict constructionists, textualism suggests that we accept the literal words as they are written. In contrast, "*purposivism*" implies that we should look behind the literal language to derive the purpose of the writing prior to interpretation; and lastly "*old intentionalism*" suggests that we examine the intent of the drafters. How would you define "bird" so as to avoid the case's ridiculous outcome?

2. Does the case deliver a lesson for drafting language involving complex numbers or financial arrangements? If so, what is it? Would it help to provide examples that included detailed calculations?

3. Does a four legged creature have two legs? Why isn't the result in *Regina v. Ojibway* perfectly reasonable?

4. Rumor has it that the case of *Regina v. Ojibway* never existed but instead was a made-up solely for jurisprudence casebooks to demonstrate the ridiculous extremes of textualism. What do you think?

§ *1.32 Legalized Mathematics*

Although rare, there is nothing more comical than a politician attempting to solve mathematical problems through the legislative process. Attempts to legislate mathematical truths by way of law (instead of visa versa) can serve up some very entertaining lunacy. It's a case of the blind attempting to lead the eagle-eyed.

You may recall, from junior high school, your first bout with π (an irrational number) which never "came out even" during your attempts to deal with circumferences and diameters. Consider the attempt by one Edwin J. Goodman to resolve the irrationality of π by proposing a different measuring standard. It seems that Ed just couldn't seem to comprehend or accept the meaning of "irrational." So Ed's Indiana House Bill #246 (1897) gave us this:[13]

> "Be it enacted by the General Assembly of the State of Indiana: It has been found that the circular area is to the quadrant of the circumference, as the area of an equilateral triangle is to the square of one side. The diameter employed as a linear unit according to the present rule in computing the circle's area is entirely wrong."

Goodman's method would cause pi to equal 9.2376, which, is of course, clearly incorrect (So who's being irrational now?). As you head-out to explore the remainder of the text, try to put aside your pessimism and remember that not every jurist is a lunatic.

Notes and Problems

1. Do you suppose that any politician who ever used the term "fuzzy math" actually understood basic mathematics? Or was the fuzziness in the mind of the politicians as opposed to the math?

2. Analyze the *Class Action Fairness Act of 2005* and develop a set of mathematical equations (inequalities) that would determine whether or not a federal court, *must, may,* or *cannot* exercise subject matter jurisdiction based on the residency of the class.

3. As lawyers, we basically do what our accountants and CPAs direct. But the bean counters, like lawyers, are far from perfect. In fact, most taxes are confusing and subject to multiple interpretations. The Internal Revenue Code, for example, has been properly termed: "The most complex body of laws in the history of the human race." There is a substantial body of law suggesting that taxing statutes should be strictly construed against the taxing power and in favor of the taxpayer, and further that all doubts should be resolved

13. See Peter Beckman, *A History of Pi* (St. Martin's Griffin, 1971).

in favor of the taxpayer.[14] Later, we will see how a clever mathematical interpretation of a mere footnote in a United Supreme Court case saved clever taxpayers (and cost the government) billions. Incredibly, the footnote was nothing more than hypothetical obiter dictum.

4. Many decades ago the state of Michigan decided to replace a multitude of business taxes (intangibles tax, inventory tax, etc.) with a so-called Single Business Tax, which was eventually repealed as of December 31, 2007. It was a gross-bearing tax that was unpopular with many. The focus for such a tax was gross receipts as opposed to net income. (i.e., you could end up with a net loss for the year and incur a tax, nevertheless). After the tax was implemented, one of several errors occurred on the tax form. A simple misuse of language caused a critical numerator to become a denominator. Thus, if you had calculated your "proper" tax liability to be $9,000, you were, according to the form only obligated to pay 1/9,000 or $0.0001111. Basically, you were required to pay nothing. Notwithstanding the "clear" directions on the form, approximately 60% of CPAs and accountants ignored the error (knowing it to be an error and directed full payment), while a substantial portion of the population simply did not pay the infamous, hated and now-repealed tax.

Years, later, I inquired as to whether there was ever going to be any follow-up for those who did not ignore the error. I was told by a state official that there would be none, "There's nothing that we [i.e., the state] could do about it except correct the mistake for future years." This is only one example of errors made by taxing authorities at all levels on a regular basis.

§ 1.40 Numbers for the Practitioner

Regularly practitioners use numbers and figures. Billing, whether hourly or contingent, requires calculations as does nearly every other aspect of a legal practice. The number of times that math is used in negotiating, drafting and litigating issues involving contracts or negotiable instruments is immeasurable. That will become more apparent as we proceed.

§ *1.41 Contracts and Negotiations*

Particular care is needed when numbers or calculations are to become material parts of a contract. Mathematical simplicity is preferred, even where sophisticated derivative transactions are involved. Regardless of whether or not quantitative issues are involved, complexity is often a sign of poor draftsmanship or an attempt by a party with devious intentions to gain an advantage. For basic contracts or negotiable instruments, the level of math needed should rarely exceed that which would be required to compute compound interest.

The plain-language movement has had an impact on many types of contracts, but simplification is not a magic bullet for the elimination of ambiguity and confusion. Throughout this chapter, and indeed, the entire book, it should become abundantly clear

14. See Steve R. Johnson, *Pro-Taxpayer Interpretation of State and Local Tax Laws*, 51 State Tax Notes (William S. Boyd School of Law, University of Nevada) 441 (2009).

that numbers present special drafting problems. A partial solution involves referencing or attaching specific examples to contracts. Fictitious numbers would be assigned to a variety of transactions. Each transaction would then be followed by the specific amounts that the parties would be entitled to receive under specified conditions.

Back in the 1970s, pursuant to a contract with the United States Navy, Rudolph Flesch and Peter Kincaid became known for developing a set of scales and tests that purported to measure how difficult or easy it was to read a given piece of text based upon the length of the words and sentences comprising it. These formulas were based upon the notion that there is a negative correlation between readability and the length of words and sentences. In the Flesch Reading Ease Test, a high score indicates that the material is relatively easy to read while lower numbers indicate greater difficulty. The formula for the Flesch Reading Ease Score (FRES) test is as follows:

$$206.835 - (1.015)\,\frac{\text{Total Words}}{\text{Total Sentences}} - (84.6)\,\frac{\text{Total Syllables}}{\text{Total Words}}$$

SCORE	SIGNIFICANCE
90–100	Easily understood by average 11-year-old students
60–70	Easily understood by average 13–15-year-old students
0–30	Best understood by university graduates

The ratios intuitively seem to make sense since shorter sentences and fewer syllables are generally easier to read and have a greater tendency to avoid ambiguity. The creators' selection of constants, however, resulted in a situation where the Score has no lower bound. It seems as if a more coherent scale would have eliminated negative values since zero does not seem to possess any particular significance. The Test, nevertheless has achieved wide acceptance; its scale has even been used to set minimum readability requirements for ballot measures. See *Greene v. Kulongoski*, 903 P.2d 366 (1995) and *Arenz v. Keisling*, 855 P.2d 1109 (1993). It has also been used to judge the coherence of questionnaires. See *Roth v. Ohio Dept. of Commerce*, 171 Ohio App.3d 326 (2007). In addition, the Test has been employed as a basis for readability requirements in contracts of insurance. See *Daly v. Paul Revere Variable Annuity Ins. Co.*, 489 A.2d 1279 (N.J. Super.L.,1984).

Cintron v. Universal Underwriters Group, 601 A.2d 1051 (Del.Super., 1990) involved an insured's action against a motor vehicle insurer, seeking to recover benefits for medical bills and lost wages resulting from an accident. With respect to the poorly drafted policy, in ruling for the insured, the court stated:

> The legislature designated the Flesch Formula for Readability as the test of readability and generally adopts a minimum score of readability of not less than 40. It requires that insurance policies be readable and understandable by a person of average intelligence and education. 18 Del.C. § 2741. The Flesch Scale is between 0 (practically unreadable) and 100 (easy for any literate person). See, R. Flesch, How to Write, Speak and Think More Effectively, (1960); R. Flesch, How to Write Plain English: A Book for Lawyers & Consumers, (1979). See, also, State

Farm Fire & Cas. Co. v. Alstadt, Cal. 4th Dist., 113 Cal.App.3d 33, 169 Cal.Rptr. 593 (1980); Deras v. Roberts, Ore.Supr., 785 P.2d 1045 (1990); Deras v. Roberts, Ore.Supr., 788 P.2d 987 (1990) (discussing application of the Flesch Formula).

Reader's Digest magazine has a readability index of approximately 65. Time magazine scores about 52. The Harvard Law Review has a general readability score in the low 30s. The highest score would go to the old *Dick and Jane* books of kindergarten fame.

It is notable that the Flesch Formula does not measure the compound use of disjunctives or conjunctives. Nor is there any reduction in readability for the inclusion of mathematical or scientific components. Where a contract involves calculations, the attorneys' common practice of attaching or including specific examples attests to the added complexity and potential ambiguity.

Negotiation is, of course, a separate skill, though closely related to contract drafting. The goal centers around the ubiquitous attempts to obtain something that you want more than that which you are giving up. It's basically a numbers game, but as we shall see later, terminology can be critical.

One particular rule of lawyer negotiation is worth mentioning because despite its importance, it seems to be unwritten. It applies to all forms of negotiation, litigation settlements in particular, since that seems to be a common numbers-related discussion. The rule is simple: Once a number is uttered by a party, it never comes off of the table, especially when a settlement judge or mediator is involved. From a plaintiff's perspective, when a number leaves the mouth, he or she will never be able to settle *above* that amount. Lawyer epithets such as "This offer is, of course, subject to client approval," or "This offer will remain on the table for a day" are pragmatically meaningless. The corollary is also true with respect to a defendant, who will never be able to settle for *less* than any number that leaves his or her lips. These are not laws of math or ethics; but they seem to be immutable laws of nature. Exceptions exist; but experience has demonstrated that they are statistically rare.

§ *1.42 Litigious Numerics*

Numbers play an uncountable variety of roles in the litigation process, both in the civil and criminal arenas. Throughout the remainder of this treatise, we can only hope to scratch a finite number of surfaces.

Basic mathematics takes us into the area of patents, boundaries and statutory distances (Chapter 2); logic and set theory are prevalent throughout all aspects of the litigation process (Chapter 3); probability theory has assumed a prominent position in the criminal evidence arena, especially since the advent of DNA match-ups (Chapter 4); statistical evidence has played an increasingly large role in criminal matters, product liability and every conceivable type of action predicated upon discrimination (Chapter 5); with the expanding use of demonstrative evidence, mathematical charts and graphs are becoming regular fare in the courtroom (Chapter 6); both classical and modern physics frequently make their appearances to explain a wide variety of physical phenomena (Chapters 7 and 8); and with respect to accounting and finance the presence of our subject is somewhat self-explanatory (Chapter 9).

§ *1.43 Quantitative Ethics*

The Model Rules of Professional Conduct of the American Bar Association require "competence." One would think that a command of basic arithmetic would be included as a "given." However, years ago, what 8th graders considered as nothing more than common sense is, today, at least in the eyes of one attorney, brilliance that justifies an otherwise excessive attorney fee. In the *Matter of Fordham*, 423 Mass. 481, 668 N.E.2d 816 (1996), attorney Laurence Fordham was accused of unethical conduct for charging (for a drunk driving defense) an excessive fee that was several times the amount normally charged in the community ($50,000). The basis for the high fee was a dismissal that resulted from the suppression of the client's breathalyzer results. Fordham argued that 0.02 (the client's blood-alcohol reading) was not *less* than 0.02 (the statutory range within which the two tests were required to fall). Thankfully, the court was not swayed by Fordham's expert witness who nevertheless termed the defense as being "very ingenious."

Serious ethical problems abound with respect to client retainer agreements, engagements letters and contingent fee agreements. In California, personal injury contingent fee agreements typically provide that an attorney's fee is to be paid from the gross recovery prior to deductions for court costs, discovery costs and medical expert fees. Client's rarely understand the egregious mathematical results, which lawyers sometimes neglect to explain. Consider the following hypothetical settlement statement where the attorney achieved a low recovery and claimed a 50% contingent fee (*Method 1*):

Gross Recovery	$200,000	
Less attorney fee	-$100,000 (50%)	
Less deposition costs	-$5,000	
Less expert witness fees	-$55,000	
Less court costs	-$2,000	
	$162,000	
Net Recovery for Client		$38,000

In an unsuccessful trial with a high priced expert witness, it is, of course conceivable that the client could end up with nothing despite a handsome attorney fee. Consider this more equitable distribution (*Method 2*):

Gross Recovery	$200,000	
Less deposition costs		-$5,000
Less expert witness fees		-$55,000
Less court costs		-$2,000
Total Costs and Fees	$62,000	
Net Recovery	$138,000	
Less attorney fee	-$69,000 (50%)	
Net Recovery for Client	$69,000	

By permitting *Method 1* to exist, do you consider the California State Bar's indifference: rational? Unconscionable? Angelic?

Separate and apart from the ethical issues plaguing contingent fee agreements are those inherent in hourly billing calculations. Aside from the usual problems involving "padding," there are legitimate concerns over the nature of the services provided and the particular level of expertise of those providing them. In Newtonian parlance, *time is time*; therefore, one might argue that 1 hour of time spent by a neophyte paralegal sitting in traffic on the way to file a pleading should be billed at the same hourly rate as the firm's chief trial attorney while in trial. Common sense and sound ethics, however, suggest the imposition of a more relativistic treatment of "time." In most situations involving hourly billing, the retainer agreement or engagement letter between the client and the firm will govern, but not always.

Disputes over the reasonableness of billing practices often arise where an attorney has been discharged, or has received an extremely large fee in a class action (which is subject to court approval regardless of the agreement among the parties) or where fees are to be shifted to the opposing party for whatever reason. Enter the so-called *Lodestar method* of billing. *Lindy Bros. Builders, Inc. v. American Radiator & Standard Sanitary Corp.*, 487 F.2d 161 (3rd Cir., 1973) gets the credit for this iconic calculus of billing that is universally recognized. In ruling on a fee dispute, the *Lindy* court coined the title by stating that: "[T]he amount thus found to constitute reasonable compensation should be the *lodestar* of the court's fee determination."

The *Lodestar* is a method for calculating attorneys' fees whereby the lawfirm multiplies a reasonable hourly rate by a reasonable number of hours expended, for each firm member, then takes the sum of the individual products. In order to calculate a fee award, the court must determine: The reasonableness of the hourly rate charged, and the reasonableness of the hours expended on the litigation. An attorney's usual billing rate is presumptively the reasonable rate, provided that this rate is in line with those prevailing in the community for similar services (by lawyers of reasonably comparable skill, experience and reputation). Mathematically speaking, assuming that there are n persons who work on a given file:

$$\text{Lodestar} = \sum(\text{Billable Time})_n =$$
$$(\text{time}_1 \text{ x rate}_1)_{\text{Member1}} + (\text{time}_2 \text{ x rate}_2)_{\text{Member2}} + \ldots (\text{time}_n \text{ x rate}_n)_{\text{Membern}}$$

According to *Perdue v. Kenny A. ex rel. Winn*, 130 S.Ct. 1662 (2010), the lodestar approach has "achieved dominance in the federal courts." Although imperfect, it has several important virtues: It produces an award that approximates the fee that the prevailing attorney would have received for representing a paying client who was billed by the hour in a comparable case; and it is readily administrable. *Gisbrecht v. Barnhart*, 535 U.S. 789, 801 (2002).

In calculating the final bill, any ambiguity in a fee agreement should be construed against the attorney as the drafter of the agreement. *In re Clark*, 405 B.R. 457 (E.D.Mich., 2009); *Hill v. Astrue*, 248 Fed.Appx. 923, (10th Cir., 2007); *Mello v. Davis*, 182 S.W.3d 622 (Mo.App. E.D., 2005); *In re Seigel*, 300 A.D.2d 668, 754 N.Y.S.2d 300 (2002); *Lane v. Wilkins*, 229 Cal.App.2d 315 (1964). Texas, however, seems to eyeball things differently. See *Anglo-Dutch Petroleum Intern., Inc. v. Greenberg Peden*, P.C., 267 S.W.3d 454 (Tex.App., 2008).

§ 1.50 Numbers for Sale

There exists an esoteric and somewhat controversial subject that has been debated for hundreds of years involving the origin of mathematics itself. There are those who believe that the mathematical structures of the universe (with emphasis upon the interrelationship of certain important constants) were put into place by some non-human force and then discovered by mankind (Platonic Theory);[15] by contrast, others claim that all of mathematics was merely invented (Constructivist Theory) instead of having been discovered. In either event, the rules of pure mathematics cannot be monopolized. No person has a right to copyright or patent the laws of mathematics; this fact is clear. Nor does anyone have a right to lay claim to the natural laws of physics. Both sets of laws are owned by Mother Nature, if anyone. As a corollary, a "principle" or abstract sense of a fundamental truth, original cause, or motive is not patentable. *Le Roy v. Tatham*, 55 U.S. 156 (1852). It may, nevertheless, surprise many to learn that equations can in fact be patented if they describe a particular process. See supra, §2.32.

§ 1.60 Numbers for Deception

Among the most important subjects covered in this text are the many ways in which the numbers and images of mathematics and physics are distorted. The rigorous reliability of the two disciplines renders the deception that much more offensive, and the lawyers' *need to know* that much more acute. According to author Dostoevsky: "Man has such a predilection for systems and abstract deductions that he is ready to distort the truth intentionally, he is ready to deny the evidence of his senses only to justify his logic." Lawyers are not the only professionals guilty of this misconduct but it seems as if they are the most notorious.

Mindful that our quasi-logical legal system is always subject to suspicion, as lawyers, we often seek to ally ourselves with scientific and mathematical systems precisely because *they* are not. We are raised to blindly trust certain sciences and the readily available experts who indulge in them even though for every expert who professes a principle, there is at least one ready to refute it.[16]

Throughout the remaining chapters, we will see how truths are twisted. I am not referring to outright fraud or unethical behavior; those companions seem less dangerous and insidious when compared to the distortions that can be achieved through our most trusted scientific disciplines. When politics, sociological studies and rhetoric are used to mislead, we think little of it; after all, fools regularly succumb to their pretended accuracies. But when mathematics and physics, the most trusted truths of our universe, speak falsely, everyone has reason to be concerned.

---

15. See http://works.bepress.com/cgi/viewcontent.cgi?article=1000&context=ashley_lipson. Lipson, "*The Intelligent Construction of the Universe: a Mathematical Proof—the Link among Science, Natural Law and Jurisprudence*," 2007.

16. The Technical Advisory Service for Attorneys (known simply as "TASA") provides experts in over 10,000 categories of scientific of expertise ranging from "adhesives" through "zippers" and "zoology."

§ 1.70 Numbers for Law Professors[17]

Long ago law professors, seizing upon the name of a far better-organized philosopher than themselves, created the so-called Socratic system of instruction, an excellent technique when used properly. But like many good ideas, undesirable myths attached themselves, contaminating the main purpose. One such myth propelled by the all-knowing Professor Kingsfield in the Oscar winning film, *Paper Chase*, suggests that: In law there are no answers, only questions. This, in turn, implies that there is no reason to construct basic systems, axioms or classification schemes, such as those employed by virtually every other logic-based endeavor. Instead, the myth suggests that we should allow our students to wade through a Grand Canyon of cases and notes, as they pretend to construct their own sign posts along the way, attempting to find their own answers, which we say don't exist.

Imagine that you're a mathematician or scientist teaching a course with little uniformity and no standard axioms, structures or definitions. You might conclude that if Science and Mathematics were taught by law professors, perhaps we would all still be calculating with an abacas, debating which beads were being unfairly discriminated against. Thus, the discipline that the law demands appears to be lacking in the very educational structure responsible for its dissemination. And this lack of discipline shows. It shows in the shortcomings of our practitioners, the inefficiency of our judicial system and the poor development of our laws and statutes.

If the goal, therefore, is to confuse the students, especially the 1Ls, then why not do it right. Many casebooks already do it "right." But for those writers preparing to write a new casebook, here are the "Dirty Dozen" rules that must be followed:

[1] Take the most important rules of law and bury them in the case notes. That way, the students who are worried only about being called upon in class (but don't have time to read the notes) will be thoroughly confused and mislead.

[2] Present a long series of cases, all aiming toward one beautiful immaculate truth; then, at the tail end of some obscure case note, cite a statute that abrogates it.

[3] In your bold-titled cases, present lengthy unedited opinions that present narrow exceptions based upon silly fact patterns; even better if the case involves a crooked bigamist attorney (such as John Mitchell of *Pennoyer v. Neff* fame).

[4] Early in the book, be sure to present cases that utilize complex procedures and terminology that even lawyers have yet to understand. Throw in some olde English cases from the 1700s.

[5] Edit out that part of the case that tells who won. Or better yet, carefully cut bits and pieces so that only incoherent, discombobbled text remains.

[6] Never include cases written by judges, who in plain English, tell us the outcomes. Make certain they contain language like this:

> This court, therefore, modifies the lower court as to issues five and seventeen, which affirmed and at the same time reversed the trial court's ruling during the second re-hearing as to issues three and six, and is therefore, modified in accordance with the remainder of this incoherent opinion. Reversed on other grounds.

17. Ashley S. Lipson, *The Dirty Dozen*, republished from *Techniques For Teaching Law II*, Gonzaga University School of Law (2011).

[7] In your Teacher's Manual, instruct professors to tell the students that *Young v. Guy* and *Ferens v. John Deere* are "easy" cases that can be understood by any neophyte student; be sure to do this to the students during pre-finals week.

[8] Before students have learned the significance of the Supreme Court or the New York Court of Appeals, give them some long winded slip opinions from a Mississippi trial court (with big bold titles, of course). If you can't find one, a Trent Lott speech will do.

[9] Never present a cohesive, logically structured outline. Instead, scatter the topics hither, thither and yon, or do as most, and scatter them Curley, Moe and Shemp.

[10] Always select cases that have several parties and entities with the same names. That way, the students won't be able to distinguish the plaintiffs from the defendants.

[11] Always let the tail (minority positions) wag the dog when it comes to selecting cases for purposes of teaching precedents.

[12] Never define a complex legal term; instead, refer the students to that "other" pillar of high priced, elegantly bound, confusion—*Black's Law Dictionary*.

Chapter 2

Math Basics

"Mathematics possesses not only truth, but also supreme beauty."

—Bertrand Russell

Replacing your fear and hatred of mathematics with an appreciation of its beauty and power does not come smoothly. It is far too easy to get bogged down with repetitive number crunching, or lost in a jumble of symbols and confusing equations.[1] Remembering early math classes often conjures up an image of the back side of a sergeant-like school teacher with her rifle arm extended toward a blackboard delivering screeching chalk noises to accompany some repetitive information. I can't recall any teacher ever lecturing about the ability of math to reach out and solve the great mysteries of the universe.

§ 2.10 Impressive Symbolism

Mathematicians, physicists, statisticians, chemists, geneticists and logicians make extensive use of scarey symbols. For them, symbols provide an efficient, effective shorthand and most importantly, they are algebraically manipulable in accordance with well-defined, universally accepted rules and laws. The rules are often very complex. Perhaps accounting is the only topic in this text that uses 10 simple digits to express its own ideas and truths.

There is a negative side to the use of symbols. Just as lawyers are often accused of shrouding their secrets and deceptive communications behind a wall of legalese, so too are scientists sometimes suspected of hiding their findings and opinions under a blanket of incomprehensible symbolic logic. There has never been a "Plain English" movement in the scientific arena. Nevertheless, we need symbolic logic.

The *trick* is simply not to be frightened, nor intimidated. In fact, a few simple tables (infra) have managed to corral all of the symbols that you should ever need. Often, the scariest looking symbols deliver the simplest messages. If, for example, someone were to hand you a small set of numbers (x_1, x_2, x_3, x_4, ... x_n), let us say "N" numbers to be exact, and ask you calculate the average, you would have very little difficulty doing so. But if that same person were to give you the same task by ordering you to calculate the "mean of a sample" using Equation (a) below, as a typical law student, you might have a seizure even though, there is no difference between Equations (a) and (b):

1. At its highest level, mathematics has the power to reveal truths about time, space and matter. Even Einstein appreciated the difference between repetitive busy-work and math's higher purpose, when he said: "*I want to know God's thoughts; the rest are details.*"

(a) $\mu = \Sigma x_n / N$ (b) Average = Sum / Total Number

In (a), the result, i.e., the average or mean is represented by the Greek symbol μ (Notice that a fancy Greek symbol is far more impressive than a simple "A" for "average"). The Σ symbol, one of the most unnecessarily intimidating symbols, means nothing more than "the sum of."

The index notation (x_n) is nothing more than a shorthand for a list of numbers.[2] This particular symbolism is worth taking the time to understand; it is used heavily in matters of probability and statistics, areas of great concern to the legal profession.[3]

Example 2-1: Let 2, 3, 3, 5, 6, 7, 9 be a list of numbers. Add these numbers; but make your work look as impressive as if you were billing someone hourly for the work.

Solution: $x_1 = 2 \quad x_2 = 3 \quad x_3 = 3 \quad x_4 = 5 \quad x_5 = 6 \quad x_6 = 7 \quad x_7 = 9$

$\Sigma x_i = x_1 + x_2 + x_3 + x_4 + x_5 + x_6 + x_7$

$\Sigma x_i = 2 + 3 + 3 + 5 + 6 + 7 + 9$

$\Sigma x_i = 35$

Example 2-2: There are other ways to express a simple sum. Another common symbolic notation might read as follows: "The sum of 7 numbers starting with the Index 1 and ending with Index 7." Notice that the index notation is nothing other than a method for assigning a name or title to the individual values.

$$\sum_{n=1}^{7} n$$

We might prefer to drop the details and formalities and use the following notation as an alternative: $\Sigma x_n = 2 + 3 + 3 + 5 + 6 + 7 + 9 = 35$

§2.11 *The Nouns and Verbs of Mathematics*

The following Table 2-1 provides a fairly complete list of all of the mathematical operations that most lawyers should ever need. They represent the "verbs" of mathematics, physics and logic or otherwise involve the performance of some calculations or computations. Combined with Table 2-2 (the nouns), you should have enough tools of the trade to keep you busy learning the mysteries of the universe, and discovering the methods by which information is twisted and misused. The importance of the symbols will become more apparent in the chapters that follow.

2. Einstein carried shorthand notation to an even higher (more abbreviated) level when he greatly shortened otherwise lengthy vectors and tensors to prove his theory of general relativity. His methodology is often referred to as the "Einstein Notation."

3. According to Oliver Wendell Holmes Sr. "All economical and practical wisdom is an extension of the following arithmetical formula: 2 + 2 = 4. Every philosophical proposition has the character of the expression a + b = c. We are mere operatives, empirics, and egotists until we learn to think in letters instead of figures." Oliver Wendell Holmes, *The Autocrat of the Breakfast-Table* (Phillips, Sampson, and Co., 1858).

Table 2-1 [Mathematical Operations]

SYMBOL	MEANING	EXAMPLE
$\neq$	Does not equal	$5 \neq 9$
$<$	Is less than	$5 < 9$
$\leq$	Is less than or equal to	$9 \leq 9$
$>$	Is greater than	$9 > 5$
$\geq$	Is greater than or equal to	$5 \geq 5$
$\Rightarrow$	Implies	$a \Rightarrow b$
$\Leftrightarrow$	If and only if (also "iff")	$(a \Leftrightarrow b) \Rightarrow (a \Rightarrow b \;\&\; b \Rightarrow a)$
$\subset$	Is a proper subset of	$\{a, b\} \subset \{a, b, c\}$
$\subseteq$	Is a subset of	$\{a, b, c\} \subseteq \{a, b, c\}$
$\supset$	Is a proper superset of	$\{a, b, c, d\} \supset \{b, c\}$
$\supseteq$	Is a superset of	$\{a, b, c, d\} \supseteq \{a, b, c, d\}$
$\not\subset$	Is not a subset of	$\{a, b, c\} \not\subset \{a, c\}$
$\not\supset$	Is not a superset of	$\{a, b\} \not\supset \{a, b, c\}$
$\cap$	Is an intersection of	$\{a, b, c\} \cap \{a, b\} = \{a, b\}$
$\cup$	Is a union of	$\{a\} \cup \{b\} = \{a, b\}$
$y = f(x)$	y is a function of x	$y = 3x$
$\surd$	Root Radical	Square root of 2
ε	Is an element of	$\{a\} \; \varepsilon \; \{a, b, c\}$
$\notin$	Is not an element of	$\{t\} \notin \{a, b, c\}$
Δ	Change in	$\Delta d = x_2 - x_1$
$\int$	Integration	Examples provided later
dy/dx	Differentiation	Examples provided later
Σ	Summation	See *supra*, Example 2-2
+ - / ×	Arithmetic operations	Elementary operations
!	Factorial	$4! = 4 \times 3 \times 2 \times 1$

As stated, Table 2-1 listed our mathematical operations (verbs); this next table presents the universe of mathematical "nouns," which unsurprisingly happens to consist primarily of numbers. At the very top of the food chain is that set of numbers that we refer to as "complex." In junior high school, they were first introduced as "imaginary numbers" because for every complex number, one of its two components included the square root of -1 as a factor. In the real number system (which is a subset of the complex number system) there is no such thing as the square root of a negative number. Attaching the label "imaginary" to complex numbers ($\mathbb{C}$), however, was a foolish trivialization of an important concept. Complex numbers not only exist, but they form an indispensable tool for both physics and mathematics. All complex numbers are of the form: $a + bi$, where a and b are real numbers and i is the square root of -1.

The set of all real numbers ($\mathbb{R}$) form a proper subset of complex numbers ($\mathbb{R} \subset \mathbb{C}$). This becomes more apparent by merely considering all elements of $\mathbb{C}$ wherein $b = 0$. With the elimination of i as a factor, we are left only with real numbers.

The set of rational numbers, represented by Q is a proper subset of the set of real numbers. Every real number is either rational or irrational. An irrational number cannot be represented by a fraction, nor can it be represented by a repeating decimal. Rational numbers are those that can be represented by a fraction or a repeating decimal. Next, the integers (Z), often referred to simply as "whole numbers" form a proper subset of the set of rational numbers (Q). And, natural numbers ($\mathbb{N}$) continue the process by forming a proper subset of the integers. Natural numbers are merely positive integers; the zero and the negatives are not included. Therefore: $\mathbb{N} \subset Z \subset Q \subset \mathbb{R} \subset \mathbb{C}$.

Table 2-2 [Numbers]

SYMBOL	SET	EXAMPLES
$\mathbb{C}$	Complex Numbers	4 + 3i, 1, 2, 3, -4, I
$\mathbb{R}$	Real Numbers	π, 1, -1, 1234.5678
Q	Rational Numbers	3.4245, -.666666667
Z	Integers	-1, -2, 1, 0, 2, 3, 4, 5, 6
$\mathbb{N}$	Natural Numbers	1, 2, 3, 4, 5, 6, 7, 8
$\aleph_0$	Countably Infinite Numbers	{$\mathbb{N}$}
$\aleph_1$	Uncountably Infinite Numbers	{$\mathbb{R}$}
∞	Infinity	Undefinably Large, $\aleph_0$ or $\aleph_1$
∅	The null set	Nothing

The $\aleph_0$ (pronounced "aleph-naught") represents the size (cardinality) of any set that is infinite but countable, meaning that each of its elements can be placed in a one-to-one correspondence (i.e., lined up) with the natural numbers. $\aleph_1$ on the other hand represents an uncountably infinite set and cannot be placed in a one-to-one correspondence with anything.

§2.12 *Special Numbers*

A "constant" is nothing more than a number. The term is often used to distinguish variables such as the infamous "x." All of the elements of all of the sets in Table 2-2 are constants. Some constants, however, comparatively few in number, happen to be very special. They form a connecting bridge between the universe of abstract mathematics and the real world. See Table 2-3.

These special constants have special properties and happen to appear more frequently in mathematics and physics than others even though they are no less "numbers," plain and simple. The number 1, known as the multiplicative identity, if multiplied by any number, will yield that same number without any alteration. Likewise, the number 0, the additive identity, can be added to any number without altering it. The number zero does, however, have mathematical meaning separate and apart from the null set {∅} referenced in Table 2-2. Unlike ∅ (an empty set), the number zero is a real number. It has real meaning, such as the net value of a $1,000 claim that is offset by an equal counterclaim, or the amount that your HMO intends to pay you if you are injured.

The constants are special because they form ratios of natural objects or tangible concepts that never change. As ratios, they do not depend upon the existence of any constructured scheme or system. If, for example, you were to pick any circular object in the universe and divide that object's circumference by its diameter, the result would always be the same here as on Mars. The magnitude of that ratio would not change regardless of whether we called it "pi," π, or "apple pie," and regardless of whether we used a decimal, binary, Arabic, Roman or alien numbering system to measure it. Pi (π) is familiar to all; this particular constant is deceptively elementary, yet extremely mysterious and impenetrable.[4] Its value in our decimal system will always equal:

3.14159265358979323846264338327952884197 . . .

The trailing dots stand as a perpetual reminder that we will never truly touch, see or write the exact number. That is because the decimal expansion continues forever and never repeats, implying that π is an irrational number and one that cannot be represented as a fraction.[5] The fact that a set of only ten digits can be written forever and never "repeat" is in itself miraculous and difficult to imagine.

The number π is not only irrational, it is also transcendental, meaning that it cannot be represented algebraically. Even if we employ a different numbering system or nomenclature, it would make no difference. The value would always equal the same constant. Whatever created the universe and created π, placed its grasp beyond our reach while waving it teasingly in our faces every time we see the sun, the moon, the rim of a cup, a sphere or any circular object.

The fourth item on your chart (Table 2-3) is an equally mysterious constant and is known simply as "*e*." Forming the base for our so-called natural (Napierian) logarithms, the number $e = 2.718281825459045235360287 4\ldots$. It is also an unchanging number. As is the case with π, we only have an approximation, because like π, the constant e is both irrational and transcendental. Raised to the variable power "x," e and only e produces a function that equals its own derivative (i.e., slope).[6] Whereas π has great "static" significance as a mathematical fixture, the constant e is a dynamic, energetic mainstay for calculus; its growth function is fundamental to the expansion of all species,[7] humans included.[8]

The constant e is also used to describe the rate of radioactive decay of a substance and, at the same time, is a mainstay for computing something as mundane and non-

4. Although this ratio was used by ancient Babylonians and Egyptians, the symbol "π" was not assigned until the 18th Century (by Euler). See Peter Beckman, *A History of Pi* (St. Martin's Griffin, 1971).

5. The fraction for pi that we first encountered in elementary school (22/7) is, of course, only a crude approximation. It is interesting to note that the ancient Pythagoreans believed that the mere existence of irrational numbers was "so horrific that it must represent some sort of cosmic error, one that should be suppressed and kept secret." Mario Livio, *The Golden Ratio* (Broadway Illustration, 2002) at 5.

6. If $y = f(x) = e^x$ then $f(x) = dy/dx\ f(x) = e^x$. Like π, Euler is credited for labeling "*e*."

7. "Besides serving as the base for the natural logarithms, *e* is a number useful everywhere in mathematics and applied science. No other mathematical constant, not even pi, is more closely connected with human affairs. In economics, statistics, probability theory or wherever exponential change is involved, it plays a significant role. It also plays an integral part in helping mathematicians describe and predict a most important natural phenomena—that of population growth." Edward Kasner and James Newman, *Mathematics and the Imagination* (Oxford University Press, 1967) 89. $Q = Q_0\, e^{k\,t}$ where Q = the population of any species at time t; k is a constant.

8. See *id.*: "A universe in which *e* and pi were lacking, would not, as some anthropomorphic soul has said, be inconceivable. One could hardly imagine that the sun would fail to rise, or the tides cease to flow for lack of pi and *e*. But without these mathematical artifacts, what we know about the sun and tides, indeed our ability to describe all natural phenomena, physical, biological, chemical or statistical, would be reduced to primitive dimensions."

celestial as compound interest.[9] It is notable that the concept of expansion, for which *e* is indispensable, ties together the physical, biological and economic universes. It is *more* than notable that the same constant used to accurately describe species contraction and extinction is likewise employed to calculate the expansion of an elemental capital asset.

The fifth constant on the chart is farther removed than even the most irrational and transcendental numbers; it requires an additional mathematical dimension for expressions that exist outside the scope of our traditional "real" number system. As previously discussed, it is the square root of minus one, that nostalgic headache first introduced in junior high school simply as "*i*."[10]

The next constant on the chart (φ), often referred to as the "golden mean" is a ratio that happens to appear frequently and naturally in the world around us. Because of its extensive use in art and architecture, it goes by many names: The mean ratio, medial section, divine proportion, divine section, golden proportion, golden cut and golden number. The golden mean is expressed in the arrangement of branches and stems of plants.

The golden ratio also manifests itself in the geometric arrangement of crystals and chemical compounds. It also applies to human body proportions. As its entry on the table indicates, the proportion (i.e., golden mean) of line a to b equals proportion of the length of the entire line to a. Notice that as is the case with π and *e*, the golden mean is also irrational.

Table 2-3 [Very Special Constants]

SYMBOL	VALUE	SIGNIFICANCE
1	1	The multiplicative identity
0	0	The additive identity
π	3.1415....	The ratio of a circle's radius to its circumference
e	2.1729....	The base of the natural logarithm
i	$\sqrt{-1}$	Basis for complex number system
φ	1.6180339887....	The Golden Mean: ________ a b (a + b) / a = a / b
h	6.626×10^{-34} (Joule Seconds)	The Planck Constant: A fundamental number for measurements involving quantum mechanics.

The last item on the list (h) is a physical, as opposed to a pure mathematical constant. Its inclusion suggests extraordinary importance. It is an indispensable number in the study of quantum mechanics. Known as the Planck Constant, it provides the means for

9. $M = m_0 e^{-at}$ (M = resulting mass, of initial mass m_0 after time *t; a* is a constant). $A = Pe^{rt}$. (A = Amount accrued after time *t* at rate *r*).

10. This constant, with a very lengthy story of its own, carries us into an expanded universe beyond our real number system. Without it, there would be no square root for any negative numbers. Such a constant is needed to complete our "complex" numbering system. In truth, our familiar *real* numbers are only but a subset of complex numbers. See generally Peter Beckman, *A History of Pi* (St. Martin's Griffin, 1971). Where a and b are real, the complex number: C = a + b*i*.

measuring the smallest possible "piece" of matter, suggesting that there is a limit to how small something can become before it is no longer divisible. Unlike classical physics, that constant suggests that neither matter nor waves are infinitely divisible.

§ 2.20 Number Foundations

This treatise is going to navigate the subjects of mathematics, physics and accounting in a manner different from most textbooks. There will be no attempt to hide or shield you from terms, topics or principles just because they happen to be "very complex" or advanced. You won't find phrases such as "beyond the scope of this text" or "for scientists only." If a concept is important enough to mention, it should be worthy of definition and presentation even though in many instances space will not permit in depth coverage of everything contained within the scope of the broad topics that are presented.

You should not be disturbed by an inability to fully comprehend all that is presented. If you were able to master everything in the text, you would not need its accompanying course. By the same token, you probably know more math than many treatises give you credit for.

§ *2.21 Simple Arithmetic*

For many of us, early experiences with math and science teachers have created a crisis of confidence. We tend to "overthink" problems or otherwise make them more difficult than they truly are. Consider the following questions, sometimes considered as nothing more than "language traps." It may surprise you know that not a single calculation is required to solve any of the problems:

1. A farmer had 20 long horn cattle. All of them except 12 died. How many were left alive?
2. If a brick and a half weighs a pound and a half, how much does a brick weigh?
3. How many 3 cent stamps are in a dozen?
4. A rope ladder 20 feet long hangs over the side of a ship. The rungs on the ladder are exactly one foot apart, with the bottom rung resting comfortably on the surface of the water. If the ship is rising at the rate of two feet per hour, how long will it take for two of the rungs to be underwater?
5. If train A leaves from D.C. to New York one hour before a different train B leaves from New York heading for D.C., which train will be closer to New York, as the two trains are passing one another?
6. Which has greatest mass, two pounds of feathers, two pounds of steel or two pounds of Paris Hilton's brain?

The examples above require nothing more than the application of simple logic.[11] Other seemingly complex problems may involve simple arithmetic. Often vague, ambiguous claims can be quickly deflated with a few simple calculations:

---

11. Credit for assembling several of these problems goes to Lederer, *April Foolishness: Don't Get Your Brain Caught In Language Traps*, Mensa Bulletin, April/May 2010 at 17. Here are the answers: (1) 12 are still alive. (2) One pound. (3) Twelve. (4) The rungs will never be under water because the ship will rise as the water rises. (5) If the trains are in the act of passing one another, they will both

Example 2-3: Included within the collective bargaining topics in a severe labor dispute among employees of a publishing company was the increasing cost of postage and its alleged effect on wages. From management's side of the table came this statement: "Since the last contract, postage rates for our publication have increased from $0.39 to $0.48. We have over 100,000 subscribers!"

Using simple math, from the employees' side came this observation:

$0.48 - $0.39 = $0.09

$0.09 × 100,000 = $9,000.00

$9,000.00 is far less than the cost of a single employee's salary.

Example 2-4: A very bitter dispute arose over the foreclosure of a marina and what turned out to be a very bad investment for several partners, all of whom were locked in litigation with one another. The matter went to trial. With respect to an ousted partner (Plaintiff), claiming a right to receive compensation for work performed, came the following cross-examination:

Q: (By Defense Counsel) On direct examination you testified that you personally spent several thousand hours, painting, fixing, improving and repairing the subject marina; is that correct?

A: (By Plaintiff) That is correct.

Q: Exactly how many hours?

A: I don't know. I did the work in my spare time and didn't ever think that I would ever need exact records. I could guess at 4,000 hours.

Q: Are you saying that you don't have any records?

A: That is correct.

Q: And I believe you testified that you were a dentist by occupation?

A: Yes.

Q: Part time or full time?

A: Full time.

Q: On average, how many days do you practice per week?

A: Monday through Saturday, about 7 or 8 hours per day.

Q: Are you like most people ... get at least 8 hours sleep per night?

A: Yes.

Q: When did you acquire the marina?

A: We had the closing for the purchase on New Years day, 2008.

Q: When was your interest foreclosed?

A: Late December of 2008.

Q: Giving you the benefit of a doubt, if you practiced dentistry for at least 7 hours per day as you say, you would have practiced close to 2200 hours during the two year period? (7 hrs × 6 days × 52 weeks = 2184 Hours).

A: If you say so.

---

be the same distance from New York, regardless of speed and regardless of amount of the head start. (6) All have equal mass.

Q: If I add the hours that you claim you spent working at the marina, it comes to 5200 hours for a year? (2200 + 4000 = 6200)?

A: Again, if you say so.

Q: But that would mean that you would have had to work close to 17 hours every day, including every holiday.... Not possible if you sleep 8 hours, even if you never eat? (6200 / 365 = 16.986).

A: (Silence)

Q: Would you like me to hand you a calculator?

§ 2.22 Ordinality v. Cardinality

Every number has two personalities.[12] The *cardinality* of a number refers to its size, while the *ordinality* pertains to its position. Unless a particular discussion concerns pure number theory, both terms (ordinal and cardinal) are best thought of adjectives. The number 9, for example, would possess a greater cardinality than the number 2 because 9 > 2. Nine tickets to the Lakers game is more than two tickets. 60,000 barrels of oil pouring into the Gulf of Mexico is more than 5,000 barrels. Therefore, we say that "9 has a greater cardinality than 2." Likewise, 60,000 has a greater cardinality than 5,000.

By contrast, the ordinal 9 describes a position as opposed to a magnitude. It might, for example, refer to the ninth person in line at Pink's Hot Dog Stand waiting to buy a chili dog. That same person might soon thereafter be patient number 27 on a waiting list to get his stomach pumped at Cedars Sinai Hospital in Beverly Hills where he might thereafter be charged the highest (number "1") bill for such services in the entire country.

Notes and Problems

1. Which would be more annoying, 9 self-styled divas dancing or a single dancing diva occupying the 9th position in a line-up of otherwise non publicity-hungry wannabees? Your answer should represent a selection of either "ordinality" or "cardinality."

2. In the song, *The Twelfth Day of Christmas*, the lyrics begin: "On the first day of Christmas my true love gave to me, a partridge in a pear tree." On the second day, he received "two turtle doves," then "three french hens" on the third day. Does this song reference ordinals or cardinals?

3. In set theory, the cardinality of a set refers to the number of elements in that set. For example, the set of integers: {1, 2, 3, 4, 5, 6, 7, 8, 9} has a cardinality of 9. The cardinality of senators who receive favors or money from lobbyists equals 100. The cardinality of Los Angeles City busses that are reliable is 0.

4. The set of all integers (i.e., natural numbers: 1, 2, 3, 4 ...) has a cardinality of infinity (See *supra* Table 2-2), which is designated: $\aleph_0$ (pronounced "aleph naught"). Interestingly, the mathematician Georg Cantor proved that all infinities are not equal. For example, $\aleph_0$ is the smallest infinity; it is sometimes referred to as "countable" infinity. By way of contrast, $\aleph_1$ represents a set whose elements are uncountably infinite, such as the set of real numbers. The term "countable" does not mean that you can physically

12. *"Number is different from quantity."* —George Bateson.

count the elements; rather, it means that all of the elements can be neatly arranged in a 1-to-1 correspondence with the natural numbers. (In other words, you can place them in a vertical list and number them.)

§ 2.23 The Metric System

The International System of Units (Abbreviated "SI"), often referred to as the "metric system" had its origins in France where it was first adopted in 1791. In essence, the Metric System is excellent for measuring things in the physical world. This internationally recognized system is based upon decimal number prefixes. Every sophisticated nation has adopted this scientifically based methodology; the United States, Burma and Liberia have not. Metric units are globally used for commercial and scientific purposes. The unit prefixes denote powers of 10, thereby simplifying calculations that are particularly large or small. Except for the scientific community, the Metric System never quite caught-on in the United States. We still measure with feet, quarts and pounds while the rest of the world uses meters, liters and grams. The mere fact that two systems exist in our shrinking world is a basis for concern. Consider the following classic exchange from the film *Pulp Fiction*:[13]

Jules: (Addressing Brad) I do love the taste of a good burger. Mmmm. Do you know what they call a Quarter Pounder with Cheese in France?

Brad: (Nervously looking up the barrel of Jules's pistol) No.

Jules: Tell him, Vincent.

Vince: A Roy'al with Cheese.

Jules: You know why they call it that?

Brad: B ... B ... Because of the Metric System?

Jules: Check out the big brain on Brad! You're a smart ____ ____ (expletive deleted)!

The fact that Jules and Vince emptied their pistols into Brad shortly after their scientific discourse should not dampen your enthusiasm about the Metric System. Dangers lurk. Historically, as might be expected, errors involving disparities between the two systems have been much more serious than re-naming hamburgers.

Many malpractice lawsuits (medical and pharmaceutical) have arisen over errors caused by overdoses and underdoses of medication as a result of mix-ups between the two systems. Medical malpractice lawsuits, for example, have risen in the United States over a confusion between a grain and a gram (1 grain = 0.06 grams). In 1999, NASA managed to lose a Mars orbiter costing $125,000,000.00 because one engineering team at NASA used metric units while another employed customary U.S. units; the mistake, as indicated, cost us a lot of customary U.S. dollars. In 1983, a Boeing 767 passenger jet ran out of fuel while airborne in the middle of a flight because of two mistakes in calculating the fuel supply of Air Canada's first aircraft to use metric measurements.

On the Internet, quick conversion information abounds. For non-scientific users, the following information should suffice:

13. Produced by Quentin Tarantino in 1994, the popular film won an Oscar for Best Original Screenplay. Samuel L. Jackson played Jules; John Travolta portrayed Vince.

Measures of Length:

1 Inch = 2.54 Centimeters = 25.4 Millimeters (mm)
1 Foot = 0.3048 Meters (m)
1 Mile = 5280 Feet = 1.609344 Kilometers (km)
1 Nautical Mile = 1.852 Kilometers (km)

Measures of Weight:

1 Ounce = 28.34952312 Grams (g)
1 Pound = 0.45359237 Kilograms (kg)
1 Stone = 14 Pounds

Measures of Area:

1 Square Foot = 929.0304 Square Centimeters
1 Square Mile = 2.58998811 Square Kilometers
1 Acre = 43,560 Square Feet
1 Section = 640 Acres

Measures of Volume:

1 U.S. fluid ounce = 29.57353 Milliliters (ml)
1 U.S. gallon = 3.78541178 Liters (l)

Temperature:

Celsius = 5 (Fahrenheit - 32) / 9

§ *2.24 Dimensional Conversions*

There is a simple, error-free trick for calculating conversions that can be used regardless of the type of system or systems involved. The trick takes advantage of the simple law of math that allows you to multiply anything by "1" without changing its value.

Example 2-5: Consider a very simple example which involves changing miles to feet. Problem: 18,480 feet equals how many miles?

Solution:

$$\frac{18{,}400\ \cancel{ft}}{1} \times \frac{1\ \text{Mile}}{5{,}280\ \cancel{ft}} = = 18{,}480 / 5280 = 3.5\ \text{Miles}$$

Step 1: Write the value that you wish to convert as a fraction with a denominator of 1. Be very careful to correctly write the units. This is, of course, perfectly legal since a denominator of 1 can never change or alter the value of anything.

Step 2: Create a second fraction equal to 1. Again, the trick is to keep the units accurate. Because 1 mile equals 5280 feet, the second fraction is equal to 1. And, because you are only multiplying by 1, you are maintaining your original value. Be certain to place the unit that you wish to cancel-out on the bottom. In our example, notice that the feet, "ft" in the two fractions will cancel one another leaving only the "miles" unit of measure.

Step 3: Perform the necessary multiplication.

SOLVED PROBLEM

Example 2-6: How many centimeters are in 2½ Miles?

Solution:

$$\frac{2.5 \text{ Mile(s)}}{1} \times \frac{5{,}280 \text{ ft}}{1 \text{ Mile(s)}} \times \frac{12 \text{ Inches}}{1 \text{ ft}} \times \frac{2.54 \text{ cm}}{1 \text{ Inch(es)}} = \frac{(2.5)(5280)(12)(2.54)\text{ (cm)}}{1}$$

$$= 402{,}336 \text{ cm}$$

Notice the manner in which all of the units cancel one another, leaving only the desired unit of measure, i.e., the centimeter (cm). Miles will cross out miles; "ft" will cross out "ft," etc. Notice also that each fraction used as a factor in the conversion is equal to 1.

Example 2-7: How many square inches are there in an acre?

Solution: Recall from § 2.23 that 1 Acre = 43,560 square feet. Also, 1 square foot consists of 144 square inches. Therefore:

$$\frac{1 \text{ Acre}}{1} \times \frac{43{,}560 \text{ ft}^2}{1 \text{ Acre}} \times \frac{144 \text{ Inches}^2}{1 \text{ ft}^2} =$$

$$\frac{1 \cancel{\text{Acre}}}{1} \times \frac{43{,}560 \cancel{\text{ft}^2}}{1 \cancel{\text{Acre}}} \times \frac{144 \text{ Inches}^2}{1 \cancel{\text{ft}^2}} = 6{,}272{,}640 \text{ Square Inches}$$

Drake v. Holstead

Court of Appeals of Texas, 757 S.W.2d 909 (1988)

BURGESS, Justice.

This is an appeal from a judgment in favor of the defendant in a personal injury suit. Plaintiff and defendant were traveling toward each other when the defendant turned left in front of plaintiff. Their cars collided. Plaintiff sued the defendant for personal injuries arising from the accident.

At trial, plaintiff testified she crossed the intersection on a green light. Defendant testified that her left-turn light was green. There was nothing to indicate the traffic lights might have malfunctioned, and each party introduced evidence to support her testimony that she had the green light. Plaintiff's witness to the accident testified defendant's light was red when defendant turned left in front of the plaintiff.

Defense witness Robert Mills testified that when plaintiff was 40 to 50 yards from the intersection, plaintiff's light turned yellow. He did not notice the color of the traffic light after that, however, so he could not testify to the color of the light when plaintiff's car entered the intersection. From Mills' testimony that plaintiff's light turned yellow 40 to 50 yards before she entered the intersection, the jury could infer the light turned red before she entered the intersection. Mills testified plaintiff was driving at around 40 to 45 miles per hour when her light turned yellow. Plaintiff contends that with the aid of judicial notice of certain facts, she planned to turn Mills' testimony to her advantage by showing that, given previously submitted evidence of the timed sequence of the traffic light and Mills' account of plaintiff's rate of speed and her distance from the intersection when the light turned yellow, she would have crossed the intersection with time to spare before her light turned red. During plaintiff's cross-examination of Mills, the following transpired:

Plaintiff Counsel:	If, as you testified, if Trina Drake, was going forty miles a [sic] hour at the time she passed you, do you know how far she would travel in 3.6 seconds when the light was yelow [sic]?
Witness:	No.
Plaintiff Counsel:	Would it surprise you to learn that in 3.6 seconds—
Defense Counsel:	I would object—
THE COURT:	Objection sustained. That is an improper question, counsel.
Plaintiff Counsel:	Your Honor, at this time, I would ask the court to take judicial notice—
THE COURT:	(interrupting) No, sir, counsel. If you are getting ready to tell me to take judicial notice of measurements of time periods in order for a car to travel—is that your? [sic]
Plaintiff's Counsel:	Yes, sir.
THE COURT:	No, sir. The answer is no. I will not take judicial notice. That takes testimony, counsel.
Plaintiff's Counsel:	All right, your Honor.

On appeal, plaintiff/appellant contends the taking of judicial notice was mandatory in this case and that the trial court's refusal to take judicial notice was harmful error. Evidently, the trial judge believed calculations of rates of speed are not susceptible to judicial notice; he indicated testimony was required to prove the facts asserted. It is true that the usual method of establishing adjudicative facts is through the introduction of evidence, ordinarily consisting of the testimony of witnesses. However, when particular facts are outside the area of reasonable controversy, this process is dispensed with as unnecessary. Fed.R.Evid. 201(a). The Texas Rule 201 is virtually identical to the federal rule:

> Rule 201 of the Texas Rules of Civil Evidence provides:
>
> (a) Scope of rule. This rule governs only judicial notice of adjudicative facts.
>
> (b) Kinds of facts. A judicially noticed fact must be one not subject to reasonable dispute in that it is either (1) generally known within the territorial jurisdiction of the trial court or (2) capable of accurate and ready determination by resort to sources whose accuracy cannot reasonably be questioned.
>
> (c) When discretionary. A court may take judicial notice, whether requested or not.
>
> (d) When mandatory. A court shall take judicial notice if requested by a party and supplied with the necessary information.
>
> (e) Opportunity to be heard. A party is entitled upon timely request to an opportunity to be heard as to the propriety of taking judicial notice and the tenor of the matter noticed. In the absence of prior notification, the request may be made after judicial notice has been taken....

The doctrine of judicial notice is applied to scientific facts and principles that are generally recognized and ought to be known by men of ordinary understanding and intelligence. Alexander v. Firemen's Ins. Co., 317 S.W.2d 752, 755 (Tex.Civ.App.—Waco 1958, no writ). Texas courts take judicial notice of natural forces as well as the primary laws of physics, mechanics, and mathematics. Searcy v. Sellers, 470 S.W.2d 103, 110 (Tex.Civ.App.—Amarillo 1971, writ ref'd n.r.e.). A fact is generally known even though it has to be processed with commonly possessed mental skills. 21 C. Wright & K. Graham, Federal Practice and Procedure sec. 5105, at 232 (1987 Supp.).

Plaintiff sought judicial notice of the indisputable results of a simple mathematical computation. The matter was well within the range of matters courts traditionally accept into evidence by judicial notice. A Texas court of civil appeals took judicial notice of just such a matter in Searcy, 470 S.W.2d at 110, where it found that a car traveling 50 miles per hour travels 100 feet in 1.36 seconds. The fact that a matter is susceptible to judicial notice does not require the court to take such notice, however. A party seeking mandatory judicial notice must supply the court with the "necessary information." In her bill of exceptions, plaintiff supplied the court with a sheet of typewritten computations which arrived at the figures she wished the court to judicially notice:

We note plaintiff's calculations erroneously conclude that 60 miles per hour divided by 60 minutes per hour is equal to one mile per hour, instead of one mile per minute:

50 MPH = 73.33 FT/SEC

60 MPH / 60 (minutes per hr) = 1 MPH

Opposing counsel and the trial judge are free to object to the computations offered by their proponent and to offer calculations of their own. No such objection was made in this case, however, and the error did not affect the ultimate figures sought to be noticed.

1 MPH 2 × 5280 (ft per mile) = 5280 Ft/min

5280 Ft/min / 60 (seconds per min) = 88 Ft/sec

60 MPH = 88 Ft/sec

5/6 × 60 MPH = 5/6 × 88 Ft/sec

50 MPH = 73.33 Ft/sec = 24.44 Yd/sec

40 MPH = 58.66 FT/SEC

4/6 × 60 MPH = 4/6 × 88 Ft/sec

40 MPH = 58.66 Ft/sec = 19.55 Yd/sec

Plaintiff did not supply the court with verification that there are, in fact, 5280 feet in a mile. However, rule 201(b) contemplates that a court may judicially notice facts generally known within the territorial jurisdiction without specific confirmation of those facts. At any rate, the number of feet per mile is easily accessible: we find the figure under the definition of "mile" in a standard desk-side dictionary. See The American Heritage Dictionary of the English Language (1975). Once the number of minutes in an hour, seconds in a minute, and feet in a mile are ascertained, it is a matter of simple mathematical computation to arrive at the number of feet traveled in a second by an object traveling at a given rate per mile. The trial court erred in refusing to take judicial notice of the matters requested by the plaintiff.

Plaintiff has not, however, shown reversible error. The charge to the jury was presented generally: the jury was asked, "Did the negligence, if any, of the persons named below proximately cause the occurrence in question?" The jury answered "no" as to the defendant and "yes" as to the plaintiff. Since no special issues were presented inquiring into the potential statutory negligence of the parties, this court has no means to ascertain what particular act or acts by the plaintiff the jury found to be negligent. And because the jury would have had to find both that defendant was negligent and that defendant's negligence was a proximate cause of the occurrence in order to answer the charge "yes" as to the defendant, we do not know that the jury did not, in fact, believe that defendant ran a red light. To be reversible, the error complained of must amount to "such a denial of the rights of the appellant as was reasonably calculated to cause and probably did cause rendition of an improper judgment in the case. We cannot say this is so under these facts. Appellant's first point of error is overruled.

Appellant also alleges the trial court erred in failing to grant a new trial because (1) the jury's failure to find liability on defendant's part was against the great weight and preponderance of the evidence and (2) there was insufficient evidence to support the jury's finding that the plaintiff's negligence was a proximate cause of the occurrence in question. In reviewing factual insufficiency points, this court considers and weighs all the evidence in the case to determine whether the verdict is supported by insufficient evidence or is so against the great weight and preponderance of the evidence as to be manifestly unjust. Pool v. Ford Motor Co., 715 S.W.2d 629, 635 (Tex.1986); see also P.T. & E. Co. v. Beasley, 698 S.W.2d 190, 193 (Tex.App.—Beaumont 1985, writ ref'd n.r.e.).

It is true that testimony was introduced indicating defendant's light was red when she turned left at the intersection. However, conflicting evidence of probative force was introduced tending to prove the existence of negligence and causation on plaintiff's part and tending to disprove negligence and causation on defendant's part. In particular, testimony was introduced to indicate plaintiff crossed the intersection while her traffic light was red.

It is the jury's function to decide credibility issues. Shwiff v. Priest, 650 S.W.2d 894, 900 (Tex.App.—San Antonio 1983, writ ref'd n.r.e.). The court of appeals will uphold the jury's finding where it would not be clearly wrong and manifestly unjust to do so. In the case at bar, the evidence was sufficient to uphold the jury's verdict. Appellant's second and third points of error are overruled.

The judgment of the trial court is affirmed.

Notes and Problems

1. Should the outcome have been different if the Appellant had supplied a formal judicial notice request to the court in advance of trial?

2. Do you think that the errors and bumbling computations would have been avoided if the calculations were performed according to the conversion methodology set forth at the beginning of this section? Do you think that methodology would be relatively easy to demonstrate for a jury? Keep in mind that 4 out of every 3 (non-sic) people have difficulty with fractions.

§2.30 The Significance of Equality

Just as the term "*equality*" carries a great deal of weight in the field of law, so too does the equality sign in the world of mathematics. With its many meanings and implications, it appears that *equality* is pervasive as both a legal and mathematical concept.

Our earliest encounter may have involved a law or definition of mathematics that taught us how to calculate a particular value, such as A = lw (The area of a rectangle equals length times width). Later, it may have involved a law of physics such as $E = mc^2$.[14]

In junior high school an altogether different brand of equation came in the form of a problem, as opposed to a statement. Those particular equations were nothing more than mixtures of constants (i.e., numbers) and unknown variables, overpopulated with the infamous "*x*," for which you were required to find the solution (i.e., value). An example: $25x^2 + 3x + 5 = 32$. By contrast, an equation might assume the form of a so-called "identity," which stated a law of mathematics that was true for all unknown or variable values. An example is the basic trigonometric identity: $\sin^2 x + \cos^2 x = 1$.

Later still in your undergraduate training, you may have encountered yet another breed of "equation" that involved nothing but constants; among them we find an equality that stands alone, above all others, as a testament to the beauty and bewilderment of mathematics because it mysteriously brings together the five most important constant numbers in the universe (1, 0, *e*, *i*, and π) in a simple eloquent formula: $e^{i\pi} + 1 = 0$. Discovered by one of the world's greatest mathematicians, Leonhard Euler, it consists only of constants that appear naturally in the real world.[15] Finally, we confront those equations that both describe and direct the operation of a specific process. They might be considered as recipes. The *Diamond* case (infra) describes such an equation.

§2.31 *The Equalities That Changed the World*

Our most frequent use of equality comes in the form of the indispensable "equation" which is dealt with in many forms throughout this text; it is not limited to pure mathematics. In Chapters 7 and 8, we consider what many claim to be the most important equations in the history of the world. Starting with basic Newtonian physics, we are led through Maxwell's equations for electromagnetic fields, eventually to Schroedinger's equation which purports to describe the movement of all of the particles in universe, big, small and subatomic.

§2.32 *Claiming Ownership of Equations*

In the world of intellectual property, it is a firm principle of law that no one can claim a patent or copyright on any of the preceding equations. Such laws of nature belong to everyone, and not to any one person to the exclusion of others. But is it possible for an equation to express a process that may be subject to a private patent?

---

14. "*Equations are more important to me, because politics is for the present, but an equation is something for eternity.*"—Albert Einstein (when asked about the importance of politics).

15. Ashley S. Lipson, *The Intelligent Construction of the Universe: A Mathematical Proof—The Link Among Science, Natural Law and Jurisprudence* (2007) http://works.bepress.com/ashley_lipson/1.

Gottschalk v. Benson

United States Supreme Court, 409 U.S. 63, 93 S.Ct. 253 (1972)

Mr. Justice DOUGLAS delivered the opinion of the Court.

Respondents filed in the Patent Office an application for an invention which was described as being related 'to the processing of data by program and more particularly to the programmed conversion of numerical information' in general-purpose digital computers. They claimed a method for converting binary-coded decimal (BCD) numerals into pure binary numerals. The claims were not limited to any particular art or technology, to any particular apparatus or machinery, or to any particular end use. They purported to cover any use of the claimed method in a general-purpose digital computer of any type. Claims 8 and 31[16] were rejected by the Patent Office but sustained by the Court of Customs and Patent Appeals, 441 F.2d 682. The case is here on a petition for a writ of certiorari. Gottschalk v. Benson, 405 U.S. 915, 92 S.Ct. 934, 30 L.Ed.2d 784. The question is whether the method described and claimed is a 'process' within the meaning of the Patent Act.[17]

A digital computer, as distinguished from an analog computer, operates on data expressed in digits, solving a problem by doing arithmetic as a person would do it by head and hand. Some of the digits are stored as components of the computer. Others are introduced into the computer in a form which it is designed to recognize. The computer operates then upon both new and previously stored data. The general-purpose computer is designed to perform operations under many different programs.

The representation of numbers may be in the form of a time series of electrical impulses, magnetized spots on the surface of tapes, drums, or discs, charged spots on cathode-ray tube screens, the presence or absence of punched holes on paper cards, or other devices. The method or program is a sequence of coded instructions for a digital computer.

The patent sought is on a method of programming a general-purpose digital computer to convert signals from binary-coded decimal form into pure binary form. A procedure for solving a given type of mathematical problem is known as an 'algorithm.' The procedures set forth in the present claims are of that kind; that is to say, they are a generalized formulation for programs to solve mathematical problems of converting one form of numerical representation to another. From the generic formulation, programs may be developed as specific applications.

The decimal system uses as digits the 10 symbols 0, 1, 2, 3, 4, 5, 6, 7, 8, and 9. The value represented by any digit depends, as it does in any positional system of notation, both on its individual value and on its relative position in the numeral. Decimal numerals are written by placing digits in the appropriate positions or columns of the numerical sequence, i.e., 'unit' (10^0), 'tens' (10^1), 'hundreds' (10^2), 'thousands' (10^3), etc. Accordingly, the numeral 1492 signifies (10^3) + 4 x (10^2) + 9 x (10^2) + 2 x (10^0).

16. They are set forth in the Appendix to this opinion (Case's Appendix omitted from this text).

17. Title 35 U.S.C. § 100(b) provides: "The term 'process' means process, art or method, and includes a new use of a known process, machine, manufacture, composition of matter, or material." Title 35 U.S.C. § 101 provides: "Whoever invents or discovers any new and useful process, machine, manufacture, or composition of matter, or any new and useful improvement thereof, may obtain a patent therefor, subject to the conditions and requirements of this title."

The pure binary system of positional notation uses two symbols as digits—0 and 1, placed in a numerical sequence with values based on consecutively ascending powers of 2. In pure binary notation, what would be the tens position is the twos position; what would be hundreds position is the fours position; what would be the thousands position is the eights. Any decimal number from 0 to 10 can be represented in the binary system with four digits or positions as indicated in the following table. Shown as the sum of powers of 2:

		2^3		2^2		2^1		2^0		
Decimal:		(8)		(4)		(2)		(1)		**Pure Binary**
0	=	0	+	0	+	0	+	0	=	0000
1	=	0	+	0	+	0	+	2^0	=	0001
2	=	0	+	0	+	2^1	+	0	=	0010
3	=	0	+	0	+	2^1	+	2^0	=	0011
4	=	0	+	2^2	+	0	+	0	=	0100
5	=	0	+	2^2	+	0	+	2^0	=	0101
6	=	0	+	2^2	+	2^1	+	0	=	0110
7	=	0	+	2^2	+	2^1	+	2^0	=	0111
8	=	2^3	+	0	+	0	+	0	=	1000
9	=	2^3	+	0	+	0	+	2^0	=	1001
10	=	2^3	+	0	+	2^1	+	0	=	1010

The BCD system using decimal numerals replaces the character for each component decimal digit in the decimal numeral with the corresponding four-digit binary numeral, shown in the right hand column of the table. Thus decimal 53 is represented as 0101 0011 in BCD, because decimal 5 is equal to binary 0101 and decimal 3 is equivalent to binary 0011.

In pure binary notation, however, decimal 53 equals binary 110101. The conversion of BCD numerals to pure binary numerals can be done mentally through use of the foregoing table. The method sought to be patented varies the ordinary arithmetic steps a human would use by changing the order of the steps, changing the symbolism for writing the multiplier used in some steps, and by taking subtotals after each successive operation. The mathematical procedures can be carried out in existing computers long in use, no new machinery being necessary. And, as noted, they can also be performed without a computer.

The Court stated in Mackay Co. v. Radio Corp., 306 U.S. 86, 94, 59 S.Ct. 427, 431, 83 L.Ed. 506 that '(w)hile a scientific truth, or the mathematical expression of it, is not patentable invention, a novel and useful structure created with the aid of knowledge of scientific truth may be.' That statement followed the longstanding rule that '(a)n idea of itself is not patentable.' Rubber-Tip Pencil Co. v. Howard, 20 Wall. (87 U.S.) 498, 507, 22 L.Ed. 410. 'A principle, in the abstract, is a fundamental truth; an original cause; a motive; these cannot be patented, as no one can claim in either of them an exclusive right.' Le Roy v. Tatham, 14 How. (55 U.S.) 156, 175, 14 L.Ed. 367.

Phenomena of nature, though just discovered, mental processes, and abstract intellectual concepts are not patentable, as they are the basic tools of scientific and technological work. As we stated in Funk Bros. Seed Co. v. Kalo Co., 333 U.S. 127, 130, 68 S.Ct. 440, 441, 92 L.Ed. 588, 'He who discovers a hitherto unknown phenomenon of nature has no claim to a monopoly of it which the law recognizes. If there is to be invention from such a discovery, it must come from the application of the law of nature to a new and useful end.' We dealt there with a 'product' claim, while the present case deals with a 'process' claim. But we think the same principle applies.

Here the 'process' claim is so abstract and sweeping as to cover both known and unknown uses of the BCD to pure binary conversion. The end use may (1) vary from the operation of a train to verification of drivers' licenses to researching the law books for precedents and (2) be performed through any existing machinery or future-devised machinery or without any apparatus.

It is argued that a process patent must either be tied to a particular machine or apparatus or must operate to change articles or materials to a 'different state or thing.' We do not hold that no process patent could ever qualify if it did not meet the requirements of our prior precedents. It is said that the decision precludes a patent for any program servicing a computer. We do not so hold. It is said that we have before us a program for a digital computer but extend our holding to programs for analog computers. We have, however, made clear from the start that we deal with a program only for digital computers. It is said we freeze process patents to old technologies, leaving no room for the revelations of the new, onrushing technology. Such is not our purpose. What we come down to in a nutshell is the following. It is conceded that one may not patent an idea. But in practical effect that would be the result if the formula for converting BCD numerals to pure binary numerals were patented in this case. The mathematical formula involved here has no substantial practical application except in connection with a digital computer, which means that if the judgment below is affirmed, the patent would wholly pre-empt the mathematical formula and in practical effect would be a patent on the algorithm itself. It may be that the patent laws should be extended to cover these programs, a policy matter to which we are not competent to speak. The President's Commission on the Patent System rejected the proposal that these programs be patentable:

> 'Uncertainty now exists as to whether the statute permits a valid patent to be granted on programs. Direct attempts to patent programs have been rejected on the ground of nonstatutory subject matter. Indirect attempts to obtain patents and avoid the rejection, by drafting claims as a process, or a machine or components thereof programmed in a given manner, rather than as a program itself, have confused the issue further and should not be permitted.
>
> The Patent Office now cannot examine applications for programs because of a lack of a classification technique and the requisite search files. Even if these were available, reliable searches would not be feasible or economic because of the tremendous volume of prior art being generated. Without this search, the patenting of programs would be tantamount to mere registration and the presumption of validity would be all but nonexistent.
>
> It is noted that the creation of programs has undergone substantial and satisfactory growth in the absence of patent protection and that copyright protection for programs is presently available.'

If these programs are to be patentable, considerable problems are raised which only committees of Congress can manage, for broad powers of investigation are needed, including hearings which canvass the wide variety of views which those operating in this field entertain. The technological problems tendered in the many briefs before us indicate to us that considered action by the Congress is needed.

Reversed.

Mr. Justice STEWART, Mr. Justice BLACKMUN, and Mr. Justice POWELL took no part in the consideration or decision of this case.

Diamond v. Diehr

Supreme Court of the United States
450 U.S. 175 (1981)

Justice Rehnquist:

We granted certiorari to determine whether a process for curing synthetic rubber which includes in several of its steps the use of a mathematical formula and a programmed digital computer is patentable subject matter under 35 U.S.C. § 101.

I.

The patent application at issue was filed by the respondents on August 6, 1975. The claimed invention is a process for molding raw, uncured synthetic rubber into cured precision products. The process uses a mold for precisely shaping the uncured material under heat and pressure and then curing the synthetic rubber in the mold so that the product will retain its shape and be functionally operative after the molding is completed.[18]

Respondents claim that their process ensures the production of molded articles which are properly cured. Achieving the perfect cure depends upon several factors including the thickness of the article to be molded, the temperature of the molding process, and the amount of time that the article is allowed to remain in the press.

It is possible using well-known time, temperature, and cure relationships to calculate by means of the Arrhenius equation[19] when to open the press and remove the cured product. Nonetheless, according to the respondents, the industry has not been able to obtain uniformly accurate cures because the temperature of the molding press could not be precisely measured, thus making it difficult to do the necessary computations to determine cure time.[20] Because the temperature inside the press has heretofore been viewed as an uncontrollable variable, the conventional industry practice has been to calculate the cure time as the shortest time in which all parts of the product will definitely be cured, assuming a reasonable amount of mold-opening time during loading and unloading.

---

18. A "cure" is obtained by mixing curing agents into the uncured polymer in advance of molding and then applying heat over a period of time. If the synthetic rubber is cured for the right length of time at the right temperature, it becomes a usable product.

19. The equation is named after its discoverer Svante Arrhenius and has long been used to calculate the cure time in rubber-molding presses. The equation can be expressed as follows:

$$\ln v = CZ + x$$

wherein ln v is the natural logarithm of v, the total required cure time; C is the activation constant, a unique figure for each batch of each compound being molded, determined in accordance with rheometer measurements of each batch; Z is the temperature in the mold; and x is a constant dependent on the geometry of the particular mold in the press. A rheometer is an instrument to measure flow of viscous substances.

20. During the time a press is open for loading, it will cool. The longer it is open, the cooler it becomes and the longer it takes to reheat the press to the desired temperature range. Thus, the time necessary to raise the mold temperature to curing temperature is an unpredictable variable. The respondents claim to have overcome this problem by continuously measuring the actual temperature in the closed press through the use of a thermocouple.

But the shortcoming of this practice is that operating with an uncontrollable variable inevitably led in some instances to overestimating the mold-opening time and overcuring the rubber, and in other instances to underestimating that time and undercuring the product.[21]

Respondents characterize their contribution to the art to reside in the process of constantly measuring the actual temperature inside the mold. These temperature measurements are then automatically fed into a computer which repeatedly recalculates the cure time by use of the Arrhenius equation. When the recalculated time equals the actual time that has elapsed since the press was closed, the computer signals a device to open the press. According to the respondents, the continuous measuring of the temperature inside the mold cavity, the feeding of this information to a digital computer which constantly recalculates the cure time, and the signaling by the computer to open the press, are all new in the art.

The patent examiner rejected the respondents' claims on the sole ground that they were drawn to nonstatutory subject matter under 35 U.S.C. § 101.[22] He determined that those steps in respondents' claims that are carried out by a computer under control of a stored program constituted nonstatutory subject matter under this Court's decision in Gottschalk v. Benson, 409 U.S. 63, 93 S.Ct. 253, 34 L.Ed.2d 273 (1972). The remaining steps—installing rubber in the press and the subsequent closing of the press—were "conventional and necessary to the process and cannot be the basis of patentability." The examiner concluded that respondents' claims defined and sought protection of a computer program for operating a rubber-molding press.

The Patent and Trademark Office Board of Appeals agreed with the examiner, but the Court of Customs and Patent Appeals reversed. In re Diehr, 602 F.2d 982 (1979). The court noted that a claim drawn to subject matter otherwise statutory does not become nonstatutory because a computer is involved. The respondents' claims were not directed to a mathematical algorithm or an improved method of calculation but rather recited an improved process for molding rubber articles by solving a practical problem which had risen in the molding of rubber products.

---

21. We note that the petitioner does not seriously contest the respondents' assertions regarding the inability of the industry to obtain accurate cures on a uniform basis.

22. Respondents' application contained 11 different claims. Three examples are claims 1, 2, and 11 which provide:

> "1. A method of operating a rubber-molding press for precision molded compounds with the aid of a digital computer, comprising: providing said computer with a data base for said press including at least, natural logarithm conversion data (ln), the activation energy constant (c) unique to each batch of said compound being molded, and a constant (x) dependent upon the geometry of the particular mold of the press, initiating an interval timer in said computer upon the closure of the press for monitoring the elapsed time of said closure, constantly determining the temperature (Z) of the mold at a location closely adjacent to the mold cavity in the press during molding, constantly providing the computer with the temperature (Z), repetitively calculating in the computer, at frequent intervals during each cure, the Arrhenius equation for reaction time during the cure, which is $\ln v = CZ + x$, where v is the total required cure time, repetitively comparing in the computer at said frequent intervals during the cure each said calculation of the total required cure time calculated with the Arrhenius equation and said elapsed time, and opening the press automatically when a said comparison indicates equivalence."
>
> "2. The method of claim 1 including measuring the activation energy constant for the compound being molded in the press with a rheometer and automatically updating said data base within the computer in the event of changes in the compound being molded in said press as measured by said rheometer."

The Commission of Patents and Trademarks sought certiorari arguing that the decision of the Court of Customs and Patent Appeals was inconsistent with prior decisions of this Court. Because of the importance of the question presented, we granted the writ. 445 U.S. 926, 100 S.Ct. 1311, 63 L.Ed.2d 758 (1980).

II.

Last Term in Diamond v. Chakrabarty, 447 U.S. 303, 100 S.Ct. 2204, 65 L.Ed.2d 144 (1980), this Court discussed the historical purposes of the patent laws and in particular 35 U.S.C. § 101. As in Chakrabarty, we must here construe 35 U.S.C. § 101 which provides: "Whoever, invents or discovers any new and useful process, machine manufacture, or composition of matter, or any new and useful improvement thereof, may obtain a patent therefor, subject to the conditions and requirements of this title."[23]

In cases of statutory construction, we begin with the language of the statute. Unless otherwise defined, "words will be interpreted as taking their ordinary, contemporary, common meaning," Perrin v. United States, 444 U.S. 37, 42, 100 S.Ct. 311, 314, 62 L.Ed.2d 199 (1979), and, in dealing with the patent laws, we have more than once cautioned that "courts 'should not read into the patent laws limitations and conditions which the legislature has not expressed.'" Diamond v. Chakrabarty, supra, at 308, 100 S.Ct., at 2207 quoting United States v. Dubilier Condenser Corp., 289 U.S. 178, 199, 53 S.Ct. 554, 561, 77 L.Ed. 1114 (1933).

The Patent Act of 1793 defined statutory subject matter as "any new and useful art, machine, manufacture or composition of matter, or any new or useful improvement [thereof]." Act of Feb. 21, 1793, ch. 11, § 1, 1 Stat. 318. Not until the patent laws were recodified in 1952 did Congress replace the word "art" with the word "process." It is that latter word which we confront today, and in order to determine its meaning we may not be unmindful of the Committee Reports accompanying the 1952 Act which inform us that Congress intended statutory subject matter to "include anything under the sun that is made by man." S.Rep.No.1979, 82d Cong., 2d Sess., 5 (1952); H.R.Rep.No.1923, 82d Cong., 2d Sess., 6 (1952), U.S.Code Cong. & Admin.News 1952, pp. 2394, 2399.

"That a process may be patentable, irrespective of the particular form of the instrumentalities used, cannot be disputed.... A process is a mode of treatment of certain materials to produce a given result. It is an act, or a series of acts, performed upon the subject-matter to be transformed and reduced to a different state or thing. If new and useful, it is just as patentable as is a piece of machinery. In the language of the patent law, it is an art. The machinery pointed out as suitable to perform the process may or may not be new or patentable; whilst the process itself may be altogether new, and produce an entirely new result. The process requires that certain things should be done with certain substances, and in a certain order; but the tools to be used in doing this may be of secondary consequence." Cochrane v. Deener, 94 U.S. 780, 787–788, 24 L.Ed. 139 (1877). Analysis of the eligibility of a claim of patent protection for a "process" did not change with the addition of that term to § 101.

Recently, in Gottschalk v. Benson, 409 U.S. 63, 93 S.Ct. 253, 34 L.Ed.2d 273 (1972), we repeated the above definition recited in Cochrane v. Deener, adding: "Transformation and reduction of an article 'to a different state or thing' is the clue to the patentability of a process claim that does not include particular machines." 409 U.S., at 70, 93 S.Ct., at 256.

---

23. The word "process" is defined in 35 U.S.C. § 100(b): "The term 'process' means process, art or method, and includes a new use of a known process, machine, manufacture, composition of matter, or material."

We think that a physical and chemical process for molding precision synthetic rubber products falls within the § 101 categories of possibly patentable subject matter. That respondents' claims involve the transformation of an article, in this case raw, uncured synthetic rubber, into a different state or thing cannot be disputed. The respondents' claims describe in detail a step-by-step method for accomplishing such, beginning with the loading of a mold with raw, uncured rubber and ending with the eventual opening of the press at the conclusion of the cure. Industrial processes such as this are the types which have historically been eligible to receive the protection of our patent laws.

III.

Our conclusion regarding respondents' claims is not altered by the fact that in several steps of the process a mathematical equation and a programmed digital computer are used. This Court has undoubtedly recognized limits to § 101 and every discovery is not embraced within the statutory terms. Excluded from such patent protection are laws of nature, natural phenomena, and abstract ideas. See Parker v. Flook, 437 U.S. 584, 98 S.Ct. 2522, 57 L.Ed.2d 451 (1978); Gottschalk v. Benson, supra, at 67, 93 S.Ct., at 255; Funk Bros. Seed Co. v. Kalo Inoculant Co., 333 U.S. 127, 130, 68 S.Ct. 440, 441, 92 L.Ed. 588 (1948). "An idea of itself is not patentable," Rubber-Tip Pencil Co. v. Howard, 20 Wall. 498, 507, 22 L.Ed. 410 (1874). "A principle, in the abstract, is a fundamental truth; an original cause; a motive; these cannot be patented, as no one can claim in either of them an exclusive right." Le Roy v. Tatham, 14 How. 156, 175, 14 L.Ed. 367 (1853). Only last Term, we explained: "[A] new mineral discovered in the earth or a new plant found in the wild is not patentable subject matter. Likewise, Einstein could not patent his celebrated law that $E = mc^2$; nor could Newton have patented the law of gravity. Such discoveries are 'manifestations of ... nature, free to all men and reserved exclusively to none.'" Diamond v. Chakrabarty, 447 U.S., at 309, 100 S.Ct., at 2208, quoting Funk Bros. Seed Co. v. Kalo Inoculant Co., supra, at 130, 68 S.Ct., at 441.

Our recent holdings in Gottschalk v. Benson, supra, and Parker v. Flook, supra, both of which are computer-related, stand for no more than these long-established principles. In Benson, we held unpatentable claims for an algorithm used to convert binary code decimal numbers to equivalent pure binary numbers. The sole practical application of the algorithm was in connection with the programming of a general purpose digital computer. We defined "algorithm" as a "procedure for solving a given type of mathematical problem," and we concluded that such an algorithm, or mathematical formula, is like a law of nature, which cannot be the subject of a patent.[24]

This definition is significantly broader than the definition this Court employed in Benson and Flook. Our previous decisions regarding the patentability of "algorithms" are necessarily limited to the more narrow definition employed by the Court, and we do not pass judgment on whether processes falling outside the definition previously used by this

---

24. (Appears as Footnote 9 in the case) The term "algorithm" is subject to a variety of definitions. The petitioner defines the term to mean:

> "1. A fixed step-by-step procedure for accomplishing a given result; usually a simplified procedure for solving a complex problem, also a full statement of a finite number of steps. 2. A defined process or set of rules that leads [sic] and assures development of a desired output from a given input. A sequence of formulas and/or algebraic/logical steps to calculate or determine a given task; processing rules." Brief for Petitioner in Diamond v. Bradley, O.T. 1980, No. 79-855, p. 6, n. 12, quoting C. Sippl & R. Sippl, Computer Dictionary and Handbook 23 (2d ed 1972).

Court, but within the definition offered by the petitioner, would be patentable subject matter.

Parker v. Flook, supra, presented a similar situation. The claims were drawn to a method for computing an "alarm limit." An "alarm limit" is simply a number and the Court concluded that the application sought to protect a formula for computing this number. Using this formula, the updated alarm limit could be calculated if several other variables were known. The application, however, did not purport to explain how these other variables were to be determined,[25] nor did it purport "to contain any disclosure relating to the chemical processes at work, the monitoring of process variables, or the means of setting off an alarm or adjusting an alarm system. All that it provides is a formula for computing an updated alarm limit." 437 U.S., at 586, 98 S.Ct., at 2523.

In contrast, the respondents here do not seek to patent a mathematical formula. Instead, they seek patent protection for a process of curing synthetic rubber. Their process admittedly employs a well-known mathematical equation, but they do not seek to preempt the use of that equation. Rather, they seek only to foreclose from others the use of that equation in conjunction with all of the other steps in their claimed process.

These include installing rubber in a press, closing the mold, constantly determining the temperature of the mold, constantly recalculating the appropriate cure time through the use of the formula and a digital computer, and automatically opening the press at the proper time. Obviously, one does not need a "computer" to cure natural or synthetic rubber, but if the computer use incorporated in the process patent significantly lessens the possibility of "overcuring" or "undercuring," the process as a whole does not thereby become unpatentable subject matter.

Our earlier opinions lend support to our present conclusion that a claim drawn to subject matter otherwise statutory does not become nonstatutory simply because it uses a mathematical formula, computer program, or digital computer. In Gottschalk v. Benson, we noted: "It is said that the decision precludes a patent for any program servicing a computer. We do not so hold." 409 U.S., at 71, 93 S.Ct., at 257.

Similarly, in Parker v. Flook, we stated that "a process is not unpatentable simply because it contains a law of nature or a mathematical algorithm." 437 U.S., at 590, 98 S.Ct., at 2526. It is now commonplace that an application of a law of nature or mathematical formula to a known structure or process may well be deserving of patent protection. (Citations omitted.)

As Justice Stone explained four decades ago:

> "While a scientific truth, or the mathematical expression of it, is not a patentable invention, a novel and useful structure created with the aid of knowledge of scientific truth may be." Mackay Radio & Telegraph Co. v. Radio of America, 306 U.S. 86, 94, 59 S.Ct. 427, 431, 83 L.Ed. 506 (1939).

Although we were dealing with a "product" claim in Funk Bros., the same principle applies to a process claim. Gottschalk v. Benson, 409 U.S. 63, 68, 93 S.Ct. 253, 255, 34 L.Ed.2d 273 (1972). We think this statement in Mackay takes us a long way toward the correct answer in this case. Arrhenius' equation is not patentable in isolation, but when

---

25. As we explained in Flook, in order for an operator using the formula to calculate an updated alarm limit the operator would need to know the original alarm base, the appropriate margin of safety, the time interval that should elapse between each updating, the current temperature (or other process variable), and the appropriate weighing factor to be used to average the alarm base and the current temperature. 437 U.S., at 586, 98 S.Ct., at 2523. The patent application did not "explain how to select the approximate margin of safety, the weighing factor, or any of the other variables."

a process for curing rubber is devised which incorporates in it a more efficient solution of the equation, that process is at the very least not barred at the threshold by § 101.

In determining the eligibility of respondents' claimed process for patent protection under § 101, their claims must be considered as a whole. It is inappropriate to dissect the claims into old and new elements and then to ignore the presence of the old elements in the analysis. This is particularly true in a process claim because a new combination of steps in a process may be patentable even though all the constituents of the combination were well known and in common use before the combination was made. The "novelty" of any element or steps in a process, or even of the process itself, is of no relevance in determining whether the subject matter of a claim falls within the § 101 categories of possibly patentable subject matter.[26]

Presumably, this argument results from the language in § 101 referring to any "new and useful" process, machine, etc. Section 101, however, is a general statement of the type of subject matter that is eligible for patent protection "subject to the conditions and requirements of this title." Specific conditions for patentability follow and § 102 covers in detail the conditions relating to novelty. The question therefore of whether a particular invention is novel is "wholly apart from whether the invention falls into a category of statutory subject matter." In re Bergy, 596 F.2d 952, 961 (Cust. & Pat.App., 1979) (emphasis deleted). See also Nickola v. Peterson, 580 F.2d 898 (CA6 1978). The legislative history of the 1952 Patent Act is in accord with this reasoning.

In this case, it may later be determined that the respondents' process is not deserving of patent protection because it fails to satisfy the statutory conditions of novelty under § 102 or nonobviousness under § 103. A rejection on either of these grounds does not affect the determination that respondents' claims recited subject matter which was eligible for patent protection under § 101.

IV.

We have before us today only the question of whether respondents' claims fall within the § 101 categories of possibly patentable subject matter. We view respondents' claims as nothing more than a process for molding rubber products and not as an attempt to patent a mathematical formula. We recognize, of course, that when a claim recites a mathematical formula (or scientific principle or phenomenon of nature), an inquiry must be made into whether the claim is seeking patent protection for that formula in the abstract. A mathematical formula as such is not accorded the protection of our patent laws, Gottschalk v. Benson, 409 U.S. 63, 93 S.Ct. 253, 34 L.Ed.2d 273 (1972), and this principle cannot be circumvented by attempting to limit the use of the formula to a particular technological environment. Parker v. Flook, 437 U.S. 584, 98 S.Ct. 2522, 57 L.Ed.2d 451 (1978). Similarly, insignificant post-solution activity will not transform an unpatentable principle into a patentable process. To hold otherwise would allow a competent draftsman to evade the recognized limitations on the type of subject matter eligible for patent

---

26. It is argued that the procedure of dissecting a claim into old and new elements is mandated by our decision in Flook which noted that a mathematical algorithm must be assumed to be within the "prior art." It is from this language that the petitioner premises his argument that if everything other than the algorithm is determined to be old in the art, then the claim cannot recite statutory subject matter. The fallacy in this argument is that we did not hold in Flook that the mathematical algorithm could not be considered at all when making the § 101 determination. To accept the analysis proffered by the petitioner would, if carried to its extreme, make all inventions unpatentable because all inventions can be reduced to underlying principles of nature which, once known, make their implementation obvious. The analysis suggested by the petitioner would also undermine our earlier decisions regarding the criteria to consider in determining the eligibility of a process for patent protection.

See, e.g., *Gottschalk v. Benson*, *supra*; and *Cochrane v. Deener*, 94 U.S. 780, 24 L.Ed. 139 (1877).

protection. On the other hand, when a claim containing a mathematical formula implements or applies that formula in a structure or process which, when considered as a whole, is performing a function which the patent laws were designed to protect (e.g., transforming or reducing an article to a different state or thing), then the claim satisfies the requirements of § 101. Because we do not view respondents' claims as an attempt to patent a mathematical formula, but rather to be drawn to an industrial process for the molding of rubber products, we affirm the judgment of the Court of Customs and Patent Appeals.

It is so ordered.

Justice STEVENS, with whom Justice BRENNAN, Justice MARSHALL, and Justice BLACKMUN join, dissenting.

Notes and Problems

1. If $\ln v = CZ + x$ is patentable, why aren't Newton's Laws?

2. Do you agree with *Diamond v. Diehr*? Why or why not?

3. Based solely upon the dicta of *Diamond v. Diehr* can you articulate a distinction between those equations that are non-patentable laws of nature and those that *are* patentable because they describe an invention or process?

4. Einstein's famous equation for special relativity, $E = mc^2$ always seems to pop up whenever the physics of big weaponry is discussed. The equation formed the underlying basis for nuclear fission, which in turn was the lynchpin for the nuclear bomb. Assuming that at the time he developed special relativity, Einstein, a German citizen, would have been accorded the rights of a U.S. citizen, could he have obtained a patent for his now-famous equation? (After the atomic devices were exploded over Hiroshima and Nagasaki, killing thousands, Einstein is reputed to have said: "If only I had known, I should have become a watchmaker.")

5. Quite often, there is a fine line between an equation that merely describes an invented or improved process and the process itself. See *Exxon Research and Engineering Co. v. U.S.*, 265 F.3d 1371 (Fed. Cir., 2001), a patent rejection battle involving improvements in the methodology of converting natural gas into liquid hydrocarbon products.

§ 2.40 The Meaning of Inequality

This subject is not one for the ACLU. There are no unfavorable connotations linked to mathematical inequality. On the contrary, the symbols: $<$, $>$, $\leq$ and $\geq$ are often used to describe ranges, lengths, areas, volumes, minimums and maximums.

For example, it is repeatedly stated that one of the requirements for establishing federal diversity jurisdiction is the amount of $75,000. That is incorrect. The amount in controversy must *exceed* $75,000. In other words, in order to satisfy 28 U.S.C. § 1332, the amount in controversy might be represented by "x" such that:

$$x > \$75{,}000$$

Would the term $x = \$75{,}000$ suffice? No. Would $x \geq \$75{,}000$ suffice? No, because "$\geq$" (greater than *or* equal) implies that "$x = \$75{,}000$" is an acceptable alternative; we know that it is not.

Our subject matter jurisdiction example is quite simple. Mathematical structures involving inequalities can, by contrast, be very complex. In fact, they can provide a great legislative tool for hiding pork barrel measures and earmarks. If the statutory language is complex enough, only the lobbyists and special interests will fully comprehend its unfairness. The Internal Revenue Code is a perfect example. If the wage-earning, taxpaying public truly understood and appreciated the Code's provisions, there would be more than mere discontent.

Example 2-8: The *Class Action Fairness Act of 2005* 28 U.S.C. § 1332(d)[27] was designed to chop down the number of large class action lawsuits that consumers might otherwise file and maintain in state courts. If the statute merely provided for federal subject matter jurisdiction for *all* class action lawsuits, the transparency and purpose would draw opposition, particularly from attorneys and consumer groups. So instead, to obfuscate the legislation, a more complex structure was created whereby three tiers of jurisdictional qualifications were employed.

A class action might consist of a single class of plaintiffs or may be comprised of several sub-classes. Let: $x_1 + x_2 + x_3 + x_4 + x_5 + x_6 + \ldots x_n$ constitute one subclass of plaintiffs and let: $y_1 + y_2 + y_3 + y_4 + y_5 + y_6 + \ldots y_m$ comprise a second sub-class. We could continue the process for each subclass, but for this example, we will presume that there are only two.

$$\text{Let } \Sigma x_i = x_1 + x_2 + x_3 + x_4 + x_5 + x_6 + \ldots x_p$$

$$\text{Let } \Sigma y_j = y_1 + y_2 + y_3 + y_4 + y_5 + y_6 + \ldots y_q$$

Let $A = \Sigma x_i \cup \Sigma y_j$ represent the aggregate plaintiffs for all of the sub-classes. Let $n(A)$ = total number of plaintiffs in the lawsuit. Let $n(C)$ = the number of all plaintiffs who are citizens of the state where the action was originally filed. According to § 1332(d)[28]: if the amount in controversy exceeds $5,000,000

27. Always be suspicious whenever words such as "fairness" are inserted into statutes. The Act could just as easily have been termed the "Class Action Limitation Act." Every plaintiff practitioner knows that the federal court system is a burial ground for state based, consumer oriented lawsuits. That is why defendants uniformly remove state cases to federal court.

28. The statute contains many other provisions; the definitions in (d)(1) are omitted. Subsection (d)(2) states: "The district courts shall have original jurisdiction of any civil action in which the matter in controversy exceeds the sum or value of $5,000,000, exclusive of interest and costs, and is a class action in which—(A) any member of a class of plaintiffs is a citizen of a State different from any defendant; (B) any member of a class of plaintiffs is a foreign state or a citizen or subject of a foreign state and any defendant is a citizen of a State; or (C) any member of a class of plaintiffs is a citizen of a State and any defendant is a foreign state or a citizen or subject of a foreign state. (3) A district court may, in the interests of justice and looking at the totality of the circumstances, decline to exercise jurisdiction under paragraph (2) over a class action in which greater than one-third but less than two-thirds of the members of all proposed plaintiff classes in the aggregate and the primary defendants are citizens of the State in which the action was originally filed based on consideration of—(A) whether the claims asserted involve matters of national or interstate interest; (B) whether the claims asserted will be governed by laws of the State in which the action was originally filed or by the laws of other States; (C) whether the class action has been pleaded in a manner that seeks to avoid Federal jurisdiction; (D) whether the action was brought in a forum with a distinct nexus with the class members, the alleged harm, or the defendants; (E) whether the number of citizens of the State in which the action was originally filed in all proposed plaintiff classes in the aggregate is substantially larger than the number of citizens from any other State, and the citizenship of the other members of the proposed class is dispersed among a substantial number of States; and (F) whether, during the 3-year period preceding the filing of that class action, 1 or more other class actions asserting the same or similar claims on behalf of the same or other persons have been filed. (4) A district court shall decline to exercise jurisdiction under paragraph (2)—(A) (I) over a class action in which—(I) greater than two-thirds of the members of all proposed plaintiff classes in the aggregate are citizens of the State in which the action was originally filed; (II) at least 1 defendant is a defendant—(aa) from whom

and if $n(A) > 100$, then with respect to the U.S. District Court's subject matter jurisdiction:

[1] If $n(C)/n(A) \leq 1/3 \Rightarrow$ Court shall have Jurisdiction.

[2] If $1/3 < n(C)/n(A) < 2/3 \Rightarrow$ Court may decline jurisdiction.

[3] If $2/3 < n(C)/n(A) \Rightarrow$ Court shall decline jurisdiction.

Notes and Problems

1. What happens if $n(C)/n(A) = 2/3$?

2. Other than the lobbyists who drafted § 1332(d), how many lawmakers do you suppose actually read it? Understood it?

§ 2.50 Determining the Unknown

Ever since junior high school, we learned that x's were everywhere. One of the elements on Table 2-1 is the ubiquitous generic function $y = f(x)$. This simple equation is, of course, a fundamental player in the universe of mathematics and physics. A single rule for solving equations seems to triumph over all others. For expediency, put aside all of the many rules of arithmetic and math that you were exposed to during your high school and college courses in algebra. We can reduce most of them to a single precept.

> **Easy (though oversimplified) Rule:** Do anything whatsoever that you want to do to one side of the equation as long as you do precisely the same thing to the other side.

This simple rule applies regardless of whether the equal sign represents an equation, an identity, a relationship, or even a process.

---

significant relief is sought by members of the plaintiff class; (bb) whose alleged conduct forms a significant basis for the claims asserted by the proposed plaintiff class; and (cc) who is a citizen of the State in which the action was originally filed; and (III) principal injuries resulting from the alleged conduct or any related conduct of each defendant were incurred in the State in which the action was originally filed; and (ii) during the 3-year period preceding the filing of that class action, no other class action has been filed asserting the same or similar factual allegations against any of the defendants on behalf of the same or other persons; or (B) two-thirds or more of the members of all proposed plaintiff classes in the aggregate, and the primary defendants, are citizens of the State in which the action was originally filed. (5) Paragraphs (2) through (4) shall not apply to any class action in which—(A) the primary defendants are States, State officials, or other governmental entities against whom the district court may be foreclosed from ordering relief; or (B) the number of members of all proposed plaintiff classes in the aggregate is less than 100. (6) In any class action, the claims of the individual class members shall be aggregated to determine whether the matter in controversy exceeds the sum or value of $5,000,000, exclusive of interest and costs. (7) Citizenship of the members of the proposed plaintiff classes shall be determined for purposes of paragraphs (2) through (6) as of the date of filing of the complaint or amended complaint, or, if the case stated by the initial pleading is not subject to Federal jurisdiction, as of the date of service by plaintiffs of an amended pleading, motion, or other paper, indicating the existence of Federal jurisdiction. (8) This subsection shall apply to any class action before or after the entry of a class certification order by the court with respect to that action." Subparagraphs... (9)–(11) are omitted.

Example 2-9: Solve the simple equation: $x + 4 = 7$.

Solution: You can solve this one in your head. But what if you wanted to subtract 2 from each side instead of 4? It would be a foolish waste of time, perhaps. But would the proposed action be incorrect? No. The equation $x + 2 = 5$ is no more accurate or inaccurate than our original problem. Eventually, you would get around to subtracting the number on the left to wind up with $x = 3$. If you obey the *Rule*, you should eventually solve the problem. The principle applies if we wish to double, triple, square or take the cube root, as long as the *same operation is performed on both sides of the equation.*

Example 2-10: Net Worth = Assets - Liabilities. This very basic Balance Sheet Equation of accounting is recognized by all (and discussed in Chapter 9). Is it true that we can increase our assets by merely going deeper into debt? Is the term "assets" misleading?

Solution: Add "Liabilities" to both sides of the equation to find out.

Net Worth = Assets - Liabilities ⇒

Net Worth + ***Liabilities*** = Assets - Liabilities + ***Liabilities*** ⇒

Net Worth + Liabilities = Assets - 0 ⇒

Assets = Net Worth + Liabilities

Notice that according to the above equation, as one goes further into debt, his or her assets increase because the liability would increase. This, of course, suggests that "Assets" can be an indicator of net worth but it can equally be an indicator of debt.

§ 2.60 Analyzing Shapes and Configurations

For the legal profession, visual mathematics in the form of trigonometry, geometry and topology are most closely associated with real property and boundary disputes. But as we shall see in the chapters that follow, the topics have an impact upon many other diverse legal disputes.

§ 2.61 Keep Your Distance

For most of us, elementary school provided an early confrontation with the Pythagorean Theorem. There is, therefore, an element of poetic justice when the Theorem is used to ensnare dope dealers as opposed to hapless math students.

People v. Robbins

5 N.Y.3d 556, 840 N.E.2d 1020 (2005)

KAYE, Chief J.

It is a class B felony to sell drugs within 1,000 feet of a school. Because we agree with the courts below that this distance is to be measured in a straight line between the nearest school and the point of sale, we uphold defendant's conviction.

Penal Law § 220.44(2) proscribes the third-degree criminal sale of a controlled substance "when such sale takes place upon school grounds." School grounds, in turn, means not only "in or on or within" school buildings or school property, but also "any area accessible to the public located within one thousand feet of the real property boundary line" of a school, as well as any vehicle parked within 1,000 feet of such line (Penal Law § 220.00). An "area accessible to the public" means "sidewalks, streets, parking lots, parks, playgrounds, stores and restaurants."

Defendant was charged with selling drugs to an undercover police officer at the northeast corner of 40th Street and Eighth Avenue in Manhattan. At trial, the People offered proof that the nearest school was the Holy Cross grade school, located on 43rd Street between Eighth and Ninth Avenues. It is undisputed that the sale took place within 1,000 feet of the school when measured in a straight line between the sale and the school's boundary line, but that existing buildings made it impossible to walk in a straight line between those two points. It is further undisputed that the shortest actual distance that a pedestrian would have to travel in order to walk from the scene of the crime to the school using city streets and publicly accessible locations exceeded 1,000 feet.

Defendant contends that the statute mandates that the requisite distance be calculated by the "pedestrian method," and that the evidence was therefore legally insufficient to support his conviction. We are unpersuaded, and hold that the proper method of measurement under the statute is the "straight-line method."

The school-grounds law was enacted in order to create a drug-free buffer zone of protection—"a corridor of safety for children coming to and from school" (Mem. of State Exec. Dept., 1986 McKinney's Session Laws of NY, at 2892).

By defining school grounds to encompass both the school property itself and the entire area "within" 1,000 feet of the property boundary line, the Legislature effectively extended the boundaries of school grounds outward in order to encompass all public areas within a 1,000-foot radius of the school.

Indeed, the 1,000-foot measurement was chosen because of the high number of narcotics-related arrests that had occurred within a two-block "radius" of elementary school grounds (see Governor's Memo approving L 1986, ch 280, 1986 McKinney's Session Laws of NY, at 3161 [because many, if not most, sales of drugs to children occur near but not inside school buildings, the bill defines school grounds to include "the area" within 1,000 feet of the real property boundary line, providing a "zone of safety" for school children]). Thus, the intent of the statute was to circumscribe a fixed geographical area, without regard to whether that area might contain obstacles around which people might have to detour. One thousand feet is "the equivalent of two city blocks."

Other jurisdictions with similar statutes have construed their laws to require a straight-line measurement. Taylor Drug Stores, Inc. v. Indiana Alcoholic Beverage Commn., 497 N.E.2d 932, 936 [Ct. App. 1986] [proximity between a church and a premises seeking a permit to distribute alcoholic beverages should be measured along the shortest straight line connecting the church and the proposed premises, regardless of intervening obstacles].

Similarly, the federal schoolyard law—which prohibits "distributing, possessing with intent to distribute, or manufacturing a controlled substance in or on, or within one thousand feet of, the real property comprising" a school (21 U.S.C. § 860)—has been universally interpreted to contemplate the straight-line method of measurement (Citations omitted).

Defendant's contrary reading would introduce uncertainty and open the statute to a charge of vagueness. Plainly, guilt under the statute cannot depend on whether a particular building in a person's path to a school happens to be open to the public or locked at the time of a drug sale. At a minimum, "[r]equiring speculation about pedestrian routes would create uncertainty in a statute which was meant to establish clear lines of demarcation" (Watson, 887 F.2d at 980–981).

Further, if the pedestrian method of measurement were used, a drug dealer could take deliberate steps to evade the reach of the statute by putting up obstructions to render the footpath to a nearby school beyond the 1,000-foot limit (see Clavis, 956 F.2d at 1088). But the intent of the legislation would be defeated "if dealers were allowed to escape prosecution by creating circuitous routes to their narcotic transactions" (Watson, 887 F.2d at 981). Thus, as the courts below properly held, the statutory distance is to be measured "as the crow flies."

Defendant's remaining contention is without merit.

Accordingly, the order of the Appellate Division should be affirmed.

Notes and Problems

1. For the *Robbins* case, the defendant was positioned at the corner of 40th Street and 8th Avenue in Manhattan. The Holy Cross Grade School is located on 43rd Street between 8th and 9th Avenues. Assume that the defendant's position is represented by the vertex of Angle <AC. Further assume that Holy Cross Grade School is located at the vertex of Angle <BC. Side a is 490 feet. Side b is 764 feet. Is $c < 1{,}000$ feet?

2. Recall that the Pythagorean Theorem states that for every right triangle, the square of the hypotenuse (c) equals the sum of the squares of the other two legs (a,b). The length of the hypotenuse, therefore, would be obtained by taking the square root of both sides of the equation:

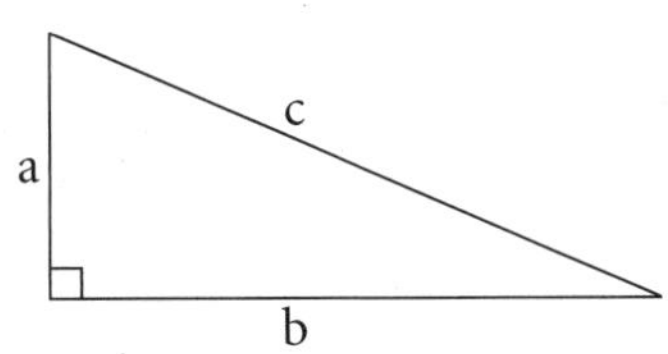

$$a^2 + b^2 = c^2$$

$$c = \sqrt{a^2 + b^2}$$

3. Pythagoras (571–496 B.C.) was a vegetarian who worshiped numbers and believed that mathematics was the answer to everything including legal issues. "Justice," for example was embodied in the number 4 because it was both even and a perfect square root. He was, on the other hand, extremely upset by irrational numbers such

as π and the square root of 2. It was rumored that Pythagoras once drowned one of his students (Hipparsus of Tarentum) for revealing the irrational nature of certain numbers to others.

In addition to his famous theorem, Pythagoras played another important role in the history of mathematics when he advanced the notion of the "theorem" by declaring that mathematical truths should be proved not just accepted. His contributions to jurisprudence, on the other hand, were marginal at best; if in doubt, try asking Hipparsus of Tarentum.

§ 2.62 Respecting Boundaries

The "science" of cartography relies on its own terminology along with a skewed use of basic trigonometric and mathematical principles. All deeds and sales contracts require some type of legal description to identify the subject property and its boundaries. There are three basic methods used to determine the exact location and boundaries of a parcel of land:

(1) **Rectangular Survey**

(2) **Metes and Bounds**

(3) **Lot and Block**

The Rectangular Survey System, created by Congress in 1785, began by marking large tracts of land received by the United States in its infancy. The rectangular survey system designates areas by using a familiar rectangular coordinate system (similar to the one created by the great philosopher Rene Descartes and employed in analytical geometry). There are principal meridians that run North and South to indicate longitude, and base lines that run East and West to designate latitude. Principal meridians and base lines partition the lands into quadrangles (squares of land with each side measuring 24 miles). The quadrangles are further subdivided into 16 townships. Each side of a township line measures 6 miles resulting in an area of 36 square miles per township. Townships are then further subdivided into 36 square mile sections. Each Section is equal to 640 acres.

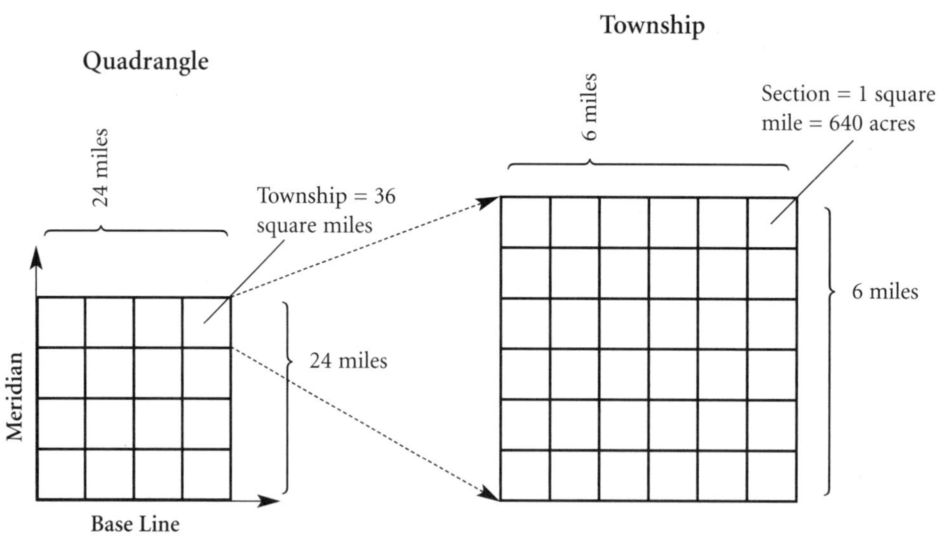

In many states, the Rectangular Survey System is supplemented with a metes and bounds description to zero—in on a particular parcel of land. A metes and bounds description uses distances and angles of boundaries starting at the Point of Beginning (POB). From the POB, the length of each boundary is described along with the angle that it forms with the previous boundary. This process continues to the Point of End (POE), which must be identical to the POB. A failure to close the boundary would topologically encompass the world.

Every legal description has to begin somewhere. This *somewhere* is normally a monument or permanent landmark, such as a large rock, tree, river, railroad, road and intersection, or another object recorded by surveyors in order to locate real property. There are also special iron pins and concrete posts scattered about by surveyors or the U.S. Corp of Engineers.

In most deeds, the bearing is *not* described in conjunction with the typical clockwise degree measurement from 0 to 360 degrees that we are all familiar with. Instead, the journey would commence with a point (POB), then a direction would be specified (N, S), followed by a degree measure ($0° < x < 90°$), then another direction (W, E), then a distance. This method has the advantage of providing the same degree measure regardless of which direction a particular boundary is being measured from; the boundary can be traversed in the opposite direction simply by exchanging N for S and E for W. In other words, "N 45° 30' W" describes the same boundary as "S 45° 30' E", but is traversed in the opposite direction.

Example 2-11: An example that combines a block and lot (Platt) reference might read as follows:

> A part of the NW 1/4 of Section 16, TIN, R9E, City of Farmington Hills, Oakland County Michigan, more particularly described As:
>
> > Commencing at the NW corner of said Section 16; thence S 02° 15' 39" E 791.71 feet along the W line of Section 16 and following Drake Road; thence N 87° 44' 21" E 50.00 to a point on the E right of way line of Drake Road... thence 02° 15' 39" E 314.11 feet to a point on the northerly right of way of Interstate 696; thence S 85° 28' 16" W 340.00 feet along said line to the easterly right of way line of Drake Road; thence N 02° 15' 39" W 118.17 feet to the point of beginning.

Up to this point we have been assuming that all boundaries are formed by nice straight lines or natural barriers. That is not necessarily the case, although it is certainly preferable. Some lines are curved. There are many ways to describe curves. Some variations result from various conventions. At least two elements of a curve are required to define a curve: usually radius and length. A third, orientation, is preferable. The descriptions can become relatively complex. Here is an example:

> ... Thence westerly along said curve 100.72 feet through a central angle of 54° 11' 10" to the beginning of a 200.00 foot radius compound/reverse curve, concave southerly (the radius point of which bears South 20° 11' 44" West)...

Problem: Using a tree as your POB, utilize metes and bounds to describe a simple square parcel of real estate consisting of 100 square yards.

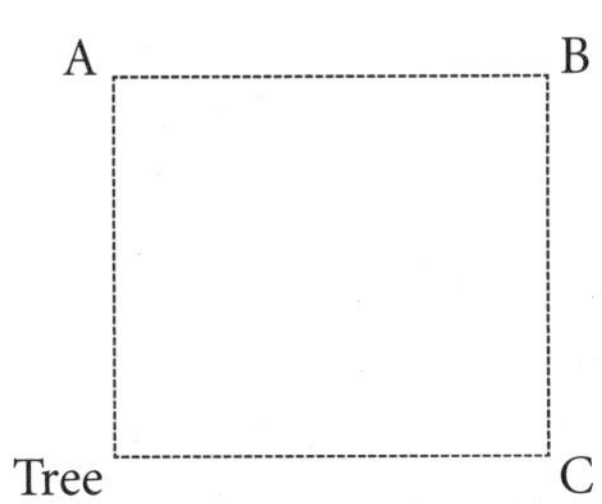

Important: As a practical matter, the most important thing for the practitioner to know is the "Rule of 4," which states that if you intend to draft a legal description or read an abstract (as opposed to advising your client to purchase title insurance) you should quadruple your malpractice coverage.

Pennsylvania Electric Company v. Waltman

Superior Court of Pennsylvania, 448 Pa.Super. 174, 670 A.2d 1165 (1995)

POPOVICH, Judge:

Pennsylvania Electric Company ("Penelec") appeals from the judgment entered in the Court of Common Pleas of Clarion County following a bench trial in a boundary dispute. After careful review, we affirm.

The factual background of this dispute revolves around ownership to a strip of approximately ten acres of land abutting the Clarion River. Penelec, an electric utility company, operates a hydroelectric plant on the Clarion River. This plant was constructed by Penelec's predecessor as part of the Piney Dam Project. Construction of the plant required approval by the Federal Power Commission, and an alleged prerequisite to this approval was that the utility acquire all property along a particular stretch of the Clarion River to an elevation of 1105 feet. The dispute arose as a result of certain structures that appellee, Vernon Waltman, constructed on what he believed was his property. Penelec contended that the structures were on its property because the land was allegedly for the dam construction project.

The parties trace their titles to one J.P. Maxwell. Penelec's title originated in an August, 1922 option that Maxwell granted to County Realty Company ("County Realty") for a section of a tract of his property situated below the 1105 foot elevation. Specifically, the option provided: "On the west by land of James Campbell heirs; being that part of my lands lying below said elevation 1105 feet above sea level; containing 10 acres more or less ..." The option was accepted by County Realty in November of 1922. The option included the following description of the property: On the north by that part of our lands lying above elevation 1105 feet above sea level; said elevation being determined from a bench mark established by the United States Geological Survey.... On the west by lands of James Campbell Heirs; being that part of my lands lying below said elevation 1105 feet above sea level; containing 10 acres more, or less ...

Maxwell conveyed the property to County Realty by deed in December of 1922. County Realty subsequently conveyed the tract to the Clarion River Power Company followed by a conveyance of the same tract to Penelec by deed in March of 1942. This deed made no mention of conveying all land below the 1105 foot elevation. Instead the description of the property conveyed was done by courses and distances. The deed specifically contained the following course descriptions:

> [T]hence with the line of other lands of J.P. Maxwell for the following fourteen Courses: (1) North 30° 19' East 184.6 feet; (2) North 33° 04' East 232.2 feet; (3) North 35° 03' East 262.2 feet; (4) North 13° 50' East 149.2 feet; (5) South 57° 56' East 61.0 feet; (6) North 70° 29 East 156.9 feet; (7) North 65° 00' East 243.9 feet; (8) North 63° 04' East 334.5 feet; (9) North 71° 24' East 190.1 feet; (10) North 76° 40' East 238.7 feet; (11) North 82° 00' East 218.1 feet; (12) South 88° 42' East 285.3 feet; (13) South 85° 40' East 293.9 feet; (14) South 79° 53' East 219.2 feet to a post and stones, a corner to other lands of J.P. Maxwell ...

Waltman's title originated in a 1978 deed from Maxwell's widow to one Jack Baum. This deed conveyed the remaining property in the Maxwell tract and contained an express exception for the property that was conveyed to County Realty. In 1984, Baum conveyed, by deed, a tract of approximately 15 acres to Waltman. This deed stated that the conveyance was subject to the property deeded to Penelec. Upon taking possession of his property, Waltman constructed a concrete boat dock, a house and several other improvements below the 1105 foot elevation. Penelec became aware of these improvements in 1986, and asked Waltman to remove them. When this request was refused, Penelec brought an action for injunctive relief requesting the removal of these structures.

The boundary dispute centers around the deed from the Maxwell's to County Realty ("County Realty deed"). Penelec alleges this deed is ambiguous. The ambiguity supposedly arises because the deed was missing two distance calls and the deed description did not close. Penelec contended that the following descriptions in the deed created ambiguity:

> ... Beginning at a point at low water mark on the Clarion River, opposite a stone on the right bank (looking downstream) of the Clarion River, said stone being located about 1600 feet upstream from the mouth of Calico and being a corner to the lands of M.M. Kaufman and Mrs. James Braden; thence 3 degrees-50 East to a stone; ... thence south 4 degrees 43 west to low water mark on the Clarion River ...

At trial, Penelec presented the testimony of surveyor Gregg Facciani, who provided support for Penelec's contention that the deed was ambiguous. Mr. Facciani based his opinion upon a survey he performed using a U.S. Geological Survey topographical map. He used this map because Penelec wanted to have its property tied into the 1105 foot elevation and a "stated plan coordinate system." The "stated plan coordinate system" is a system on which the entire Commonwealth is based. This system establishes monuments from the United States Geological Survey and apparently can be used to tie in deed calls to establish property lines. It does not establish deed bearings.

Penelec believed that it should be able to introduce extrinsic evidence to clarify the descriptions in the deed because of these alleged ambiguities. Waltman, on the other hand, contended that the Maxwell to County Realty deed was unambiguous. Harold Frank Scott testified as a surveying expert for Waltman. Mr. Scott was familiar with the property because of a 1978 lot development survey he had prepared on the Maxwell tract. He stated that it was possible to calculate the courses and distances in the deed even though some of these were now underwater. Waltman's structures were on his own property according to Mr. Scott. Mr. Ron Scott also presented expert testimony for Waltman and stated that he established the missing corner of the deed using information from adjoining deeds. Mr. Scott further testified that Penelec was mistaken as to its property line and that he was able to establish Penelec's property using the course and distance calls in the County Realty Deed.

William McGarvey, a surveyor, also testified for Waltman and stated that the description in the deed provided sufficient information to ascertain the exact property conveyed to Penelec. McGarvey also opined that, as a result of his various surveys and computer simulations, Waltman's improvements did not encroach upon Penelec's property. Further testimony provided that it was possible to calculate, with mathematical certainty, where the property in the County Realty deed was located.

The Honorable Paul Greiner heard this evidence at a trial in May of 1994, and subsequently entered judgment on an August 16, 1994, verdict in favor of Waltman. The verdict stated that Penelec did not meet its burden of proving the boundary line in dispute to be at an elevation of 1105 feet. Judge Greiner also adopted Waltman's proposed findings of fact and conclusions of law. This entry of judgment was also based upon a November 17, 1994, order denying Penelec's post-trial motion.

This timely appeal followed.

Penelec now presents the following questions for review: Did the lower court err as a matter of law in disregarding extrinsic evidence to determine the intent of the grantor in conveying property by deed to Penelec's predecessor in title? Did the lower court err as a matter of law in determining that Waltman owned the property lying below the 1105 elevation where the undisputed evidence was that the conveyance to Waltman's predecessor in title expressly excepted all land lying below the 1105 elevation?

In the alternative, if the deed to Penelec's predecessor in title is not ambiguous, did the lower court err as a matter of law in disregarding the prevailing natural monuments and adjoining calls in determining the boundary line of Penelec's property? In the alternative, did the lower court err as a matter of law in failing to find that Penelec owned the disputed property by de facto taking?

Penelec, in its first argument, contends that the lower court improperly refused to consider extrinsic evidence to aid in interpreting the ambiguities in the Maxwell to County Realty deed.

The court determines whether a written instrument is ambiguous. Metzger v. Clifford Realty Corp., 327 Pa.Super. 377, 476 A.2d 1 (1984). We stated in Wysinski v. Mazzotta, 325 Pa.Super. 128, 472 A.2d 680 (1984): In the absence of fraud, accident or mistake, the nature and quantity of the real estate interest conveyed must be ascertained from the deed itself and cannot be shown by parol. When the language of the deed is clear and free from ambiguity, the intent of the parties must be determined from the language of the deed. With respect to unambiguous deeds, a court must ascertain what is the meaning of the words used, not what may have been intended by the parties as shown by parol. To permit a variation of a deed description which is complete and unambiguous on its face, there must be evidence of a mutual mistake which is clear, precise and convincing (citations omitted).

However, the boundaries to property are not always clear from the plain reading of a written deed. Doman v. Brogan, 405 Pa.Super. 254, 592 A.2d 104 (1991) (citations omitted). Where a deed is uncertain because of vague or ambiguous language, extrinsic evidence may be used to explain, but not vary, the writing in the deed. Id. at 109 (citation omitted). "[T]he primary function of a court faced with a boundary dispute is to ascertain and effectuate the intentions of the parties at the time of the original subdivision."

Here, we find that the lower court did not err in finding that the deed was free from ambiguity and refusing to consider extrinsic evidence. Waltman presented three surveying experts, Mr. Frank Scott, Mr. Ronald Fox and Mr. William McGarvey. Mr. Scott prepared a survey for Waltman's predecessor in title, Mr. Jack Baum, of the original J.P. Maxwell tract. The northern and eastern boundaries of this tract were established by Mr. Scott by the use of certain physical evidence. Mr. Scott also stated that courses and distances could be established even though they were under water.

Surveying expert, Mr. Fox added support to Mr. Scott's testimony with his testimony that lost corners or lost points in a deed could be reestablished. The corner in this dispute could be re-established using deed information from deeds of adjoining properties. Mr. Fox further stated that he found the northern boundary and that the Penelec line of monuments were further away from the water's edge than the Penelec deed stated. Evidence also was presented that the Penelec line was not meant to be at an elevation of 1105 feet.

Finally, Mr. McGarvey testified that modern computer simulation could establish the acreage of a tract within an inch. He further stated that he was able to establish the Maxwell to County Realty tract of land accurately by using the courses and distances in the deed and by using deeds for surrounding conveyances. In fact, Mr. McGarvey determined that Penelec was claiming ownership of approximately twenty-four acres which contradicted the ten and one-half acres that the Maxwell to County Realty deed conveyed. Finally, review of the condemnation proceedings for the Piney Dam Project reveals that land was not condemned at an 1105 foot elevation. Instead, it was condemned using course and distance descriptions.

Penelec, through its expert Gregory Facciani, did not prove that it owned the disputed property. Mr. Facciani, stated that he used a United Stated Geological Survey and attempted to tie in Penelec's property line with this survey. Mr. Facciani admitted that the courses and distances in this map did not close and contained no reference to the 1105 foot elevation. Moreover, he ran his survey at an 1105 foot elevation because Penelec believed that its property line was at this elevation. In essence, Mr. Facciani appears to have established that Penelec owned at the 1105 foot elevation by assuming that its property line was at that elevation. The trial court, in its November 17, 1994, order stated, "[Gregory] Facciani's testimony did not support Plaintiff's burden of proving it's [sic] claim by the fair weight or preponderance of the evidence." Our review of the record leads us to conclude that Judge Greiner did not err in his conclusions.

Penelec also argues that the deed is ambiguous because it differs from the option agreement. This argument fails in light of the doctrine of merger by deed which provides "as a general rule an agreement of sale merges into the deed and no recovery may be had based upon an earlier agreement." Valvano v. Galardi, 363 Pa.Super. 584, 592, 526 A.2d 1216, 1220 n. 2 (1987), citing Stoever v. Gowen, 280 Pa. 424, 124 A. 684 (1924); See also, Doman v. Brogan, supra. Accordingly, Penelec's first allegation of error fails.

Second, Penelec contends that the lower court erred as a matter of law in determining that Waltman owned the property lying below the 1105 elevation where the undisputed evidence was that the conveyance to Waltman's predecessor in title expressly excepted all land lying below the 1105 foot elevation.

When anything is excepted from a deed all things that are dependent upon it and necessary for obtaining it are also excepted. Walker v. Forcey, 396 Pa. 80, 151 A.2d 601 (1959). Where a doubtful reservation or exception exists in a deed, the deed will be construed most strongly in favor of the grantee. (Citations omitted) "This rule applies with special force to a reservation or exception which amounts to a cutting down of the grant ..." (citation omitted).

In this case, we find that the exception did not establish that Penelec owned all the land below the 1105 foot elevation. Contrary to Penelec's assertion, the evidence was not undisputed that Waltman's predecessor in title excepted all land lying below the 1105 foot elevation. Waltman's expert, Mr. Fox, even though conceding that he did not run his survey with the exception, stated that he ran his survey using the course and distance calls in the unambiguous deed. By so doing, he was able to establish that Waltman's structures were not on Penelec property. Thus, this exception was doubtful and must be construed in favor of Waltman. Heidt, supra. Moreover, Federal Energy Regulatory Commission proceedings for the original Piney Dam Project showed that the dam was to be established at an elevation of 1085 feet. Accordingly, we find appellant's argument meritless.

Penelec next contends that if the deed to its predecessor in title is not ambiguous, the lower court erred as a matter of law in disregarding the prevailing natural monuments and adjoining calls in determining the boundary line of Penelec's property. We stated, in Baker v. Roslyn Swim Club, 206 Pa.Super. 192, 213 A.2d 145 (1965):

> The general rules to be applied in boundary cases were set forth in Walleigh v. Emery, 193 Pa.Super. 53, 58, 163 A.2d 665, as follows: "Courses and distances

in a deed must give way to monuments on the ground. Merlino v. Eannotti, 177 Pa.Super. 307, 110 A.2d 783.

"Parol evidence is admissible to establish the existence of monuments. New York State Natural Gas Corp. v. Roeder, 384 Pa. 198, 120 A.2d 170. Boundaries may be established by circumstantial as well as direct evidence. Hostetter v. Commonwealth, 367 Pa. 603, 80 A.2d 719; Guerra v. Galatic, 185 Pa.Super. 385, 137 A.2d 866."

The general rule for establishing the order of precedence between inconsistent calls in a deed is found in 12 Am.Jur.2d Boundaries § 65, at page 603, as follows: "Where the calls for the location of boundaries to land are inconsistent, other things being equal, resort is to be had first to natural objects or landmarks, next to artificial monuments, then to adjacent boundaries (which are considered a sort of monument), and thereafter to courses and distances." However, the same paragraph continues with the following language: "Where, however, it is apparent that a mistake exists with respect to the calls, an inferior means of location may control a higher one. In the last analysis the call adopted as the controlling one should be that most consistent with the apparent intent of the grantor."

Likewise, the same general rule and the same exceptions are cited in 11 C.J.S. Boundaries § 51, as follows: "The rule that artificial monuments control courses and distances in case of conflict is not an imperative and exclusive one, but is a rule of construction to ascertain, or to aid in determining, the intention of the parties; and it is not followed where strict adherence to the call for a monument would lead to a construction plainly inconsistent with such intention.

"Accordingly, courses and distances will prevail over monuments where absurd consequences might ensue by giving controlling influence to a call for the latter, or where, in any given case, a consideration of all the facts and circumstances shows a call for distance to be the more reliable or certain, or where the call for the monument was inserted by mistake or inadvertence." Baker, 213 A.2d at 148–149.

Here, Penelec contends that a stream, which is a natural monument, existed at the 1105 foot elevation. A stream did exist, but according to Mr. McGarvey's survey the stream was not at the 1105 foot elevation as Penelec contends. Mr. McGarvey ran the course and distance calls in the County Realty deed and the stream coincided with these calls. However, Mr. McGarvey found that the stream was not at the 1105 foot elevation but instead at a lower elevation. Thus, he concluded that Waltman's structures were not on Penelec's property. Contrary to Penelec's contentions, the monument was not ignored but was taken into account by Mr. McGarvey when he established the calls in the deed. Because there was doubt as to the placement of the stream, Waltman's experts, using the courses and distances in the County Realty deed, established the proper boundaries of Penelec's and Waltman's property. As the court in Baker stated, courses and distances will prevail over monuments when there is doubt as to the monument. Baker, 213 A.2d at 150. Accordingly, Penelec's argument must fail. See, Baker, supra (holding that a stone monument mentioned in three deeds did not prevail over courses and distances in the deeds because the monument was not free from doubt and could lead to an absurd result); Yoho v. Stack, 373 Pa.Super. 77, 540 A.2d 307 (1988) (holding that monument must yield to deed description where expert testimony pertaining to the deed description was more credible and that courses and distances will prevail over monuments where the monuments are doubtful); Howarth v. Miller, 382 Pa. 419, 115 A.2d 222 (1955).

Penelec's final contention is that the lower court erred as a matter of law in failing to find that Penelec owned the disputed property by de facto taking. We disagree.

The requirements of a de facto taking can be summarized as follows:

> A de facto taking occurs when an entity, clothed with the power of eminent domain, exercises that power causing damages to the property owner which are the immediate, necessary and unavoidable consequences of that exercise. Harborcreek Township v. Ring, 48 Pa.Commonwealth Ct. 542, 410 A.2d 917 (1980). To find a de facto taking, there must be exceptional circumstances which have substantially deprived the property owner of the use and enjoyment of his or her property. Jacobs Appeal, 55 Pa.Commonwealth Ct. 142, 423 A.2d 442 (1980), appeal dismissed, 499 Pa. 337, 453 A.2d 336 (1982). Hence, in determining whether a de facto taking occurred, each case must be examined on its own factual situation. McGaffic v. Redevelopment Authority of City of New Castle, 120 Pa.Commonwealth Ct. 199, 548 A.2d 653 (1988), appeal denied, 523 Pa. 644, 645, 565 A.2d 1168, 1169 (1989).

Appeal of D.R.E. Land Developing, Inc., 149 Pa.Cmwlth. 290, 294, 613 A.2d 96, 98 (1992).

Here, Penelec contends that a de facto taking resulted because it had marked off the property with wires. This fact was disputed at trial by Waltman, and there was only evidence of one small piece of wire through a tree. In fact, Penelec presented no evidence of a de facto taking at trial. This contention was raised for the first time in its Closing Argument/Trial Brief/Proposed Findings of Fact. Penelec in making this argument is basically conceding that it did not obtain the property legitimately by deed. Such an argument, would support Waltman's view that he is the legitimate owner of the disputed property. Accordingly, Penelec's final allegation of error fails.

Penelec relies on the case of Fulmer v. White Oak Borough, 146 Pa.Cmwlth. 473, 606 A.2d 589 (1992), to support its contention that there was a de facto taking. In Fulmer there was a clear taking of land by White Oak Borough. Id. at 592. Moreover, the main issues addressed in Fulmer were whether the Fulmer's action was properly brought as an action in trespass or whether it should have been brought under the Eminent Domain Code and what a proper remedy would be. Id. at 590. Obviously, Fulmer is inapposite to this case. There was not a clear taking of property by Penelec. Furthermore, Penelec is trying to shift the burden to Waltman by claiming there was a de facto taking.

Judgment affirmed.

Notes and Problems

1. How difficult would it be to establish the missing line that would close the loop in the Waltman case? Simple trigonometry combined with a little bit of geometry? Does the fact that the part of the land is underwater create any insurmountable problems? Are two-dimensional descriptions adequate for describing three-dimensional landscapes?

2. Assuming the absence of any curved borders, what geometric figure must be described by every proper legal description?

3. As might be expected from the length and complexity of most legal descriptions, many are "off" or simply incorrect. Homeowners and landowners come to accept natural borders such as streets, sidewalks and driveways. All prospective purchasers, however, are wise to have a survey conducted (complete with little poles and red ribbons) prior to closing. I have yet to see a survey precisely match the expectations of the parties. In one instance, involving a dispute between hostile neighbors, it was discovered that the defending neighbor owned the other's driveway. Do brokers or sellers owe a duty to a prospective buyer to insure accuracy?

4. Burton Abstract and Title Company is a large, old, well respected Detroit entity that is in the business of insuring title for purchasers of real estate. Benny (known in college as "Benny the Babbler") purchased a farm in the northern part of Michigan. After I had been practicing for about three years, Benny telephoned to tell me that he had carefully measured the boundaries of his farm and the boundaries of his neighbor's farm from the metes and bounds descriptions contained in the abstracts and the title insurance policies issued by Burton; coincidentally (but not surprisingly) Burton had insured the titles for both adjacent landowners. It appeared that there was a triangular portion that, according to Burton's policies, was owned by both of the neighbors; i.e., Burton had simultaneously insured that two different people owned the same parcel of land. Pursuant to Benny's instructions, I opened a file (*Benny v. Burton*) to pursue an insurance claim against Burton. My first reaction to the call, however, was: "*What kind of a shmendrick has time to make such measurements.*" I learned the answer when, after months of negotiation with Burton and hours of work, Benny had sold the farm without telling me. And, who do you think insured the title for the hapless new owner? You guessed it—Burton Abstract Company.

§ 2.70 The Calculus

Often, seemingly small advances in the field of mathematics produce extraordinary and far reaching consequences. The discovery of calculus (originally termed "*The Infinitesimal Calculus*") by Sir Isaac Newton is ranked as one of the greatest advances in the realms of both mathematics and physics, and yet, it is concerned with the very tiniest of measurements. It is difficult to imagine that this indispensable mainstay of modern science resulted from nothing more than the discovery of a method for measuring the slope of a curved line.

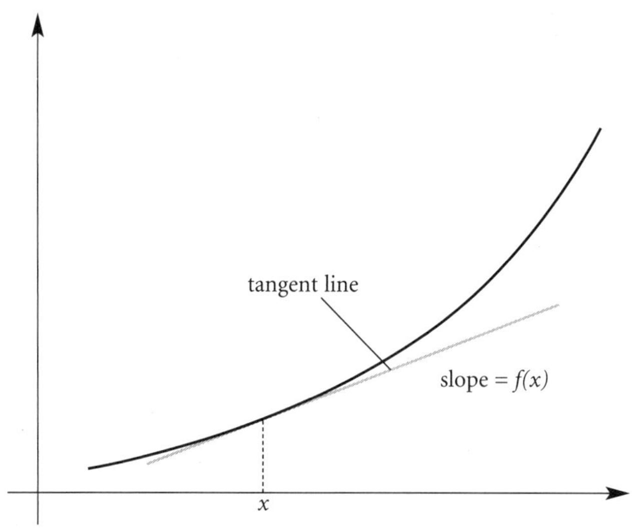

The slope of a curve, sometimes referred to as the "tangent line" to the curve, other times termed the "derivative," is formally defined as:

$$f' = \frac{\text{Change in y}}{\text{Change in x}} = \frac{\Delta y}{\Delta x} = \lim_{\Delta x \to 0} \frac{f(x + \Delta x) - f(x)}{\Delta x} = \frac{dy}{dx}$$

The Run represents the distance along the x Axis while the Rise is the distance along the y Axis. Visualize a 4 foot long board laying on the ground (run); it's slope would equal 0. Now lift one end of the board exactly 2 feet (rise) as the other end remains on the ground. The board would now have a slope of 2/4 = 0.5. Notably, it would then form a 30 degree angle (whose sine is also equal to 0.5).

Only for those craving specificity: Let f(x), sometimes referred to only as f, be a function and let x be a point in the domain (i.e., x-Axis) of f(x). The point (x, f(x)) is the point at which the vertical dotted line touches the curve. Let Δx be a number very close to 0. Then $x + \Delta x$ is a number very close to x. Therefore $(x + \Delta x, f(x + \Delta x))$ is close to $(x, f(\Delta x))$. The slope (f') of the line (referred to as a "secant line") between these two points is the measurement that we are aiming for; therefore, f' is the slope of the secant line between points $(x, f(\Delta x))$ and $(x + \Delta x, f(x + \Delta x))$. The secant line is theoretically an approximation of the behavior of the function at the point x; it does not account for what happens between x and $x + \Delta x$. Also, it is not possible to discover the behavior at x by setting x to zero because this would require division by zero, which is impossible. The derivative is, therefore, defined by taking the "limit" as Δx tends to zero, meaning that it considers the behavior of f for all small values of Δx and extracts a consistent value for the case when Δx equals zero. Geometrically, the derivative (f') is the slope of the tangent line to the curve of f at x.

Differential calculus, predicated upon the preceding discussions and its inverse, "integral calculus" are brought together by one of the most important equations in the history of the human race, known modestly as the "*Fundamental Theorem of Calculus.*"

$$\int_a^b f(x)dx = F(b) - F(a)$$

Sir Isaac Newton takes center stage with respect to the discovery of differential calculus. Followers of Newton and Gottfried Leibniz still battle over which of the two men discovered integral calculus, a field of mathematics that permits calculation of curved areas and volumes.

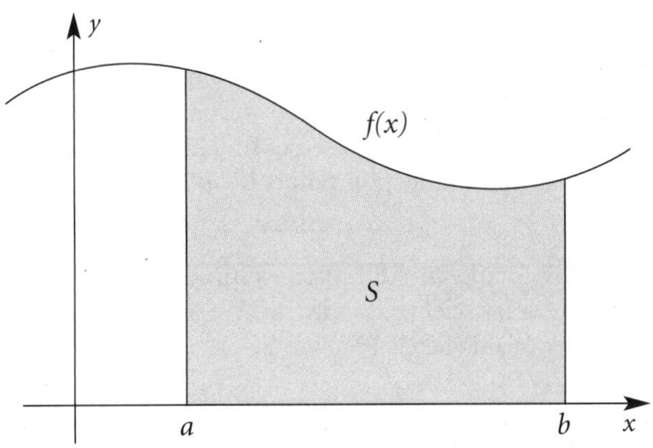

Overall, calculus (differential and integral) occupies an important place in virtually every branch of the physical sciences, actuarial sciences, engineering, electronics, thermodynamics, business theory, statistics, economics, medicine, probability theory, demography, and every other field where a physical or conceptual problem can be mathematically modeled. It allows computation and deliberation to advance from static to dynamic rates of change.

Physics in particular relies heavily on calculus; every concept in classical mechanics and electromagnetism is interrelated through calculus. The mass of a given object, for example, or the moment of inertia of objects, as well as the total energy of a system within a field can be determined more quickly and accurately by the use of calculus.

A common example of the use of calculus in mechanics involves Newton's Second Law of Motion (considered more thoroughly in Chapter 7), which defines "force" as the "rate of change" of the momentum of a body multiplied by that body's mass. Commonly expressed today as Force = Mass × acceleration, the equation involves differential calculus because acceleration is the time derivative of velocity.

To obtain a more accurate or instantaneous measurement, the classic equation, $F = ma$ (which will assume greater significance for us in Chapter 7) could be expressed as:

$$F = m\ dv/dt$$

The examples could continue indefinitely. Maxwell's theory of electromagnetism and Einstein's theory of general relativity are also expressed in the language of differential calculus. Chemistry uses calculus in determining reaction rates and radioactive decay (discussed in Chapter 8). In biology, population dynamics start with a consideration of reproduction and death rates in order to model population changes.[29]

In the realm of medicine, calculus can be used to find the optimal branching angle of a blood vessel so as to maximize flow. From the decay laws for a particular drug's elimination from the body, it is used to derive dosing laws. In nuclear medicine, it is used to build models of radiation transport in targeted tumor therapies. In economics, calculus allows for the determination of maximum profit by providing a way to easily calculate both marginal cost and marginal revenue. In fact any time that problems involve the calculation of maximums or minimums, calculus is almost certain to play an important role. Suffice it to say that any witness who takes the stand and professes to be an expert in any scientific or economic subject should be considered as fair game for cross examination involving basic calculus.

Calculus is often called upon in conjunction with other mathematical disciplines. For example, it can be used with linear algebra to find the "best fit" linear approximation for a set of points in a domain. Or it can be used in probability theory to determine the probability of a continuous random variable from an assumed density. In analytic geometry, the study of graphs of functions, calculus is used to find high points and low points (maxima and minima), concavity and inflection points in addition to slopes.

The subject of calculus could easily fill many volumes since it is essential for higher mathematics and virtually every scientific endeavor. We will use it on a *need to know* basis throughout the remainder of this text.

29. In § 2.12, we considered the significance of the constant *e* and its status as the only number, which when raised to the power of x will equal its own derivative. One of the most significant biological equations employing "*e*" will forecast whether or not a species will survive:

If Q = the future population of the species, then $Q = Q_0\, e^{kt}$ where Q = the population at time t; k is a constant.

We might note that the term "calculus" is often used (some say "mis-used") in a more generic sense to denote the quantification of a broad range of problems that do not involve derivatives or integrals. Jurists often, for example, refer to the *calculus* of damages even though nothing more than simple addition is involved.

In another example, the so-called "*Calculus of Negligence*" (the "*Hand Rule*") was derived by Judge Learned Hand in the case of *United States v. Carroll Towing*, 159 F.2d 169 (2nd Cir., 1947) to determine whether a duty of care had been breached in a tort claim. It appeared that an improperly secured barge drifted away from its pier and smashed into other boats. Judge Hand's "calculus" for determining negligence consisted of 3 elements:

P = Probability that the barge would break-away

L = The gravity of the potential loss in the event of a break-away

B = The burden or cost of taking adequate precautions

According to Judge Hand, the defendant would have breached his duty of care only if, in conjunction with a principle of economics:

$$B < P \times L$$

The Calculus of Negligence is predicated upon The *Coase Theorem*, a precept of Economics that applies mathematics to some fundamental notions about tort law in general. According to Coase, the tort system implies the pre-existence of a contract between the relevant parties in a dispute, under the assumption that a rational, cost-minimizing individual should not be required to spend money for safety or precautionary measures if those measures are more expensive than the costs of the harm that they prevent. In other words, says the theory, it would be more cost efficient to pay for the consequences of the harm than to take steps to prevent it.[30] The concept of punitive damages and the juries' repeated willingness to award them, however, has not been kind to the Coase Theorem.

§ 2.80 Further Exploration

In a vast mathematical landscape, we can barely scratch the surface. The Internet, however, is packed with free lectures for anyone wishing to dig deeper.[31]

30. In other words, if it costs more to fix a defective gas tank on an automobile than it costs to pay tort claims resulting from the defect, what the heck—let 'em burn. See *Grimshaw v. Ford Motor Company*, 119 Cal.App.3d 757 (1981). But in fairness to Ford (which was alleged to have applied the Coase Theorem when refusing to repair its tricky exploding gas tanks), see also Gary T. Schwartz, *The Myth of the Ford Pinto Case*, 43 Rutgers L. Rev., 1013 (1991).

31. As an educational effort supported by the Bill and Melinda Gates Foundation, the Khan Academy provides a very wide range of excellent, very specific mathematical short courses; most of the lectures are approximately 12 minutes in duration. The Institute's excellent purpose is to provide free educational services to the world: http://www.khanacademy.org/#browse.

Chapter 3

Logic and Set Theory

"Logic is not satisfied with assertion. It cares nothing for the opinions of the great—nothing for the prejudices of the many, and least of all for the superstitions of the dead."

—Robert G. Ingersoll

§ 3.10 Fundamental Logic

For diversity-based, subject matter federal jurisdiction to exist, we know that the plaintiff and defendant must be from different states. On the other hand, in order for any court to exercise territorial (e.g., in personam) jurisdiction, there must be presence in the state by the defendant. In order to maintain a diversity-based federal lawsuit, both subject matter jurisdiction and territorial jurisdiction must exist. Why, then, is this next statement false?

> *"A plaintiff can never sue a corporation in his own state based on diversity because if there is* **no** *presence by that company in the state, territorial jurisdiction will fail; on the other hand, if the corporation* **is** *in the state, the parties will not be diverse."*

Faulty logic abounds in the legal arena particularly where litigation is involved. As is the case in the above example, many logic-based failures are attributable to multiple meanings often attached to the words and phrases that we use. Experienced lawyers know this to be true because so much time is spent arguing about the scope and meaning of words. In our preceding example, the deceptively simple term "presence" creates the problem. In particular, thinking the problem through, you may recall that "presence" for purposes of diversity in reality means "citizenship," a status of which may involve at most two out of 50 states (state of incorporation and state where the principle place of business is located). For territorial jurisdiction, however, "presence" does not require citizenship (although it may include same), instead only minimal contacts. *International Shoe v. Washington*, 326 U.S. 310 (1945). Therefore, a corporation *can* clearly have minimal contacts with a state and simultaneously *not* be a citizen of that state. "Presence" is but one of many abstract legal concepts that invite confusion and faulty logic.

Before becoming acquainted with formal logic, it may help to be aware of its shortcomings, its limitations and above all, its ability to be twisted and misused.

Example 3-1 (Most words in the English language have multiple meanings):

A. All cardinals are birds.

B. All cardinals are high ranking members of the Catholic Church.

C. Therefore, some birds are high ranking members of the Catholic Church.

Words and phrases with multiple meanings are not the only impediments to sound logic. Another note of suspicion should be directed at so-called **truisms**. We lawyers love them. They are statements that "sound" far too logical to be untrue.[1] They can be very false and very dangerous when blindly misapplied. Consider, for example, defending a person accused of murder. Often times there will be extenuating circumstances in a homicide. Some such as self-defense or lack of capacity may exonerate the defendant completely, while others such the *battered wife syndrome* or *heat of passion* may serve to mitigate the severity of the charged offense. Even if these factors are potentially available, the defendant may, nevertheless, wish to deny that he or she committed the offense altogether, and at the same time perhaps demonstrate that the victim "deserved what he got."

Assume that a severely battered wife who feared for her life killed her husband. In planning the defense, consider the truism: "*You can't claim battered wife syndrome and at the same time claim that you didn't do it.*" The statement sounds unimpeachable, but in truth, a properly planned defense would manage to maintain the innocence of the defendant, while introducing into evidence the violent nature of someone who may implicitly "deserve killing." In recent years, several high profile, successful defenses managed to claim that the defendant "did not do it," while planting in the mind of the jurors that any doubts should be resolved in favor of the defendant precisely because the victim, in any event, may have deserved what he or she got.[2]

Generally, before engaging in any deep or protracted logic-based analysis, the practitioner should carefully analyze the words, terms and language structure employed in any strategic debate. Often, the most authoritative sounding arguments are among the most defective.

Historically, we tend to trace the study of logic back to ancient Greece and its most memorable stars, Aristotle and Plato. At that time, philosophy and logic were quite closely related. Aristotle, they say, drew his deductions and conclusions from the real world around him, while his teacher Plato was more interested in abstract mathematical truths. Aristotle frequently questioned Plato's teachings. If recent scientific history is any indicator, it would appear that Plato should have the upper hand. For if general relativity and quantum mechanics have taught us anything, it is the unreliability of our own perceptions about nature. In the face of scientific experimentation, our intuition may be worthless. Worse, the very instruments of our scientific measurements are subject to variation and distrust.

Applying logic to the discipline known as "law" has been a difficult road. There is often wide disagreement as to the manner in which laws should be applied and interpreted, not to mention disagreement as to the meaning of very fundamental concepts.[3]

---

1. Most truisms assume the form of innocent, generalized sayings and quotations such as: *Going to law is losing a cow for the sake of a cat; What large print giveth, small print taketh away; Actions speak louder than words; There is nothing to fear but fear itself; A fool and his money are soon separated*; and *Two plus two must always equal four* (Not true for 3-element cyclical subgroups).

2. See, for example, *People of the State of California v. Robert Blake*, LA 040377 (Cal. Superior Ct.).

3. See generally, Hanks, Herz and Nemerson, *Elements of the Law*, Anderson Publishing, 1994.

§ 3.11 Basic Definitions

As is the case with mastering most formal disciplines, half of the battle involves learning basic definitions and otherwise becoming familiar with the particular language employed by those who use and teach it. Logic is no different.

Deductive reasoning (Aristotelian Logic) is the application of a general statement or proposition to a specific instance. In conjunction with deductive reasoning, a **syllogism** is an argument composed of at least one **premise** (sometimes referred to as an **hypothesis**) followed by a **conclusion**. If the conclusion is true in all instances, then the conclusion is valid. If the conclusion is untrue in at least one instance, then it is **invalid**.

Example 3-2 (Syllogism for Deductive reasoning):

A. Premise: All patent infringement lawsuits must be commenced in federal court.

B. Premise: Plaintiff is going to file a patent infringement lawsuit.

C. Valid conclusion: Plaintiff must file his lawsuit in federal court.

Example 3-3:

A. Premise: Anyone who commits a crime is a criminal.

B. Premise: Joe committed a crime.

C. Valid conclusion: Joe is a criminal.

Example 3-4:

A. Premise: Perry Mason never lost a case.

B. Premise: *The Case of the Restless Redhead* was a Perry Mason case.

C. Conclusion: Perry Mason won *The Case of the Restless Redhead.*

A single counterexample will destroy the validity of the conclusion. In fact, avid fans of the popular television series know that Perry did lose one of his trials (*The Case of the Terrified Typist*). This does not, however, change the fact that he won *The Case of the Restless Redhead.* In truth, therefore, even though Premise A of Example 3-4 was false, the conclusion was, nevertheless true.

Inductive reasoning involves a general conclusion that is derived from specific premises or observations. In a sense, it is the opposite of deductive reasoning. But unlike deductive logic, inductive reasoning provides a conclusion that can never be guaranteed.

Example 3-5 (Inductive Reasoning):

A. Premise: Dirty Harry's first partner was blown up and killed.

B. Premise: Dirty Harry's second partner was shot and killed.

C. Inductive Conclusion: All of Dirty Harry's partners will be killed.

Example 3-6 (Inductive Reasoning):

A. Premise: Lee Marvin was a film actor.

B. Premise: Lee Marvin's character suffered a dramatic violent death in the film *Violent Saturday.*

C. Premise: Lee Marvin's character suffered a dramatic violent death in the film *The Man Who Shot Liberty Valence.*

D. Inductive Conclusion: The character that Lee Marvin portrays will always suffer a violent death.

As is the case with any deductive conclusion, it only takes one counterexample to destroy the validity of an inductive conclusion. Lee Marvin's character, Kid Shaleen, in *Cat Balou* did not die. In fact, Lee Marvin appeared in many films where he didn't die.

Abductive reasoning is the least reliable. Sometimes referred to as a "hunch," it involves a general conclusion that is drawn from several seemingly unrelated premises. Unlike deduction, and in some sense induction, abduction can produce results that are incorrect within its own formal system.

Example 3-7 (Abductive Reasoning—Objective Conclusion):

A. Premise: Benny Batts and Tommy argued intensely at a bar in Brooklyn.

B. Premise: Benny Batts insulted Tommy by calling him a shoeshine boy.

C. Premise: Tommy called Benny Batts a "*@$$^$!!!*" (expletives deleted).

D. Premise: Tommy attempted to physically attack Benny Batts but was restrained by Jimmy Conway, Henry Hill, and Frankie Carbone.

E. Premise: Benny Batts permanently disappeared.

F. Conclusion: Tommy caused Benny's disappearance.

The validity of an abductively reasoned conclusion, such as the one pertaining to Benny's disappearance, cannot be conclusively determined to be valid or invalid. The fact Tommy was later exterminated as revenge for Benny's disappearance may slightly add to the likelihood that Conclusion F is true, but it does not conclusively establish either the truth or validity of that conclusion.

Example 3-8 (Abductive Reasoning: Subjective Conclusion):

A. Premise: Cass Avenue in Detroit was a widely recognized distribution area for crack cocaine.

B. Premise: Officer O'Malley observed defendant walking down Cass Avenue.

C. Premise: When the defendant saw Officer O'Mally's uniform, he stopped suddenly, and then continued walking.

D. Premise: Defendant appeared to be nervous.

E. Conclusion: There is probable cause to believe that the defendant possessed contraband in the form of a concealed weapon or a controlled substance.

In general, the element of "probable cause" (which is purely abductive) is subject to definitional interpretation and is difficult to uniformly apply since most determinations are based upon unique facts. When considering the admissibility of various forms of scientific evidence, both inductive and abductive reasoning will surface. Both are to be treated with suspicion.

The preceding examples illustrate **conditionals**. Simply defined, a conditional is a compound statement of the form: "a implies b." If a implies b and b implies a, then we say that a and b are **equivalent**. Each of the conditionals are composed of statements.

A **compound** statement is one that combines a premise and a conclusion. The following is a compound statement: "If the case was defended by Perry Mason, then the defense won." A slightly more complex compound statement, but a compound statement nevertheless would be: "If the case was defended by Perry Mason, and it was not *The Case of the Terrified Typist*, then the defense won."

The **converse** of a conditional statement, merely swaps the premise and the conclusion. The **inverse** of a conditional statement negates the premise and the conclusion. If we both swap the premise with the conclusion *and* negate the premise and the conclusion, we obtain the **contrapositive**.

Example 3-9:

A. Premise: All lobbyists bribe congressmen to effect legislation.

B. Converse: All people who bribe congressmen to effect legislation are lobbyists.

C. Inverse: If you are not a lobbyist, then you do not bribe congressmen to effect legislation.

D. Contrapositive: All people who do not bribe congressmen to effect legislation are not lobbyists.

Whether or not you agree with the statements is irrelevant. For purposes of our discussions and the conventions of logic, we often presume that the premise is true. Whether or not the conclusion that follows the premise is true or false will depend upon the validity of the statements. By definition, if the premise is true, the conclusion may be either true or false. If, however, the premise (or hypothesis) is false, then the conclusion is, by definition automatically true. Example: "*If the moon is made of green cheese then I'm a monkey's uncle*" is a true statement, assuming that the moon is *not* made of green cheese.

The movement of our discussions into the realm of symbolic logic should not change the basic complexity of the subject matter. It does, however, lessen the amount of time spent writing and in many instances reduces lengthy complex premises to a manageable format. As Table 3-1 (infra, § 3.12) indicates, the symbol "→" means nothing more than: "**implies.**"

"Water is pouring from the sky → It's raining." In another example: "It's sunny → ~ Rain," the "~" means the negative, or more formally, "the **negation** of."

Example 3-10: Assume that the Premise is correct. Which of B, C and D are true?

A. Premise: *Drill baby drill* → *Spill baby spill.*

B. Converse: *Spill baby spill* → *Drill baby drill.*

C. Inverse: ~ *Drill baby drill* → ~ *Spill baby spill.*

D. Contrapositive: ~ *Spill baby spill* → ~ *Drill baby drill.*

Solution: In fairness to the petroleum industry, not every instance of off-shore drilling results in an oil leak or spill; but for our purposes, we are required to presume that our initial premise (A) is, nevertheless, true. That being the case, is the converse true? The converse implies that if there is a spill, it must have resulted from drilling; stated another way, the converse implies that all spills come from drilling. We know that is not true. In one instance, a spill resulted from the negligence of drunken Captain Hazelwood of the Exxon Valdez. (B) is therefore, untrue due to this counterexample.

The inverse is also negated by the Exxon Valdez incident. C concludes that if there is no drilling, there will be no spilling. This statement is neither true, nor is it a valid conclusion flowing from Premise A. (Arguably, however, it is probable that drilling somewhere produced the subject oil.)

The contrapositive (D) may or may or may not be true, but it *is* a valid conclusion that flows from Premise A. A implies that every time there is drilling, there will be spilling. Therefore, the lack of a spill would be sufficient evidence that there was no drilling.

Example 3-11: Following the 9/11 terrorist attack on the World Trade Center, airport screeners working for the Transportation Safety Administration (TSA) have had the authority to remove prospective passengers from their flights who were *thought* to be suspicious. Keep in mind, however, that TSA employees may not engage in racial profiling. Consider which combinations of the following premises would warrant the conclusion that a passenger should be pulled from a flight:

A. *Passenger's name is Mohammed B. Rubenstein*

B. *The name Mohammed B. Rubenstein appeared on the No-Fly List.*

C. *The name Mohammed P. Rubenstein appeared on the No-Fly List.*

D. *Passenger is an 80 year old grandmother.*

E. *Passenger's face is covered with a ski mask.*

F. *Passenger's tee shirt, which reads: "Death to the infidel dogs," has four sticks of dynamite duct taped to it.*

In law, we rarely deal with certainties, especially where some of our most critical concepts are involved. For example, the concept of probable cause, a fundamental cornerstone of constitutional protection from governmental intrusion, rests upon a principle which, by its very definition lacks certainty as does the abductive reasoning upon which it is predicated. The next case presents a classic example of abductive logic and its constitutional limitations.

Terry v. Ohio

United States Supreme Court, 392 U.S. 1 (1968)

MR. CHIEF JUSTICE WARREN delivered the opinion of the Court.

This case presents serious questions concerning the role of the Fourth Amendment in the confrontation on the street between the citizen and the policeman investigating suspicious circumstances.

Petitioner Terry was convicted of carrying a concealed weapon and sentenced to the statutorily prescribed term of one to three years in the penitentiary. Following the denial of a pretrial motion to suppress, the prosecution introduced in evidence two revolvers and a number of bullets seized from Terry and a codefendant, Richard Chilton, by Cleveland Police Detective Martin McFadden.

At the hearing on the motion to suppress this evidence, Officer McFadden testified that while he was patrolling in plain clothes in downtown Cleveland at approximately 2:30 in the afternoon of October 31, 1963, his attention was attracted by two men, Chilton and Terry, standing on the corner of Huron Road and Euclid Avenue. He had never seen the two men before, and he was unable to say precisely what first drew his eye to them. However, he testified that he had been a policeman for 39 years and a detective for 35 and that he had been assigned to patrol this vicinity of downtown Cleveland for shoplifters and pickpockets for 30 years. He explained that he had developed routine habits of observation over the years and that he would "stand and watch people or walk and watch people at many intervals of the day." He added: "Now, in this case when I looked over they didn't look right to me at the time."

His interest aroused, Officer McFadden took up a post of observation in the entrance to a store 300 to 400 feet away from the two men. "I get more purpose to watch them when I seen their movements," he testified. He saw one of the men leave the other one and walk southwest on Huron Road, past some stores. The man paused for a moment and looked in a store window, then walked on a short distance, turned around and walked back toward the corner, pausing once again to look in the same store window. He rejoined his companion at the corner, and the two conferred briefly. Then the second man went through the same series of motions, strolling down Huron Road, looking in the same window, walking on a short distance, turning back, peering in the store window again, and returning to confer with the first man at the corner.

The two men repeated this ritual alternately between five and six times apiece—in all, roughly a dozen trips. At one point, while the two were standing together on the corner, a third man approached them and engaged them briefly in conversation. This man then left the two others and walked west on Euclid Avenue. Chilton and Terry resumed their measured pacing, peering, and conferring. After this had gone on for 10 to 12 minutes, the two men walked off together, heading west on Euclid Avenue, following the path taken earlier by the third man. By this time Officer McFadden had become thoroughly suspicious. He testified that after observing their elaborately casual and oft-repeated reconnaissance of the store window on Huron Road, he suspected the two men of "casing a job, a stick-up," and that he considered it his duty as a police officer to investigate further. He added that he feared "they may have a gun."

Thus, Officer McFadden followed Chilton and Terry and saw them stop in front of Zucker's store to talk to the same man who had conferred with them earlier on the street corner. Deciding that the situation was ripe for direct action. Officer McFadden approached the three men, identified himself as a police officer and asked for their names. At this point his knowledge was confined to what he had observed. He was not acquainted with any of the three men by name or by sight, and he had received no information concerning them from any other source. When the men "mumbled something" in response to his inquiries, Officer McFadden grabbed petitioner Terry, spun him around so that they were facing the other two, with Terry between McFadden and the others, and patted down the outside of his clothing. In the left breast pocket of Terry's overcoat Officer McFadden felt a pistol. He reached inside the overcoat pocket, but was unable to remove the gun. At this point, keeping Terry between himself and the others, the officer ordered all three men to enter Zucker's store. As they went in, he removed Terry's overcoat completely, removed a .38-caliber revolver from the pocket and ordered all three men to face the wall with their hands raised. Officer McFadden proceeded to pat down the outer clothing of Chilton and the third man, Katz. He discovered another revolver in the outer pocket of Chilton's overcoat, but no weapons were found on Katz.

The officer testified that he only patted the men down to see whether they had weapons, and that he did not put his hands beneath the outer garments of either Terry or Chilton until he felt their guns. So far as appears from the record, he never placed his hands beneath Katz' outer garments. Officer McFadden seized Chilton's gun, asked the proprietor of the store to call a police wagon, and took all three men to the station, where Chilton and Terry were formally charged with carrying concealed weapons.

On the motion to suppress the guns the prosecution took the position that they had been seized following a search incident to a lawful arrest. The trial court rejected this theory, stating that it "would be stretching the facts beyond reasonable comprehension" to find that Officer McFadden had had probable cause to arrest the men before he patted them down for weapons. However, the court denied the defendants' motion on the ground that Officer McFadden, on the basis of his experience, "had reasonable cause to believe ... that the defendants were conducting themselves suspiciously, and some interrogation should be made of their action." Purely for his own protection, the court held, the officer had the right to pat down the outer clothing of these men, who he had reasonable cause to believe might be armed. The court distinguished between an investigatory "stop" and an arrest, and between a "frisk" of the outer clothing for weapons and a full-blown search for evidence of crime. The frisk, it held, was essential to the proper performance of the officer's investigatory duties, for without it "the answer to the police officer may be a bullet, and a loaded pistol discovered during the frisk is admissible."

After the court denied their motion to suppress, Chilton and Terry waived jury trial and pleaded not guilty. The court adjudged them guilty, and the Court of Appeals for the Eighth Judicial District, Cuyahoga County, affirmed. State v. Terry, 5 Ohio App. 2d 122, 214 N. E. 2d 114 (1966). The Supreme Court of Ohio dismissed their appeal on the ground that no "substantial constitutional question" was involved. We granted certiorari, 387 U.S. 929 (1967), to determine whether the admission of the revolvers in evidence violated petitioner's rights under the Fourth Amendment, made applicable to the States by the Fourteenth. Mapp v. Ohio, 367 U.S. 643 (1961). We affirm the conviction.

I.

The Fourth Amendment provides that "the right of the people to be secure in their persons, houses, papers, and effects, against unreasonable searches and seizures, shall not be violated...." This inestimable right of personal security belongs as much to the citizen on the streets of our cities as to the homeowner closeted in his study to dispose of his secret affairs. For, as this Court has always recognized,

"No right is held more sacred, or is more carefully guarded, by the common law, than the right of every individual to the possession and control of his own person, free from all restraint or interference of others, unless by clear and unquestionable authority of law." Union Pac. R. Co. v. Botsford, 141 U.S. 250, 251 (1891).

We have recently held that "the Fourth Amendment protects people, not places," Katz v. United States, 389 U.S. 347, 351 (1967), and wherever an individual may harbor a reasonable "expectation of privacy," id., at 361 (MR. JUSTICE HARLAN, concurring), he is entitled to be free from unreasonable governmental intrusion. Of course, the specific content and incidents of this right must be shaped by the context in which it is asserted. For "what the Constitution forbids is not all searches and seizures, but unreasonable searches and seizures." Elkins v. United States, 364 U.S. 206, 222 (1960). Unquestionably petitioner was entitled to the protection of the Fourth Amendment as he walked down the street in Cleveland. Beck v. Ohio, 379 U.S. 89 (1964); Rios v. United States, 364 U.S. 253 (1960); Henry v. United States, 361 U.S. 98 (1959); United States v. Di Re, 332 U.S. 581 (1948); Carroll v. United States, 267 U.S. 132 (1925). The question is whether in all the circumstances of this on-the-street encounter, his right to personal security was violated by an unreasonable search and seizure.

We would be less than candid if we did not acknowledge that this question thrusts to the fore difficult and troublesome issues regarding a sensitive area of police activity—issues which have never before been squarely presented to this Court. Reflective of the tensions involved are the practical and constitutional arguments pressed with great vigor on both sides of the public debate over the power of the police to "stop and frisk"—as it is sometimes euphemistically termed—suspicious persons. On the one hand, it is frequently argued that in dealing with the rapidly unfolding and often dangerous situations on city streets the police are in need of an escalating set of flexible responses, graduated in relation to the amount of information they possess. For this purpose it is urged that distinctions should be made between a "stop" and an "arrest" (or a "seizure" of a person), and between a "frisk" and a "search." Thus, it is argued, the police should be allowed to "stop" a person and detain him briefly for questioning upon suspicion that he may be connected with criminal activity.

Upon suspicion that the person may be armed, the police should have the power to "frisk" him for weapons. If the "stop" and the "frisk" give rise to probable cause to believe that the suspect has committed a crime, then the police should be empowered to make a formal "arrest," and a full incident "search" of the person. This scheme is justified in part upon the notion that a "stop" and a "frisk" amount to a mere "minor inconvenience and petty indignity," which can properly be imposed upon the citizen in the interest of effective law enforcement on the basis of a police officer's suspicion.

On the other side the argument is made that the authority of the police must be strictly circumscribed by the law of arrest and search as it has developed to date in the traditional jurisprudence of the Fourth Amendment. It is contended with some force that there is not—and cannot be—a variety of police activity which does not depend solely upon the voluntary cooperation of the citizen and yet which stops short of an arrest based upon probable cause to make such an arrest. The heart of the Fourth Amendment, the argument runs, is a severe requirement of specific justification for any intrusion upon protected personal security, coupled with a highly developed system of judicial controls to enforce upon the agents of the State the commands of the Constitution. Acquiescence by the courts in the compulsion inherent in the field interrogation practices at issue here, it is urged, would constitute an abdication of judicial control over, and indeed an encouragement of, substantial interference with liberty and personal security by police officers whose judgment is necessarily colored by their primary involvement in "the often competitive enterprise of ferreting out crime." Johnson v. United States, 333 U.S. 10, 14 (1948). This, it is argued, can only serve to exacerbate police-community tensions in the crowded centers of our Nation's cities.

In this context we approach the issues in this case mindful of the limitations of the judicial function in controlling the myriad daily situations in which policemen and citizens confront each other on the street. The State has characterized the issue here as "the right of a police officer ... to make an on-the-street stop, interrogate and pat down for weapons (known in street vernacular as 'stop and frisk')." But this is only partly accurate. For the issue is not the abstract propriety of the police conduct, but the admissibility against petitioner of the evidence uncovered by the search and seizure. Ever since its inception, the rule excluding evidence seized in violation of the Fourth Amendment has been recognized as a principal mode of discouraging lawless police conduct. Thus its major thrust is a deterrent one, and experience has taught that it is the only effective deterrent to police misconduct in the criminal context, and that without it the constitutional guarantee against unreasonable searches and seizures would be a mere "form of words." Mapp v. Ohio, 367 U.S. 643, 655 (1961). The rule also serves another vital function—"the imperative of judicial integrity." Elkins v. United States, 364 U.S. 206, 222 (1960). Courts which sit under our Constitution cannot and will not be made party to lawless invasions of the constitutional rights of citizens by permitting unhindered governmental use of the fruits of such invasions. Thus in our system evidentiary rulings provide the context in which the judicial process of inclusion and exclusion approves some conduct as comporting with constitutional guarantees and disapproves other actions by state agents. A ruling admitting evidence in a criminal trial, we recognize, has the necessary effect of legitimizing the conduct which produced the evidence, while an application of the exclusionary rule withholds the constitutional imprimatur.

The exclusionary rule has its limitations, however, as a tool of judicial control. It cannot properly be invoked to exclude the products of legitimate police investigative techniques on the ground that much conduct which is closely similar involves unwarranted intrusions upon constitutional protections. Moreover, in some contexts the rule is ineffective as a deterrent. Street encounters between citizens and police officers are incredibly rich in diversity. They range from wholly friendly exchanges of pleasantries or mutually useful information to hostile confrontations of armed men involving arrests, or injuries, or loss of life. Moreover, hostile confrontations are not all of a piece. Some of them begin in a friendly enough manner, only to take a different turn upon the injection of some unexpected element into the conversation.

Encounters are initiated by the police for a wide variety of purposes, some of which are wholly unrelated to a desire to prosecute for crime. Doubtless some police "field interrogation" conduct violates the Fourth Amendment. But a stern refusal by this Court to condone such activity does not necessarily render it responsive to the exclusionary rule. Regardless of how effective the rule may be where obtaining convictions is an important objective of the police, it is powerless to deter invasions of constitutionally guaranteed rights where the police either have no interest in prosecuting or are willing to forgo successful prosecution in the interest of serving some other goal.

Proper adjudication of cases in which the exclusionary rule is invoked demands a constant awareness of these limitations. The wholesale harassment by certain elements of the police community, of which minority groups, particularly Negroes, frequently complain, will not be stopped by the exclusion of any evidence from any criminal trial. Yet a rigid and unthinking application of the exclusionary rule, in futile protest against practices which it can never be used effectively to control, may exact a high toll in human injury and frustration of efforts to prevent crime. No judicial opinion can comprehend the protean variety of the street encounter, and we can only judge the facts of the case before us. Nothing we say today is to be taken as indicating approval of police conduct outside the legitimate investigative sphere. Under our decision, courts still retain their traditional responsibility to guard against police conduct which is overbearing or harassing, or which trenches upon personal security without the objective evidentiary justification which the Constitution requires. When such conduct is identified, it must be condemned by the judiciary and its fruits must be excluded from evidence in criminal trials. And, of course, our approval of legitimate and restrained investigative conduct undertaken on the basis of ample factual justification should in no way discourage the employment of other remedies than the exclusionary rule to curtail abuses for which that sanction may prove inappropriate.

Having thus roughly sketched the perimeters of the constitutional debate over the limits on police investigative conduct in general and the background against which this case presents itself, we turn our attention to the quite narrow question posed by the facts before us: whether it is always unreasonable for a policeman to seize a person and subject him to a limited search for weapons unless there is probable cause for an arrest. Given the narrowness of this question, we have no occasion to canvass in detail the constitutional limitations upon the scope of a policeman's power when he confronts a citizen without probable cause to arrest him.

II.

Our first task is to establish at what point in this encounter the Fourth Amendment becomes relevant. That is, we must decide whether and when Officer McFadden "seized" Terry and whether and when he conducted a "search." There is some suggestion in the use of such terms as "stop" and "frisk" that such police conduct is outside the purview of the Fourth Amendment because neither action rises to the level of a "search" or "seizure" within the meaning of the Constitution. We emphatically reject this notion. It is quite plain that the Fourth Amendment governs "seizures" of the person which do not eventuate in a trip to the station house and prosecution for crime—"arrests" in traditional terminology.

In this case there can be no question, then, that Officer McFadden "seized" petitioner and subjected him to a "search" when he took hold of him and patted down the outer surfaces of his clothing.

We must decide whether at that point it was reasonable for Officer McFadden to have interfered with petitioner's personal security as he did. And in determining whether the seizure and search were "unreasonable" our inquiry is a dual one—whether the officer's action was justified at its inception, and whether it was reasonably related in scope to the circumstances which justified the interference in the first place.

III.

If this case involved police conduct subject to the Warrant Clause of the Fourth Amendment, we would have to ascertain whether "probable cause" existed to justify the search and seizure which took place. However, that is not the case. We do not retreat from our holdings that the police must, whenever practicable, obtain advance judicial approval of searches and seizures through the warrant procedure or that in most instances failure to comply with the warrant requirement can only be excused by exigent circumstances, see, e.g., Warden v. Hayden, 387 U.S. 294 (1967) (hot pursuit); cf. Preston v. United States, 376 U.S. 364, 367–368 (1964). But we deal here with an entire rubric of police conduct—necessarily swift action predicated upon the on-the-spot observations of the officer on the beat—which historically has not been, and as a practical matter could not be, subjected to the warrant procedure. Instead, the conduct involved in this case must be tested by the Fourth Amendment's general proscription against unreasonable searches and seizures.

Nonetheless, the notions which underlie both the warrant procedure and the requirement of probable cause remain fully relevant in this context. In order to assess the reasonableness of Officer McFadden's conduct as a general proposition, it is necessary "first to focus upon the governmental interest which allegedly justifies official intrusion upon the constitutionally protected interests of the private citizen," for there is "no ready test for determining reasonableness other than by balancing the need to search [or seize] against the invasion which the search [or seizure] entails." Camara v. Municipal Court, 387 U.S. 523, 534–535, 536–537 (1967). And in justifying the particular intrusion the police officer must be able to point to specific and articulable facts which, taken together with rational inferences from those facts, reasonably warrant that intrusion.

The scheme of the Fourth Amendment becomes meaningful only when it is assured that at some point the conduct of those charged with enforcing the laws can be subjected to the more detached, neutral scrutiny of a judge who must evaluate the reasonableness of a particular search or seizure in light of the particular circumstances. And in making that assessment it is imperative that the facts be judged against an objective standard: would the facts available to the officer at the moment of the seizure or the search "warrant a man of reasonable caution in the belief" that the action taken was appropriate? Anything less would invite intrusions upon constitutionally guaranteed rights based on nothing more substantial than inarticulate hunches, a result this Court has consistently refused to sanction. If subjective good faith alone were the test, the protections of the Fourth Amendment would evaporate, and the people would be 'secure in their persons, houses, papers, and effects,' only in the discretion of the police."

Applying these principles to this case, we consider first the nature and extent of the governmental interests involved. One general interest is of course that of effective crime prevention and detection; it is this interest which underlies the recognition that a police officer may in appropriate circumstances and in an appropriate manner approach a person for purposes of investigating possibly criminal behavior even though there is no probable cause to make an arrest.

It was this legitimate investigative function Officer McFadden was discharging when he decided to approach petitioner and his companions. He had observed Terry, Chilton, and Katz go through a series of acts, each of them perhaps innocent in itself, but which taken together warranted further investigation. There is nothing unusual in two men standing together on a street corner, perhaps waiting for someone. Nor is there anything suspicious about people in such circumstances strolling up and down the street, singly or in pairs. Store windows, moreover, are made to be looked in. But the story in quite different where, as here, two men hover about a street corner for an extended period of time, at the end of which it becomes apparent that they are not waiting for anyone or anything; where these men pace alternately along an identical route, pausing to stare in the same store window roughly 24 times; where each completion of this route is followed immediately by a conference between the two men on the corner; where they are joined in one of these conferences by a third man who leaves swiftly; and where the two men finally follow the third and rejoin him a couple of blocks away. It would have been poor police work indeed for an officer of 30 years' experience in the detection of thievery from stores in this same neighborhood to have failed to investigate this behavior further.

The crux of this case, however, is not the propriety of Officer McFadden's taking steps to investigate petitioner's suspicious behavior, but rather, whether there was justification for McFadden's invasion of Terry's personal security by searching him for weapons in the course of that investigation. We are now concerned with more than the governmental interest in investigating crime; in addition, there is the more immediate interest of the police officer in taking steps to assure himself that the person with whom he is dealing is not armed with a weapon that could unexpectedly and fatally be used against him. Certainly it would be unreasonable to require that police officers take unnecessary risks in the performance of their duties.

American criminals have a long tradition of armed violence, and every year in this country many law enforcement officers are killed in the line of duty, and thousands more are wounded. Virtually all of these deaths and a substantial portion of the injuries are inflicted with guns and knives.

In view of these facts, we cannot blind ourselves to the need for law enforcement officers to protect themselves and other prospective victims of violence in situations where they may lack probable cause for an arrest. When an officer is justified in believing that the individual whose suspicious behavior he is investigating at close range is armed and presently dangerous to the officer or to others, it would appear to be clearly unreasonable to deny the officer the power to take necessary measures to determine whether the person is in fact carrying a weapon and to neutralize the threat of physical harm.

We must still consider, however, the nature and quality of the intrusion on individual rights which must be accepted if police officers are to be conceded the right to search for weapons in situations where probable cause to arrest for crime is lacking. Even a limited search of the outer clothing for weapons constitutes a severe, though brief, intrusion upon cherished personal security, and it must surely be an annoying, frightening, and perhaps humiliating experience. Petitioner contends that such an intrusion is permissible only incident to a lawful arrest, either for a crime involving the possession of weapons or for a crime the commission of which led the officer to investigate in the first place. However, this argument must be closely examined.

Petitioner does not argue that a police officer should refrain from making any investigation of suspicious circumstances until such time as he has probable cause to make an arrest; nor does he deny that police officers in properly discharging their investigative function may find themselves confronting persons who might well be armed and dangerous. Moreover, he does not say that an officer is always unjustified in searching a suspect to discover weapons. Rather, he says it is unreasonable for the policeman to take that step until such time as the situation evolves to a point where there is probable cause to make an arrest. When that point has been reached, petitioner would concede the officer's right to conduct a search of the suspect for weapons, fruits or instrumentalities of the crime, or "mere" evidence, incident to the arrest. There are two weaknesses in this line of reasoning, however. First, it fails to take account of traditional limitations upon the scope of searches, and thus recognizes no distinction in purpose, character, and extent between a search incident to an arrest and a limited search for weapons. The former, although justified in part by the acknowledged necessity to protect the arresting officer from assault with a concealed weapon, Preston v. United States, 376 U.S. 364, 367 (1964), is also justified on other grounds, ibid., and can therefore involve a relatively extensive exploration of the person. A search for weapons in the absence of probable cause to arrest, however, must, like any other search, be strictly circumscribed by the exigencies which justify its initiation. Warden v. Hayden, 387 U.S. 294, 310 (1967) (MR. JUSTICE FORTAS, concurring). Thus it must be limited to that which is necessary for the discovery of weapons which might be used to harm the officer or others nearby, and may realistically be characterized as something less than a "full" search, even though it remains a serious intrusion.

A second, and related, objection to petitioner's argument is that it assumes that the law of arrest has already worked out the balance between the particular interests involved here—the neutralization of danger to the policeman in the investigative circumstance and the sanctity of the individual. But this is not so.

Our evaluation of the proper balance that has to be struck in this type of case leads us to conclude that there must be a narrowly drawn authority to permit a reasonable search for weapons for the protection of the police officer, where he has reason to believe that he is dealing with an armed and dangerous individual, regardless of whether he has probable cause to arrest the individual for a crime. The officer need not be absolutely certain that the individual is armed; the issue is whether a reasonably prudent man in the circumstances would be warranted in the belief that his safety or that of others was in danger. And in determining whether the officer acted reasonably in such circumstances, due weight must be given, not to his inchoate and unparticularized suspicion or "hunch," but to the specific reasonable inferences which he is entitled to draw from the facts in light of his experience. Cf. Brinegar v. United States supra.

IV.

We must now examine the conduct of Officer McFadden in this case to determine whether his search and seizure of petitioner were reasonable, both at their inception and as conducted. He had observed Terry, together with Chilton and another man, acting in a manner he took to be preface to a "stick-up." We think on the facts and circumstances Officer McFadden detailed before the trial judge a reasonably prudent man would have been warranted in believing petitioner was armed and thus presented a threat to the officer's safety while he was investigating his suspicious behavior.

The actions of Terry and Chilton were consistent with McFadden's hypothesis that these men were contemplating a daylight robbery—which, it is reasonable to assume, would be likely to involve the use of weapons—and nothing in their conduct from the time he first noticed them until the time he confronted them and identified himself as a police officer gave him sufficient reason to negate that hypothesis. Although the trio had departed the original scene, there was nothing to indicate abandonment of an intent to commit a robbery at some point. Thus, when Officer McFadden approached the three men gathered before the display window at Zucker's store he had observed enough to make it quite reasonable to fear that they were armed; and nothing in their response to his hailing them, identifying himself as a police officer, and asking their names served to dispel that reasonable belief. We cannot say his decision at that point to seize Terry and pat his clothing for weapons was the product of a volatile or inventive imagination, or was undertaken simply as an act of harassment; the record evidences the tempered act of a policeman who in the course of an investigation had to make a quick decision as to how to protect himself and others from possible danger, and took limited steps to do so.

The manner in which the seizure and search were conducted is, of course, as vital a part of the inquiry as whether they were warranted at all. The Fourth Amendment proceeds as much by limitations upon the scope of governmental action as by imposing preconditions upon its initiation. Compare Katz v. United States, 389 U.S. 347, 354–356 (1967). We need not develop at length in this case, however, the limitations which the Fourth Amendment places upon a protective seizure and search for weapons. These limitations will have to be developed in the concrete factual circumstances of individual cases. See Sibron v. New York, post, p. 40, decided today. Suffice it to note that such a search, unlike a search without a warrant incident to a lawful arrest, is not justified by any need to prevent the disappearance or destruction of evidence of crime. The sole justification of the search in the present situation is the protection of the police officer and others nearby, and it must therefore be confined in scope to an intrusion reasonably designed to discover guns, knives, clubs, or other hidden instruments for the assault of the police officer.

The scope of the search in this case presents no serious problem in light of these standards. Officer McFadden patted down the outer clothing of petitioner and his two companions. He did not place his hands in their pockets or under the outer surface of their garments until he had felt weapons, and then he merely reached for and removed the guns. He never did invade Katz' person beyond the outer surfaces of his clothes, since he discovered nothing in his pat-down which might have been a weapon. Officer McFadden confined his search strictly to what was minimally necessary to learn whether the men were armed and to disarm them once he discovered the weapons. He did not conduct a general exploratory search for whatever evidence of criminal activity he might find.

V.

We conclude that the revolver seized from Terry was properly admitted in evidence against him. At the time he seized petitioner and searched him for weapons, Officer

McFadden had reasonable grounds to believe that petitioner was armed and dangerous, and it was necessary for the protection of himself and others to take swift measures to discover the true facts and neutralize the threat of harm if it materialized. The policeman carefully restricted his search to what was appropriate to the discovery of the particular items which he sought. Each case of this sort will, of course, have to be decided on its own facts.

We merely hold today that where a police officer observes unusual conduct which leads him reasonably to conclude in light of his experience that criminal activity may be afoot and that the persons with whom he is dealing may be armed and presently dangerous, where in the course of investigating this behavior he identifies himself as a policeman and makes reasonable inquiries, and where nothing in the initial stages of the encounter serves to dispel his reasonable fear for his own or others' safety, he is entitled for the protection of himself and others in the area to conduct a carefully limited search of the outer clothing of such persons in an attempt to discover weapons which might be used to assault him. Such a search is a reasonable search under the Fourth Amendment, and any weapons seized may properly be introduced in evidence against the person from whom they were taken.

Affirmed.

MR. JUSTICE BLACK, MR. JUSTICE HARLAN, and MR. JUSTICE WHITE, wrote concurring opinions.

MR. JUSTICE DOUGLAS, dissenting.

I agree that petitioner was "seized" within the meaning of the Fourth Amendment. I also agree that frisking petitioner and his companions for guns was a "search." But it is a mystery how that "search" and that "seizure" can be constitutional by Fourth Amendment standards, unless there was "probable cause" to believe that (1) a crime had been committed or (2) a crime was in the process of being committed or (3) a crime was about to be committed.

The opinion of the Court disclaims the existence of "probable cause." If loitering were in issue and that was the offense charged, there would be "probable cause" shown. But the crime here is carrying concealed weapons; and there is no basis for concluding that the officer had "probable cause" for believing that that crime was being committed. Had a warrant been sought, a magistrate would, therefore, have been unauthorized to issue one, for he can act only if there is a showing of "probable cause." We hold today that the police have greater authority to make a "seizure" and conduct a "search" than a judge has to authorize such action. We have said precisely the opposite over and over again. In other words, police officers up to today have been permitted to effect arrests or searches without warrants only when the facts within their personal knowledge would satisfy the constitutional standard of probable cause. At the time of their "seizure" without a warrant they must possess facts concerning the person arrested that would have satisfied a magistrate that "probable cause" was indeed present.

The term "probable cause" rings a bell of certainty that is not sounded by phrases such as "reasonable suspicion." Moreover, the meaning of "probable cause" is deeply imbedded in our constitutional history.... There have been powerful hydraulic pressures throughout our history that bear heavily on the Court to water down constitutional guarantees and give the police the upper hand. That hydraulic pressure has probably never been greater than it is today.

Yet if the individual is no longer to be sovereign, if the police can pick him up whenever they do not like the cut of his jib, if they can "seize" and "search" him in their discretion, we enter a new regime. The decision to enter it should be made only after a full debate by the people of this country.

Notes and Problems

1. Is it possible to formulate a set of objective criteria to determine who should be deemed "suspicious" and who should not? If so, what criteria would you employ? Should your efforts be discouraged by the Court's position that there is no "ready test" for determining reasonableness other than by balancing the need to search against the invasion which the search entails? Does a "balancing" requirement further complicate the problem? Do you think that the definition of "suspicious activity" is different in our post-9/11 world than it was when *Terry* was decided (in 1968)?

2. Quantifying the key evidentiary burdens of jurisprudence, Reasonable Suspicion (RS), Probable Cause (PC), Preponderance of Evidence (PE), Clear and Convincing Evidence (CCE), and Beyond a Reasonable Doubt (BRD) would yeild the following inequality:

$$RS < PC < PE < CCE < BRD$$

3. The choice of language that lawyers employ at suppression hearings is particularly interesting. In *Terry*, the Court mentioned that there is nothing suspicious about people "*strolling* up and down the street (Emphasis supplied)." In the next sentence, however, Terry wasn't just strolling, but was "hovering." Likewise, there is nothing wrong with "looking" or "window shopping." The same activity, however, becomes more sinister if we refer to it as "casing."

4. The logical "chicken-egg" dilemma faced by the court is particularly interesting. The defendant, citing good authority (i.e., the clear language of the Fourth Amendment) claimed that no search could be warranted without probable cause, whereas, the prosecution noted that the search revealed enough facts to establish probable cause. Do you agree with the majority opinion or the dissent?

5. The popularity of books teaching Blackjack players how to count cards and beat the dealers has been a mixed blessing for Las Vegas Casinos. On the one hand, the resulting increase in business has been remarkable as neophyte card-counters rushed to Las Vegas to test their skills; this caused Blackjack to supercede Craps as the world's most popular game of chance. Most of the amateurs performed as normal gamblers, i.e., they lost money thereby increasing the casinos' revenues dramatically. On the other hand, a few highly skilled counters could win significant amounts of money. The casinos, therefore, found it necessary to identify and chase away "suspicious" players (who might be serious counters) before they could win substantial sums. Without intelligent standards for determining "suspicious," the management might do the unthinkable and turn away some serious losers. The problem, therefore, involved the identification of winning professional counters and distinguishing them from the losers. Abductive criteria used to identify the counters led to its own set of books and courses on how to avoid looking suspicious. In summary, if you want to look like you're *not* a counter, be: Drunk, clueless, superstitious, stupid, with a distractive spouse or date. The safety of these so-called countermeasures, however, can be quickly erased by the most significant "suspicious activity"—winning.

§ 3.12 *Symbolic Logic*

For the solution to more complex logic problems, symbols often help. At the very least, they make the endeavor appear more impressive. Because of their common usage, some of the symbols listed below may appear in other tables or examples throughout the text. Simple letters (most commonly p, q, r, a and b) represent statements. A statement can either be a premise or a conclusion.

[Logic Symbols]

SYMBOL	DEFINITION	EXAMPLES
$\rightarrow$	Implies (Conditional)	$a \rightarrow b$
$\wedge$	Conjunctive (And)	$a \wedge b$
$\vee$	Disjunctive (Or)	$a \vee b$
$\sim$ (__)	Negation (Not)	$\sim a$
$\therefore$	Hence (Therefore)	$a \therefore b$
$\leftrightarrow$ ($\equiv$)	Equivalent Statements (Biconditional)	a if and only if b

Example 3-12: Assume that statements p, q, s, and t are true. Knowing that Al Capone hired his own gang members to wipe out Bugsy Moran's gang on St. Valentine's day, first, express the statements in (a–g) in words, then determine which of those statements are true:

p: Al was in Miami when Bugsy's gang was assassinated in Chicago.
q: Al had an alibi.
s: Al hired some killers to wipe out Bugsy's gang.
t: Al is innocent of all crimes relating to the St. Valentine's Day massacre.

(a) $\sim s \vee s$
(b) $s \rightarrow \sim t$
(c) $q \rightarrow t$
(d) $p \rightarrow q$
(e) $\sim p \rightarrow \sim q$
(f) $\sim q \wedge q$
(g) $p \vee q \rightarrow t$

Solution: (a) is always true. A "statement or its negation" must always be true. Statements that are always true are sometimes referred to as **tautologies**. (b) would seem to be true since the conspiracy to commit a crime is an act of guilt. (c) is not necessarily true for the reasons set forth in (b). (d) is true because regardless of Al's guilt or innocence, he *does* have an alibi assuming that he was Miami. (e) is false; just because Al was not in Miami does not mean that he lacks an alibi. He may have been in Detroit. (f) is always false. A statement and its negation can never both be true at the same time; such statements constitute an **inconsistency**. (g) p and q are relatively equivalent; however, as suggested by s, the statement isn't necessarily true.

Example 3-13: (Hypothetically travel back to Chicago in the 1930s)

Statement p: Al Capone will kill Bugsy.

Statement q: Bugsy will kill Al Capone.

Which is more probable? (a) $p \vee q$ or (b) $p \wedge q$

Solution: (a) is more probable since it is more likely that either "Al Capone will kill Bugsy" or "Bugsy will kill Al Capone." Proposition (b) will not occur unless both men happen to shoot one another at the same time. (Such an event allegedly occurred in the Glen Ford/Broderick Crawford film *The Fastest Gun Alive.*)

Example 3-14: Consider the following federal lawsuit, then determine which statements would permit a federal court to exercise subject matter jurisdiction over the parties:

p: The plaintiff and the defendant are citizens of different states.
q: The lawsuit involves claims that arise under federal law.
r: The amount in controversy exceeds $75,000.

(a) $\sim r \wedge p$
(b) $\sim r \wedge q$
(c) $\sim r \wedge \sim q$
(d) $p \wedge q \wedge r$
(e) $\sim q$
(f) $p \wedge r$

Solution: (a) No; the amount in controversy must be greater than $75,000; (b) Yes; q and q alone is enough to establish jurisdiction; (c) No; there is no basis for jurisdiction; (d) Yes; both predicates for federal jurisdiction are present; (e) No; no basis for jurisdiction is stated; (f) Yes; p and r represent the two predicates for diversity jurisdiction.

Notes and Problems

1. Disjunctives and conjunctives used incorrectly often destroy the value of a statement and they can totally destroy the usefulness of a deposition. A basic purpose of every discovery deposition is to obtain admissions by the opposition that can later be used at trial or to support a motion for summary judgment. If multiple interpretations of a Q-and-A sequence are possible, the answers may be worthless. Consider the following two identical questions with completely different meanings:

Q.1: Did you *stop* by the drugstore or the grocery store on the way home to pick up your prescription? A. Yes.

Q.2: Did you stop by the drugstore *or* the grocery store on the way home to pick up your prescription? A. The drugstore.

The careless use of disjunctives and conjunctives in pleadings can also cause problems; just ask the District Attorney in *U.S. v. Johnson*, 87 F.3d 133 (5th Cir., 1996) after prosecuting the defendant for possession of cocaine while "using *and* carrying a firearm." Guilt, according to the applicable statute only required that the defendant either use *or* carry a firearm. The indictment, however, used the conjunctive ("and") instead of the disjunctive. The defendant appealed his conviction after the trial court charged the jury that defendant could be convicted if he "used *or* carried" a firearm. Luckily for the careless prosecutor, the appellate court ruled that the *disjunctive* statute could be pleaded *conjunctively* and proved *disjunctively*.

2. Boolean Logic—It's the stuff that computers are made of. Unlike the real number system which contains infinite elements, Boolean logic consists of only two: "False" and "True," mathematically represented as 0 and 1 respectively. Applied to circuitry, the two

elements translate into two values: Off (0) and On (1). This single distinction is the only thing that the computer is capable of recognizing. Most users are not aware that computers are that stupid. The preceding principles (particularly those involving disjunctives, conjunctives and negations) evolved into the underlying basis for all of our computer software and microchip circuitry.

3. The Latin term: "*Expressio Unius Est Exclusio Alterius*" represents an important concept for contractual, testamentary and statutory interpretation. Literally translated as: "The expression of one thing is the exclusion of another." The precept refers to those situations where a writing lists certain items but excludes others. If, for example, a will provided a bequest to several named individuals, there would arise an implication that all others (who were not named) were intended to be excluded.

If we were to allow U to represent a collection or set, and then take the trouble to list, in writing, certain elements (A) of that set, it would be fair to presume that all of the elements of U that were *not* listed (A'), were intended by the drafter to be excluded. A' is the **compliment** of A. In other words:

A → ~ A' (The inclusion of A implies the *exclusion* of everything else)

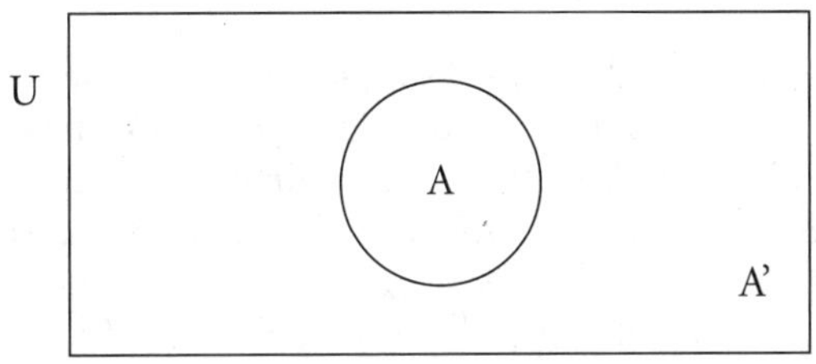

§ 3.13 De Morgan's Laws

There exist a set of rules and operations that might be performed on the symbolic expressions that are listed below. An obvious example is the so-called double negative, which yields the original statement. For example, the statement: "I'm not going to fail to attend the mud wrestling exhibition," is equivalent to "I'm going to the mud wrestling exhibition." Symbolically speaking: $\sim \sim p \equiv p$. A fellow named De Morgan, someone obviously with a lot of time on his hands, and negativity on his mind, gave us a set of laws that bear his name:

Law 1. $\sim (\sim p) \equiv p$.
Law 2. $\sim (p \wedge q) \equiv \sim p \vee \sim q$.
Law 3. $\sim (p \vee q) \equiv \sim p \wedge \sim q$.
Law 4. $\sim (p \rightarrow q) \equiv p \wedge \sim q$.

Example: In order to give concrete meaning to the abstract symbols, we might best choose a familiar setting. For Law # 2, Let p = diversity of citizenship and let q = an amount in controversy that exceeds $75,000. Knowing that both p and q are needed for diversity subject matter jurisdiction, we might assume that the statement $\sim (p \wedge q)$ tells us that jurisdiction is lacking. We would also readily accept that this is equivalent (≡) to saying that either p or q (or perhaps both) is lacking; it is represented by the expression: $\sim p \vee \sim q$.

For Law #3, we might continue our example by reading $\sim (p \vee q)$ as meaning that neither p nor q exist implying that both of the requirements for diversity jurisdiction have failed. As Law #3 indicates, this dual failure is represented by the equivalent expression: $\sim p \wedge \sim q$.

The fourth law is different from the other three because it, unlike the others, involves a conditional in its equivalence. The left side of the equivalence, "$\sim (p \rightarrow q)$" simply states that p does not imply q. In our jurisdiction example, we recognize this to be a true phrase because the mere fact that parties may be diverse does not imply anything whatsoever about the amount in controversy.

The second part of the equivalence is quite tricky. First, we must automatically assume that p is true. But how can this be possible? The answer lies in our prior determination that whenever the premise is false, the conclusion must be true (See supra, § 3.10). Therefore, it is not possible to have a negative premise with a false conclusion. The only part of the expression that can be false, therefore, is q.

§ 3.14 Truth Tables

Despite common platitudes about the "simple truth," "absolute truth," and "whole truth," no matter how far back in time we travel, it seems as if the *truth* was never simple, nor absolute nor whole. Today, in our multi-dimensional era of parallel universes and quantum dominance, the truth is at best relative and statistical. Nevertheless throughout the eons, the discipline known as *Logic* required some "absolutes" to establish its basic definitions and axioms. Today, the three basic suppositions of logic are not concerned with relativity or shades of grey: *The Law of Identity*—claims that if something is true, it is absolutely true and will not vary or change; *The Law of Contradiction*—states that if something is true, its opposite is false and visa versa; and *The Law of the Excluded Middle* says that there is no in between. As archaic and contrary to jurisprudence as these concepts may be to some, they form the basic premise for Boolean Algebra, which, in turn is precisely how our computers reason and seem to dominate our existence.

Simple statements in the Boolean world are simply measured by a "True" or a "False." As the number of compound statements increase, the variety and complexity of the conclusions increase exponentially. If you were to create a matrix, listing the permutations and combinations, the number of rows would equal 2^N where N = The number of individual statements.

Such a matrix is referred to as a **Truth Table**. The purpose of a truth table is to present all of the possible combinations of individual statements as being true and false in order to determine their effect on the applicable conclusions. Start with the simplest example: One statement—two possibilities:

Table 3-2 [Basic Truth Table]

p
True
False

Table 3-3 [Negation]

p	~ p
True	False
False	True

Notice the manner in which the table itself defines the concept of negation (~ p). Where p is true, the negation of p is false. And, where p happens to be false, its negation is true. The logic and the definitions associated with it are very black-and-white with no shades of grey to cause confusion. This is precisely the manner in which computers operate and think.

Table 3-4 [Disjunction and Conjunction]

		OR	AND
p	q	p ∨ q	p ∧ q
True	True	True	True
True	False	True	False
False	True	True	False
False	False	False	False

The table reflects what we already know about the connectors "or" and "and." With some relatively simple inductive reasoning, we can safely conclude that in order for a set of compound disjunctives to be *true*, only one of the statements needs to be true. For a set of compound conjunctives to be true, *all* of the statements must be true.

Example 3-15:

Statement p: Myron has a high LSAT Score.

Statement q: Myron has a high college grade point.

In order to gain acceptance to law school, Myron's chances will depend upon whether the school to which he is applying requires both (∧) high LSAT's *and* high college grades, or whether it requires only one of the two (∨).

This oversimplification of the process assumes that the term "high" is clearly definable. It also ignores law schools that require neither high LSAT's nor high grades (i.e., Phoenix and Abe Lincoln School of Law).

As the number of statements increase, the number of rows in our truth table will increase geometrically, as previously discussed.

Example 3-16: Our most familiar combination of disjunctives and conjunctives involves the belabored requirements for federal subject matter jurisdiction.

Statement p: The plaintiff and defendant are diverse.

Statement q: The amount in controversy (> $75,000) is satisfied.

Statement r: The case involves a federal question.

Required for Jurisdiction: $J \equiv r \vee (p \wedge q)$.

Table 3-5 [Disjunction and Conjunction]

p	q	r	p ∧ q	J
True	True	True	True	True
True	True	False	True	True
True	False	True	False	True
True	False	False	False	False
False	True	True	False	True
False	True	False	False	False
False	False	True	False	True
False	False	False	False	False

Conditionals are also represented by truth tables. Notice that whenever the hypothesis (p) is false, the conclusion (q) is always true. Let p and q be any statements:

p	q	p → q
True	True	True
True	False	False
False	True	True
False	False	True

The relationship of "equivalence," is represented as follows:

p	q	p ↔ q
True	True	True
True	False	False
False	True	False
False	False	True

Symbolic logic is most useful when the number and complexity of the compound statements increase to the point where words and sentences become clumsy. Determining the value of each box is greatly simplified by merely starting at the left and completing each column before moving to the column to the immediate right.

Notice, in Example 3-17, the manner in which each of the values build upon one another during the course of the rightward movement. The "~ q" column, for example, is generated by merely negating (i.e., "taking the opposite of") the second column.

Example 3-17:

p	q	r	~ q	p ∨ ~ q	(p ∨ ~ q) → r
True	True	True	False	True	True
True	True	False	False	True	False
True	False	True	True	True	True
True	False	False	True	True	False
False	True	True	False	False	True
False	True	False	False	False	True
False	False	True	True	True	True
False	False	False	True	True	False

Notes and Problems

1. It is important to keep in mind that the examples in this section illustrate the validity of conclusions that are predicated upon premises that are presumed to be true. Validity, however, does not insure that any given statement is either true or false.

2. Recall from our prior discussions that a tautology is always true. In trigonometry, such statements are referred to as identities. Is the following statement a tautology: (p ∧ q) → (p ∨ q)? Prove your answer through a set of truth tables.

p	q	p ∧ q	p ∨ q		(p ∧ q) → (p ∨ q)
True	True	True	True		?
True	False	False	True		?
False	True	False	True		?
False	False	False	False		?

State v. Burkhardt

1989 WL 155741 (Ohio.App., 1989)

PUTMAN, Presiding Judge.

This criminal case involves freedom to publish. The Constitutions protect publications that do not violate contemporary community standards. Miller v. California (1973), 413 U.S. 15.

The only witness in this case denied any knowledge of contemporary community standards. Accordingly, we reverse the judgment of conviction and sentence and remand this case for further proceedings according to law. Urbana, ex rel. Newlin v. Downing (1989), 43 Ohio St.3d 109.

Appellate courts must make an independent determination of the question of First Amendment constitutional coverage in reviewing convictions for selling publications. The contemporary community standards in this case having been unproved, undisclosed and unrecorded, meaningful appellate affirmance of this conviction upon review has been precluded by the way this case was tried. No proof, no conviction.

We now proceed to obey the mandate of App.R. 12(A) to determine each of the separate assigned errors giving written reasons. This case arose when on March 15, 1989, a plain clothes Canton police officer entered a retail store at 5th Street and North Market, the business district of downtown Canton in Stark County, Ohio, opened and went behind the double swinging Dutch-doors clearly marked "you must be 21 to enter," selected the magazine in evidence, paid the clerk (appellant here), departed, examined the magazine, returned, arrested the clerk and charged him with violating R.C. 2907.32(A)(2) by knowingly selling obscenity. No other crime was charged.

The arresting officer, the sole witness to appear (there was no defense), opined that the magazine was "obscene" but gave no testimony as to the elements of obscenity. Upon cross-examination, he stated that he knew nothing of any case law on the subject or any three-pronged United States Supreme Court test for obscenity and expressly denied any knowledge of what was being sold in the community. He explained that he had been busy on cruiser duty until he made this arrest.

Although there was no evidence to support it, the prosecutor argued vigorously to the jury that the type of material in evidence could not be purchased anywhere else in the community and named several specific well-known stores where he said he knew this type of material could not be found. He acknowledged that no children were present at the sale but urged the jury to convict because the magazine might fall into the hands of children and challenged the jury to "send the community a message." The jury returned a guilty verdict, and this appeal followed the sentence.

We turn now to the assignments of error. Appellant raises the following nine assignments of error on appeal:

[All Assignments of Error other than Assignment III, are omitted]

III.
The Trial Court Erred in Denying Appellant's Motion for Mistrial Based on Illegible, Unintelligible and Confusing Jury Instructions.

We sustain the third assignment of error on the ground that the jury instructions were confusing and misleading with respect to the elements of obscenity.

Publications are protected by the free press and speech provisions of the Constitution unless all of three elements appear together or coalesce. Those elements relate generally to: 1) appealing to prurient interest in sex; 2) being patently offensive to customary limits of openness or candor with which sexual matters are portrayed or depicted; and 3) lacking in political, historical, scientific, literary or artistic merit. Miller v. California (1973), 413 U.S. 15, 24.

These elements of obscenity are elements of the crime charged that the prosecution must prove. They are raised by the plea of not guilty to the charge. Communities may measure the first two elements by existing local standards, but national standards control the third element. Smith v. United States (1977), 431 U.S. 291, 293, 301. Absent proof of any one of the above elements, publications cannot be stripped of constitutional protection by the conclusion of obscenity. Accordingly, the United States Supreme Court told the state of Georgia that it could not ban the Jack Nicholson/Candice Bergen movie "Carnal Knowledge," because it had artistic merit judged by national standards no matter how offensive it was to Georgia's standards with respect to the first two elements of obscenity. In other words, local standards may be used to measure "appeal" and "offensiveness," but "merit" is measured by national standards. Stated otherwise, with respect to the third element regarding "merit," communities may not secede from the union.

State v. Burgun (1978), 56 Ohio St.2d 354, requires that in addition to the statutory definition of obscenity under R.C. 2907.01(F), the trial judge must also give the jury the limiting test for obscenity set forth by the United States Supreme Court in Miller v. California (1973), 413 U.S. 15. Regarding what constitutes obscene material, the trial judge instructed the jury expressly that: Any material is obscene if any of the following apply ...

> (1) Create, reproduce, or publish any obscene material, when the offender knows that the material is to be used for commercial exploitation or will be publicly disseminated or displayed, or when the offender is reckless in that regard;
>
> (2) Promote or advertise for sale, delivery, or dissemination; sell, deliver, publicly disseminate, publicly display, exhibit, present, rent, or provide; or offer or agree to sell, deliver, publicly disseminate, publicly display, exhibit, present, rent, or provide, any obscene material;
>
> (3) Create, direct, or produce an obscene performance, when the offender knows that it is to be used for commercial exploitation or will be publicly presented, or when the offender is reckless in that regard;
>
> (4) Advertise or promote an obscene performance for presentation, or present or participate in presenting an obscene performance, when the performance is presented publicly, or when admission is charged;
>
> (5) Buy, procure, possess, or control any obscene material with purpose to violate division (A)(2) or (4) of this section.
>
> Thereafter, he repeated the five separate disjunctives contained in R.C. 2907.01(F). Then, after some definitions, he told the jury:
>
> The test to determine whether the material, is obscene, in addition to the statutory definition which I just have given you, is a three-pronged test that has been determined by the U.S. Supreme Court:
>
> (1) Whether the average person, applying contemporary community standards, would find that the work, taken as a whole, appeals to prurient interest.
>
> (2) Whether the work depicts or describes in a patently offensive way, sexual conduct as has been defined.
>
> (3) Whether the work, taken as a whole, lacks serious literary, artistic, political or social value.

The misleading quality of this instruction is that it told the jury that they could make a finding of obscenity when any single one of the five first named elements standing alone is proved beyond a reasonable doubt. Those elements were expressly stated to be in the disjunctive. The jury was never instructed that in order for material to be obscene, all three elements of the Miller test must be proved beyond a reasonable doubt. Moreover, the jury was never told that if there was a reasonable doubt as to any one of the three elements of the Miller test, the material is not obscene no matter how many other elements articulated in R.C. 2907.01(F) may be present. Absent any one of the Miller prongs, the jury must find the defendant not guilty. They were never told that.

The jury had these written instructions in front of them. The above-quoted language was the only instruction that the jurors received concerning the Miller test. The mere use of the word "prong" is not grammatically sufficient clearly to instruct the jury that the state must prove beyond a reasonable doubt all three of these "prongs" before they can find the material to be obscene and the defendant guilty. The jury was never instructed that if they had a reasonable doubt as to any one of the "prongs," that they must find the defendant not guilty. Jurors are not law students. They are laymen who need plain language direction as to their duty depending upon what facts they find.

Furthermore, the trial court failed to instruct the jury that, as in the case of the first element regarding "appeal," the second element regarding "offensiveness" was also to be measured by contemporary community standards. Additionally there was no instruction that the "merit" element must be measured by national standards. In Smith v. United States (1977), 431 U.S. 291, the Supreme Court discussed the elements of Miller test and stated:

The phrasing of the Miller test makes clear that contemporary community standards take on meaning only when they are considered with reference to the underlying questions of fact that must be resolved in an obscenity case. The test itself shows that appeal to the prurient interest is one such question of fact for the jury to resolve. The Miller opinion indicates that patent offensiveness is to be treated in the same way.... Literary, artistic, political or scientific value, on the other hand, is not discussed in Miller in terms of contemporary community standards.

For the foregoing reasons, the judgment of the Canton Municipal Court, Stark County, Ohio, is reversed, and this cause is remanded to that court for further proceedings according to law.

Notes and Problems

1. Fill out a truth table that would determine the guilt or innocence of *Burkhardt.* Let the letters a, b, c, d and e be used to represent the disjunctive requirements; assign x, y, and z to the conjunctive requirements of *Miller v. California.* A "T" represents a jury determination that a condition has been satisfied by the prosecutor; an "F" means that the condition has not been established. **Caution!** Do not try to complete *all* of the possible combinations until you first determine how many additional rows you would be required to add (See infra, §4.40).

$P \wedge Q = T \Rightarrow$ Defendant is Guilty
$P \wedge Q = F \Rightarrow$ Defendant is Not Guilty

a	b	c	d	e	$P = a \vee b \vee c \vee d \vee e$	x	y	z	$Q = x \wedge y \wedge z$	$P \wedge Q$
T	T	T	T	T	T	T	T	T	T	T
T	T	T	T	F	T	T	T	T	T	T

2. In the landmark case used to determine the admissibility of expert testimony, *Daubert v. Merrell Dow* (See *supra*, §1.22), several factors are provided. Are those requirements disjunctive, conjunctive or neither? Explain.

§ 3.20 Set Theory

If you happened to grow up in the 1970s, you may have been a student of the so-called "new math," which ironically was designed to improve upon "old math," skills, which in truth did not need improving upon. The new math, however, was not without its benefits. It helped develop skills involving set theory which, in turn, provided a better knowledge base for higher mathematics.

Understanding set theory is not difficult. We are already familiar with most of its concepts. The key to learning this area of mathematics essentially involves a grasp of the symbolic language that accompanies it.

§ *3.21 Basic Definitions*

A **Set** is nothing more than a collection of objects or things. These objects, which are referred to as **Elements**, can consist of virtually anything—numbers, letters, words, people, investment bankers or a collection of bad actors. A set whose elements are objectively determinable (easy to determine) is said to be **Well Defined**.

The set of crooked politicians in Washington would not be well defined. The set of Louisiana governors currently serving time in prison would be well defined. Just because a set is large does **not** mean that it cannot be well defined. The **Cardinality** of a set is the number of elements in the set. The symbol "∅" represents "literally nothing." The symbol "{∅}" represents the null set. The null set is presumed to be an element of every set.

Capital letters are most commonly used to denote sets; any letter will suffice. A Capital "U" is often used to denote the **Universal** set. We typically use brackets "{ }" to encompass the elements of a set. To denote the number of elements in the set S as opposed to the set itself, we use the term n(S). Here are some examples of the manner in which symbols are used:

Example 3-18: U = {a, b, c, d, e, f}.

Translation: U is a set consisting of six letters, in particular: a, b, c, d, e, and f.

Example 3-19: B = {George, Paul, John, Ringo}.

Translation: The set B consists of the original four Beatles; the elements are: George, Paul, John and Ringo.

Example 3-20: B = {George, Paul, John, Ringo}. n(B) = 4

Translation: The set of original Beatles consists of four elements or members.

Example 3-21: If U = {All athletes who cheated on their wives}:

O.J. Simpson ∉ U.

Congressman John Edwards ∉ U.

Tiger Wood ε U.

U = U.

∅ ε U.

Translation: O.J. Simpson is not an element of the defined set since there is no evidence that he ever cheated on his wife. (He may have murdered her, but he probably never cheated on her.) Congressman John Edwards is not an element of the set because he is not an athlete. Tiger Woods is both an athlete and a confessed member of the set.

U = U because every set equals itself regardless of whether that set is well defined or not. Finally, the null set {∅} is a member of every set; it is an empty set.

Sets may also be constructed from our numbering system. Because of their frequent use, certain sets have their own special symbols, thereby dispensing with the brackets. From Table 2-2, recall that:

Z = The set of integers

$\mathbb{N}$ = The set of natural number (i.e., positive integers)

$\mathbb{R}$ = The set of real numbers

$\mathbb{C}$ = The set of complex numbers.

Each of the above four sets are infinite. Not all infinite sets are equal. The set $\mathbb{N}$ which is said to be countably infinite forms the smallest infinite set. It's cardinality is represented by $\aleph_0$ meaning that all of its elements can be stretched out, lined up and placed into a one-to-one correspondence with the set of integers (i.e., 1, 2, 3, 4, 5 ...). On the other hand, the set $\mathbb{R}$ cannot be is placed in a one to one correspondence with the set of integers; it is, therefore, uncountably infinite. It's cardinality is denoted as: $\aleph_1$.

§3.22 Subsets and Supersets

A set "A" is a **Subset** (⊆) of set S if every element of A is also an element of S. In other words: A ⊆ S. The null set {∅} is, by definition a subset of every set. So too is each set a subset of itself. A is a **Proper Subset** of S (⊂) if it is a subset that does *not* equal S and does not equal the null set {∅}. The funny little "ε" means: "is an element of," while the symbol "∉" means: "is not an element of."

The term **Superset** (⊇) should be intuitive. It is simply the reverse of the subset. If A happens to be a subset of S, then S is a superset of A. The same concepts that apply to proper subsets also apply to proper (⊃) and improper (⊇) supersets. In addition, the same principles apply with respect to the null set and sets that equal themselves.

Example 3-22: If U = {b, c, d} and V = {b, c} ⇒ V ⊂ U

Translation: If U is a set consisting of three letters b, c, and d; and if V is a set consisting of letters b and c then V is a proper subset of U.

Example 3-23: If U = {b, c, d} and V = {c, d, b} ⇒ V ⊆ U

Translation: If U is a set consisting of three letters b, c, and d; and if V is a set consisting of letters c, d and b, then V is a subset of U (not a proper subset). Notice that the order or arrangement of the elements is irrelevant.

From the preceding definitions, there are a few simple conclusions to draw. First, it is apparent that every set is a subset of itself and at the same time a superset of itself. Although, by definition a set would never be a "proper" subset or superset of itself, we rarely ever use the term "improper" when referring to a subset or superset. Instead, the preferred use involves a simple dropping of the term "proper." Next, notice from the preceding example that the order of the elements in the set are irrelevant. Let's consider some more examples:

Example 7: S = {George, Paul, John, Ringo}
L = {Paul, Ringo}
$L \subset S$

Translation: The set S consists of the original four Beatles. The remaining living Beatles form a subset, L, consisting of Paul and Ringo.

Example 3-24: If S = {George, Paul, John, Ringo}, and:
If L = {∅, John}
$L \subset S$

Translation: The set S consists of the original four Beatles. The null element and John form a subset, L.

Example 3-25: If S = {George, Paul, John, Ringo}, then:
Paul ε S
Peter ∉ S

Translation: The set S consists of the original four Beatles. Paul is an element of S; Peter is not an element of S. Had we redefined the set of original Beatles to include the ejected drummer Peter Best (the drummer who preceded Ringo), the conclusion would be different.

It is common for subsets to be constructed from the infinite number sets $\mathbb{I}$, $\mathbb{N}$, $\mathbb{R}$ and $\mathbb{C}$. We introduce the symbol "|" which translates to: "such that." That symbol is commonly used with our inequality symbols in the following manner:

Example 3-26: $A = \{x \ \varepsilon \ \mathbb{I} \mid x > 0\}$

Translation: A equals the set x, of integers, such that x is greater than zero.

It is interesting to note that, as defined above, the set $A = \mathbb{N}$. Moreover, here is an excellent example of **Nested Subsets:** $\mathbb{N} \subset \mathbb{I} \subset \mathbb{R} \subset \mathbb{C}$. Here are two more examples before moving on:

Example 3-27: $A = \{x \ \varepsilon \ R \mid 100 > x > 0\}$

Translation: A equals the set, x, of real numbers that lie between zero and one hundred.

Example 3-28: Assume that A = {a, b, c, d, e, f}

If B = {a, b} then $B \subset A$, because a ε B and a ε A. The same applies to b.

(A would be a proper superset of B)

If C = {∅} then $C \subseteq A$, because the null set is a subset of every set.

If D = {a, b, c, d, e, f} then A = D which means that $D \subseteq A$ and $A \subseteq D$.

If E = {a, b, c, p} then $E \not\subset A$ because p ∉ A.

Lawyers are perceived as analysts, superior thinkers. A competent lawyer is presumed to be precise and specific, often measured by an ability to find tiny loopholes. The next case provides an interesting contrast of expectations for lawyers v. mathematicians. As a prelude to *Fordham*, consider the following question: Does the term "within" suggest a *subset* or a *proper subset?* Or, stated symbolically: Given: $S = \{x \in R \mid x < .02\}$. Is $.02 \in S$?

Matter of Fordham

The Supreme Judicial Court of Massachusetts,
423 Mass. 481, 668 N.E.2d 816 (1996)

[An attorney discipline action against Laurence Fordham alleged that he charged an excessive fee, $50,000, for defending a client who was prosecuted for operating a motor vehicle under the influence of intoxicating liquor (OUI). The Board of Bar Overseers accepted the recommendation of hearing committee and dismissed the petition. Bar counsel appealed. The appellate court reversed and directed public censure even though Fordham had successfully defended his client with a "novel and successful motion to suppress Breathalyzer test results."]

O'CONNOR, Justice.

After five days of hearings, and with "serious reservations," the hearing committee concluded that Fordham's fee was not substantially in excess of a reasonable fee and that, therefore, the committee recommended against bar discipline. Bar counsel appealed from that determination to the board. By a vote of six to five, with one abstention, the board accepted the recommendation of the hearing committee and dismissed the petition for discipline. Bar counsel then filed in the Supreme Judicial Court for Suffolk County a claim of appeal from the board's action. We summarize the hearing committee's findings. On March 4, 1989, the Acton police department arrested Timothy, then twenty-one years old, and charged him with OUI, operating a motor vehicle after suspension, speeding, and operating an unregistered motor vehicle. At the time of the arrest, the police discovered a partially full quart of vodka in the vehicle. After failing a field sobriety test, Timothy was taken to the Acton police station where he submitted to two Breathalyzer tests which registered .10 and .12 respectively.

Subsequent to Timothy's arraignment, he and his father, Laurence Clark (Clark) consulted with three lawyers, who offered to represent Timothy for fees between $3,000 and $10,000. Shortly after the arrest, Clark went to Fordham's home to service an alarm system which he had installed several years before. While there, Clark discussed Timothy's arrest with Fordham's wife who invited Clark to discuss the case with Fordham. Fordham then met with Clark and Timothy.

At this meeting, Timothy described the incidents leading to his arrest and the charges against him. Fordham, whom the hearing committee described as a "very experienced senior trial attorney with impressive credentials," told Clark and Timothy that he had never represented a client in a driving while under the influence case or in any criminal matter, and he had never tried a case in the District Court. The hearing committee found that "Fordham explained that although he lacked experience in this area, he was a knowledgeable and hard-working attorney and that he believed he could competently represent Timothy. After the meeting, Clark hired Fordham to represent Timothy.

According to the hearing committee's findings, Fordham filed four pretrial motions on Timothy's behalf, two of which were allowed. One motion, entitled "Motion in

Limine to Suppress Results of Breathalyzer Tests," was based on the theory that, although two Breathalyzer tests were exactly .02 apart, they were not "within" .02 of one another as the regulations require. See 501 Code Mass.Regs. § 2.56(2) (1994). The hearing committee characterized the motion and its rationale as "a creative, if not novel, approach to suppression of Breathalyzer results." Although the original trial date was June 20, 1989, the trial, which was before a judge without jury, was held on October 10 and October 19, 1989. The judge found Timothy not guilty of driving while under the influence.

The bills totaled $50,022.25, reflecting 227 hours of billed time, 153 hours of which were expended by Fordham and seventy-four of which were his associates' time. Clark did not pay the first two bills when they became due and expressed to Fordham his concern about their amount. Clark paid Fordham $10,000 on June 20, 1989. At that time, Fordham assured Clark that most of the work had been completed "other than taking [the case] to trial." Clark did not make any subsequent payments. Fordham requested Clark to sign a promissory note evidencing his debt to Fordham and, on October 7, 1989, Clark did so. In the October 13, 1989, bill, Fordham added a charge of $5,000 as a "retroactive increase" in fees. On November 7, 1989, after the case was completed, Fordham sent Clark a bill for $15,000.

Bar counsel and Fordham have stipulated that all the work billed by Fordham was actually done and that Fordham and his associates spent the time they claim to have spent. They also have stipulated that Fordham acted conscientiously, diligently, and in good faith in representing Timothy and in his billing in this case.

The board dismissed bar counsel's petition for discipline against Fordham because it determined, relying in large part on the findings and recommendations of the hearing committee, that Fordham's fee was not clearly excessive. Pursuant to S.J.C. Rule 3:07, DR 2-106(B), "a fee is clearly excessive when, after a review of the facts, a lawyer of ordinary prudence, experienced in the area of the law involved, would be left with a definite and firm conviction that the fee is substantially in excess of a reasonable fee." The rule proceeds to list eight factors to be considered in ascertaining the reasonableness of the fee:

> "(1) The time and labor required, the novelty and difficulty of the questions involved, and the skill requisite to perform the legal service properly. (2) The likelihood, if apparent to the client, that the acceptance of the particular employment will preclude other employment by the lawyer. (3) The fee customarily charged in the locality for similar legal services. (4) The amount involved and the results obtained. (5) The time limitations imposed by the client or by the circumstances. (6) The nature and length of the professional relationship with the client. (7) The experience, reputation, and ability of the lawyer or lawyers performing the services.(8) Whether the fee is fixed or contingent."

In concluding that Fordham did not charge a clearly excessive fee, the board adopted, with limited exception, the hearing committee's report. The board's and the hearing committee's reasons for dismissing the petition are as follows: Bar counsel and Fordham stipulated that Fordham acted conscientiously, diligently, and in good faith in his representation of the client and his billing on the case.

Bar counsel contends that the board's decision to dismiss the petition for discipline is erroneous on three grounds: First, "[t]he hearing committee and the Board committed error by analyzing only three of the factors set out in DR 2-106(B)(1)-(8), and their findings with regard to these criteria do not support their conclusion that the fee in this case was not clearly excessive"; second, the board "misinterpreted [DR 2-106's] prohibition against charging a clearly excessive fee by reading into the rule a 'safe harbor' provision"; and third, "by allowing client acquiescence as a complete defense."

The first factor listed in DR 2-106(B) requires examining "[t]he time and labor required, the novelty and difficulty of the questions involved, and the skill requisite to perform the legal service properly." Although the hearing committee determined that Fordham "spent a large number of hours on [the] matter, in essence learning from scratch what others ... already know," it "[did] not credit Bar Counsel's argument that Fordham violated DR 2-106 by spending too many hours."

Four witnesses testified before the hearing committee as experts on OUI cases. One of the experts, testifying on behalf of bar counsel, opined that "the amount of time spent in this case is clearly excessive." He testified that there were no unusual circumstances in the OUI charge against Timothy and that it was a "standard operating under the influence case." The witness did agree that Fordham's argument for suppression of the breathalyzer test results, which was successful, was novel and would have justified additional time and labor. He also acknowledged that the acquittal was a good result; even with the suppression of the breathalyzer tests, he testified, the chances of an acquittal would have been "[n]ot likely at a bench trial." The witness estimated that it would have been necessary, for thorough preparation of the case including the novel breathalyzer suppression argument, to have billed twenty to thirty hours for preparation, not including trial time.

A second expert, testifying on behalf of bar counsel, expressed his belief that the issues presented in this case were not particularly difficult, nor novel, and that "[t]he degree of skill required to defend a case such as this ... was not that high." He did recognize, however, that the theory that Fordham utilized to suppress the breathalyzer tests was impressive and one of which he had previously never heard. Nonetheless, the witness concluded that "clearly there is no way that [he] could justify these kind of hours to do this kind of work." He estimated that an OUI case involving these types of issues would require sixteen hours of trial preparation and approximately fifteen hours of trial time. The witness explained, however, that that case had involved a second offense OUI and that it was a case of first impression, in 1987, concerning new breathalyzer equipment and comparative breathalyzer tests.

An expert called by Fordham testified that the facts of Timothy's case presented a challenge and that without the suppression of the breathalyzer test results it would have been "an almost impossible situation in terms of prevailing on the trier of fact." He further stated that, based on the particulars in Timothy's case, he believed that Fordham's hours were not excessive and, in fact, he, the witness, would have spent a comparable amount of time. The witness later admitted, however, that within the past five years, the OUI cases which he had brought to trial required no more than a total of forty billed hours, which encompassed all preparation and court appearances.

He explained that, although he had not charged more than forty hours to prepare an OUI case, in comparison to Fordham's more than 200 expended hours, Fordham nonetheless had spent a reasonable number of hours on the case in light of the continuance and the subsequent need to reprepare, as well as the "very ingenious" breathalyzer suppression argument, and the Clarks' insistence on trial. In addition, the witness testified that, although the field sobriety test, breathalyzer tests, and the presence of a half-empty liquor bottle in the car placed Fordham at a serious disadvantage in being able to prevail on the OUI charge, those circumstances were not unusual and in fact agreed that they were "[n]ormal circumstances."

The fourth expert witness, called by Fordham, testified that she believed the case was "extremely tough" and that the breathalyzer suppression theory was novel. She testified that, although the time and labor consumed on the case was more than usual in defending an OUI charge, the hours were not excessive. They were not excessive, she explained, because the case was particularly difficult due to the "stakes [and] the evidence." She conceded, however, that legal issues in defending OUI charges are "pretty standard" and that the issues presented in this case were not unusual. Furthermore, the witness testified that challenging the breathalyzer test due to the .02 discrepancy was not unusual, but the theory on which Fordham proceeded was novel. Finally, she stated that she thought she may have known of one person who might have spent close to one hundred hours on a difficult OUI case; she was not sure; but she had never heard of a fee in excess of $10,000 for a bench trial.

In considering whether a fee is "clearly excessive" within the meaning of S.J.C. Rule 3:07, DR 2-106(B), the first factor to be considered pursuant to that rule is "the novelty and difficulty of the questions involved, and the skill requisite to perform the legal service properly." DR 2-106(B)(1). That standard is similar to the familiar standard of reasonableness traditionally applied in civil fee disputes. See Society of Jesus of New England v. Boston Landmarks Comm'n, 411 Mass. 754, 759 n. 10, 585 N.E.2d 326 (1992) (reasonable hours, when calculating fees pursuant to 42 U.S.C. § 1988 [1994], requires consideration as to "whether that time was reasonable in light of the difficulty of the case and the results achieved"). Based on the testimony of the four experts, the number of hours devoted to Timothy's OUI case by Fordham and his associates was substantially in excess of the hours that a prudent experienced lawyer would have spent. According to the evidence, the number of hours spent was several times the amount of time any of the witnesses had ever spent on a similar case. We are not unmindful of the novel and successful motion to suppress the breathalyzer test results, but that effort cannot justify a $50,000 fee in a type of case in which the usual fee is less than one-third of that amount.

In charging a clearly excessive fee, Fordham departed substantially from the obligation of professional responsibility that he owed to his client. The ABA Model Standards for Imposing Lawyer Sanctions § 7.3 (1992) endorses a public reprimand as the appropriate sanction for charging a clearly excessive fee. We deem such a sanction appropriate in this case. Accordingly, a judgment is to be entered in the county court imposing a public censure. The record in this case is to be unimpounded.

So ordered.

Notes and Problems

1. For us, the issue is not whether Fordham overcharged his client but rather, whether or not the answer to the problem initially posed (Is $.02 \in \{x \in R \mid x < .02\}$?) happens to be so non-obvious as to warrant such an aberrational attorney fee. Is it? Clearly .02 is not less than .02. That seems to be obvious. But are attorneys so deficient in mathematics that the correct answer for *them* is novel or brilliant?

2. The *Fordham* discussions concerning fee factors for Disciplinary Rule 2-8 of DR 2-106(B) have been omitted. Suffice it to say that the Supreme Judicial Court determined that the proper elements to be considered were sufficient to outweigh the benefits derived from Fordham's "novel" approach. Without Fordham's novel argument, the fee would have been clearly excessive. Do you agree or disagree with the decision?

3. Fordham's witnesses described the case against Timothy as extremely tough. One witness referred to Fordham's breathalyzer suppression argument as "very ingenious." Do

you agree with this last statement? If so, examine Footnote 2 of Chapter 1 and ask yourself: Would an eighth grader from the 1800s be considered "very ingenious" for knowing that .02 was not less than .02?

4. As to recognizing the issue of whether or not readings of .10 and .12 were "within .02," is there a rule that we could formulate that might guide a lawyer as to when he or she should view a statutory requirement from a verbal as opposed to a mathematical perspective? The term "within" might be considered ambiguous by attorneys and laypersons. But as to whether or not $.02 < (.12 - .10)$, no mathematician (Or eighth grader from the 1800s) would give it a second thought.

5. Criminal defense attorneys, it is argued, should always impose the strictest mathematically precise standards upon the prosecution, whose burden of proof is beyond a reasonable doubt. Assume that a typical state statute for *Breaking and Entering* reads as follows: "It shall be unlawful to break and enter an occupied dwelling with intent to commit a felony therein." Assume that your defendant was caught in the act of smashing a window at an occupied dwelling while carrying a bag to haul away the jewelry from the house. He confessed his intentions to the police. What defense should instantly come to mind?

§ 3.23 Unions, Intersections and Compliments

The joining of two sets is referred to as a **Union** of sets and is represented by this symbol: $\cup$. The **Intersection** of sets is denoted by $\cap$. For a given single set, U and a subset S, the **Compliment** of S (denoted as S′) consists of all elements of U that are not contained in S.

1. **Definition of Union:** $A \cup B = \{ x \mid x \in A \text{ or } x \in B\}$
2. **Definition of Intersection:** $A \cap B = \{ x \mid x \in A \text{ and } x \in B\}$
3. **Definition of Compliment:** $A' = \{ x \mid x \in U \text{ and } x \notin A\}$

For the mathematical definition of *compliment,* note that if you are not given a specific set and subset, but are only provided with a single set, it is assumed that such a single set is a subset of the Universal Set U. Be careful so as not to confuse U and $\cup$.

Example 3-29: Let U = {b, c, d, e, f, g, h}
let S = {b, c, d, e} and
let T = {e, f, g}.
Calculate: (1) $S \cup T$ (2) $S \cap T$ (3) S′ and (4) T.′

Solution: (1) $S \cup T = \{b, c, d, e\} \cup \{e, f, g\} = \{b, c, d, e, f, g\}$. Notice that in arriving at our solution, we do not count the same element (e) more than once.
(2) $S \cap T = \{b, c, d, e\} \cap \{e, f, g\} = \{e\}$.
(3) $S' = U - S = \{b, c, d, e, f, g, h\} - \{b, c, d, e\} = \{f, g, h\}$.
(4) $T' = U - T = \{b, c, d, e, f, g, h\} - \{e, f, g\} = \{b, c, d, h\}$.

As common sense may suggest, the union of two sets will always be equal to, or greater than either set, and by contrast, the intersection of two sets will always be equal to, or smaller than either set. Also, it should be clear, if not intuitive that the union of any set with its complement should always equal the Universal Set. And, the intersection of any set with its compliment should always equal the null set.

$$A \cup A' = U \qquad A \cap A' = \{\varnothing\}$$

There exist a set of rules for determining the number of elements (cardinality) in sets that result from one or more of the operations that we have been discussing:

Rules of Cardinality:

1. **Cardinality of a Union:** $n(A \cup B) = n(A) + n(B) - n(A \cap B)$

2. **Cardinality of an Intersection:** $n(A \cap B) = n(A) + n(B) - n(A \cup B)$

3. **Cardinality of a Complement:** $n(A') = n(U) - n(A)$

4. **De Morgan's Law** states that the complement of an intersection is equal to the union of the complements. In other words: $(A \cap B)' = A' \cup B'$

Example 3-30: In the city of Wayne, the political grouping is as follows:

U = {All registered voters in the city of Wayne} = $D \cup G \cup I$

D = {Voters registered as Democrats}

G = {Voters registered as Republicans}

I = {Voters registered as Independents} = $P \cup C \cup F$

P = {10 Pinkos} $\subseteq I$

C = {22 Commies} $\subseteq I$

F = {30 Fascists} $\subseteq I$

It might be natural to assume that $D \cap I = \varnothing$, $G \cap I = \varnothing$, and $D \cap G = \varnothing$ since the voters would normally be expected to register as a Democrat, Republican or Independent, but not more than one of the three. We can verify compliance with DeMorgan's Law by determining that both sides of his equation equal U in the example below (Note that the compliment of $\varnothing$ always equals the Universe):

$(D \cap G) = \varnothing \Rightarrow (D \cap G)' = U.$

$D' = (G \cup I).$

$G' = (D \cup I).$

$(G \cup I) \cup (D \cup I) = U$

As for the Pinkos, Commies and Fascists, apply the different Cardinality equations to convince yourself that DeMorgan is correct.

§ 3.40 Tying It All Together

There are obvious common links among the studies of philosophy, logic, set theory and computer circuitry (Boolean Algebra). The simple concepts: Implications, negations, conjunctives and disjunctives work their way through abstract philosophical discussions and through the logical analyses, on their way to more precise mathematical inquiries, and finally into the realm of the microchip. Boolean logic plays an extremely important role in conducting legal research and searching for information in databases. All of the disciplines, as complex as they may seem, place heavy reliance upon two simple pairs that vary only slightly in meaning: Yes/No = Truth/Falsity = Validity/Invalidity = 1/0. They all utilize the same basic concepts.

§ 3.50 Venn Diagrams

It is not uncommon for set theorists to use so-called "Venn diagrams" in order to visualize the union and intersection of various sets. They also become handy when solving simple problems involving deductive logic. Here is a simple example:

Figure 3-1:

{J} =

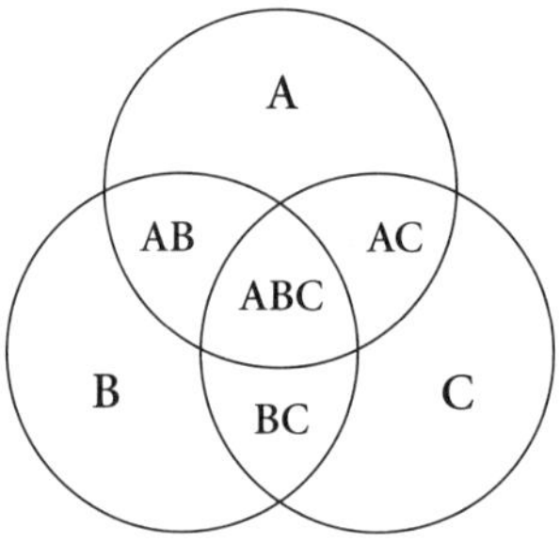

Assume that a particular jury pool consisted of a set of individuals {J} possessing the following sets of non-exclusive traits:

A = {Those who expressed anti-police attitudes during voir dire}

B = {Blue Collar Workers}

C = {College Graduates}

Figure 3-1 might represent the overlapping characteristics of the jury pool. The members in the center slice, for example ("ABC"), would have all three of the listed traits, while the other members would possess the trait or traits indicated.

Venn diagrams may be used to help illustrate or explain a wide range of complex problems to the judge or jury. See *Tecklenburg v. Appellate Div. of Superior Court*, 169 Cal.App.4th 1402 (2009); *Judulang v. Holder*, 132 S.Ct. 476 (2011); *Viscontini v. Secretary of Health and Human Services*, 2011 WL 5842577 (Fed.Cl., 2011); *George v. Junior Achievement of Central Indiana, Inc.*, 2011 WL 2960716 (S.D.Ind., 2011); *Paone v. Microsoft Corp.*, 771 F.Supp.2d 224 (E.D.N.Y., 2011); *U.S. v. Rajaratnam*, 753 F.Supp.2d 299 (S.D.N.Y., 2010); and *State v. Gupta*, 998 A.2d 1085 (Conn., 2010).

Chapter 4

Probability Theory

"A reasonable probability is the only certainty."

—E. W. Howe

Probability plays a role in nearly every area of legal endeavor. The worlds of investment, finance, litigation and dispute resolution could not function without it. Every time a jury retires to deliberate, it must consider the probability that each witness is lying or telling the truth. For the legal profession, the science of probability is by no means limited to esoteric matters involving fingerprinting, DNA and jury selection even though such matters often seem to dominate the media spotlight.

§4.10 Basic Rules and Terminology

Probability theory is not very complex. The trick is to master a few definitions along with the symbols and rules associated with them.

§4.11 Basic Definitions

We begin with our basic **Universe**, which, similar to our set theory universe, consists of all of the elements that we wish to consider. This universe might consist of the population of the world, a bunch of fruit flies or perhaps a simple deck of 52 playing cards.

A **Sample Space** (S) is defined as the set of all possible **outcomes** or results of a given experiment or trial conducted on the elements of our universe. An **experiment** or trial is the process by which an outcome is observed. An experiment may consist of the simple toss of a coin or spinning of a roulette wheel, or the measurement or description of a simple object. The specific result or outcome of such an experiment is called an "event." As such, an **event** represents a **subset** of the sample space. For example, the set of all possible outcomes that might result from the selection of a card from a complete deck of cards might represent our sample space (as it frequently does in these discussions); that set would contain 52 outcomes. The random selection of a club from the deck would represent an event. That event (i.e., 1 of 13 possible clubs) would constitute a subset of the sample space.

We will let "P" represent the **probability** that a given event might occur, which may sometimes be referred to as a **success**. Likewise, an outcome wherein the event does not occur might be termed a **failure**. The value of P will always range between 0 (representing a zero probability) to 1 (meaning that the event is certain to occur).

If n(E) is the number of potential successful events in sample space S, and n(S) is the number of total events in the sample space, then the Probability (P) that a given event (E) will occur, where all of the events in S have an equal likelihood of occurrence, is as follows:

Definition of Probability: P(E) = n(E) / n(S)

Example 4-1: Assuming that you are to pull a single card blindly from a deck of 52 cards, what is the probability of pulling an Ace (assuming that you're not David Copperfield)?

Solution: P (Ace) = n(Aces) / n(Cards in Deck).
P (Ace) = 4 / 52 = .0769 (1 in 13).

Simplified: There are 4 Aces in a deck of 52 cards. Therefore: P = 4/52.

---

§4.12 *Axioms*

A fellow by the name of Kolmogorov is given credit for developing the following axioms for probability theory. As is the case with most rules designated as "axioms," they are fundamental to the development of mathematical theorems and rules, while at the same time being somewhat obvious.

1. **P (∅) = 0** The probability of selecting nothing from a selected sample.
2. **P (S) = 1** The probability of selecting something (i.e., anything) from a selected sample.
3. **P (E) + P (E′) = 1** The probability of selecting from E ∪ E′ (Recall that S = E ∪ E′).
4. **0 ≤ P(E) ≤1** The probability of selecting a specific outcome from a sample.

As a more detailed example, consider the Senate of the United States with its 100 senators as our universe (U). Assume that there are 50 Republican Senators and 50 Democratic Senators. A typical event (E) might consist of the selection of a Republican Senator at random. If that is the case, then our sample space (S) would equal all of the possible outcomes. Each of the 100 possible outcomes (S) would consist of the *selection* of a senator. The number of possible outcomes for the sample space n(S) = 100.

Given that our event consisted of the selection of *any* senator, the probability of *not* selecting a senator is zero. Stated differently: P (∅) = 0. Likewise, the probability of selecting a senator after it has been already been determined that we *will* select a senator is absolute (100%) or simply: P (S) = 1.

The probability that we are going to select someone who is either a Republican (E) or a non-Republican (E′) is, once again 100% since everyone in the Senate is one or the other. Recall from basic set theory that E′ is simply the compliment of E. Therefore, without regard to randomness, we know that: P (E) + P (E′) = 1. We also know that the probability of any event must, by definition, consist of a real number between 0 (The Event can't happen) and number 1 (The Event is certain to happen). In other words: 0 ≤ P(E) ≤1.

In our hypothetical and in all of the problems that follow, unless otherwise indicated, it is to be presumed that the probability of each event in the sample space is equal to the probability of each of the remaining events in the sample space. In other words, the probability of selecting a senator from California is the same as the probability of a picking a senator from Michigan; each state has 2 senators regardless of size or population. P(Barbara Boxer) = P(Carl Levin).

Back to our original problem—the random selection of a Republican Senator. Because our sample space consists of 100 possible events, n(S) = 100. Because there are exactly 50 Republican Senators, n(E) = 50. Therefore, the probability of selecting a Republican Senator is simply calculated as follows:

P(E) = n(E) / n(S)

P(E) = 50 / 100 = .50

Let's a consider a slightly more difficult problem. Assume that we are attempting to locate a politician who is *not* having an extramarital affair. More specifically, we want to determine the probability of selecting such a non-cheating person at random. The sample space that we choose becomes extremely significant. We might, for example, select the U.S. Senate (100 senators) as our sample space, in which case n(S) = 100. Or we might select a smaller sample space consisting only of a selection of married senators; after all, every unmarried senator would automatically constitute "success" because such a senator could *not* by definition engage in an extramarital affair. (As used in our example, a "cheater" must be married.)

Assume that E = The selection of a non-cheating Senator. Then E′ = cheating Senators. The first two axioms are somewhat obvious. If we select a senator at random, there is 0 probability that we have *not* selected a senator. Likewise, the probability that we have made the selection is absolute (1.0). So too is the probability that we must either select a cheater or non-cheater because E ∪ E′ = The entire pool of senators.

Senator David Vitter (the Louisianan), Senator Larry Craig (the bathroom enthusiast) and former Senator and democratic presidential candidate John Edwards (the love-child guy) would all be considered as "failures" in the probabilistic sense because of the manner in which we defined a successful event; i.e., all three were cheaters. Florida Congressmen Mark Foley would not constitute a success *or* failure simply because he was a member of the House of Representatives, *not* the Senate. Therefore, he could not even be considered as a member of our sample space (S).

As previously suggested, we can expand or contract the sample space depending upon our purpose. Restrictions, of course, tend to shrink the sample space. If, for example, we were seeking to randomly select married "*family-value preaching*" senators who are *not* having an extramarital affair, our sample space might be smaller than our initial example. (Or would it?)

Let's assume that S = 100 senators and we will further assume for convenience that ½ of the senators are married. Of the married senators, we will optimistically presume that a full 5 senators are actually faithful to their wives. What then is the probability of selecting a senator at random who is *not* having an extramarital affair?

Start with our basic definition of probability: P(E) = n(E) / n(S).

Next: n(E) = 50 (Singles Senators) +5 (Faithful Senators) = 55.

We know that: n(S) = 100.

Therefore: P(E) = 55/100. Therefore, P(E) = .55.

Jed Martin and Clem Coy, both unrelated residents of notorious Harlan County, Kentucky (a dry county), ran a respectable moonshine operation in Cumberland (a wet city). Jed got liquored up one night and shot up Clem's barn and still along with a few of his stray kinfolk. Jed was charged with a variety of crimes. At trial, the jury pool consisted of 100 potential jurors. As it turned out, 40 of the potential jurors were related to Jed, 20 were related to Clem and another 10 were related to both (don't ask) of them. Of the pool, 50 prospective jurors had close business relationships with either Jed or Clem.

Before calculating the probability of any given event P(E) relating to the selection of an impartial jury, we should re-acquaint ourselves with some of the basic set theory rules from the preceding chapter. We know, for example, that if JR is a relative of Jed and CR is a relative of Clem, then:

$$\{JR\} \cup \{CR\} = 70 \text{ and } \{JR\} \cap \{CR\} = 10$$

But we do not know what overlap, if any, there may be with respect to relatives and persons with a close business relationship {B}. In particular: $\{JR\} \cap \{B\} = ?$ $\{CR\} \cap \{B\} = ?$

If $\{JR\} \cap \{CR\} \cap \{B\} = \{\varnothing\}$ then the probability of selecting an unbiased juror P(E) = 0. If, however $\{JR\} \cup \{CR\} = \{B\}$ then, in such an instance, the close business relationships would not effect the selection probabilities.

§ 4.13 *Calculating the Odds*

The terms "odds" and "probability" are closely related to one another and are often freely transposed. Whereas "probability" is expressed as a single decimal number (ranging from 0 to 1), the term **odds** is a ratio used to denote the likelihood or success of an event as compared to its failure. By definition: **Odds = n(E) : n(E′)**. The odds that a true coin toss will come up heads is 1:1 (even odds).

For example, early in the 2010 football season, the Las Vegas odds that the Detroit Lions would win the Super Bowl were 2,000 to 1. A successful event, therefore, would yield $2,000 for an otherwise "losing" $1 bet.

> **Example 4-2:** The notorious *Three Penny Opera* line : "Five will get you ten ole' Mackie's back in town," implies that the odds that Mackie is *not* back are 2 to 1. Don't be fooled by the reverse order of the odds statement. That statement is essentially setting the odds at 10:5 (not the reverse). Rearranged, the statement is saying: "The odds that Mackie is not back are 10 to 5." What is the probability that Mackie is back (assuming that the pay-off fairly represents the probability)?
>
> **Solution:** Start with the basic definition of probability. P(E) = n(E) / n(S).
>
> We know that n(E) = 5
>
> We know that for the event: Mackie is *not* back, n(E′) = 10.
>
> Our third basic rule of probability (*supra*) tells us that: P(E) + P(E′) = 1.
>
> (This is based upon the logical certainty that either he's back or he's not)
>
> If n(E) = 5 and n(E′) = 10 then our sample space (S) would contain 15 events.
>
> Therefore, P(E) = 10/15 = .333333.

The preceding example, in truth, combines three very closely related concepts: Probability, Expectation (see § 4.30) and Odds. The Expectation does not necessarily reflect the probability. In fact, because of the number of dead bodies that suddenly popped up around town, the speaker is suggesting a much higher probability that Mackie is actually back.

Because the odds are 2 to 1, there is a natural temptation to assume that the number .50 will figure into the probability, but that is clearly not the case.[1] In any event, people who talk of "odds" are normally gamblers and therefore losers (crap shooters and pony players), whereas those who speak in terms of probabilities (blackjack card counters) tend to be more successful. The remainder of our discussions, will therefore, be limited to probabilities.

---

§ *4.14 Algebraic Rules*

The examples that follow in the next section involve simple single events such the ones pertaining to the senators. Thereafter, a consideration of multiple outcomes will require additional rules, some of which are listed below. Assume that E and F are both events that are subsets sample space S.

Rules of Probability For Multiple Outcomes:

1. $P(E \cup F) = P(E) + P(F)$ for events that are mutually exclusive.
2. $P(E \cup F) = P(E) + P(F) - P(E \cap F)$ for events that are not mutually exclusive.
3. $P(E' \cap F') = P((E \cup F)')$ DeMorgan's Law.

§ 4.20 Measuring Event Probability

To learn and fully understand the important concepts presented in this chapter, we will employ seemingly benign sample spaces such as decks of cards, roulette wheels and jury pools. But make no mistake; the concepts to be derived from them are not simplistic. For the justice system, probabilities are crucial. It is important, for example, to know the probability that a polygraph result will be correct, that an eyewitness identification has an indicia of reliability, or that a DNA match is not misleading. The odds that a parolee will commit further crimes is another of many instances where probability theory plays a direct and critical role in the American justice system.

For instructional purposes, unless otherwise stated, all events, whether single or compound, have a probability of success that is equal to all other events in the given sample space. This notion of equality is periodically reinforced by terms such as "fair" dice or "true" coin. Loaded dice are not fair because their results are not truly random. Other non-true (i.e., non-random) events include slot machines and "fixed" roulette wheels.

---

1. Caution: New estate planners frequently make a serious similar mistake. Assume that the testator instructs you to draft a will or trust, stating: "I want to leave everything to my two children, Joe and Jim; Joe is to receive twice the amount that Jim is to receive." What percentage should each son receive? If your answer is 2/3 and 1/3 you are correct. If your answer was "one half" and "one quarter," then who gets the other 25%?

§ 4.21 Single Trials

The most elementary item for us to measure is the single event. Determining its probability requires only the two factors listed in definition of probability: The number of elements in the sample space and the number of elements of the event. Recall: P(E) = n(E) / n(S).

A. Mathematical Probability of Single Outcomes

The best place to start is with the simple toss of a fair coin. The sample space (S) consists of a total of only two events: Heads and Tails. Therefore: n(S) = 2. If we wish to determine the probability that a random toss will produce the observable outcome known as "Heads," we simply define: Heads as the event E. This will automatically imply that its compliment E′ = Tails. Recall from the prior discussions pertaining to set theory that, by definition, a subset and its compliment must equal the universal set. By definition, therefore, $E \cup E' = S$. Since there is only 1 Heads on any given coin, n(E) = n(Heads) = 1.

Example 4-3: What is the probability that the fair toss of a fair coin will come up heads?

Solution: P(E) = n(E) / n(S)
P(Heads) = n(Heads) / n(S)
P(Heads) = n(1) / n(2)
P(Heads) = 1 / 2
Therefore, P(Heads) = .50. It's that simple.

If we elect to use a deck of cards as our sample space, we assume that the number of events in the space (S) is 52 (We don't use Jokers). A subset of that space might consist of a given suit, such as clubs; such a subset would contain 13 items. In a prior example, a specific card, such as an Ace, would carry the number: n(Ace) = 4.

Example 4-4: What is the probability of drawing a single card from a deck of cards such that the selected card turns out to be a spade?

Solution: n(S) = 52 because there are 52 cards in a deck.
n(E) = 13 because there are 13 spades in the deck.
Therefore:
P(E) = 13/52 = 0.25
There is a 1 in 4 chance of pulling a spade.

B. Mathematical Probability of Multiple Outcomes

There are circumstances where we may wish to measure more than a single attribute from the event that has been randomly selected. In the preceding example, we only measured the probability of pulling a card of a particular suit. But we know that every playing card has several attributes, including: numerical value, suit, royalty, and color.

Mutual Exclusivity:

Two or more events are deemed to be **Mutually Exclusive** if only one of the events is possible during any given trial. In other words, the occurrence of one event will automatically exclude the others. The probability of *both* events occurring during a single event = $P(E \cap F) = 0$. Therefore, when considering mutually exclusive events, we are normally concerned with a disjunctive analysis. A coin, for example, cannot come both heads *and* tails as the result of a single toss.

Assume that Murry is going draw a single card from a deck of 52 cards. The probability that he will draw an Ace is, of course, 4/52. Likewise, the probability that he will draw a King is also 4/52. We know that a single card cannot be both an ace and a king at the same time. The two events, E = Ace and F = King are, therefore, mutually exclusive. We may, however, want to know the probability that if Manny selects a single card, that card will be *either* an Ace or a King.

From Rule (1.) of our Algebraic Rules, we know that, where the events are mutually exclusive: **P (E ∪ F) = P(E) + P(F).**

Therefore:

P (Ace or King) = P(Ace) + P(King).

P (Ace or King) = P(4/52) + P(4/52) = 8/52 = .1538.

Non-Mutual Exclusivity:

Certain events have multiple consequences that are said to be non-mutually exclusive. If, for example, we were to draw a card at random, there is a probability of 1/13 that the card would be an Ace. During the same event, there is a probability of 1/4 that the selected card would be a spade. The two events are non-mutually exclusive because both can occur at the same time, upon a single drawing of a card. In other words, the Ace of spades is a single card that would satisfy both conditions. Therefore, the "suit" and "card value" attributes are said to be non-mutually exclusive. By way of contrast, the attributes of "red" and "black" are mutually exclusive. There is a probability of .50 that a randomly drawn card will be a member of a red suit (hearts or diamonds) and an equal probability that the card would be of a black suit (spades or clubs). But the probability of drawing a single card that is **both** red and black is 0.

Example 4-5: What is the probability of drawing a single card from a deck of cards such that the selected card turns out to be *either* a spade or an Ace?

Solution: From Algebraic Rule 2 (supra), we know that where the events are non-mutually exclusive:

P (E ∪ F) = P(E) + P(F) - P(E ∩ F).
P(E) = Probability of selecting a spade = 1/4.
P(F) = Probability of selecting an Ace = 4/52 = 1/13.
P(E ∩ F) = Probability of selecting an Ace that is also a spade = 1/52.
Therefore, P (E ∪ F) = 1/4 + 1/13 - 1/52.
P (E ∪ F) = .25 + .0769 - .0192 = .3077

Example 4-6: Product liability generator Mozzarella Motors is the maker of the Apperson Jackrabbit automobile. 5% of the automobiles have defective accelerators that suddenly and without driver intervention take off at high speeds. 3% of the manufactured automobiles have brakes that suddenly fail for no apparent reason. And, 7% of all Apperson Jackrabbits have at least one of these two defects. Determine the probability that one of these cars, selected at random: (a) will have both defects; (b) will have neither defect.

Solution: Let E = Accelerator defect. Let F = Brake defect.
P(E) = .05. P(F) = .03. P (E ∪ F) = .07

(a) We need to determine P(E ∩ F).
Using Algebraic Rule 2 (supra): P (E ∪ F) = P(E) + P(F) - P(E ∩ F).
Plugging in our known values: .07 = .05 + .03 - P(E ∩ F)
Therefore: .07 = .05 + .03 - P(E ∩ F) ∩ .07 = .08 - P(E ∩ F) ⇒ - P(E ∩ F) = - .01 ∩ P(E ∩ F) = .01.

Therefore, 1% have both defects.

(b) The probability that the car has neither defect = $P(E' \cap F')$.
Using DeMorgan: $P(E' \cap F') = P((E \cup F)')$.
$P((E \cup F)') + P(E \cup F) = 1$ (Recall: A set + its compliment = 1)
Therefore: $P((E \cup F)') = 1 - P(E \cup F) = 1.0 - .07 = .93$
Therefore: 93% of the automobiles are safe.[2]

A typical U.S. casino roulette wheel has 38 numbered slots consisting of the numbers 1 to 36, a zero, and a double-zero. The betting layout consists of every individual number as well as a host of other combinations of numbers. After the players make their bets, the dealer spins the wheel and a ball; after several seconds the ball will land in one of the numbered slots. There are 18 red numbers and 18 black numbers.

During a single roll of the roulette wheel, it is typical that many different players will have bets down for a variety different types of pay-offs. Some, for example, may bet that a red number will hit; others may place money on the "1 to 12" box (With this bet, any one of 12 numbers will win). Still other gutsy players may put all of their capital on a single number, hoping for a large pay-off. There are a total of 38 possible numbers or events that could result from a single spin. Of those numbers, 18 are red, 18 are black; 0 and 00 are both colored green. 18 of the numbers are odd and 18 are even. There is no odd/even designation for 0 or 00.

Example 4-7: In order to insure an excellent chance of winning something, Murry places a one dollar chip on "red," and at the same time places a one dollar chip on the "1 to12" box. What is the probability that Murry will win something?

Solution: E (Red) and F (1–12) are non-mutually exclusive events, meaning that both events can occur in a single spin.

From Rule 2.$P(E \cup F) = P(E) + P(F) - P(E \cap F)$.
$P(E)$ = Probability of Red = 18/38
$P(F)$ = Probability of "1 to 12" = 12/38
$P(E \cap F)$ = Probability of a Red "1 to 12" = 6/38.
Therefore:
$P(E \cup F) = 18/38 + 12/38 - 6/38$.
$P(E \cup F) = 24/38$
$P(E \cup F) = .6316$
Therefore:
The probability that Murry will win something = .6316

Notes and Problems

1. With a .6316 (greater than 50%) probability of success, in the preceding example should you leave your current job, go to Las Vegas and play this strategy forever?

2. While playing roulette, you are paid even money if you bet "red," meaning that you are gambling that the ball will fall into one of the 18 red numbers. Is this a good bet (meaning that the payoff exceeds the probability of a success) a bad bet (meaning that the payoff is less than the probability of a success) or a fair bet (meaning that the payoff equals probability of a success)?

---

2. Based on the facts provided, a simpler solution involves merely subtracting 7% from 100% to arrive at the same 93% solution.

3. Why is the probability of the null set (∅) equal to zero, and the probability of the sample space equal to 1?

4. The jury pool consists of 100 venire men. There are 20 females, 78 males, and two hermaphrodites. Half of the venire men (i.e., venire people) are employed. What is the probability that the first person called will be an employed male?

5. Is poker a game of chance or a game of luck? Obviously, it is a combination of both. For state anti-gambling statutes that distinguish games of skill and games of chance, the balance between the two may prove to be quite sensitive. In *Bell Gardens Bicycle Club v. Department of Justice*, 36 Cal.App.4th 717, 42 Cal.Rptr.2d 730 (1995), the California appellate court reversed a ruling by a Los Angeles Superior Court that had declared: "The issue to be decided is whether Jackpot Poker is an illegal lottery under Penal Code section 319, or an adjunct to poker, requiring a comparable level of skill to poker, and therefore lawful on a local option basis under Penal Code section 330. This court concludes that jackpot poker is not a lottery and will enter judgment for plaintiffs." The plaintiffs were clubs that promoted Jackpot Poker, which unlike traditional poker, provided for a special jackpot that was unrelated to any skill. In determining that Jackpot Poker's involved chance, the appellate court noted:

> "The money in the jackpot accumulates until a player in a lowball game achieves a hand with a particular combination of cards under specified conditions which make him the winner of all the funds in the jackpot. As explained to us the best possible lowball hand is a five-four-three-two-ace, often called a 'wheel' and the second best hand is a six-four-three-two-ace, often called a 'sixty-four'. Under the jackpot poker rules the winner of the jackpot is the player who holds a 'sixty-four' hand when another player has a 'wheel' and thus wins the pot. The jackpot feature does not interfere with the regular lowball poker play except for the withdrawal from each pot for the jackpot...."
>
> "When the 7 poker players 'played' 5,500,000 games, there were 1,625 jackpots hit, or 1 jackpot hit about every 3,385 games played among the 7 players at the table. [Witness] Caro explained that any one 'individual would have a much less chance of being the recipient of that jackpot' because the 3,385 jackpots-won figure was a total 'per table' figure; he concluded that an individual's chance of winning the jackpot when 5,500,000 games (consisting of 38,500,000 hands) are played is about 'one in 23,692'; and when there are 8 players who play each game to its completion, he estimated that an individual's chance of winning the jackpot is about 1 chance in 20,307."

C. Relative Frequency

We have just calculated the theoretical or mathematical probability of several events using a number of different axioms and equations. Our ability to do so results from our confidence that our coins, cards and roulette wheels are true and fair, which in turn, implies that the subject events are truly random.

Another way to determine the probability of an event might be to repeat an experiment a large number of times. For example, draw a card from the deck, record its suit and its value, then put it back in the deck, shuffle and repeat the process. This methodology, which avoids the use of equations, would likely yield a number different from, but close to, the calculated probability. This number is termed: "**Relative Frequency.**"

The **Law of Large Numbers** suggests that as the number of trials becomes large, the relative frequency should approach the mathematically calculated probability. In other words, if for example you were to toss a coin twice, there is no guarantee that Heads will come up once or that Tails would come up once, even though we know that P(Heads) = P(Tails) = .50. But if we continue to toss the coin many times, the relative frequency will get closer and closer to .50. This methodology, of course, is not as accurate as if we were to employ pure mathematics, but for many scientific studies, there may be no other choice.

The mathematical principles permit us to deductively calculate our probabilities with near perfect results, whereas relative-frequency experimentation provides inductively reasoned results which must lack such reliability.

Before a new drug can be approved by the Food and Drug Administration (FDA) for consumer use, certain probabilities must first be determined. There are no mathematical equations to decide the probability that the drug's benefits will outweigh its probabilistic risks. The only way to test the drug's safety is through the relative frequency methodologies referred to above.

For our legal systems, both civil and criminal, the probabilistic challenges well exceed the scope of inductive logic. A criminal defendant normally gets a single trial, not a large frequency of them. Jurors at most number twelve. And, until recently, through DNA studies, there was never any method that would conclusively determine the accuracy of our court systems.

Many treasured safeguards are designed to see to it that although some criminals may go free, never should an innocent person be falsely convicted. To secure our peace of mind in that regard and shift the odds in favor of the accused, we have the Hearsay Rule, the "Beyond a reasonable doubt" burden of proof, and a litany of landmark cases that suppress or otherwise exclude questionable or illegally obtained evidence. *Miranda v. Arizona*, 384 U.S. 436 (1966) will toss out coerced and uninformed confessions; *Crawford v. Washington* 541 U.S. 36 (2004) will exclude certain statements by co-conspirators even though they would otherwise constitute admissible declarations against interest; *and Mapp v. Ohio*, 367 U.S. 643 (1961) will see to it that illegally obtained and discovered evidence is excluded. In order to further protect the accused, *Powell v. Alabama*, 287 U.S. 45 (1932), Gideon v. *Wainwright*, 372 U.S. 335, and *Escobedo v. Illinois*, 378 U.S. 478 (1964) will see to it that the accused is provided with legal representation regardless of whether he or she can afford it.

Unfortunately, however, all of the safeguards apparently do not work. The too-high probability that innocent people go to prison (or worse) still exists.[3]

---

3. If DNA studies are any indication, the term: "*Beyond a reasonable doubt*" does not imply the high probability of guilt that we might otherwise expect. According to the Innocence Project, a nonprofit organization dedicated to exonerating falsely convicted felons, there are more innocent people wrongly convicted than we had ever imagined. See http://www.innocenceproject.org. Visited May 20, 2013. The greatest causes for the false convictions were: Mistaken identifications, incorrect blood tests, police misconduct and prosecutorial misconduct. In the first 239 exonerations, there were 175 eyewitness misidentification cases. 62% involved a single witness, 25% involved two witnesses and in 13% of the cases, three separate witnesses misidentified the innocent defendant.

People v. Mountain

Court of Appeals of New York, 66 N.Y.2d 197, 486 N.E.2d 802 (1985)

WACHTLER, Chief Judge.

The defendant, a former police officer, has been convicted of raping and sodomizing a woman being held for arraignment in a detention cell at a police station. The Appellate Division affirmed the conviction and the defendant has appealed by leave of a Judge of this court.

The defendant raises a number of issues, most of which relate to the admission or exclusion of evidence at the trial. The primary issue concerns efforts by the prosecutor to establish a relationship between the defendant's blood type and the blood type found in the assailant's sperm, a procedure universally accepted in other jurisdictions but previously not allowed in this State in cases, such as this, where the blood type is a common one.

In the early morning hours of Saturday, September 12, 1981 the Schenectady Police arrested a young woman on a minor charge. She was transported to a police station where she was held in a detention cell for arraignment on Monday morning. On the day of the arrest the defendant was a police officer assigned to that station as a telephone operator. Although his duties did not require him to care for the female detainees, who were in the charge of a matron, he approached the complainant several times during the first day of her detention. He initially inquired about her bail and then suggested he could help her or indicated that he was attempting to raise money for her release. He also called her and gave her a telephone number at which he could be reached. On another occasion that day he gave her a note asking her to call him at a certain number if she was released.

On Sunday afternoon the defendant again came to the complainant's cell. He unlocked the door and asked her if she was ready and when she said she was, he pointed to something behind her. As she turned the defendant grabbed her throat, cupped his hand over her mouth, told her to pull down her pants and warned her that he would kill her if she screamed. He then forced her to her knees and proceeded to rape her. After experiencing some difficulty, he told her to perform oral sodomy on him. When that act was completed the defendant left and the complainant spit his sperm into a paper cup from her lunch tray.

She did not inform the matron then in charge what had occurred. However, when that matron was relieved she soon informed the new matron that she had been raped and that she had a "specimen." When she left her cell the following morning she crushed the cup and put it in her pocket. At her arraignment the charge against her was dismissed because the person who had made the complaint against her did not appear. She then informed the court that she had been raped and had a specimen. At the court's suggestion she reported the incident to a police officer in the building. After her release she went to her girlfriend's house and then immediately to St. Clare's Hospital where she again reported the rape and produced the cup. The hospital examination revealed sperm in her vagina. A laboratory examination of the cup showed that it contained sperm. A blood test of the sperm indicated that it contained type A blood.

The defendant was indicted for rape, sodomy, and official misconduct. The prosecutor applied for and was granted an order directing the defendant to submit to a blood test and provide a handwriting sample. Based on the handwriting sample an expert concluded that the note the complainant had received in the detention cell was written by the defendant. The police investigation also revealed that the telephone number on the note was registered to the security office in St. Clare's Hospital where the defendant was employed part time as a security guard.

In his opening statement the prosecutor informed the jury that they were going to hear testimony with regard to the blood type of the sperm found in the cup, as well as the blood type of the defendant. The defendant objected to the introduction of this evidence but the court sustained the objection only to the extent of excluding evidence of the defendant's blood type. The blood type of the sperm in the cup and the fact that a blood test was performed on the defendant, together with the other evidence set forth above, was submitted to the jury. The defendant did not testify on his own behalf.

During their deliberations the jury asked the court whether the defendant's blood type was in evidence and were informed that it was not. The jury found the defendant guilty of rape, sodomy, and two counts of official misconduct. The Appellate Division unanimously affirmed (105 A.D.2d 494, 481 N.Y.S.2d 449).

On this appeal the defendant urges that the court erred in permitting the prosecutor to introduce evidence that the assailant's sperm contained type A blood, and allowing "references" to the defendant's blood type. He relies on People v. Robinson, 27 N.Y.2d 864, 317 N.Y.S.2d 19, 265 N.E.2d 543.

In the Robinson case this court stated at page 865, 317 N.Y.S.2d 19, 265 N.E.2d 543: "Proof that the defendant had type 'A' blood and that the semen found in and on the body of the decedent was derived from a man with type 'A' blood was of no probative value in the case against defendant in view of the large proportion of the general population having blood of this type and, therefore, should not have been admitted." That case is distinguishable because there the relationship between the defendant's blood type and the assailant's was established. In the case now before us the connection was not made because the defendant successfully objected to the introduction of evidence of his blood type. It may be, as the defendant suggests, that the jury was confused by the evidence of blood testing and typing. The evidence was so nearly complete that an inattentive jury might have been misled into believing that the relationship had been established. On the other hand, the jury did ask the court during deliberations whether the defendant's blood type was in evidence and was informed that it was not. Thus it appears that the defendant was not prejudiced by the proof because the jury detected the gap and realized that the People had not proven that the defendant had the same blood type as the rapist.

In essence the defendant urges that the Robinson rule should be extended to make it reversible error whenever the court admits evidence that the assailant had a common blood type. However, thorough consideration of the Robinson rule shows that it is not well founded and should no longer be followed.

The Robinson decision itself is a two-line memorandum in which the court stated the rule quoted above but nevertheless affirmed the conviction on harmless error grounds ... It was followed in People v. Macedonio, 42 N.Y.2d 944, 397 N.Y.S.2d 1002, 366 N.E.2d 1355, where the conviction was reversed with the court noting in another brief memorandum, the introduction of the blood evidence and the fact that the prosecutor had introduced a confession obtained in violation of the defendant's constitutional rights. More recently in Matter of Abe A., 56 N.Y.2d 288, 299, 452 N.Y.S.2d 6, 437 N.E.2d 265, where the blood type at issue was found in less than 1% of the population, Robinson was found inapplicable. The court stated at page 299, 452 N.Y.S.2d 6: "The scientific validity and reliability of tests used to identify the type of blood a particular individual carries and to determine whether the blood of one person matches that of another are well recognized in both the medical and legal communities ... Moreover, the relative rarity of the assailant's type of blood relegates arguments as to remoteness to the realm of weight rather than admissibility."

The Robinson rule has been criticized by noted scholars (see, e.g., McCormick, Evidence § 205, at 619 [3d ed]) and in other jurisdictions has been universally rejected (see, People v. Lindsey, 84 Cal.App.3d 851, 149 Cal.Rptr. 47; Admissibility, Weight, and Sufficiency of Blood-Grouping tests in Criminal Cases, Ann., 2 ALR4th 500, 511–512). For a time it appeared that Iowa would accept the rule but now it is clear that New York stands alone (see, State v. Mark, 286 N.W.2d 396, 412–413 [Iowa]; McCormick, Evidence § 205, at 619, n.31[3d ed]).

The basic problem with the rule is the premise on which it is founded. Although blood grouping may only serve to show that the defendant and the assailant are part of a large group having that particular characteristic, it does not follow that such proof completely lacks probative value. When identity is in issue, proof that the defendant and the perpetrator share similar physical characteristics is not rendered inadmissible simply because those characteristics are also shared by large segments of the population. For instance, evidence that the person who committed the crime was white would not be excluded although that may include 80% of the population. Similarly, evidence of a person's sex, which would include roughly 50% of the general population, is routinely accepted as having some probative value with respect to identification. Proof of such common characteristics, of little value individually, may acquire great probative value when considered cumulatively (McCormick, Evidence § 171, at 364–365 [1954]). Thus, the Robinson rule goes too far in holding that evidence showing that the defendant and the assailant have type A blood, is of no probative value merely because that includes 40% of the population. It appears that the only justification for excluding such evidence is a fear that the jury may accord it undue weight, beyond its probative value, because of its scientific basis (cf. Richardson, Evidence § 147 [10th ed]). That, however, can generally be avoided by instructions, where requested, emphasizing the fact that it is only circumstantial evidence and noting, perhaps, the percentage of the population involved. In cases where the defendant can show that the potential prejudice outweighs the probative value the court may in its discretion exclude the evidence.

In sum, there is no showing that the defendant was prejudiced by the blood test evidence in this case, and in future cases such evidence should be admissible unless prejudice is shown ...

Accordingly, the order of the Appellate Division should be affirmed.

§ 4.22 Multiple Trials

If we toss a coin once, we know that the probability of obtaining Heads is .50 and if we continue to toss the coin, that probability will not change. Each toss is, for all practical purposes, a single event, separate and apart from all other tosses or trials. In contrast to single events, there arise circumstances where we are required to determine the relationship, if any, among multiple trials or events. We are frequently concerned with the manner in which multiple experiments relate to one another and the resulting probabilistic consequences.

Where there are multiple trials or attempts to secure a single successful result, as common sense suggests, the probability of obtaining success will increase with the number of trials or "chances" permitted. On the other hand, where a series of events must all register as a success, the chance of such an outcome diminishes very quickly as the number of required events increases. That is why horse racing trifecta's and football betting parlays pay such large returns for very small bets.

To deal with multiple trials, we use the term: **P(E | F)**, meaning "The probability of event E, given that event F will occur." Do not become distressed if we elect to use subscripts instead of consecutive letters. The term $P(E_1 \mid E_2)$, for our purposes means the same thing as P(E | F).

A. Independent Trials

The events E and F are said to be **independent** if P(E | F) = P(E). In such an instance, P(E) is totally unaffected by event F, as the equation indicates.

Upon the toss of a coin, for example, the probability of Heads is unaffected by any prior toss. The fact that prior tosses came up Heads 10 times in a row would not change the fact that the probability of the next toss coming up Tails would still be equal to .50.

If there exist several events, we may wish to know the probability that two or more of them might occur in sequence. If, for example, I were to toss a fair coin once, we know that the probability of it coming up heads is: $P(E_{Heads}) = P(E_1) = 1/S = ½$ or 0.5. It does not matter how many times that the coin is tossed, the probability of it coming up heads will always be 0.5. If we assume that E_1 is the first toss, that E_2 is the second toss, on and on with E_n being the nth toss, then, even though the actual result may be different, the probabilities do not change: $P(E_1) = P(E_2) = P(E_3) = P(E_4) =\ldots. .= P(E_n) = 0.5$.

To calculate the probability that every toss, in a series of tosses, is going to come up Heads, or more generally—To calculate the probability that several independent events are *all* going to occur, we employ:

The Product Rule For Independent Events: P(E | F) = P(E) × P(F)

If more than two trials are involved, we may wish to employ a generalized version of the Product Rule:

$$P(E_2 \mid E_1 \ldots \mid E_{n)} = P(E_1) \times P(E_2) \times \ldots. P(E_n)$$

Example 4-8: Having already tossed the coin once (It came up heads), what is the probability of getting tails on the next toss?

Solution: This is an independent event. The conditional nature of the toss is irrelevant, since the first toss and the second toss are independent events.

$P(E_2 \mid E_1) = P(E_2)$
$P(E_2) = 0.5$

Example 4-9: What is the probability of tossing two heads in a row using a true quarter?

Solution: These are two successively independent events.

$P(E_1) = 0.5$
$P(E_2) = 0.5$
$P(E_2 \mid E_1) = 0.5 \times .05 = 0.25$ or 1/4.

Example 4-10: What is the probability of selecting a King from a deck of cards while blindfolded, then immediately placing the card back in the deck, shuffling the deck and then drawing any spade from the deck?

Solution: This problem presents two unrelated characteristics which again are successively independent events.

$P(E_1)$ = Selection of a King
$P(E_2)$ = Selection of a spade
$P(E_1) = 4/52$ and $P(E_2) = 1/4$
$P(E_2 \mid E_1) = 4/52 \times 1/4 = .0192$

Wilson v. State

370 Md. 191, 803 A.2d 1034 (2002)

RAKER, Judge.

The primary question we address in this appeal is whether the trial court abused its discretion in permitting the State to use statistical data and a product rule computation to prove the improbability of two Sudden Infant Death Syndrome ("SIDS") deaths in a single family. In 1989, the National Institute of Child Health and Human Development published the following definition of Sudden Infant Death Syndrome (SIDS): "The sudden death of an infant under 1 year of age, which remains unexplained after a thorough case investigation, including performance of a complete autopsy, examination of the death scene, and review of the clinical history."

We shall hold that because the evidence did not satisfy the test we adopted in Reed v. State, 283 Md. 374, 391 A.2d 364 (1978), which guides the admissibility of expert testimony in Maryland, the trial court abused its discretion in admitting the evidence.

Garrett Eldred Wilson, petitioner, was convicted by a jury in the Circuit Court for Montgomery County of first degree premeditated murder of his infant son, Garrett Michael Wilson. He was sentenced to a term of imprisonment of life without the possibility of parole. The Court of Special Appeals affirmed his conviction. Wilson v. State, 136 Md.App. 27, 764 A.2d 284 (2000). We granted Wilson's petition for writ of certiorari to consider the following questions:

> "1. May the State use statistical data and a product rule computation to prove the improbability of two SIDS deaths in a family where such evidence lacks an adequate foundation and is highly susceptible to misuse by the jury?
>
> 2. Did the trial judge take inadequate corrective action when the State's Attorney argued to the jury that a statistical computation he performed accurately represented the 1 in 10 million probability of petitioner's innocence?
>
> 3. May State experts in forensic pathology tell the jury that they drew incriminatory inferences from the defendant's purchase of life insurance on his infant children and that they made credibility assessments of witnesses in the case?
>
> 4. Did the trial judge err in prohibiting the defense pathologist from explaining why, in his opinion, the defendant's purchase of life insurance on his children is irrelevant to an expert pathologist's opinion as to matter of death?
>
> 5. Did the trial judge err in admitting evidence of Appellant's alleged murder of his infant daughter six years before the alleged murder in this case?"

I.

On February 25, 1981, Deborah Oliver Fennell, then petitioner's wife, gave birth to a daughter, Brandi Jean Wilson. After Brandi's birth, petitioner purchased two life insurance policies, worth a total of $40,000, on Brandi's life. Petitioner was the primary beneficiary of these policies, Ms. Fennell the contingent beneficiary. On April 30, 1981, Brandi died. After an autopsy, her death was labeled as a SIDS death.

On March 22, 1987, Mary Anastasi, petitioner's wife as of March 1986, gave birth to a son, Garrett Michael Wilson. After Garrett's birth, petitioner purchased two life insurance policies, worth a total of $150,000, on his son's life. As with his daughter Brandi's insurance policies, petitioner was the primary beneficiary and his wife the contingent beneficiary. On August 13, 1987, Garrett died. After an autopsy, his death was also attributed to SIDS.

On May 28, 1998, the Grand Jury for Montgomery County indicted petitioner for the murder of Garrett Michael Wilson. Wilson proceeded to trial before a jury in the Circuit Court for Montgomery County.

At trial, Ms. Fennell testified that the night Brandi died was the first and only night that petitioner took care of the child. Soon after Brandi's death, petitioner filed claims with the two insurance companies from which he had purchased the policies on Brandi's life, and he collected the insurance proceeds. Like Ms. Fennell, Ms. Anastasi testified that the night of Garrett's death was the first night that petitioner alone took care of their baby. Petitioner collected the money from insurance policies he had taken out on Garrett's life soon after the infant's death.

At trial, the State presented testimony from the doctors who performed autopsies on Brandi and Garrett. Dr. Ann Dixon performed the autopsy on Brandi. Dr. Dixon testified that she changed her opinion as to the cause of death in Brandi's case to "probable suffocation" and the manner of death to "undetermined" on the basis of information provided by the police, including witnesses' statements and information about the life insurance policies on the two children taken out by petitioner.

Dr. Charles Kokes performed the autopsy on Garrett. He testified that he changed his opinion as to the cause of death in Garrett's case to smothering and the manner of death to homicide. Dr. Kokes changed his opinion based on additional investigative information provided by the State, including the facts surrounding Brandi's death and interviews with Garrett's mother.

Dr. John Smialek, the chief medical examiner for the State of Maryland, reviewed the original autopsy reports on Brandi and Garrett. In addition, the State provided Dr. Smialek with statements by the children's parents, family friends, and information regarding the life insurance policies. Dr. Smialek changed his opinion as to the cause of death in Brandi's case to suffocation and the manner of death to undetermined. Focusing largely on swelling in Garrett's brain, Dr. Smialek changed his opinion in Garrett's case to suffocation and the manner of death to homicide.

Dr. Linda Norton did not conduct the autopsy on Brandi or Garrett. She was hired by the State to review the children's deaths. Dr. Norton concluded that the cause of death in Brandi's case should be changed to suffocation and the manner of death to homicide. She focused on the similarity between Brandi and Garrett's deaths and pictures that suggested Brandi's face was pushed into the mattress of her crib. Dr. Norton also concluded that the cause of death in Garrett's case was suffocation and that the manner of death was homicide. Her opinion was based on the fact that petitioner was caring for both children at the time they died, and that these occasions were the only times petitioner had cared for either child during the night. She also considered statements from Garrett's mother and the insurance policies taken out by petitioner.

Two of the experts, Dr. Kokes and Dr. Norton, also relied on statistics, utilizing the product rule as a basis for their opinion and in calculating the probability that Garrett had not died of SIDS. In general terms, the product rule has been defined as follows: that "the probability of the joint occurrence of a number of mutually independent events is equal to the product of the individual probabilities that each of the events will occur." People v. Collins, 68 Cal.2d 319, 66 Cal.Rptr. 497, 438 P.2d 33, 36 (1968).

Dr. Kokes testified that "[t]he death rate from Sudden Infant Death Syndrome back in 1987 was somewhere between 1 to 2 deaths for every 1,000 live births." He also noted that Garret had cerebral swelling, a condition that effects less than one percent of children who die from SIDS. Employing the product rule, Dr. Kokes multiplied the probability of a child's dying of SIDS and the probability of a SIDS death involving cerebral swelling. He concluded that the "the mathematical possibility of having a SIDS death occurring with cerebral swelling would be 1 in 100,000 live births." Dr. Kokes then took into account the fact that Garrett was the second child in the family to die of SIDS. He multiplied the probability of Garrett's dying from SIDS, 1 in 100,000, by the probability of Brandi's dying of SIDS, 1 in 1,000. He concluded that the probability that Garrett died from SIDS was 1 in 100,000,000.

Dr. Norton also testified as to the probability that Garrett died of SIDS. Dr. Norton relied on different statistics that indicated that SIDS occurs in 1 infant out of every 2,000 live births. Dr. Norton employed the product rule and concluded that the probability of two SIDS deaths occurring in one family is 1 in 2,000 multiplied by 1 in 2,000, or 1 in 4,000,000.

So that I know that the child at least has been autopsied; that the child is not of an inappropriate age, you know, 12 months, 15 months, something of that nature; that toxicology of a reasonable nature has been done; that microscopic examination has been done; and that we can be reasonably assured that at least the death does fall within what is supposed to be the rules, as it were, before you can call a death SIDS. So, the information or the statistic that I use is one that is generally accepted as that which is produced by a good medical examiner system."

Petitioner moved in limine to exclude the expert testimony regarding the probability that a single family would suffer two SIDS deaths as well as the evidence that he was involved in Brandi's death. The trial judge denied both motions.

During rebuttal closing argument, the State's Attorney referred to the statistics that the experts relied on in forming their opinion that Garrett's death was criminal homicide, and argued the probability of petitioner's innocence. The State's Attorney did not merely argue that there was a low probability that two SIDS deaths would occur in one family; he argued that there was a low probability that petitioner was innocent. He told the jury, "[i]f you multiply his numbers, instead of 1 in 4 million, you get 1 in 10 million that the man sitting here is innocent. That was what a doctor, their expert, told you." Defense counsel's motion for a mistrial was denied and, instead, the court gave a curative instruction.

II.

We turn first to petitioner's argument that the trial court erred in admitting expert testimony that relied on the product rule to calculate the probability that petitioner's children died of SIDS. It is the general rule that the admissibility of expert testimony is within the sound discretion of the trial judge and will not be disturbed on appeal unless clearly erroneous. In this regard, the trial judge has wide latitude in determining whether expert testimony is sufficiently reliable to be admissible. See In Re Adoption—Guardianship No. CCJ14746, 360 Md. 634, 759 A.2d 755 (2000). Maryland Rule 5-702, addressing the admissibility of expert testimony, provides that such testimony is admissible "if the court determines that the testimony will assist the trier of fact to understand the evidence or to determine a fact in issue."

Testimony concerning an unreliable scientific process, technique or unreliable opinion is of little value to a jury. In Reed v. State, 283 Md. 374, 391 A.2d 364 (1978), this Court adopted the standard set forth in Frye v. United States, 293 F. 1013 (D.C.Cir.1923), for determining the admissibility of scientific evidence and expert testimony. Thus, before a scientific expert opinion may be received in evidence, the basis of that opinion must be shown to be generally accepted as reliable within the expert's particular scientific field.

A brief discussion of SIDS and the product rule is helpful in understanding how Frye-Reed applies to the case before us. Approximately fifty years ago, the medical community began a search to understand and prevent SIDS. See Committee on Child Abuse and Neglect, Distinguishing Sudden Infant Death Syndrome From Child Abuse Fatalities, 107 Pediatrics 437 (2001). Today, understanding of the etiology of SIDS still is incomplete. SIDS remains a "diagnosis of exclusion," meaning that a "diagnosis of SIDS reflects the clear admission by medical professionals that an infant's death remains completely unexplained."

Medical studies consistently have identified the following risk factors for SIDS: prone sleep position, sleeping on a soft surface, maternal smoking during pregnancy, overheating, late or no prenatal care, young maternal age, prematurity and/or low birth weight, and male sex. See Task Force on Infant Positioning and SIDS, Changing Concepts of Sudden Infant Death Syndrome: Implications for Infant Sleeping Environment and Sleep Position, 105 Pediatrics 650 (2000). African Americans and American Indians have consistently higher rates, two to three times the national average. Id. Because the cause of SIDS remains unknown, none of those risk factors are of help in calculating the probability that a child will die of SIDS. Robert M. Reece, Fatal Child Abuse and Sudden Infant Death Syndrome: A Critical Diagnostic Decision, 91 Pediatrics 423 (1993).

Beyond these commonly accepted risk factors, there is little agreement as to the causes of SIDS. This is particularly true with regard to the role of genetics. Some, including the State, argue that it is generally accepted that there is no genetic defect or condition that can be tied to SIDS. See id. (noting that "[t]he issue of recurrent SIDS within a family raises the possibility of genetically determined conditions.... But when SIDS occurrences among siblings of SIDS cases were compared with those among non-SIDS siblings in maternal age-and birth rank matched-control families, there was no statistically significant difference in SIDS rates.... Thus, the notion that having a SIDS baby makes having another more likely was dispelled.").

In contrast, a recent article in the Journal of the American Medical Association presents a study suggesting that SIDS may result from a genetic condition. See Michael J. Ackerman et al., Postmortem Molecular Analysis of SCN5A Defects in Sudden Infant Death Syndrome, JAMA 2264 (2001). This study draws into question the assertion that SIDS deaths within a single family are independent or unrelated events. Similarly, in the March 2000 edition of Pediatrics, the Task Force on Infant Positioning and SIDS, chaired by John Kattwinkel, M.D., expressed uncertainty as to the risk of SIDS among siblings.

With this background in mind, we now turn to the product rule. The product rule "states that the probability of the joint occurrence of a number of mutually independent events is equal to the product of the individual probabilities that each of the events will occur." People v. Collins, 68 Cal.2d 319, 66 Cal.Rptr. 497, 438 P.2d 33, 36 (1968) (emphasis in original). See also Armstead, 342 Md. at 69-70, 673 A.2d at 236 (noting that the product rule is a probability principle that establishes that "the probability of two events occurring together is equal to the probability that event one will occur multiplied by the probability that event two will occur."). In Armstead, we explained that "[t]he classic illustration is coin tossing; the probability of finding heads on two successive coin tosses is equal to the probability of heads on the first toss, 50%, times the probability of heads on the second toss, 50%, equaling 25%." Id. at 70, 673 A.2d at 236. We noted that the product rule may be applied only to events that are unrelated to one another, or are independent. We explained:

> "In the coin toss example, this means that the outcome of the first coin toss does not affect the outcome of the second coin toss, which is a valid assumption. By comparison, assume we wish to calculate the probability of having both a checking account and a loan from a particular bank. This is an example of non-independent or linked events. We can not calculate the probability of having both a loan and a checking account at the same bank by multiplying together the individual probabilities under the product rule because a person is more likely to obtain a loan from the bank where he maintains a checking account. To illustrate nonindependence as it applies to human characteristics (although not genetic characteristics), assume we wish to determine the probability a man will have both a beard and a moustache. Also assume that the probability of having a beard is 1/20, and the probability of having a moustache is 1/10. It would be incorrect to infer that the probability of having both a beard and a moustache, applying the product rule, is 1/200, because it is likely that these are non-independent events; men who have beards are probably more likely than others to also have moustaches."

In the case sub judice, petitioner contends that the product rule should not have been used to calculate the likelihood that both of his children died of SIDS because it is not generally accepted in the medical field that SIDS deaths within a single family are independent. The State argues that the statistical evidence introduced at trial was generally accepted and reliable, and further, that because the risk factors for SIDS are independent, it was appropriate for the experts to utilize the product rule and to multiply the probability of one child dying of SIDS times the probability of a second child also dying of SIDS. The State posits that it is accepted universally that SIDS is not caused by any genetic defect, and that the testimony based on the product rule was therefore permissible. The State relies heavily on the conclusion of the Court of Special Appeals that SIDS deaths are independent events. The Court of Special Appeals reasoned as follows:

> "[W]e merely note that appellant's argument that SIDS deaths are interrelated is not totally accurate. It is accurate to assert that the deaths labeled as SIDS deaths may indeed have a connection. The inaccuracy arises once that connection is discovered, because the deaths should no longer be identified as SIDS deaths. We must keep in mind that SIDS is a diagnosis of exclusion that is subject to change if an actual cause is uncovered. The scientific literature has shown that, from what is currently known, SIDS risk factors are not interrelated but are independent. This is particularly true of multiple SIDS deaths in one family, where the likelihood of recurrence is less than one percent."

The State and the Court of Special Appeals cite articles that suggest that there is no genetic component to SIDS. Both fail to acknowledge that most of the articles reflect that it is unknown whether there is a genetic component to SIDS. For example, the intermediate appellate court cited the Task Force on Infant Positioning and SIDS, Changing Concepts of Sudden Infant Death Syndrome: Implications for Infant Sleeping Environment and Sleep Position, 105 Pediatrics 650 (2000), for the proposition that SIDS deaths are independent. Wilson, 136 Md.App. at 69, 764 A.2d at 306. The court omitted the author's discussion of genetics as a risk factor.

A strong argument that there is no genetic component to SIDS is found in the article by Robert M. Reece, Fatal Child Abuse and Sudden Infant Death Syndrome: A Critical Diagnostic Decision, 91 Pediatrics 423 (1993). The author states "The issue of recurrent SIDS within a family raises the possibility of genetically determined conditions. It also provokes questions of a forensic nature. In a 14-year study of subsequent siblings of SIDS victims in Norway, and in a Washington State study over 16 years, the SIDS sibling risk was seen to be almost four times that of the SIDS risk among births at large. But when SIDS occurrences among siblings of SIDS cases were compared with those among non-SIDS siblings in maternal age- and birth rank-matched control families, there was no statistically significant difference in SIDS rates or in total infant mortality rates in families with a history of SIDS compared with families with no SIDS. Thus, the notion that having a SIDS baby makes having another more likely was dispelled. With the exclusion from the SIDS statistics of some of the deaths now thought to be due to inborn errors of metabolism, the chances for subsequent SIDS in families seems even less likely."

As demonstrated above, few authors in this area exhibit such confidence that there is no genetic component to SIDS. "Several studies that have evaluated SIDS among siblings have found that having a sibling who died of SIDS is a significant risk factor. However, others have failed to find such a relationship or have found that siblings of infants who have died of SIDS are at risk for all causes of infant death, not just SIDS. In addition, most of the studies reporting familial SIDS have the limitation of having been conducted during a period when case and scene investigations were not routine and assignment of the SIDS diagnosis may have been flawed. Thus, the true risk is unknown." The Court of Special Appeals also relied on a student note for the proposition that multiple SIDS deaths in single family are unrelated. Wilson, 136 Md.App. at 70, 764 A.2d at 304 (quoting Goldenberg, Comment: Sudden Infant Death Syndrome as a Mask for Murder: Investigating and Prosecuting Infanticide, 28 Sw. U.L.Rev. 599, 606 (1999)). Both the State's brief before this Court and the Court of Special Appeals' opinion omit the designation "Comment" in the citation of this student note, thereby failing to indicate that it is a student-written piece. See The Bluebook: A Uniform System of Citation R. 16.6.2(a), at 121 (Columbia Law Review Ass'n et al. eds., 17th ed.2000) (noting that signed and titled notes, comments, projects, etc. are cited in the same manner as any other signed article in a law review, except that the designation of the piece should appear before the title of the work to indicate that it is student-written).

We hold that the trial court erred in admitting expert testimony based on the product rule because a condition necessary to the proper application of the product rule was lacking: there was inadequate proof of the independence of Brandi and Garrett's deaths. As evidenced by the authorities above cited, there is not general agreement in the scientific community as to the relationship between SIDS deaths within a single family. Stated another way, there is not general agreement in the medical community that multiple SIDS deaths in a single family are genetically unrelated. The literature continues to reflect a lively debate concerning the role of genetics in SIDS. Moreover, the recent study in the

Journal of the American Medical Association suggests that there may well be a genetic component to SIDS. See Michael J. Ackerman et al., Postmortem Molecular Analysis of SCN5A Defects in Sudden Infant Death Syndrome, 286 JAMA 2264 (2001). If there is any consensus in the field, it is that more research into the question is necessary before general acceptance is reached. One article, which is reflective of thinking in the field, states:

III.

Petitioner argues that the trial court took inadequate corrective action when the State's Attorney, in closing argument, calculated the statistical probability of petitioner's innocence. Prior to closing arguments, the court reviewed proposed jury instructions with counsel. The State's Attorney desired a particular instruction to guide the jury's consideration of the statistical evidence. The following discussion ensued:

> "STATE'S ATTORNEY: I mean we shouldn't stand up ... and say there is a one-in-200-million chance that this man is innocent. We shouldn't be able to say that. What we should be able to say is-and this limits it.... 'Ladies and gentlemen, Dr. Kokes thinks that the chances of SIDS happening the second time is one in 200 million. And you-that is the only purpose for which we can argue and for which they can consider. They can't and shouldn't be able to perhaps-
>
> COURT: And you are afraid they are going to misuse that evidence.
>
> STATE'S ATTORNEY: They might. And if this instruction is given-and the jurors are presumed to follow the Court's instructions-then there could be no prejudice inured from the use of statistics at all during the course of this trial, because the only purpose for the statistics to be introduced was to assist-was because the medical experts relied on those statistics."

Despite his knowledge that he was prohibited from using the statistics to calculate the probability of petitioner's innocence, the State's Attorney argued, in closing, as follows:

> "But what he also told you in terms of the statistics we have talked about, the doctors, relying on the statistics of SIDS, he told you that in his numbers that it was 3 in 1,000 certain SIDS deaths.
>
> The second time, the death could be attributed to SIDS. There is 3 in 1,000 live births that would be a SIDS. Well, let us use his numbers and be conservative.
>
> Assuming it is 3 in 1,000 for the first. He also told you that less than 1 percent of SIDS deaths had the brain swelling, the edema.
>
> If you multiply his numbers, instead of 1 in 4 million, you get 1 in 10 million that the man sitting here is innocent. That was what a doctor, their expert, told you."

Petitioner moved for a mistrial on the grounds that the State improperly used the statistical evidence to calculate the probability that petitioner was innocent. The court denied the motion, and petitioner requested a curative instruction. Defense counsel stated: "I would ask you to ... tell [the jury] that you can disregard it. You can never ever, ever, use statistics and compare that to the burden of proof or reasonable doubt. They have no place in this case, and that is what I am asking." The court declined to instruct the jury as requested by defense counsel and, instead, essentially reiterated the earlier instruction.

The State's Attorney's comment constituted error. The statistics stated by the experts in this case were admitted as evidence to satisfy the State's burden that Garrett's death was a homicide, not that petitioner was the person who committed the homicide. The State's Attorney, however, improperly used the statistics to argue that there was only a minuscule possibility that the defendant was innocent. The State's Attorney was well aware that the statistical evidence could not be used to calculate the probability of petitioner's innocence. The colloquy at the bench makes this crystal clear. His argument was improper.

The Court of Special Appeals found the State's Attorney's argument "perplexing and disturbing," Wilson, 136 Md.App. at 73, 764 A.2d at 309, but held that the improper remark was not so egregious as to have undermined the presumption that appellant was innocent. Id. at 78, 764 A.2d at 311.

The courts that have considered this issue have concluded that it is impermissible to assign a number to the probability of guilt or innocence. See e.g., People v. Collins, 68 Cal.2d 319, 66 Cal.Rptr. 497, 438 P.2d 33, 40 (1968). In Collins, the Supreme Court of California stated:

> "Confronted with an equation which purports to yield a numerical index of probable guilt, few juries could resist the temptation to accord disproportionate weight to that index; only an exceptional juror, and indeed only a defense attorney schooled in mathematics, could successfully keep in mind the fact that the probability commuted by the prosecution can represent, at best, the likelihood that a random couple would share the characteristics testified to by the People's witnesses- not necessarily the characteristics of the actually guilty couple."

Moreover, the trial judge's supplemental instruction did not cure the prejudice caused by the State's Attorney's remark. The instructions directed the jurors to use the statistical evidence only to evaluate the expert's testimony, but, given the powerful nature of statistics, did nothing to cure the prosecutor's invitation to misuse the evidence to conclude that there was a 1 in 10,000,000 chance that petitioner was innocent. The trial court's instructions did not cure the prejudice the State's comment engendered.

We turn next to petitioner's contention that the trial court erred in admitting evidence of the alleged murder of his infant daughter, Brandi, six years before the alleged murder in this case. The State sought a pre-trial ruling regarding the admissibility of evidence related to Brandi's death. The court considered certain exhibits introduced by the State, facts agreed to by both parties as uncontested and the arguments of counsel. In a memorandum opinion and order, the trial court ruled that the evidence was admissible.

Before this Court, petitioner argues that the evidence of Brandi's death was inadmissible other crimes or bad act evidence under Md. Rule 5-404(b) because it lacked special relevance, there was not clear and convincing evidence of his involvement in Brandi's death, and the probative value of the evidence was outweighed by unfair prejudice. The State argues that the evidence was admissible on several bases-for identity, motive, absence of accident and intent. The State maintains that petitioner's involvement in Brandi's death was established by clear and convincing evidence, and that the probative nature of the evidence outweighed the prejudice.

We shall not address this issue because in light of our ruling as to the expert's reliance on the product rule, it is unclear what the substance of the testimony will be in any future proceeding. The expert testimony on the cause and manner of Brandi's death will bear on the determination of whether her death is viewed properly as a prior bad act and, consequently, whether the admissibility of evidence surrounding her death is subject to Maryland Rule 5-404(b). The expert testimony will also affect if, and how, the evidence of Brandi's death may be used to prove the corpus delicti.

For all the reasons stated herein, the judgment of conviction is reversed. Judgment of the court of special appeals reversed, case remanded to that court with directions to reverse the judgment of the circuit court for Montgomery County and remand the case to the Circuit Court for a new trial. Costs in this court and the court of special appeals to be paid by respondent.

People v. Collins

Supreme Court of California, 68 Cal.2d 319, 438 P.2d 33, (1968)

SULLIVAN, Justice.

We deal here with the novel question whether evidence of mathematical probability has been properly introduced and used by the prosecution in a criminal case. While we discern no inherent incompatibility between the disciplines of law and mathematics and intend no general disapproval or disparagement of the latter as an auxiliary in the fact-finding processes of the former, we cannot uphold the technique employed in the instant case. As we explain in detail Infra, the testimony as to mathematical probability infected the case with fatal error and distorted the jury's traditional role of determining guilt or innocence according to long-settled rules. Mathematics, a veritable sorcerer in our computerized society, while assisting the trier of fact in the search for truth, must not cast a spell over him. We conclude that on the record before us defendant should not have had his guilt determined by the odds and that he is entitled to a new trial. We reverse the judgment.

A jury found defendant Malcolm Ricardo Collins and his wife defendant Janet Louise Collins guilty of second degree robbery (Pen.Code, ss 211, 211a, 1157). Malcolm appeals from the judgment of conviction. Janet has not appealed. (Hereafter, the term 'defendant' is intended to apply only to Malcolm, but the term 'defendants' to Malcolm and Janet.)

On June 18, 1964, about 11:30 a.m. Mrs. Juanita Brooks, who had been shopping, was walking home along an alley in the San Pedro area of the City of Los Angeles. She was pulling behind her a wicker basket carryall containing groceries and had her purse on top of the packages. She was using a cane. As she stooped down to pick up an empty carton, she was suddenly pushed to the ground by a person whom she neither saw nor heard approach. She was stunned by the fall and felt some pain. She managed to look up and saw a young woman running from the scene. According to Mrs. Brooks the latter appeared to weigh about 145 pounds, was wearing 'something dark,' and had hair 'between a dark blond and a light blond,' but lighter than the color of defendant Janet Collins' hair as it appeared at trial. Immediately after the incident, Mrs. Brooks discovered that her purse, containing between \$35 and \$40, was missing.

About the same time as the robbery, John Bass, who lived on the street at the end of the alley, was in front of his house watering his lawn. His attention was attracted by 'a lot of crying and screaming' coming from the alley. As he looked in that direction, he saw a woman run out of the alley and enter a yellow automobile parked across the street from him. He was unable to give the make of the car. The car started off immediately and pulled wide around another parked vehicle so that in the narrow street it passed within six feet of Bass. The latter then saw that it was being driven by a male Negro, wearing a mustache and beard. At the trial Bass identified defendant as the driver of the yellow automobile. However, an attempt was made to impeach his identification by his admission that at the preliminary hearing he testified to an uncertain identification at the police lineup shortly after the attack on Mrs. Brooks, when defendant was beardless. In his testimony Bass described the woman who ran from the alley as a Caucasian, slightly over five feet tall, of ordinary build, with her hair in a dark blond ponytail, and wearing dark clothing. He further testified that her ponytail was 'just like' one which Janet had in a police photograph taken on June 22, 1964.

On the day of the robbery, Janet was employed as a housemaid in San Pedro. Her employer testified that she had arrived for work at 8:50 a.m. and that defendant had picked her up in a light yellow car about 11:30 a.m. (Other witnesses variously described the car as yellow, as yellow with an off-white top, and yellow with an egg-shell white top. The car was also described as being medium to large in size. Defendant drove a car at or near the times in question which was a Lincoln with a yellow body and a white top.)

On that day, according to the witness, Janet was wearing her hair in a blonde ponytail but lighter in color than it appeared at trial. There are inferences which may be drawn from the evidence that Janet attempted to alter the appearance of her hair after June 18. Janet denies that she cut, colored or bleached her hair at any time after June 18, and a number of witnesses supported her testimony.

There was evidence from which it could be inferred that defendants had ample time do drive from Janet's place of employment and participate in the robbery. Defendants testified, however, that they went directly from her employer's house to the home of friends, where they remained for several hours.

In the morning of June 22, Los Angeles Police Officer Kinsey, who was investigating the robbery, went to defendants' home. He saw a yellow Lincoln automobile with an off-white top in front of the house. He talked with defendants. Janet, whose hair appeared to be a dark blonde, was wearing it in a ponytail. Malcolm did not have a beard.

The officer explained to them that he was investigating a robbery specifying the time and place; that the victim had been knocked down and her purse snatched; and that the person responsible was a female Caucasian with blonde hair in a ponytail who had left the scene in a yellow car driven by a male Negro. He requested that defendants accompany him to the police station at San Pedro and they did so. There, in response to police inquiries as to defendants' activities at the time of the robbery, Janet stated, according to Officer Kinsey, that her husband had picked her up at her place of employment at 1 p.m. and that they had then visited at the home of friends in Los Angeles. Malcolm confirmed this. Defendants were detained for an hour or two, were photographed but not booked, and were eventually released and driven home by the police.

Late in the afternoon of the same day, Officer Kinsey, while driving home from work in his own car, saw defendants riding in their yellow Lincoln. Although the transcript fails to disclose what prompted such action Kinsey proceeded to place them under surveillance and eventually followed them home. He called for assistance and arranged to meet other police officers in the vicinity of defendants' home. Kinsey took a position in the rear of the premises. The other officers, who were in uniform and had arrived in a marked police car, approached defendants' front door. As they did so, Kinsey saw defendant Malcolm Collins run out the back door toward a rear fence and disappear behind a tree. Meanwhile the other officers emerged with Janet Collins whom they had placed under arrest. A search was made for Malcolm who was found in a closet of a neighboring home and also arrested. Defendants were again taken to the police station, were kept in custody for 48 hours, and were again released without any charges being made against them.

Officer Kinsey interrogated defendants separately on June 23 while they were in custody and testified to their statements over defense counsel's objections based on the decision in Escobedo and our first decision in Dorado. According to the officer, Malcolm stated that he sometimes wore a beard but that he did not wear a beard on June 18 (the day of the robbery), having shaved it off on June 2, 1964. Evidence as to defendant's beard and mustache is conflicting. Defense witnesses appeared to support defendant's claims that he had shaved his beard on June 2. There was testimony that on June 19 when defendant appeared in court to pay fines on another matter he was bearded. By June 22 the beard had been removed.

He also explained two receipts for traffic fines totaling $35 paid on June 19, which receipts had been found on his person, by saying that he used funds won in a gambling game at a labor hall. Janet, on the other hand, said that the $35 used to pay the fines had come from her earnings.

On July 9, 1964, defendants were again arrested and were booked for the first time. While they were in custody and awaiting the preliminary hearing, Janet requested to talk with Officer Kinsey. There followed a lengthy conversation during the first part of which Malcolm was not present. During this time Janet expressed concern about defendant and inquired as to what the outcome would be If it appeared that she committed the crime and Malcolm knew nothing about it. In general she indicated a wish that defendant be released from any charges because of his prior criminal record and that if someone must be held responsible, she alone would bear the guilt. The officer told her that no assurances could be given, that if she wanted to admit responsibility disposition of the matter would be in the hands of the court and that if she committed the crime and defendant knew nothing about it the only way she could help him would be by telling the truth. Defendant was then brought into the room and participated in the rest of the conversation. The officer asked to hear defendant's version of the matter, saying that he believed defendant was at the scene. However, neither Janet nor defendant confessed or expressly made damaging admissions although constantly urged by the investigating officer to make truthful statements. On several occasions defendant denied that he knew what had gone on in the alley. On the other hand, the whole tone of the conversation evidenced a strong consciousness of guilt on the part of both defendants who appeared to be seeking the most advantageous way out. Over defense counsel's same objections based on Escobedo and Dorado, some parts of the foregoing conversation were testified to by Officer Kinsey and in addition a tape recording of the entire conversation was introduced in evidence and played to the jury.

At the seven-day trial the prosecution experienced some difficulty in establishing the identities of the perpetrators of the crime. The victim could not identify Janet and had never seen defendant. The identification by the witness Bass, who observed the girl run out of the alley and get into the automobile, was incomplete as to Janet and may have been weakened as to defendant. There was also evidence, introduced by the defense, that Janet had worn light-colored clothing on the day in question, but both the victim and Bass testified that the girl they observed had worn dark clothing.

In an apparent attempt to bolster the identifications, the prosecutor called an instructor of mathematics at a state college. Through this witness he sought to establish that, assuming the robbery was committed by a Caucasian woman with a blond ponytail who left the scene accompanied by a Negro with a beard and mustache, there was an overwhelming probability that the crime was committed by any couple answering such distinctive characteristics.

The witness testified, in substance, to the 'product rule,' which states that the probability of the joint occurrence of a number of Mutually independent events is equal to the product of the individual probabilities that each of the events will occur.[4] Without presenting any statistical evidence whatsoever in support of the probabilities for the factors selected, the prosecutor then proceeded to have the witness Assume[5] probability factors for the various characteristics which he deemed to be shared by the guilty couple and all other couples answering to such distinctive characteristics.

Although the prosecutor insisted that the factors he used were only for illustrative purposes—to demonstrate how the probability of the occurrence of mutually independent factors affected the probability that they would occur together, he nevertheless attempted to use factors which he personally related to the distinctive characteristics of defendants. In his argument to the jury he invited the jurors to apply their own factors, and asked defense counsel to suggest what the latter would deem as reasonable. The prosecutor himself proposed the individual probabilities set out in the table below. Although the transcript of the examination of the mathematics instructor and the information volunteered by the prosecutor at that time create some uncertainty as to precisely which of the characteristics the prosecutor assigned to the individual probabilities, he restated in his argument to the jury that they should be as follows:

Characteristic	Individual Probability
A. Partly yellow automobile	1/10
B. Man with mustache	1/4
C. Girl with ponytail	1/10
D. Girl with blond hair	1/3
E. Negro man with beard	1/10
F. Interracial couple in car	1/1000

In his brief on appeal defendant agrees that the foregoing appeared on a table presented in the trial court. Applying the product rule to his own factors the prosecutor arrived at a probability that there was but one chance in 12 million that any couple possessed the distinctive characteristics of the defendants.

Accordingly, under this theory, it was to be inferred that there could be but one chance in 12 million that defendants were innocent and that another equally distinctive couple actually committed the robbery. Expanding on what he had thus purported to suggest as a hypothesis, the prosecutor offered the completely unfounded and improper testimonial assertion that, in his opinion, the factors he had assigned were 'conservative estimates' and that, in reality 'the chances of anyone else besides these defendants being there, having every similarity, is somewhat like one in a billion.'

Objections were timely made to the mathematician's testimony on the grounds that it was immaterial, that it invaded the province of the jury, and that it was based on unfounded assumptions. The objections were 'temporarily overruled' and the evidence admitted subject to a motion to strike. When that motion was made at the conclusion of the direct examination, the court denied it, stating that the testimony had been received only

4. In the example employed for illustrative purposes at the trial, the probability of rolling one die and coming up with a '2' is 1/6, that is, any one of the six faces of a die has one chance in six of landing face up on any particular roll. The probability of rolling two '2s' in succession is 1/6 1/6, or 1/36, that is, on only one occasion out of 36 double rolls (or the roll of two dice), will the selected number land face up on each roll or die.

5. His argument to the jury was based on the same gratuitous assumptions or on similar assumptions which he invited the jury to make.

for the 'purpose of illustrating the mathematical probabilities of various matters, the possibilities for them occurring or re-occurring.'

Both defendants took the stand in their own behalf. They denied and knowledge of or participation in the crime and stated that after Malcolm called for Janet at her employer's house they went directly to a friend's house in Los Angeles where they remained for some time. According to this testimony defendants were not near the scene of the robbery when it occurred. Defendant's friends testified to a visit by them 'in the middle of June' although she could not recall the precise date. Janet further testified that certain inducements were held out to her during the July 9 interrogation on condition that she confess her participation.

Defendant makes two basic contentions before us: First, that the admission in evidence of the statements made by defendants while in custody on June 23 and July 9, 1964, constitutes reversible error under the rules announced in the Escobedo and Dorado decisions; and second, that the introduction of evidence pertaining to the mathematical theory of probability and the use of the same by the prosecution during the trial was error prejudicial to defendant. We consider the latter claim first.

As we shall explain, the prosecution's introduction and use of mathematical probability statistics injected two fundamental prejudicial errors into the case: (1) The testimony itself lacked an adequate foundation both in evidence and in statistical theory; and (2) the testimony and the manner in which the prosecution used it distracted the jury from its proper and requisite function of weighing the evidence on the issue of guilt, encouraged the jurors to rely upon an engaging but logically irrelevant expert demonstration, foreclosed the possibility of an effective defense by an attorney apparently unschooled in mathematical refinements, and placed the jurors and defense counsel at a disadvantage in sifting relevant fact from inapplicable theory.

We initially consider the defects in the testimony itself. As we have indicated, the specific technique presented through the mathematician's testimony and advanced by the prosecutor to measure the probabilities in question suffered from two basic and pervasive defects—an inadequate evidentiary foundation and an inadequate proof of statistical independence. First, as to the foundation requirement, we find the record devoid of any evidence relating to any of the six individual probability factors used by the prosecutor and ascribed by him to the six characteristics as we have set them out in footnote 10, Ante. To put it another way, the prosecution produced no evidence whatsoever showing, or from which it could be in any way inferred, that only one out of every ten cars which might have been at the scene of the robbery was partly yellow, that only one out of every four men who might have been there wore a mustache, that only one out of every ten girls who might have been there wore a ponytail, or that any of the other individual probability factors listed were even roughly accurate.

We seriously doubt that such evidence could ever be compiled since no statistician could possibly determine after the fact which cars, or which individuals, 'might' have been present at the scene of the robbery; certainly there is no reason to suppose that the human and automotive populations of San Pedro, California, include all potential culprits—or, conversely, that all members of these populations are proper candidates for inclusion. Thus the sample from which the relevant probabilities would have to be derived is itself undeterminable. (See generally, Yaman, Statistics, An Introductory Analysis (1964), ch. I.)

The bare, inescapable fact is that the prosecution made no attempt to offer any such evidence. Instead, through leading questions having perfunctorily elicited from the witness the response that the latter could not assign a probability factor for the characteristics involved,[6] the prosecutor himself suggested what the various probabilities should be and these became the basis of the witness' testimony.

It is a curious circumstance of this adventure in proof that the prosecutor not only made his own assertions of these factors in the hope that they were 'conservative' but also in later argument to the jury invited the jurors to substitute their 'estimates' should they wish to do so. We can hardly conceive of a more fatal gap in the prosecution's scheme of proof. A foundation for the admissibility of the witness' testimony was never even attempted to be laid, let alone established. His testimony was neither made to rest on his own testimonial knowledge nor presented by proper hypothetical questions based upon valid data in the record. (See generally: 2 Wigmore on Evidence (3d ed. 1940) ss 478, 650–652, 657, 659, 672–684; Witkin, Cal. Evidence (2d ed. 1966) s 771; McCormick on Evidence pp. 19–20; Evidence: Admission of Mathematical Probability Statistics Held Erroneous for Want of Demonstration of Validity (1967) Duke L.J. 665, 675–678, citing People v. Risley (1915) 214 N.Y. 75, 85, 108 N.E. 200; State v. Sneed (1966) 76 N.M. 349, 414 P.2d 858.)

In the Sneed case, the court reversed a conviction based on probabilistic evidence, stating: 'We hold that mathematical odds are not admissible as evidence to identify a defendant in a criminal proceeding so long as the odds are based on estimates, the validity of which have (sic) not been demonstrated.' (Italics added.) (414 P.2d at p. 862.)

But, as we have indicated, there was another glaring defect in the prosecution's technique, namely an inadequate proof of the statistical independence of the six factors. No proof was presented that the characteristics selected were mutually independent, even though the witness himself acknowledged that such condition was essential to the proper application of the 'product rule' or 'multiplication rule.'[7]

To the extent that the traits or characteristics were not mutually independent (e.g. Negroes with beards and men with mustaches obviously represent overlapping categories[8]), the 'product rule' would inevitably yield a wholly erroneous and exaggerated result even if all of the individual components had been determined with precision. (Siegel, Nonparametric Statistics for the Behavioral Sciences (1956) 19; see generally Harmon, Modern Factor Analysis (1960).)

In the instant case, therefore, because of the aforementioned two defects—the inadequate evidentiary foundation and the inadequate proof of statistical independence—

6. The prosecutor asked the mathematics instructor: "Now, let me see if you can be of some help to us with some independent factors, and you have some paper you may use. Your specialty does not equip you, I suppose, to give us some probability of such things as a yellow car as contrasted with any other kind of car, does it? I appreciate the fact that you can't assign a probability for a car being yellow as contrasted to some other car, can you? A. No, I couldn't."

7. It is there stated that: "A trait is said to be independent of a second trait when the occurrence or non-occurrence of one does not affect the probability of the occurrence of the other trait. The multiplication rule cannot be used without some degree of error where the traits are not independent." (Citing Huntsberger, Elements of Statistical Inference (1961) 77; Kingston & Kirk, The Use of Statistics in Criminalistics (1964) 55 J.Crim.L., C. & P.S. 516.) (Note, *supra*, Duke L.J. fn. 25, p. 670.)

8. Assuming Arguendo that factors B and E, were correctly estimated, nevertheless it is still arguable that most Negro men with beards Also have mustaches (exhibit 3 herein, for instance, shows defendant with both a mustache and a beard, indeed in a hirsute continuum); if so, there is no basis for multiplying 1/4 by 1/10 to estimate the proportion of Negroes who wear beards And mustaches. Again, the prosecution's technique could Never be meaningfully applied, since its accurate use would call for information as to the degree of interdependence among the six individual factors. (See Yamane, op. cit. *supra*.) Such information cannot be compiled, however, since the relevant sample necessarily remains unknown.

the technique employed by the prosecutor could only lead to wild conjecture without demonstrated relevancy to the issues presented. It acquired no redeeming quality from the prosecutor's statement that it was being used only 'for illustrative purposes' since, as we shall point out, the prosecutor's subsequent utilization of the mathematical testimony was not confined within such limits.

We now turn to the second fundamental error caused by the probability testimony. Quite apart from our foregoing objections to the specific technique employed by the prosecution to estimate the probability in question, we think that the entire enterprise upon which the prosecution embarked, and which was directed to the objective of measuring the likelihood of a random couple possessing the characteristics allegedly distinguishing the robbers, was gravely misguided. At best, it might yield an estimate as to how infrequently bearded Negroes drive yellow cars in the company of blonde females with ponytails.

The prosecution's approach, however, could furnish the jury with absolutely no guidance on the crucial issue: Of the admittedly few such couples, which one, if any, was guilty of committing this robbery? Probability theory necessarily remains silent on that question, since no mathematical equation can prove beyond a reasonable doubt (1) that the guilty couple In fact possessed the characteristics described by the People's witnesses, or even (2) that only One couple possessing those distinctive characteristics could be found in the entire Los Angeles area.

As to the first inherent failing we observe that the prosecution's theory of probability rested on the assumption that the witnesses called by the People had conclusively established that the guilty couple possessed the precise characteristics relied upon by the prosecution. But no mathematical formula could ever establish beyond a reasonable doubt that the prosecution's witnesses correctly observed and accurately described the distinctive features which were employed to link defendants to the crime. (See 2 Wigmore on Evidence (3d ed. 1940) s 478.) Conceivably, for example, the guilty couple might have included a light-skinned Negress with bleached hair rather than a Caucasian blonde; or the driver of the car might have been wearing a false beard as a disguise; or the prosecution's witnesses might simply have been unreliable. In the instant case, for instance, the victim could not state whether the girl had a ponytail, although the victim observed the girl as she ran away.

The witness Bass, on the other hand, was sure that the girl whom he saw had a ponytail. The demonstration engaged in by the prosecutor also leaves no room for the possibility, although perhaps a small one, that the girl whom the victim and the witness observed was, in fact, the same girl.

The foregoing risks of error permeate the prosecution's circumstantial case. Traditionally, the jury weighs such risks in evaluating the credibility and probative value of trial testimony, but the likelihood of human error or of falsification obviously cannot be quantified; that likelihood must therefore be excluded from any effort to assign a Number to the probability of guilt or innocence. Confronted with an equation which purports to yield a numerical index of probable guilt, few juries could resist the temptation to accord disproportionate weight to that index; only an exceptional juror, and indeed only a defense attorney schooled in mathematics, could successfully keep in mind the fact that the probability computed by the prosecution can represent, At best, the likelihood that a random couple would share the characteristics testified to by the People's witnesses—Not necessarily the characteristics of the actually guilty couple.

As to the second inherent failing in the prosecution's approach, even assuming that the first failing could be discounted, the most a mathematical computation could ever yield would be a measure of the probability that a random couple would possess the distinctive features in question. In the present case, for example, the prosecution attempted to compute the probability that a random couple would include a bearded Negro, a blonde girl with a ponytail, and a partly yellow car; the prosecution urged that this probability was no more than one in 12 million. Even accepting this conclusion as arithmetically accurate, however, one still could not conclude that the Collinses were probably the guilty couple. On the contrary, as we explain in the Appendix, the prosecution's figures actually imply a likelihood of over 40 percent that the Collinses could be 'duplicated' by at least One other couple who might equally have committed the San Pedro robbery. Urging that the Collinses be convicted on the basis of evidence which logically establishes no more than this seems as indefensible as arguing for the conviction of X on the ground that a witness saw either X or X's twin commit the crime.

Again, few defense attorneys, and certainly few jurors, could be expected to comprehend this basic flaw in the prosecution's analysis. Conceivably even the prosecutor erroneously believed that his equation established a high probability that No other bearded Negro in the Los Angeles area drove a yellow car accompanied by a ponytailed blonde. In any event, although his technique could demonstrate no such thing, he solemnly told the jury that he had supplied mathematical proof of guilt.

Sensing the novelty of that notion, the prosecutor told the jurors that the traditional idea of proof beyond a reasonable doubt represented 'the most hackneyed, stereotyped, trite, misunderstood concept in criminal law.' He sought to reconcile the jury to the risk that, under his 'new math' approach to criminal jurisprudence, 'on some rare occasion an innocent person may be convicted.' 'Without taking that risk,' the prosecution continued, 'life would be intolerable because there would be immunity for the Collinses, for people who chose not to be employed to go down and push old ladies down and take their money and be immune because how could we ever be sure they are the ones who did it?'

In essence this argument of the prosecutor was calculated to persuade the jury to convict defendants whether or not they were convinced of their guilt to a moral certainty and beyond a reasonable doubt. (Pen.Code, s 1096.) Undoubtedly the jurors were unduly impressed by the mystique of the mathematical demonstration but were unable to assess its relevancy or value. Although we make no appraisal of the proper applications of mathematical techniques in the proof of facts (see People v. Jordan (1955) 45 Cal.2d 697, 707, 290 P.2d 484; People v. Trujillo (1948) 32 Cal.2d 105, 109, 194 P.2d 681; in a slightly differing context see Whitus v. State of Georgia (1967) 385 U.S. 545, 552, fn. 2, 87 S.Ct. 643, 17 L.Ed.2d 599; Finkelstein, The Application of Statistical Decision Theory to the Jury Discrimination Cases (1966) 80 Harv.L.Rev. 338, 338–340), we have strong feelings that such applications, particularly in a criminal case, must be critically examined in view of the substantial unfairness to a defendant which may result from ill conceived techniques with which the trier of fact is not technically equipped to cope. (See State v. Sneed, supra, 414 P.2d 858; Note, supra, Duke L.J. 665.) We feel that the technique employed in the case before us falls into the latter category.

We conclude that the court erred in admitting over defendant's objection the evidence pertaining to the mathematical theory of probability and in denying defendant's motion to strike such evidence. The case was apparently a close one. The jury began its deliberations at 2:46 p.m. on November 24, 1964, and retired for the night at 7:46 p.m.; the parties stipulated that a juror could be excused for illness and that a verdict could be reached by the remaining 11 jurors; the jury resumed deliberations the next morning at 8:40 a.m. and returned verdicts at 11:58 a.m. after five ballots had been taken. In the light of the closeness of the case, which as we have said was a circumstantial one, there is a reasonable likelihood that the result would have been more favorable to defendant if the prosecution had not urged the jury to render a probabilistic verdict. In any event, we think that under the circumstances the 'trial by mathematics' so distorted the role of the jury and so disadvantaged counsel for the defense, as to constitute in itself a miscarriage of justice. After an examination of the entire cause, including the evidence, we are of the opinion that it is reasonably probable that a result more favorable to defendant would have been reached in the absence of the above error. (People v. Watson (1956) 46 Cal.2d 818, 836; 299 P.2d 243.) The judgment against defendant must therefore be reversed.

In view of the foregoing conclusion, we deem it unnecessary to consider whether the admission of defendants' extrajudicial statements constitutes error under the rules announced in Escobedo and Dorado. Upon retrial, the admissibility of these or any other extrajudicial statements sought to be introduced by the prosecution must be determined in the light of the rules set forth in Miranda v. State of Arizona (1966) 384 U.S. 436, 86 S.Ct. 1602, 16 L.Ed.2d 694. (People v. Doherty (1967) 67 A.C. 1, 4, 9–13, 59 Cal.Rptr. 857, 429 P.2d 177.) As we have pointed out, the trial herein took place between our first and second Dorado decisions. Although defense counsel was commendably alert in basing objections to the admission of the statements upon the decisions in Escobedo and Dorado, he of course did not have the benefit of our numerous decisions beginning with the second Dorado decision expounding various facets of the exclusionary rule. In the event any extrajudicial statements made by defendant are offered in evidence on retrial, the parties will have an opportunity to make a record on pertinent issues subject to prior determination by the court in the light of Miranda rules before such statements are received in evidence. It would be fruitless for us to essay such a task at this point when such record does not yet exist.

The judgment is reversed.

APPENDIX (*People v. Collins*)

If 'Pr' represents the probability that a certain distinctive combination of characteristics, hereinafter designated 'C,' will occur jointly in a random couple, then the probability that C will Not occur in a random couple is (1-Pr). Applying the product rule, the probability that C will occur in None of N couples chosen at random is (1-Pr)N, so that the probability of C occurring in At least one of N random couples is (1-(1-Pr) N).

Given a particular couple selected from a random set of N, the probability of C occurring in that couple (i.e., Pr), multiplied by the probability of C occurring in none of the remaining N-1 couples (i.e., (1-Pr)N-1), yields the probability that C will occur in the selected couple and in no other. Thus the probability of C occurring in any particular couple, and in that couple alone, is ((Pr) (1-Pr)N-1). Since this is true for each of the N couples, the probability that C will occur in precisely One of the N couples, without regard to which one, is ((Pr) (1-Pr)N-1) added N times, because the probability of the occurrence of one of several Mutually exclusive events is equal to the Sum of the individual probabilities. Thus the probability of C occurring in Exactly one of N random couples (Any one, but Only one) is ((N) (Pr) (1-Pr)N-1).

By subtracting the probability that C will occur in Exactly one couple from the probability that C will occur in At least one couple, one obtains the probability that C will occur in More than one couple: (1 - (1-Pr)N-((N) (Pr) (1-Pr)N-1). Dividing this difference by the probability that C will occur in at least one couple (i.e., dividing the difference by (1-(1-Pr)N)) then yields The probability that C will occur more than once in a group of N couples in which C occurs at least once.

Turning to the case in which C represents the characteristics which distinguish a bearded Negro accompanied by a ponytailed blonde in a yellow car, the prosecution sought to establish that the probability of C occurring in a random couple was 1/12,000,000—i.e., that Pr = 1/12,000,000.

Treating this conclusion as accurate, it follows that, in a population of N random couples, the probability of C occurring Exactly once is ((N) (1/12,000,000) (1-1/12,000,000)N-1). Subtracting this product from (1-(1-1/12,000,000)N), the probability of C occurring in At least one couple, and dividing the resulting difference by (1-(1-1/12,000,000)N), the probability that C will occur in at least one couple, yields the probability that C will occur more than once in a group of N random couples of which at least one couple (namely, the one seen by the witnesses) possesses characteristics C. In other words, the probability of Another such couple in a population of N is the quotient A/B, where A designates the numerator (1 - (1 - 1/12,000,000)N)-((N) 1/12,000,000) (1-1/12,000,000)N-1), and B designates the denominator (1-(1-1/12,000,000)N).

N, which represents the total number of all couples who might conceivably have been at the scene of the San Pedro robbery, is not determinable, a fact which suggests yet another basic difficulty with the use of probability theory in establishing identity.

One of the imponderables in determining N may well be the number of N-type couples in which a single person may participate. Such considerations make it evident that N, in the area adjoining the robbery, is in excess of several million; as N assumes values of such magnitude, the quotient A/B computed as above, representing the probability of a second couple as distinctive as the one described by the prosecution's witnesses, soon exceeds 4/10.

Indeed, as N approaches 12 million, this probability quotient rises to approximately 41 percent. We note parenthetically that if 1/N = Pr, then as N increases indefinitely, the quotient in question approaches a limit of (e-2) / (e-1), where 'e' represents the transcendental number (approximately 2.71828) familiar in mathematics and physics.

Hence, even if we should accept the prosecution's figures without question, we would derive a probability of over 40 percent that the couple observed by the witnesses could be 'duplicated' by at least one other equally distinctive interracial couple in the area, including a Negro with a beard and mustache, driving a partly yellow car in the company of a blonde with a ponytail.

Thus the prosecution's computations, far from establishing beyond a reasonable doubt that the Collinses were the couple described by the prosecution's witnesses, imply a very substantial likelihood that the area contained more than one such couple, and that a couple other than the Collinses was the one observed at the scene of the robbery. (See generally: Hoel, Introduction to Mathematical Statistics (3d ed. 1962); Hodges & Leymann, Basic Concepts of Probability and Statistics (1964); Lindgren & McElrath, Introduction to Probability and Statistics (1959).)

Notes and Problems

1. Do you agree with the *Collins* decision? As a matter of hindsight, if you were the prosecuting attorney, how would alter your statements to the jury (pertaining to the Product Rule) so that the impact would have been as great but the reversal avoided?

2. What if it happened that all interracial couples drove yellow automobiles, or what if only blonds wore ponytails in the 1960s? The events in Collins would not be independent.

3. A local resident of Blyth, California (a town that is basically in the middle of nowhere) robbed the only liquor store in that city. The store clerk did not know the robber personally but told police that the thief had red hair, was a male with a large beard. In Blyth, half of the city is male, 1/9 of the population has red hair and 1/5 of the population has a beard. The police arrested Teddy the Termite, the first person they could grab who satisfied the three requirements. What is the probability that they pinched the right redhead?

4. More recently, the so-called "ceiling principle" (described in § 4.51), which purports to provide a more conservative match-up, has acted to modify and, in some instances, replace the product rule. Thus, for some jurisdictions, the ceiling principle method of calculating the odds of a random DNA match is generally preferred. Such jurisdictions hold that DNA evidence is admissible if supporting statistics are offered that were calculated using the ceiling principle. See *People v. Castro*, 144 Misc.2d 956, 545 N.Y.S.2d 985 (1989); *State v. Vandebogart*, 136 N.H. 365 (1992). But this "acceptance" of the ceiling principle is not without controversy. See *People v. Watson*, 965 N.E.2d 474 (Ill.App., 2012). See also J. R. Slimowitz and J. E. Cohen, *Violations of the Ceiling Principle: Exact Conditions and Statistical Evidence,* The Journal of Human Genetics, August 1993 (The authors claim to provide explicit analytical formulas to describe when the ceiling principle fails and why it is not always mathematically reliable.).

5. There is a fine line between providing the jury with probabilities of "guilt" and probabilities of events suggesting that guilt exists. The former may entail potential constitutional violations of due process and rights of confrontation, not to mention evidentiary problems. A prosecution witness came dangerously close to that fine line in *Rachals v. State*, 184 Ga.App. 420 (1987), involving a nurse who was found guilty (but mentally ill) of aggravated assault with intent to murder a patient. The case matter involved Terri Rachels, a nurse who experienced an unusually high number of cardiac arrests and patient deaths while she was on duty. The conviction was affirmed on appeal, but not without concern over the testimony of Dr. Franks, an epidemiologist from the U.S. Centers for Disease Control, on loan to the Georgia Department of Human Resources; he had apparently suggested during the trial that "Terri Rachals was, by odds of five out of nine, probably guilty."

B. Dependent Trials

Thus far, we have been considering allegedly independent events. E and F are said to be independent when $P(E \mid F) = P(E)$ on the other hand, if $P(E \mid F) \neq P(E)$, then events E and F are said to be **dependent**.

Most of the casino games are predicated upon these independent events, enabling the casinos to carefully calculate the probabilities and pay-offs, which are always in favor of the "house." So-called systems to beat the house-odds for roulette or craps (i.e., dice) simply do not work. In the absence of cheating or some other external influence that would render the events "not equally probable," each roll of the dice or spin of the wheel is independent of everything that transpired previously. Therefore, if someone were to repeatedly roll a string of 7s at the dice table, the probability of a 7 on the next roll would still be the same as if the that person never rolled a 7 previously in his or her life.

Unlike craps, roulette and a coin toss, several popular card games are *not* based upon independent trials. When cards are not shuffled, subsequent hands or events become dependent upon whatever has previously transpired. If, for example, we were to calculate the probability of pulling two Queens from a deck of cards without a shuffle, the probability of pulling the second would be dependent upon the status of the deck after the first Queen were pulled. We need new definitions and rules.

Definition of Conditional Probability: P(E | F) = n(E ∩ F) / n(F)

Bayes Theorem: P(E | F) = P(F | E) x P(E) / P(F)

The probability of event E (given the occurrence of event F) equals the number of events that are common to both E and F, divided by the number of potential events of F. It can be algebraically demonstrated that the above definition is equivalent to:

The Product Rule For Dependant Events: P(E ∩ F) = P(E | F) x P(F)

Example 4-11: What is the probability of pulling two sequential Queens from the deck without a shuffle (above example)?

Solution: The probability of drawing the 1st Queen and the 2nd Queen = P(E ∩ F).

Probability of drawing the 1st Queen = P(F) = 4/52.

Probability of drawing 2nd Queen, given the 1st Queen = P(E | F) = 3/51

(After the first Queen is played and removed from the deck, only 3 queens and 51 cards remain)

P(E ∩ F) = P(E | F) × P(F) = 4/52 x 3/51 = .0004

Example 4-12: There is an opaque bowl of 6 hidden jelly beans; 3 of them are red and 3 are black. What is the probability of selecting 2 red jelly beans sequentially (assuming that the first jelly bean to be selected is not returned to the bowl)?

Solution: The probability of selecting the 1st red jelly bean and the 2nd red jelly bean = P(E ∩ F).

Probability of selecting the 1st red jelly bean = P(F) = 3/6.

Probability of selecting the 2nd red jelly bean,

given the 1st jelly bean = P(E | F) = 2/5

P(E ∩ F) = P(E | F) x P(F) = 3/6 x 2/5 = .20

Notes and Problems

1. Several decades ago, a mathematician named Edward Thorp turned the gambling world upside down when he released what was to become a best selling book tiled: *Beat the Dealer.*[9] Unlike roulette or craps, Thorp discovered that Blackjack (a.k.a. "21") was *not* based upon independent trials. Because the casino dealers did not reshuffle the cards after each game, a player willing to keep track of the cards that were played (i.e., count the cards) could gain a slight advantage over the house, a feat that was previously unheard of. By counting the detrimental cards that were "out of play," the player could know when it was best to increase or decrease the size of his or her wager. When the deck became favorable (lots of 10's, picture cards and aces), the player would double or triple his or her bet; on the other hand, when the deck favored the house (lots of 5's and small cards), the player would decrease the bet. In summary, the game of Blackjack was discovered to be based on dependent as opposed to independent trials.[10] As is the case with Blackjack, in any series of events where $P(E_2)$ is influenced by

9. Edward O. Thorp, *Beat the Dealer: A Winning Strategy for the Game of Twenty-One* (Vintage, 1966).

10. Generally, aces, picture cards and 10's are favorable to the player, while the smaller cards benefit the house. Statistically, the single most significant card in the deck is the 5, which is generally bad news for the player because it is most likely to guarantee that the dealer will not bust. In a lawsuit

the outcome of E_1 we say that $P(E_2)$ is a **Dependent Event**, that depends upon the outcome of E_1.

2. There's a saying in poker that "only a sucker pulls to an inside straight." If you were dealt a Two, Three, Four, Six and King, what is the probability of throwing in the King and drawing a Five?

3. At a very famous poker game, Wild Bill Hickok was shot in the back while holding two pair, Aces and Eights, all black cards (ubiquitously known as the "Dead man's hand"). What is the probability of being dealt a dead man's hand?

Kammer v. Young

Court of Special Appeals of Maryland, 73 Md.App. 565 (1988)

BISHOP, Judge.

A Baltimore City jury found appellant, Thomas Robert Kammer, to be the father of the child of Christine J. Young, appellee. From the decree of the circuit court based on this verdict, Kammer appeals.

FACTS

Appellee gave birth to a child on May 24, 1982. She alleged that appellant was the only man with whom she had had sexual intercourse in the year immediately preceding the child's birth. Appellant responded that his sexual relations with appellee had ended more than 15 months before the birth.

The court admitted into evidence, over objection, the results of the blood tests of appellant, appellee and the child, obtained in accordance with MD.FAM.LAW CODE ANN. Section 5-1029. Additional facts will be given in the course of the discussion of the issues.

I.
Admissibility of Blood Test Evidence

(a) Admission of Opinion Evidence Concerning the Interpretation of the Blood Tests

In his attack on the admissibility of the interpretation of the blood tests, appellant claims that:

> (1) the expert witnesses appellee called at trial were not qualified to perform statistical analysis, to interpret the results or to express an opinion on the ultimate issue of the "statistical probability" of the appellant's paternity; and
>
> (2) the number presented to the jury was predicated upon a scientifically invalid formula and was not a competent "statistical probability of paternity" thereby rendering the laboratory report inadmissible at trial.

Appellant also asks that we reconsider our decision in Haines v. Shanholtz, 57 Md.App. 92, 468 A.2d 1365, cert. denied 300 Md. 90, 475 A.2d 1201 (1984), in which we held that by virtue of legislative enactment, blood test results which meet certain threshold requirements are admissible. We decline to do so. This controversy is governed by § 5-1029 of the MD.FAM.LAW CODE ANN. (1984) which provides in pertinent part:

---

against a Del Webb Casino, a big loser claimed that extra 5s were illegally stuffed into the Blackjack decks.

Blood Tests

(a) In general.—On the motion of a party to the proceeding or on its own motion, the court shall order the mother, child, and alleged father to submit to blood tests to determine whether the alleged father can be excluded as being the father of the child.

(b) Approved laboratory required.—The blood tests shall be made in a laboratory selected by the court from a list of laboratories provided by the Administration.

(c) Form of results.—The laboratory shall report the results of each blood test in writing and in the form the court requires.

(d) Copies of results. A copy of the results of each blood test shall be provided to the parties or their counsel in the manner that the court directs.

(e) Results as evidence.

(1) The results of each blood test shall be received in evidence if:

(I) definite exclusion is established; or

(ii) the testing is sufficiently extensive to exclude 97.3% of alleged fathers who are not biological fathers, and the statistical probability of the alleged father's paternity is at least 97.3%.

(2) A laboratory report is prima facie evidence of the results of a blood test.

(3) If a laboratory report is admitted in evidence, the laboratory technician who made the test is subject to cross-examination by any party to the proceeding. (The omitted subsections, (f) and (g), deal with failure to submit to the tests and payment of costs for the tests.)

We begin our analysis of the applicability of this section to the appellant's claims by setting out the following facts which are undisputed by the parties:

(1) Appellee's witnesses are employed by an approved blood testing laboratory and are qualified in the field of paternity testing, but are not trained statisticians. Appellant's witness is a trained statistician and although he is not employed by a blood testing laboratory he does consult, as part of his employment, with the paternity testing laboratory at Johns Hopkins Hospital.

(2) The blood tests were performed by an approved laboratory and the testing was done for blood group markers in seven different systems, six in the red cell laboratory and for HLA in the white cell laboratory (HLA is the abbreviation for human leucocyte antigen).

(3) The combination of the HLA and red cell tests is sufficiently extensive to exclude more than "97.3% of alleged fathers who are not biological fathers."

(4) The "paternity index" in this case is 460 to 1, i.e., it is 460 times more likely that appellant could produce the single sperm carrying the necessary genetic information than a random man in the population. (A paternity index is a ratio which expresses the odds that the accused is the father based on his chance of producing a sperm that carries all the necessary genetic information in conjunction with the mother's genetic information as compared to finding such a sperm in the random population.)

(5) The prior probability used by the laboratory in converting the paternity index to a percentage was .5, and that number is the standard used for such tests (Prior probability is expressed as a number, in the form of a percentage, which repre-

sents the accumulated non-genetic evidence that tends to indicate the accused man's paternity).

(6) Using a prior probability of .5, the percentage which results from application of Bayes' Theorem to the paternity index of 460 to 1 is 99.78%. (Bayes' Theorem is a mathematical technique accepted in the blood testing field for calculating conditional probabilities. See Ellman and Kaye, Probabilities and Proof: Can HLA and Blood Group Testing Prove Paternity? 54 N.Y.U.L.Rev. 1131, 1149 (1979). See infra p. 941 and Appendix, p. 946.)

At this point, however, the dispute is centered on §5-1029(e)(1)(ii), which requires that "the statistical probability of the alleged father's paternity [be] at least 97.3%." Appellant argues that this phrase should be construed strictly and that the method used to arrive at the percentage should be scrutinized according to the technical definition given by statisticians to the term "statistical probability". This argument forms the basis for his claims that: (1) appellee's witnesses, who lack training in statistics, are incompetent to interpret and to present the results of the blood test as a "statistical probability" of paternity; and (2) the methodology used to arrive at the statistical probability of paternity is based on a scientifically invalid formula and therefore it is not a true "statistical probability" of paternity.

In Haines v. Shanholtz, supra, we held that by virtue of the adoption of Chapter 784, Acts of 1982, the legislature expressly recognized the reliability of genetic testing. Chapter 784 amended Art. 16 Chancery §66G, the predecessor of §5-1029, and permitted the use of blood test results which are sufficiently extensive to exclude 97.3% of putative fathers who are not biological fathers where the statistical probability of paternity is at least 97.3%. The statute was amended again in 1984 to require the introduction of blood test evidence which meets the statutory standards instead of merely allowing it. We found the legislative intent "clear" and stated:

> [G]enetic testing is valid, subject to cross-examination of those responsible for performing and evaluating the tests. The trial judge has the discretion of admitting testimony that may have a bearing on the weight of the testimony offered as to the genetic testing, but not, by reason of the statute, as to the admissibility vel non of genetic testing to establish paternity. Id. at 98, 468 A.2d 1365 (emphasis in original).

The question that arises in the case sub judice is therefore: whether appellant's two-pronged attack goes to the admissibility of genetic testing or to the weight of the testimony offered, the former being impermissible as a result of the statute.

We conclude that the first prong of appellant's attack, regarding the qualifications of appellee's witnesses, goes to the weight of the testimony offered, but we find it unpersuasive. Our review of the legislative history of §5-1029(e)(1)(ii) does not support appellant's claim that the legislature intended that the results of the blood test be expressed in terms of a "true" "statistical probability."

Neither the Judicial Proceedings Committee of the Maryland State Senate nor the Judiciary Committee of the Maryland House of Delegates received testimony from a statistician, and the legislative history is devoid of any indication that there was testimony regarding the precise scientific formula required in order to render a "true" statistical probability. We hold that the legislature used the term "statistical probability" within the context and according to the meaning contained in the testimony of the experts from whom it heard: those in the blood testing community, including appellee's witness Margaret Brooks (Mrs. Brooks testified before the Senate committee only).

Mrs. Brooks testified at trial that the terms plausibility and probability are interchangeable in the blood testing community. We must assume the legislature was aware of this and we therefore reject the "scientifically literal" interpretation of § 5-1029(e)(1)(ii) urged by appellant. To the blood testing community, the terms plausibility and probability carry with them a distinction without a difference. (Even appellant's expert witness, Dr. Gary Chase, admitted that the paternity testing community does not use the same terms within the same context as the statistician's community, of which he is a member.) To express the paternity index ratio, which in this case is 460 to 1, as percentage requires the use of a statistical formula. The legislature called the resulting percentage a "STATISTICAL PROBABILITY." This choice of terms makes sense, because that is how it is expressed in the community that conducts the tests, and because the percentage is reached only after application of a statistical formula.

Based on the foregoing, we conclude that appellee's expert witnesses are qualified, within the relevant community, to conduct the calculations necessary to express the 460 to 1 paternity index as a percentage as is required by the statute. We note that appellant's witness, Dr. Chase, acknowledged at trial that the methods and terms used by the blood testing laboratory are universal in the field of paternity testing.

Appellant's second challenge, regarding the admissibility of the laboratory report, is permissible only to the extent that it goes to the weight of the testimony offered. As we pointed out in Haines, once appellee meets the requirements of § 5-1029(e)(1)(ii), the statute "mandates the admissibility of the test results." 57 Md.App. at 100–01, 468 A.2d 1365. Appellant attempts to circumvent this holding by attacking the genetic testing procedure used and by arguing that the intent of the legislature was hardly to sanction an inaccurate or unreliable scientific methodology which results in an incorrect determination of paternity.

We acknowledged in Haines that the Legislature stopped short of adopting "any specific tests to establish the percentages necessary to include or exclude putative fathers." We surmised that the reason for this was that the "General Assembly recognized that new technology may become available and, by not addressing any specific combination of tests, laboratories will be allowed to utilize the most effective tests without further legislative change". Id. The fact that the application of Bayes' Theorem, with a prior probability of .5, to the paternity index does not yield a "true" statistical probability of paternity does not invalidate the procedure or the result. The procedure used in this case is universally accepted by the relevant community of paternity blood testers and is the same procedure that was being used in 1984 when the Legislature amended the blood testing statute to require the introduction of blood test evidence. Indeed, the laboratory that performed these tests and computed these results is the same one that supplied the experts who testified in regard to the 1982 statutory amendment and who testified in Haines.

By contrast, the interpretation urged upon us by appellant would effectively nullify the statute. If we were to agree with his expert, Dr. Chase, no blood testing laboratory could produce results sufficient to satisfy the "statistical probability" requirement of the statute since they would not have access to the relevant non-genetic evidence needed to assign an "accurate" prior probability of paternity to Bayes' Theorem. This interpretation runs counter to the legislative goal which, as expressed in Haines, was "to protect illegitimate children through court ordered support based upon sophisticated and reliable genetic testing." (emphasis added).

(b) Admission of the Blood Test as a Violation of Due Process

Appellant contends that even if appellee properly complied with the blood test statute, the admission of any opinion evidence concerning the father's alleged "statistical probability" of paternity violates the due process provisions of the United States and Maryland Constitutions. He also claims that apart from the due process violation, the introduction of statistical and mathematical evidence on the ultimate issue before the jury effectively deprived him of a fair and impartial trial by jury as guaranteed by Article 23 of the Maryland Declaration of Rights.

Appellant's due process challenge is actually a two-pronged attack. He asks whether: (1) due process of law permits a paternity index to be converted to a statistical probability of paternity through the use of Bayes' Theorem with an assigned prior probability; and (2) the limited reference population used in this case can be used to compute the "paternity index" used in the equation. Appellant's blood markers were compared against a gene frequency chart which shows how often his specific gene pattern arose in the United States in tests of 12,000 persons of his same race.

Once appellant's paternity index has been calculated, it must then be converted to a statistical probability in order to determine whether it meets the 97.3% benchmark for admissibility set forth in § 5-1029(e)(1)(ii). The mathematical equation used in making this transformation is Bayes' Theorem, a basic formula of probability theory. It is expressed as follows:

$$P(F \mid M) = 1/\ ((1\text{-}f) + (f/P(F)))$$

$$\text{where } f = P(M \mid \text{not-}F)\ /\ P(M \mid F)$$

In this formula, P(F) represents the prior probability and it is factored into the equation in order to account for the non-genetic evidence of the accused man's paternity. The remaining variables are defined in the attached Appendix.

Ideally, the prior probability should reflect an assessment of all the relevant non-genetic evidence in each individual case and should take into consideration such factors as the accused man's opportunity for access to the child's mother during the period of conception, the evidence linking the child's mother to other men during that time period and the credibility of the witnesses. In the case sub judice, however, an assumed prior probability of fifty percent (.5) was used in converting appellant's paternity index to a statistical probability. He contends that this was a violation of his due process rights. We disagree.

Appellant claims that the methodology employed by all of appellee's experts assumes a fact not in evidence (the prior probability) and it is therefore scientifically inaccurate and invalid. We disagree. The evidence presented at trial makes plain that within the relevant community of blood testers, the paternity probability calculations in the present case were based upon scientific methods, accepted world-wide, which incorporate both Bayes' Theorem and the .5 prior probability. Because the use of the .5 as a basis for prior probability meets the requirement of general acceptance in the relevant community, it satisfies the test for admission of scientific evidence set out in Frye v. United States, 293 F. 1013 (D.C.Cir.1923) and adopted by the Court of Appeals in Reed v. State, 283 Md. 374, 391 A.2d 364 (1978).

All of the experts who testified at trial agreed that the paternity index, representing only the purely genetic evidence (the HLA and the red cell tests), was 460 to 1. Furthermore, using Bayes' Theorem and a prior probability of .5, all of the experts (including appellant's expert Dr. Chase) calculated the same probability percentage of 99.78%. The mathematical basis for the .5 figure was explained by Dr. Chase in the following manner:

> If you know nothing about a set of possibilities then the principle of insufficient reason says you are entitled to assign equal prior probability to those alternatives. [T]hat's what's used to get this .5. This .5 comes from a principle that says we don't know whether he was the father or somebody else was the father. So we are

> going to split the probability fifty-fifty between the two possibilities. We are going to load him with 50 percent of it and give the other percent to all the other men.

When calculating a statistical probability of paternity, the blood testing laboratory does not have before it any of the individualized non-genetic evidence, and it therefore uses a standard figure (.5) whenever it makes its calculations. This fact neither renders the methodology "scientifically invalid" nor does it deny appellant an opportunity to present the non-genetic evidence. If the laboratory's calculations meet the prerequisites for admissibility set forth in § 5-1029(e)(1)(ii) the blood test evidence is admitted, but it is neither conclusive nor does it create a presumption, it is merely one piece of evidence that is put before the jury.

Appellant was free to, and in fact did, put on non-genetic evidence which not only disputed generally his paternity but, in effect, was an attack upon the use of the .5 prior probability figure. This allowed him an opportunity to counterbalance appellee's introduction of the blood test results and the prior probability on which they were based and served to protect his due process rights.

We decline to respond to appellant's second contention, regarding the sufficiency of the reference population used in determining the paternity index, on the basis of Maryland Rule 1031 c 5 and this Court's holding in Van Meter v. State, 30 Md.App. 406, 352 A.2d 850 (1976). Nowhere in the record or in the briefs does appellant offer any argument in support of his claim that the reference population used was inadequate. In his brief, appellant merely states:

> The record also raises a serious question about whether the "reference population" used by the Baltimore RH Laboratory, consisting in total of a mere 12,000 persons, is legally sufficient upon which to predicate the appellant's paternity index" of 460:1. Appellant respectfully submits that there is no legally sufficient evidence to support the assumption made by the Baltimore RH Laboratory that this data base is adequate to use with Bayes' Theorem to extrapolate the statistical "plausibilities" made in this case, or that it is a proper "reference population" at all.

Yet, he offers no proof to support his conclusion. From this bald assertion, appellant leaps to the conclusion that his due process rights have been violated because of the size of the reference population used. Appellant, by failing to include in his brief any argument to support his claim that the reference population was insufficient, provides us with nothing to consider as a basis for his contention.

Appellant does provide some argument on his right to a fair and impartial jury trial claim; however, we are not persuaded. He argues that the "statistical approach to the case distracted the jurors from weighing the evidence and determining the ultimate credibility of the plaintiff and defendant ..." and that the jury "unquestionably was overwhelmed" by the testimony that the "plausibility" of the appellant's paternity was "99.78%", and therefore he was denied a right to a fair and impartial trial. The "statistical approach" to this case is no different from any other paternity case, and it is precisely what is required by the statute. The fact that the percentage obtained, 99.78%, may be perceived as "overwhelming" because of the extent to which it reaches 100% conclusiveness does not in any way prevent appellant from receiving a fair trial. The "statistical probability of paternity" required by the statute comprises only one part of the evidence that is put before the jury concerning the alleged father's probable paternity. It is not the only evidence and it does not create an irrebuttable presumption of paternity. Appellant was allowed to put on evidence in support of his claim that he is not the child's father. All of the evidence went to the jury and there is no evidentiary support for his claim that the jurors improperly relied on the statistical probability as being dispositive of the issue ...

JUDGMENT AFFIRMED; COSTS TO BE PAID BY APPELLANT.

APPENDIX (*Kammer v. Young*)

P(F) = The prior probability

P(F/M) = The statistical probability of paternity

f = The quotient yielded, when the random man's chance of fathering the child is divided by the accused's chance of fathering the child

P(M/not-F) = The random man's chance of fathering the child based upon the value calculated for each of the seven blood systems tested using a hypothetical blood type and the mother's blood type

P(M/F) = The accused's chance of fathering the child based upon the value calculated for each of the seven blood systems tested using his blood type and the mother's blood type

Notes and Problems

1. Assume that you are a male walking down the street on a crisp Fall day. A car pulls along side of you and the female driver asks you to get in. You tell her your name, address and enough other information to enable her to obtain detailed personal intelligence about you. The two of you have never previously met, yet she has decided that you are the father of her three month-old-child. When the accusation of fatherhood is directed your way in a paternity action, she says "Yes," you say "No." To apply Bayes' Theorem, what number should the court use as a prior probability for the purpose of admitting blood type results?

2. With respect to *Kammer*, explain how a paternity index of 460 to 1 complies with Maryland's *Code of Civil Procedure*, which required at least a probability of 97.3% to establish a foundation for the admissibility of the defendant's blood type evidence.

3. In *Kammer*, the Court observed: "The fact that the application of Bayes' Theorem, with a prior probability of .5, to the paternity index does not yield a 'true' statistical probability of paternity does not invalidate the procedure or the result. The procedure used in this case is universally accepted by the relevant community of paternity blood testers ..." Do you think that the use of a .5 prior probability is fair or merely expedient? Do you think the stated statutory goal ("To protect illegitimate children through court ordered support based upon sophisticated and reliable genetic testing") played a role in the selection of a .5 prior probability?

§ 4.30 Probability Expectation

We encounter the challenges of probability expectation every day of our lives. Whenever the subject arises, the most common thought turns to the daily lottery. But that is a far too narrow view of a function that we confront second by second. Every time we leave the safety of our home, take a flight, or run a yellow light, we are weighing probabilities, expectations and, of course, consequences. Miss your turn-off on the Pacific Coast Highway in Malibu and you have a choice between a very high risk, deadly u-turn or driving half way to San Francisco before you can safely double back.

The sciences of fingerprinting, ballistics and serology are recognized as being worthy of recognition by the realm of criminal jurisprudence. Because of their high probability of success, they have a high expectation of accuracy. In other words, we expect that a fingerprint match-up yields an extremely high probability that the latent print (taken from the scene of the crime) belongs to the suspect that produced a matching sample. Nothing is conclusive; the match-up alone doesn't establish guilt. A high probability of accuracy, however, permits the evidence to be admitted. If "high probability" alone were enough to convict, there would be no need for ever conducting any trial for the murder of a married victim. They might as well just arrest and hang the spouse; in all "*probability*," justice would thereby be served, and very efficiently.

Practitioners confront concrete probability expectations on a regular basis. More than 95% of civil lawsuits are settled because the plaintiffs have determined that the probability of a monetary expectation from a jury, less the costs to be incurred, does not exceed a present offer of settlement.[11] Rising court costs combined with the uncertainties of jury verdict's tend to increase the desire for settlement.

§ 4.31 The Zero Sum Game

Most of the discussions thus far have involved what are commonly referred to as "Zero Sum" games. A zero sum game is simply a game wherein the amount won by the winner (or winners) must equal the amount lost by the loser or losers. Here is a more impressive definition (The x_L terms will, of course, constitute negative values):

$$\Sigma\,(x_W + x_L) = 0$$

The rule is basically one of common sense. If someone wins money, someone else has to lose it. This concept is fundamental to all games of chance and forms the underlying predicate for the prosecution of insider traders by the Securities and Exchange commission.

Not all "games" are zero sum. In the world of collective bargaining, the threat of a strike is the primary game weapon of labor. But if it is deployed and a strike is called, everyone (both union and management) loses; clearly, this is a non-zero sum game.

Litigation combines the concepts of a zero-sum and non-zero sum game. If a case is to be settled quickly, zero-sum principles apply. If, however, litigation is to proceed to a jury trial, and perhaps even appeals, the costs of litigation (money that is lost to unrelated third parties) render the exercise—non-zero. In fact, whenever a verdict is rendered for a defendant in a civil action, everyone loses.

The fact that all of our stock market purchases and investments are predicated upon the *zero sum game* concept is particularly significant with respect to insider trading laws. No one can derive a gain from the market unless someone else suffers a loss.

11. As the costs of litigation (such as expert witnesses fees, deposition costs and court costs) increase, and the probability expectations become less certain, settlements become more attractive. Economists frequently get into the act. The notorious Kaldor-Hicks Criterion defines a potential economic move as "efficient" if, after the move, the winners gain more than enough to compensate the losers.

§ 4.32 *Calculating Expectations*

In order to reduce our probabilistic expectations to something that is mathematically measurable, we are going to presume that our results are reducible to dollars and cents. Therefore, if P(M) represents the probability of receiving a specific amount of money, and M represents the amount of money, then the mathematical expectation of receiving the money is provided by the following:

Definition of Mathematical Expectation: $E(M) = P(M) \times M$

For circumstances involving multiple events (i.e., payoffs), a more complex equation is required. But rest assured that it is nothing more than a generalization of the simple definition stated above:

$$E(M) = P_1(M)\,M_1 + P_2(M)\,M_2 + P_3(M)\,M_3 + \ldots + P_n(M)\,M_n = \Sigma_n\,P_n(M)\,M_n$$

Example 4-13: You place a \$10 bet on "red" at the roulette wheel. Recall that there 18 slots that are red and 18 slots that are black in addition to 0 and double-0 (which are counted as losers). After the ball has begun to spin, what is the Mathematical Expectation of your \$10 bet. (It would also be correct to refer to the "value" of your bet.)

Solution: The Mathematical Expectation $E(M) = P(M) \times M$. $P(M) = 18/38$.

M = \$20 (You will receive \$20.00 if you win)

Therefore $E(M) = 18/38 \times \$20 = .4737 \times \$20 = \$9.47$

(You would be correct to conclude that a \$10 wager whose value is only \$9.47 is not a good bet.)

Example 4-14: Kenny the Worm is planning to knock over a bank (i.e., make a withdraw of funds that were never deposited). If he robs BA (Bank of America), there is a 20% chance that he will be pinched (caught); if, on the other hand, he hits Well Fargo, there is a 40% chance that he will be caught. The average amount of money in the teller's cash drawer at BA is \$10,000; At Wells Fargo (WF), the average amount is \$25,000. Where should Kenny pull the heist?

Solution: Calculate the Mathematical Expectation as follows:

$E(M) = P(M) \times M$.

For Bank of America, P(M) = .80 and M = \$10,000

And therefore, E(M for BA) = \$8,000.

For Wells Fargo, E(M) = .60 and M = \$25,000

And therefore, E(M for WF) = \$15,000

Therefore, the Mathematical Expectation is higher for Wells Fargo (The problem, however, does not consider the penal expectation in the event of capture).

Example 4-15: If you purchase a Super Macho Power Ball Lottery Ticket, you can win a first prize of \$50,000 or a second prize of \$20,000. The probabilities of winning the first and second prizes are .001 and .003 respectively. In order to make a favorable investment decision, what is the most that you should pay for the ticket?

Solution: The Mathematical Expectation:

$E(M) = P_1(M)\,M_1 + P_2(M)\,M_2$

$P_1(M)\,M_1 = (.001)\,(\$50{,}000) = \$50$

$P_2(M)\,M_2 = (.003)\,(\$20{,}000) = \$60$

$E(M) = P_1(M)\,M_1 + P_2(M)\,M_2 = \$110$

Fennell v. Southern Maryland Hosp. Center, Inc.

320 Md. 776, 580 A.2d 206 (1990)

CHASANOW, Judge.

Appellants, Robert N. Fennell, Katrina Lynn Fennell, and Robin Tracy Fennell (the Fennells) are the spouse, children, personal representative, and heirs at law of decedent Cora L. Fennell (Mrs. Fennell). The Fennells filed a medical malpractice claim against Appellee, Southern Maryland Hospital Center, Inc. (the Hospital). Both the Health Claims Arbitration Board and the Circuit Court for Prince George's County granted summary judgment in favor of the Hospital. The Fennells appealed, and this Court granted certiorari prior to consideration by the Court of Special Appeals. Pursuant to Maryland Rule 8-413, the parties agreed to and filed a statement of the case in lieu of the entire record. The relevant facts agreed to by the parties are as follows:

"Early on the morning of July 14, 1981, decedent, Cora L. Fennell, woke up with a severe headache. Appellant, Robert N. Fennell, took his wife to the Emergency Room of [the Hospital]. The Emergency Room records indicate that Cora L. Fennell arrived there at 2:27 a.m. on the morning of July 14, 1981. [Mrs.] Fennell was then seen, examined and treated in the Emergency Room by Roy Kring, M.D. After an initial evaluation, Dr. Kring believed that Mrs. Fennell was suffering from a neurosurgical emergency. He contacted by telephone Ronald Uscinski, M.D., a neurosurgeon, for consultation. Dr. Kring conceded that he was aware at the time he evaluated [Mrs.] Fennell in the Emergency Room that her condition was critical. Drs. Kring and Uscinski were primarily concerned about the possibility of an intracranial bleed, and it was agreed by these doctors that the patient be transferred to Radiology for a CT Scan. The Hospital chart does not indicate the time that the CT Scan was performed, but it was concluded by Dr. Bach, Plaintiff's expert, that the scan was performed at approximately 3:30 a.m. on the morning of July 14, 1981. Pursuant to the Hospital records, Cora L. Fennell was admitted to the Hospital at 4:00 a.m. on July 14, 1981, and the records note Dr. Uscinski as the admitting physician. The Hospital records further indicate that [Mrs.] Fennell was admitted to the Intensive Care Unit at 5:00 a.m. on the same date. [Mrs.] Fennell arrested at 7:40 a.m. on July 14, 1981, and it was determined by Ronald B. Landman, M.D., who saw her at that time, that she was brain dead. However, she was maintained on life support systems until the early morning hours of July 15, 1981 when she arrested for a second time and was pronounced dead. Although Dr. Uscinski was noted as the admitting doctor on the Hospital chart, [Mrs.] Fennell was not seen by him at any time before she arrested at 7:40 a.m. on the morning of July 14, 1981. [Mrs.] Fennell was not seen by any treating physician at any time between the time she was admitted to the Hospital at 4:00 a.m., and the time she was first seen by internist, Dr. Landman, at approximately 7:30 a.m. on the morning of July 14, 1981, just minutes before she suffered a fatal arrest. The above referenced CT Scan which, according to Dr. Bach, was taken at approximately 3:30 a.m. on the morning of July 14, 1981, ruled out an intracranial bleed and suggested inflammatory process. It was ultimately determined on autopsy that Cora L. Fennell was suffering from meningitis."

In their opposition to the Hospital's motion for summary judgment, the Fennells filed the affidavit of Dr. Michael C. Bach, an infectious disease expert. For the purpose of the motion for summary judgment, the Hospital did not dispute the affidavit. The affidavit established that decedent was suffering from bacterial meningitis.

A complication of bacterial meningitis is swelling of the brain, a life threatening condition requiring aggressive treatment. Dr. Bach stated that the CT Scan taken at approximately 3:30 a.m. revealed brain swelling and that proper medical treatment required that a lumbar puncture be completed within one-half hour thereafter. He further stated that had decedent been diagnosed and treated in accordance with the appropriate standard of care, she would have had a 40% chance of survival. He concluded that the failure to follow-up the CT Scan with a lumbar puncture and the failure to immediately and aggressively reduce the swelling of the brain that was detected by the CT Scan were violations of the standard of care, and that as a result, the progress of the disease was irreversible. In effect, Dr. Bach established that decedent had a 40% chance of surviving the meningitis, but that the chance was lost as a result of the Hospital's negligence.

We held that, in a wrongful death case, a plaintiff must prove to a reasonable medical probability that death was proximately caused by the defendant's negligence.

The Hospital filed a motion for summary judgment citing testimony of plaintiffs' expert that decedent, even if treated in accordance with the appropriate standard of care, still had only a 40% chance of survival. The motion for summary judgment was granted by the Health Claims Arbitration panel chairman, and in the Circuit Court for Prince George's County, summary judgment in favor of the defendant was again entered on both the wrongful death and survival counts. The Fennells, recognizing that Weimer is dispositive of the wrongful death claim, have raised on appeal only the summary judgment in the survival claim, contending that "the Maryland courts have left open the issue of whether loss of chance is compensable in a survival action where the degree of proof that death was caused by negligence does not meet the 'more likely than not' standard."

Loss of chance may include loss of chance of a positive or more desirable medical outcome, loss of chance of avoiding some physical injury or disease, or a loss of chance to survive. Because the instant case involves a loss of chance to survive, when we refer to "loss of chance," we mean decreasing the chance of survival as a result of negligent treatment where the likelihood of recovery from the pre-existing disease or injury, prior to any alleged negligent treatment, was improbable, i.e., 50% or less.

Negligent treatment resulting in a loss of chance of survival may or may not eliminate all chance of survival or recovery. If the chance of recovery is 40%, as in the instant case, the risk of non-recovery must be 60%; and the loss of the 40% chance of recovery increased the risk of non-recovery to 100%. Thus, the loss of a 40% chance of recovery in this case eliminated all chance of recovery. It is also conceivable that negligent treatment may result in loss of a chance of survival without eliminating all chance of survival. For example, if the patient had a 40% chance of recovery and negligent treatment reduced the patient's chance of survival to 10%, then the actual loss of chance of survival would be 30%. By loss of chance, we mean the net loss of chance of survival directly attributable to the negligence.

Loss of chance medical malpractice actions have been recognized in several other jurisdictions, but the cases do not always clearly state the basis for recognizing the cause of action. A number of jurisdictions have adopted the loss of chance doctrine, while others have not clearly resolved whether to recognize the doctrine. Still other jurisdictions have refused to recognize the loss of chance doctrine.

Although we have not yet had the issue squarely before us, speculation has arisen whether Maryland will depart from common law principles and recognize a claim for loss of chance in survival actions. Cooper v. Hartman, 311 Md. 259, 261 (1987) (even if loss of chance doctrine was to be recognized, facts of case did not warrant its application); Weimer v. Hetrick, 309 Md. 536 (1987) (loss of chance damages not recoverable in wrongful death action), reversing 67 Md.App. 522, 508 A.2d 522 (1986) (recognizing doctrine); Chudson v. Ratra, 76 Md.App. 753 (1988), cert. denied, 314 Md. 628, 552 A.2d 894 (1989) (doctrine considered but found inapplicable); Waffen v. U.S. Dept. of Health & Human Services, 799 F.2d 911, 917 (4th Cir.1986) (pre- Weimer dictum that, under Maryland law, "the loss of a substantial chance of survival is a cognizable harm"); Kroll v. U.S., 694 F.Supp. 1210, 1213 (D.Md.1988) (post- Weimer prediction that loss of chance will be recognized as element of damages in medical malpractice cases). The doctrine has also received attention in the law reviews. See Feldman, Chances as Protected Interests: Recovery For the Loss of a Chance and Increased Risk, 17 U.Balt.L.Rev. 139, 158 (1987) (urging recognition of chances as protected interests in Maryland).

In Weimer v. Hetrick, supra, this Court held that loss of chance damages were not recoverable in an action brought pursuant to the Wrongful Death Act. In order to recover under that statute, the plaintiff must prove, by a preponderance of the evidence, that the death was caused by the wrongful act of the defendant. Since the issue was not before us, we declined to consider whether loss of chance damages might be recoverable in a survival action. In Cooper v. Hartman, supra, we again declined to address the issue of whether loss of chance damages were recoverable in a survival action, but Judge Adkins writing for the Court identified the divergent approaches to the doctrine, stating: "The various courts that have adopted these ['loss of chance'] rules ... have taken different approaches. Some have simply relaxed the standards regarding causation and allowed full compensation for an injury or death where the plaintiff demonstrated less than a 50% chance of recovery. Others have left the traditional rules of causation intact but have viewed the loss of a chance as a way of approaching damages." Cooper, 311 Md. at 265, 533 A.2d at 1297.

We shall examine both the relaxed causation/new cause of action approach, as well as the new form of damages approach. However, the relaxed causation/new cause of action approach, allowing full recovery for the death, was not advocated by the Fennells.

RELAXED CAUSATION/NEW CAUSE OF ACTION APPROACH

In the instant case, the circuit court refused to allow loss of chance recovery either as a new cause of action by relaxing traditional rules of causation or, in the alternative, by establishing a new form of damages for loss of chance. On appeal, Appellants do not challenge the refusal to create a new cause of action by relaxing traditional rules of causation. Instead, they argue for loss of chance as a new form of damages. Indeed, they expressly reject the idea of changing traditional principles of causation stating:

> "It does not quite seem to make sense to impose upon a Defendant doctor the full responsibility for a patient's death when the patient had a less than even chance of surviving with proper treatment.... A rule allowing recovery for loss of chance ... leaves intact Maryland's traditional principles relating to causation, e.g., no recovery would be allowed for the death itself in a case where Plaintiff is unable to prove, by a preponderance of the evidence, that death was proximately caused by the negligence of the Defendant...."

For that reason, the relaxed causation/new cause of action approach is not before this Court. It deserves, however, at least a brief discussion.

Courts adopting the relaxed causation/new cause of action approach continue to award "all or nothing" damages, and when the plaintiff establishes a "substantial possibility" that the doctor's negligence caused the death, there is full recovery. The result is that relaxing the rules of causation merely improves the plaintiff's odds of receiving all rather than nothing. The relaxed causation approach in medical malpractice cases has often been attributed to dictum in Hicks v. United States, 368 F.2d 626 (4th Cir.1966), where Judge Sobeloff writing for the Court stated:

> When a defendant's negligent action or inaction has effectively terminated a person's chance of survival, it does not lie in the defendant's mouth to raise conjectures as to the measure of the chances that he has put beyond the possibility of realization. If there was any substantial possibility of survival and the defendant has destroyed it, he is answerable. Rarely is it possible to demonstrate to an absolute certainty what would have happened in circumstances that the wrongdoer did not allow to come to pass. The law does not in the existing circumstances require the plaintiff to show to a certainty that the patient would have lived had she been hospitalized and operated on promptly." Hicks, 368 F.2d at 632.

Hicks was extensively discussed by this Court in Weimer v. Hetrick, supra, where we pointed out that Judge Sobeloff's language was at best dictum since there was testimony by two experts that the defendant's negligence was the proximate cause of the decedent's death.

In addition to the Hicks dictum, Section 323 of the Restatement (Second) of Torts, which would impose liability where the defendant's negligence increases the risk of harm to another, has sometimes been extended by courts as a justification for relaxing the traditional preponderance standard for proving proximate cause in loss of chance of survival cases. Restatement (Second) of Torts § 323 (1965) provides:

> One who undertakes, gratuitously or for consideration, to render services to another which he should recognize as necessary for the protection of the other's person or things, is subject to liability to the other for physical harm resulting from his failure to exercise reasonable care to perform his undertaking, if (a) his failure to exercise such care increases the risk of such harm, or (b) the harm is suffered because of the other's reliance upon the undertaking.

We are unwilling to relax traditional rules of causation and create a new tort allowing full recovery for causing death by causing a loss of less than 50% chance of survival. In order to demonstrate proximate cause, the burden is on the plaintiff to prove by a preponderance of the evidence that "it is more probable than not that defendant's act caused his injury." Peterson v. Underwood, 258 Md. 9, 264 A.2d 851 (1970).

We might also note that in 1986 the General Assembly enacted Md.Code (1974, 1989 Repl.Vol.), Courts and Judicial Proceedings Article, § 3-2A-04(b)(1)(I), which requires dismissal of a malpractice action "if the claimant fails to file a certificate of a qualified expert with the Director attesting to the departure from standards of care, and that the departure from standards of care is the proximate cause of the alleged injury ..." (emphasis added). This statute may have manifested a legislative intent that there be a preliminary showing of traditional causation as a prerequisite to filing a medical malpractice claim.

THE DAMAGE APPROACH TO LOSS OF CHANCE

Appellants urge this Court to adopt the loss of chance/damages approach first suggested by Professor King in his law review article King, Causation, Valuation and Chance in Personal Injury Torts Involving Pre-existing Conditions and Future Consequences, 90 Yale L.J. 1353 (1981).

In Cooper v. Hartman, supra, this Court described the damage approach as leaving "the traditional rules of causation intact," but treating "the loss of a chance as a way of approaching damages." Cooper, 311 Md. at 265, 533 A.2d at 1297. The Court quoted a passage from Professor King's law review article which, by coincidence, describes the same "chance of survival" percentage shown in the instant case:

> "To illustrate, consider a patient who suffers a heart attack and dies as a result. Assume that the defendant-physician negligently misdiagnosed the patient's condition, but that the patient would have had only a 40% chance of survival even with a timely diagnosis and proper care. Regardless of whether it could be said that the defendant caused the decedent's death, he caused the loss of a chance, and that chance-interest should be completely redressed in its own right. Under the proposed rule, the plaintiff's compensation for the loss of the victim's chance of surviving the heart attack would be 40% of the compensable value of the victim's life had he survived (including what his earning capacity would otherwise have been in the years following death). The value placed on the patient's life would reflect such factors as his age, health, and earning potential, including the fact that he had suffered the heart attack and the assumption that he had survived it. The 40% computation would be applied to that base figure." King at 1382.

Judge McAuliffe, in his concurring opinion in Weimer, supra, outlined the damage approach:

> "[A] claim under that theory does not involve the creation of a new tort, but rather involves a redefinition of damages involved in the claim. Traditional principles of law relating to duty, breach, causation, and burden of proof remain the same-what changes is the acceptance of the concept that damages may be recovered for the loss of a chance of survival where that chance is substantial and can be identified and quantified (and thus valued) without resort to conjecture or speculation." Weimer, 309 Md. at 556, 525 A.2d at 653 (McAuliffe, J., concurring).

Weimer and Cooper discussed, but neither endorsed nor rejected, Professor King's suggestion for loss of chance damages in survival actions. Several other states have adopted Professor King's damage approach.

A good argument can be made that damages ought to be recoverable when, due to a doctor's negligence, a patient loses a substantial, though less than probable, chance of survival. There are also some arguments against expanding damages in medical malpractice cases to include damages for loss of chance of survival.

Because loss of chance recovery is based on statistical probabilities, it might be appropriate to examine the statistical probabilities of achieving a "just" result with loss of chance damages. In Brennwald, Proving Causation in "Loss of a Chance" Cases: A Proportional Approach, 34 Cath.U.L.Rev. 747, 779 n. 254 (1985), the author, citing Orloff and Stedinger's article A Framework for Evaluating the Preponderance-of-the-Evidence Standard, 131 U.Pa.L.Rev. 1159 (1983), attempted to analyze statistically the errors produced using traditional tort recovery as compared to the errors produced by loss of chance recovery.

To compare the two rules, assume a hypothetical group of 99 cancer patients, each of whom would have had a 33 1/3% chance of survival. Each received negligent medical care, and all 99 died. Traditional tort law would deny recovery in all 99 cases because each patient had less than a 50% chance of recovery and the probable cause of death was the pre-existing cancer not the negligence. Statistically, had all 99 received proper treatment, 33 would have lived and 66 would have died; so the traditional rule would have statistically produced 33 errors by denying recovery to all 99.

The loss of chance rule would allow all 99 patients to recover, but each would recover 33 1/3% of the normal value of the case. Again, with proper care 33 patients would have survived. Thus, the 33 patients who statistically would have survived with proper care would receive only one-third of the appropriate recovery, while the 66 patients who died as a result of the pre-existing condition, not the negligence, would be overcompensated by one-third. The loss of chance rule would have produced errors in all 99 cases.

Re-defining loss of chance of survival as a new form of damages so that the compensable injury is not the death, but is the loss of chance of survival itself, may really be an exercise in semantics. Loss of chance of survival in itself is not compensable unless and until death ensues. Thus, it would seem that the true injury is the death.

While we should not award damages if there is no injury, the logical extension of the loss of chance damages theory arguably should allow loss of chance damages for negligence, even when the patient miraculously recovers. For example, if a doctor negligently treats a person with a 40% chance of recovery and the doctor's negligence reduces the patient's chance of recovery to only 10%, whether the patient lives or dies, the doctor's negligence cost the patient a 30% loss of chance of survival. If the patient dies, the probable cause of death was the pre-existing disease or injury; it is unlikely that the negligence caused the death. If the patient lives, the negligence clearly did not cause the death. In both scenarios, there was negligence resulting in a 30% loss of chance of survival. If courts are going to allow damages solely for the loss of chance of survival, logically there ought to be recovery for loss of chance regardless of whether the patient succumbs to the unrelated pre-existing medical problem or miraculously recovers despite the negligence and unfavorable odds.

Since loss of chance damages are only permitted when the patient dies, it is also arguable that, when we strip away the rhetoric, damages are really being awarded for the possibility that the negligence was a cause of the death. Maryland law clearly does not allow damages based on mere possibilities. Pennwalt Corp. v. Nasios, 314 Md. 433, 444, 550 A.2d 1155, 1161 (1988).

Another factor weighing against adoption of a loss of chance damages approach is its practical application in civil jury trials. Probabilities and statistical evidence comprise a substantial portion of the evidence submitted to the trier of fact in loss of chance actions. This evidence will generally be in the form of opinions based on statistics that show chance of survival of other individuals similarly situated to the victim. The use of statistics in trials is subject to criticism as being unreliable, misleading, easily manipulated, and confusing to a jury. When large damage awards will be based on the statistical chance of survival before the negligent treatment, minus the statistical chance of survival after the negligent treatment, times the value of the lost life, we can imagine the bewildering sets of numbers with which the jury will be confronted, as well as the difficulties juries will have in assessing the comparative reliability of the divergent statistical evidence offered by each side.

Traditional tort law is based on probabilities. If a patient had a 49% chance of dying from an injury or disease and if the patient was negligently treated and dies, full recovery will be permitted because, absent the negligence, it was more likely than not that the patient would have survived. Based on the 51% probability of surviving the injury or disease, we exclude the injury or disease as the cause of death. Damages are not reduced by the fact that there was a strong possibility that the patient would have died absent the negligence. Conversely, if the patient had a 51% chance of dying from an injury or disease, and was negligently treated and died, it was probably the pre-existing medical condition, not the negligence, that killed the patient, and there is no recovery. Damages must be proven by a preponderance of the evidence. Damages are not proven when it is more likely than not that death was caused by the antecedent disease or injury rather than the negligence of the physician.

When urged to adopt new theories of causation and/or new elements of tort damages, this Court should be "concerned with the substantial expansion of tortfeasor liability and the accompanying societal costs that will be imposed by this new cause of action." Gaver v. Harrant, 316 Md. 17, 31, 557 A.2d 210, 217 (1989).

If loss of chance damages are permitted they would under present case law have to be awarded under a survival action, not a wrongful death action. Weimer, supra. Survival action damages currently include conscious pain and suffering as well as medical expenses, but exclude future loss of earnings, solatium damages, and damages which result to other persons from the death. See Rhone v. Fisher, 224 Md. 223, 167 A.2d 773 (1961); Stewart v. United Electric Light & Power Co., 104 Md. 332, 65 A. 49 (1906). The determination of whether to recognize loss of chance damages should also include a determination of whether ordinary survival action damages should be expanded in loss of chance cases. These issues are more properly resolved by the legislature.

If loss of chance damages are to be recognized, amendments to the wrongful death statute should also be considered. As a class, medical malpractice plaintiffs benefit from the fact that they are entitled to recover 100% of their damages from a defendant whose negligence caused only 51% of their loss because it is more probable than not that the defendant's negligence caused the loss. Reciprocally, a defendant whose negligence caused less than 50% of a plaintiff's loss pays nothing because it is probable that the negligence did not cause the loss. If a plaintiff whose decedent had a 49% chance of survival, which was lost through negligent treatment, is permitted to recover 49% of the value of the decedent's life, then a plaintiff whose decedent had a 51% chance of survival, which was lost through negligent treatment, perhaps ought to have recovery limited to 51% of the value of the life lost. The latter result would require a change in our current wrongful death statute.

Loss of chance damages did not exist at common law. In Gaver v. Harrant, supra, Chief Judge Murphy noted that this Court has modified the common law when necessary due to changing circumstances or increased knowledge. But, "[a]doption of a new cause of action involves serious public policy concerns." Id. at 29, 557 A.2d at 216. "Declaration of the public policy of Maryland is normally the function of the General Assembly." Harrison v. Montgomery County Board of Education, 295 Md. 442, 460, 456 A.2d 894, 903 (1983). Recognizing loss of chance damages in a survival action would involve serious public policy concerns. We are not convinced that such a change should be initiated by this Court.

Appellants have not identified any "changing circumstances or increased knowledge" that would necessitate changing traditional common law survival action damage principles. Although there has been a significant increase in medical malpractice litigation in recent years, there is no evidence to suggest that this is due to a significant increase in malpractice by physicians. There is also no reason to believe that health care providers are systematically failing to treat or are improperly treating the critically ill. The legislature is better able to ascertain the reasons for the increase in malpractice claims, and what, if any, remedial action is necessary.

Recognition of loss of chance damages would allow a new form of damages as well as allow medical malpractice claims by an entirely new class of plaintiffs who traditionally have had no cause of action at common law. Patients whose chances of surviving their pre-existing injuries or diseases were 50% or less had no cause of action for negligent treatment under traditional tort principles. Although their chances of survival were decreased, survival was unlikely; and therefore, actual demonstrable harm, in all probability, did not occur. Recognition of this new form of medical malpractice damages for loss of a chance would undoubtedly cause an increase in medical malpractice litigation, as well as result in an increase in medical malpractice insurance costs.

Since the 1976 enactment of the Health Care Malpractice Claims Act, every subsequent legislative session has devoted considerable time to legislation aimed at managing the soaring costs of malpractice litigation. There is obvious legislative recognition of a public policy of reducing medical malpractice litigation and insurance costs. Professor Paul Weiler in his forthcoming book, Legal Policies for Medical Injuries (to be published in 1991 by Harvard University Press), points out that expenditures for medical liability insurance in the United States have increased from about sixty million in 1960, to more than seven billion in 1988. The number of tort claims has also risen dramatically from approximately one tort claim per hundred doctors around 1960, to more than one tort claim per ten doctors at the end of the eighties.

Consequently, we are not persuaded that the benefits of allowing loss of chance damages in a survival action offset the detriments of a probable increase in medical malpractice litigation and malpractice insurance costs.

JUDGMENT OF THE CIRCUIT COURT FOR PRINCE GEORGE'S COUNTY AFFIRMED. COSTS TO BE PAID BY APPELLANT.

ADKINS, Judge, dissenting.

Chiefly because of a theory of damages based on loss of a chance does not meet its test of mathematical precision, the majority rejects that approach to the calculation of damages. I respectfully dissent. Tort law is not about mathematical niceties; it has to do with fairness to fault-free victims who have suffered harm by reason of the tortious acts or omissions of others. It is a basic principle of our tort system that those who can prove they have been so harmed should be compensated. Why should we reject that principle when the harm is loss of a chance of survival that is less than 51%? Is it fair that one who can show that a doctor's negligence had a 51% possibility of producing the harm complained of can recover full damages whereas if the proof is only of a 50% possibility, there can be no recovery whatsoever? Obviously not.

For the reasons stated by Professor King, King, Causation, Valuation and Chance in Personal Injury Torts Involving Pre-existing Conditions and Future Consequences, 90 Yale L.J. 1353 (1981). I would allow loss of chance damages in survival actions. The non-mathematical arguments offered by the majority in support of its holding are equally unpersuasive.

The use of statistical evidence is not limited to loss of chance cases. A great deal of evidence, especially medical expert testimony, is based on less than the certainty the majority seems to desire. The elitist notion that Maryland juries would be unable to cope with loss of chance evidence, 320 Md. at 790-791, 580 A.2d at 213-214, surely would surprise those who adopted the second paragraph of Article 23 of our Declaration of Rights. "The right of trial by jury of all issues of fact in civil proceedings in the several Courts of Law in this State, where the amount in controversy exceeds the sum of five hundred dollars, shall be inviolably observed."

As to speculation about " 'the substantial expansion of tortfeasor liability and the accompanying societal costs that will be impressed by this new cause of action,' " Id. at 792,

580 A.2d at 214 (quoting Gaver v. Harrant, 316 Md. 17, 31, 557 A.2d 210, 217 (1989)), the majority should not be so quick to adopt the parade of horribles so facilely conjured up by the defense bar. It is not inevitable that "societal costs" (i.e. insurance premiums) would be increased should this approach to damages be applied, or that new floodgates of litigation will open. See Feldman, 17 U.Balt.L.Rev. at 150-151. And what of the social cost of failing to compensate those who have been harmed by tortious deprivation of loss of a chance of survival of less than 51%? Even if "social costs" may increase, "that fact alone is not reason enough to deny recovery to a class of plaintiffs that deserves a remedy consistent with the public policy of this State." Gaver, 316 Md. at 45, 557 A.2d at 225 (Adkins, J., dissenting).

And despite the majority's assertion to the contrary, 320 Md. at 794, 580 A.2d at 215, adoption of the loss of a chance theory of damages is not contrary to legislative policy. The thrust of legislative policy in the tort claim area has been largely directed to a cap on damages (Courts Article § 11-108), and the elimination of frivolous claims (Courts Article § 3-2A-04(b)). To allow recovery for a properly supported loss of a chance does not conflict with these measures.

Nor is this matter one that must be left solely to the legislature. The majority, as it must, recognizes our power to modify the common law, and suggests that we may do so only when "necessary due to changing circumstances or increased knowledge." 320 Md. at 793, 580 A.2d at 214. In Adler v. American Standard Corp., 291 Md. 31, 42-43, 432 A.2d 464, 471 (1981) we modified the common law by recognizing a cause of action for abusive discharge. We did so largely because of a perceived need to protect employees at will from unfair firing- i.e. on the ground of fairness. Id. at 42, 432 A.2d at 470. The only changed circumstance we mentioned was that "a growing number of jurisdictions ... have recognized wrongful discharge as a new cause of action." Id. at 43, 432 A.2d at 471.

Both of those factors exist here. I would permit a damages claim based on loss of a chance of survival at least above the de minimis level. Consequently, I would reverse the judgment of the Circuit Court for Prince George's County.

Notes and Problems

1. In Blackjack, if the dealer's exhibited card happens to be an Ace, the players are provided an opportunity to "insure" that the dealer does *not* have a 10 or face card as her second, hidden card (A 10 or face card underneath would mean that everyone at the table without a Blackjack automatically loses the hand.) The insurance bet will pay 2 to 1. Calculate whether or not an insurance bet is basically a "sucker" bet.

2. **Non-Zero Sum Game:** Following a robbery-homicide at a secluded all night liquor store, the police arrest Thelma and Louise. There are no eye witnesses but the police, based on the M.O. (Modus operandi), are fairly convinced about the identity of the two perps (perpetrators). Instantly, the women are placed in two separate holding cells and questioned. Without a confession, no one will be convicted and both women will serve 0 years (an optimal outcome wherein both women win). If convicted, however, either or both will fry in the electric chair (i.e., both lose). The interrogating officers will offer each one a deal (using outdated Mickey Spillane cliches). First, Officer Krupke talks privately to Thelma: "The guy is dead as hell" (From *Vengeance is Mine*, 1950), and: "You're going to fry in the big chair unless you rat on Louise." If, however, you sing (i.e. confess and im-

plicate Louise), you will only "do life." If you reject the deal, and Louise rats on you, then you will fry. Only one person will be offered the deal. What is Thelma's optimal strategy?

3. Among the many brainless offerings of daytime television was a show called "Deal or No Deal," wherein a player (normally an over-exuberant noisemaker of limited capacity) selects a suitcase from among 26. Hosted by Howie Mandel, "Deal or No Deal" was a hit game show where contestants play and deal for a top prize of $1 million. Each night, the game of odds unfolds when a contestant is confronted with 26 sealed suitcases full of varying amounts of cash, ranging from a penny to a million dollars. Without knowing the amount in each briefcase, the contestant picks one that is his to keep until its unsealing at game's end. The risk element kicks in when the player must then instinctively eliminate the remaining 25 cases which are opened and the amount of cash inside revealed.

The pressure mounts in each round, after a pre-determined number of cases are opened the participant is tempted by a mysterious voice from a second story window enclosure (that resembles a New Orleans house of prostitution) known only as "the Banker" to accept an offer of cash in exchange for what might be contained in the contestant's closed briefcase. As each case is opened, the likelihood of the player having a valuable cash amount in his or her own case decreases or increases. The contestant knows that as long as the larger cash prizes haven't been opened, the Banker's deals will only get higher. If, however, the conflicted contestant accidentally opens a case with a bigger cash value, the Banker's offer could suddenly evaporate. The show clearly panders to an intellect incapable of figuring out that the Banker's offers are always too low.

Assume that the suitcases containing the following amounts of money have yet to be opened and that one of those suitcases belongs to you:

$1,000,000.00
$1,000,000.00
$50,000.00
$25,000.00
$25.00
$.10
$100.00

Based on a calculation of your expectation, what is the least amount of money that you should accept to surrender your suitcase and end the play?

§ 4.40 Combinatorics

Combinatorics is a branch of mathematics that concerns the study of finite or countable discrete sets and subsets. The topic is important with respect to probability theory. Many experiments in probability involve determining a number of possible outcomes and combinations of outcomes. Until now, we may have taken for granted our ability to quickly count or determine the number of events in a given set or subset. When dealing with combinations and permutations, things can quickly get out hand without some formulaic assistance.

It's time to introduce another important symbol commonly used in the subject of probability combinatorics and in the world of mathematics generally. The exclamation "!" which is termed "factorial," is used to denote the product of an incrementally declining series of integers ending with 1. Let's start with a simple definition.

Definition of Factorial (!): n! = (n) (n-1) … 1

Simply defined, the factorial of a positive integer n, denoted (n!) is the product of all integers less than or equal to n. For example:

$$5! = 5 \times 4 \times 3 \times 2 \times 1 = 120.$$

Notice how the factorial product (!) increases rapidly as the number n increases. 3! = 6, 10! = 3,628,000. 20! = 2,432,902,008,176,640,000. **By definition 0! = 1.**

Example 4-16: Five horses, all of equal ability are going to race at Churchill Downs. What is the probability of picking the exact order of finish for the entire field?

Solution: Using The Product Rule For Dependent Events (extended):

P(Horse 1): 1/5
P(Horse 2): 1/4
P(Horse 3): 1/3
P(Horse 4): 1/2
P(Horse 5): 1/1 (Position is predetermined if prior picks are successful)

$P(E) = 1 / 5! = 1/ (5 \times 4 \times 3 \times 2 \times 1) = 1/120 = .0083$

$$P(E) = 1 / 5! = \frac{1}{5 \times 4 \times 3 \times 2 \times 1} =$$

$P(E) = 1/120 = .0083$

Consider the set of 5 horses from the prior example, or any set of distinct finite elements. A rearrangement of the elements of a given set is referred to as a **permutation.** When elements are selected and extracted without replacement from a set, we sometimes refer to the remaining elements as a **combination.** For any set of n members or elements:

If P_n = The number of permutations of n elements:

$$P_n = n!$$

If nPr = The number of permutations (P) of r items selected from n elements (Order Counts):

$$_nP_r = \frac{n!}{(n-r)!}$$

If $_nC_r$ = The number of Combinations (C) of r items Picked from n elements (Order Unimportant):

$$_nC_r = \frac{n!}{r!(n-r)!}$$

To further abbreviate to symbols, the following notation is often used for the calculation of combinations:

$$_nC_r = \binom{n}{r}$$

Example 4-17: How many permutations are there in a set consisting of 3 elements {a, b, c}? What are they?

Solution: $P_n = n!$

$P_n = 3! = 3 \times 2 \times 1 = 6$

They are: {a b c},
{a c b},
{b a c},
{b c a},
{c a b},
and {c b a}.

Example 4-18: The *Disease of the Month Club* maintains a set of 50 DVDs relating to weird and contagious diseases. New members could select any 5 DVDs. From how many different combinations of 5-DVD selections may a new member select?

Solution: ${}_nC_r = n! / r!(n-r)!$
n = 50 and r = 5

$${}_nC_r = \frac{50!}{5!\,(50\text{-}5)!} = \frac{50!}{5!\,(45!)} = \frac{50 \cdot 49 \cdot 48 \cdot 47 \cdot 46\,(45!)}{5!\,(45!)} =$$

$$\frac{50 \cdot 49 \cdot 48 \cdot 47 \cdot 46}{5!} = \frac{50 \cdot 49 \cdot 48 \cdot 47 \cdot 46}{5 \cdot 4 \cdot 3 \cdot 2 \cdot 1} =$$

$${}_nC_r = 2{,}118{,}760$$

Notes

Sentencing criminals in merry olde England was simple in the 1600s and it was consistent; everyone was hanged, so there was no problem with multiple offenders. Today, however, in the U.S., sentencing can be quite complex due to multiple-offense statutes, *concurrent v. consecutive* sentencing and other laws and rules that superimpose penalties upon penalties. The net effect is the micro-management of judicial discretion through the use of variations in the number of years to be served. The result is not always desirable or consistent.

18 U.S.C.A. § 924(c) involves sentencing for the possession of a firearm in furtherance of a drug trafficking offense or a crime of violence. The variations and permutations in the number of years imposed are many; the factors relating to the firearm alone include: Was it fired? Or brandished but not fired? Short-barreled? Rifle or shotgun? Semiautomatic? Assault weapon? Machinegun or a destructive device? Equipped with a firearm silencer or firearm muffler? In a dispute over the application of § 924(c), the court in *U.S. v. Jefferson*, 302 F.Supp.2d 1295 (M.D., Ala., 2004), commenting upon an earlier case, *United States v. Johnson*, 25 F.3d 1335 (6th Cir.1994) observed:

> Despite the differences between the challenges to § 924(c) in Johnson and the question this court faces today, Johnson is illustrative of how § 924(c) can lead to arbitrary sentences by forcing courts to engage in an exercise in **combinatorics (a branch of mathematics founded on the formulas for permutations**

and combinations), as opposed to an exercise in judgment. As the Johnson court noted: "The absurdity is that a defendant with one firearm and one marijuana joint, one rock of crack and one Dilaudid would receive fifteen years (or forty-five years, under the government's theory) for the gun in addition to the sentence on the drug charges, assuming the drugs were possessed with intent to distribute, while a drug kingpin with ten kilos of crack and the same firearm would only receive one five-year sentence, consecutive to his sentence for the drug offense."

§ 4.50 Forensic Misuse of Probabilistic Theory

Without a doubt, statistics are the weapons of choice for the heavyweight liars of the world. No one would doubt that for a moment, as the next chapter will shortly reveal. Consider this section as warm-up for the welterweights. The distinction between probabilistic theory and statistics is fuzzy at best. Statistics normally involves numbers that are not quite as precise as those in the probability arena, thereby allowing more room for shenanigans. For probabilistic theorists, however, there is also room for chicanery.

§ *4.51 Dazzling Numbers for the Jury*

Probability theory plays its most controversial role when used to determine the guilt or innocence of someone charged with a crime that he or she denies committing. In *People v. Collins, supra* we witnessed how six relatively innocent characteristics can quickly add up to a "12,000,000 to one" shot that the defendant is guilty. Both of our product rules (See, *supra*, § 4.22) for multiple trials (both dependent and independent) cause relatively small numbers to become large very quickly. If you consider, for example, just three independent personal traits, each shared by 10% of the population, a defendant with all three traits could face a convincing argument by the prosecutor that the odds are 1,000 to one in favor of guilt even though there is no other corroborating evidence.

When numbers grow ridiculously large, they may tend to unfairly prejudice the defendant. It has been argued that a jury, which probably does not understand probabilistic theory to begin with, tends to place a great deal of trust in large numbers. Such trust, it is argued may rule out any exculpatory evidence that might otherwise exist. In face of the following numbers, a million is probably considered as chump change:

1 Billion = 1,000,000,000 = 10^9

1 Trillion = 1,000,000,000,000 = 10^{12}

1 Quadrillion = 1,000,000,000,000,000 = 10^{15}

1 Quintillion = 1,000,000,000,000,000,000 = 10^{18}

1 Sextillion = 1,000,000,000,000,000,000,000 = 10^{21} = A Million Trillion

Where DNA evidence is involved, the numbers get extremely large and impressive. *Snowden v. State*, 574 So.2d 960 (Alabama Criminal Appeals, 1990) provided a case with an alleged range of 1 in 2.5 billion to 27 trillion (Quite a "range"). *People v. Johnson*, 139 Cal. App. 4th 1135 (2006) dealt with 1 in 240 quadrillion. *U.S. v. Jenkins* 887 A.2d 1013 (D.C., 2005) placed the probability that someone other than defendant committed the offense at 1 in 870 quintillion assuming a Caucasian population and 1 in 4 sextillion for the Southwestern Hispanic population. Finally, there is *State v. Watts*, 616 S.E.2d 290 (North Carolina, 2005) with a million trillion.

At this juncture, arguments that very high probabilities would be unfairly prejudicial to the defendant, have been unsuccessful. *U.S. v. Morrow*, 374 F.Supp.2d 51 (D.D.C., 2005). Despite the lack of a successful attack on the use of monstrous numbers, it is interesting to note that the likelihood of a test false-positive or a mix-up in the samples, or some other laboratory error is far more probable than the tenth-to-the-whatever-power numbers tossed around for the benefit of the jurors.

Humans will always make errors; the probability of mislabeling and misinterpretation will always be significantly greater than zero. In *Armstead v. State*, 342 Md. 38, 673 A.2d 221 (Md., 1996), the defendant was convicted of first-degree rape and other offenses following a jury trial; he appealed and certiorari was granted to consider the admissibility of DNA evidence. The general laboratory error rate of 0.7% for Cellmark Diagnostics, promulgator of the damning probability match, was shown to be based on two errors identified in proficiency tests conducted earlier (Since that time, Cellmark claims that its protocol has been revised). In affirming the conviction, said the court:

> "Armstead was tried before a jury in the Circuit Court for Howard County. At trial, the State called several expert witnesses who presented both the product rule and ceiling principle calculations to the jury, explaining the rationale for each method. The witnesses explained that the product rule calculation yielded odds of a random match between Armstead's DNA and the DNA recovered from the victim of one in 480 million, while the ceiling principle calculation yielded odds of a random match of one in 800,000. The jury was also informed of the laboratory error rates. Petitioner did not call any expert witnesses at trial to challenge the State's DNA evidence; he did, however, cross-examine the State's experts and elicited testimony regarding the controversy over the proper method of calculating match probabilities."

In a *nuts-to-the-bothersome-numbers* approach, a more unabashed Maryland court in *Young v. State*, 388 Md. 99, 879 A.2d 44 (2005), held that when a DNA method analyzes genetic markers at sufficient locations to arrive at an *infinitesimal* random match probability, expert opinion testimony of a match and of the source of the DNA evidence is admissible *without* accompanying contextual statistics.

In a more recent case, *People v. Watson*, 965 N.E.2d 474 (Ill.App., 2012), one might detect a pull-back from the attitude of invincibility that seems to surround DNA matching. In sharp contrast to *Young*, the *Watson* court not only found DNA evidence challengeable, but it went so far as to find that defense counsel's failure to challenge a forensic scientist's conclusions about a partial-profile DNA comparison established an ineffective-assistance-of-counsel claim. It is, however, notable that the DNA match-up was the only evidence against Watson (who was charged with, and convicted of burglary).

The Ceiling Principle: Notice how the "ceiling principle," mentioned in *Armstead*, yields a match-up that is more conservative than the product rule; in fact, it is about 600 times more conservative. That is because, unlike the product rule, the ceiling principle places a limit or ceiling on the scarcity of the factors. Under the ceiling principle, statements endowing traits (i.e., factors) with a "1 in a billion members of the population" are attenuated or limited to a less dramatic ratio, with a smaller denominator. Stated another way, consider the following dicta from *Armstead* involving a debate regarding The Ceiling Principle v. The Product Rule (As used in the discussion that follows, assume that the term "allele" refers to a single genetic trait):

> In 1992, the National Research Council issued a report on forensic DNA testing to attempt to resolve this dispute. Although the report recommended using a conservative modification of the product rule known as the ceiling principle, it did not ultimately reject the product rule. Instead, it merely suggested that until data could be collected to confirm or refute the existence of significant population substructuring effects, the interim approach should be to incorporate several conservative assumptions into the product rule calculation. The effect of these conservative assumptions is to maximize the likelihood of random matches, thus decreasing the power of the DNA results to some extent. (The ceiling principle "assumes the worst" about the amount of variation in allele frequencies across subpopulations. Under the product rule, the odds of a random match at one allele are determined by comparison with the frequency of occurrence of that allele in the population. Therefore, if a particular allele that shows up in the suspect's sample occurs in 1% of the population, there is a 1% chance of a random match.)
>
> Suppose, instead, that the allele in question occurs in a low percentage of Caucasians, but in a much higher percentage of Hispanics. The ceiling principle proponents at one point suggested that, in case such variability existed (although data did not then exist to confirm or refute this), we should assume a much higher maximum frequency than is likely, so that the possibility of error due to a random match is minimized. The NRC report recommended using an "interim" population allele frequency of 10%, to be replaced by a figure of 5% once preliminary data was collected. These rates were considerably higher than the likely maximum allele frequency for any subgroup, based even on then-existing data. See, e.g., B. Devlin et al., Statistical Evaluation of DNA Fingerprinting: A Critique of the NRC's Report, 259 Science 748, 749 (1993).
>
> Other than incorporating these maximum allele frequencies, the product rule would still be applied as discussed above ... Ultimately, the studies have generally concluded that use of the ceiling principle is unnecessary.

---

§4.52 The Prosecutor's Transposition Fallacy

As we witnessed in Chapter 1, when the English language, with all of its ambiguities is used to define or describe quantitative matters, errors are bound to abound. Miss-stating probabilities can be accidental or intentional; either way, the mathematics are just as incorrect.

Recall from our earlier discussion of Bayes' Theorem (*supra*, §4.22(B)) that the probability of E given F = P(E | F) = P(F | E) x P(E) / P(F). Notice from the equation that unless P(E) happens to equal P(F), that P(F | E) ≠ P(E | F). In other words, the probability of E given F does not necessarily equal the probability of F given E.

Despite the obvious inequality, prosecutors presenting match evidence (most commonly DNA) during the course of a criminal trial or a summation to the jury, will switch the term: [1] "probability that the match is random" with [2] "random match probability." The two terms are no more equal than the inequality represented above.

> [1] Probability that the match is random: The probability (F) that the individual tested has been selected at random ***given that*** (E) the individual has the requisite genotype. This statement = P(F).
>
> [2] Random match probability: The probability that (E) the requisite genotype is in the sample taken from the individual tested ***if*** (F) the individual tested has been selected at random. This statement = P(E).

In order to appreciate the nature of the transposition, let L = The random selection of a lawyer. Let J = The random selection of a judge. There would be a huge discrepancy between the probability of selecting a judge who happens to be a lawyer P(J | L), as compared to selecting a lawyer who happens to be a judge P(L | J). Assume that 2% of lawyers are judges. We can further assume that at least 99% of all judges are lawyers.

> P(J | L) = .0200
>
> P(L | J) = .9900
>
> Therefore, clearly: P(J | L) ≠ P(L | J)

§4.53 The Criminal Defendant's Favorite Fallacy

The defendant confronted with DNA evidence is faced with a choice of explaining how his genetic material managed to affix itself to the victim's property, or otherwise discredit its probabilistic conclusions. Difficulty breeds some interesting statements. Consider the following argument to the jury:

> PROSECUTOR: There is only one person in a million that possesses the same DNA and genetic markers as those of the defendant. In particular, the probability that the genotype would be in the sample taken from the defendant at random would be 1 in a million.
>
> DEFENSE : Even if we were to assume what the prosecution says is true, there are at least 12 million people in this city and its surrounding area. That would mean there are 11 other people with the matching genetic material. This means that the odds against the defendant having committed the crime are 12 to 1.

The fallacy, of course, involves the defendant's assumption that the only link between the defendant and the crime scene was the DNA match. The fallacious argument ignores the fact that there must have been probable cause for the defendant's arrest. Presumably other evidence would connect the defendant. There is, however, the probability of a "cold hit," where a person is selected from a database of offenders, and thereafter becomes the subject of an investigation; in such an instance, the defendant's "fallacy" might not be far off the mark in the absence of corroborating evidence.

A similar misuse of probability theory occurred in the notorious multiple-murder case of *People v. O.J. Simpson*, California Superior Ct., No.: BA097211 (1992). In charging the former football player with the murder of his ex-wife, the prosecution presented a lengthy history of physical, domestic abuse by the defendant directed at the victim, a matter which was certainly relevant. The defense team argued that there are many millions of women abused and battered every year, including 1992. Some studies go so far as to claim that 1/3 of all of the women in the world have been the victim of violence at some time in their lives. See Heise, L., Ellsberg, M. and Gottemoeller, M., *Ending Violence Against Women*, Population Reports, Series L, No. 11 (1999, December). The defense, argued, based on various studies, that only 1 woman out of 2,500 ended up as a murder victim, and therefore, there was only a .04 % chance that O.J. was guilty. It was almost as if the defense was claiming, due to the low odds, that O.J. was *less* likely to have committed the murder because he was a wife beater. In truth, the probabilities presented were irrelevant; the mere fact that O.J. was a wife beater certainly did not *reduce* the significance of other evidence of guilt (i.e., DNA match-ups, bloody footprints and bloody gloves). A different situation would have existed if there had been no other evidence against the defendant. Nevertheless, the defense strategy must have worked, since O.J. was acquitted by a jury, which according to prosecuting attorney Marsha Clark, consisted of "moon stones."

More recently, arrested, hapless Drew Walter Peterson, a former Illinois police sergeant had four wives who either disappeared or went missing. A defense attorney might start an argument by pointing out that unless spouses happen to die simultaneously (not a common occurrence), one will *always* outlive the other. Many survivors will then remarry, with a close-to-50% probability that the prior widow or widower will again be the survivor. For the third event, a prosecutor might argue that the odds are a million to one that all of the deaths were random; the response, of course, would be to point out that there are many *tens of millions* of couples. "Drew was merely the unlucky one." By the time we get to the fourth missing wife, however, even creative statisticians might have trouble keeping a straight face. For Drew, it was the disappearance of the fourth spouse that prompted forensic re-examination of missing wife number 3, for which he was eventually found guilty of murder.

§ 4.54 False Conclusions

Many assume that a match-up of any kind is tantamount to a conclusion of guilt or innocence. For a prosecutor to state such a conclusion would be improper. It would not only dismiss all other relevant evidence, but it would also invade the province of the jury.

§ 4.55 False Positives

There is always the possibility of a false-positive laboratory result or an error with respect to labeling samples, or a host of other mistakes that might have intervened between the crime scene and the laboratory. In the first O.J. Simpson trial (*People v. Simpson*), the jury was apparently convinced that the defendant's DNA was planted on the victims.

§ 4.56 False Explanations

Even if the numbers are reliable and the mathematics happens to be correct, attempting to explain those concepts to jurors can present major problems. A slip of the tongue, the misuse of a single word or worse, an attempt to explain the corrections can all lead to disasters and mistrials.

U.S. v. Massey

594 F.2d 676 (8th Cir., Minn., 1979)

LAY, Circuit Judge.

Carl Massey was convicted of bank robbery in violation of 18 U.S.C. ss 2113(a) and (d), receiving a sentence of 12 years. On appeal Massey challenges (1) the sufficiency of the evidence and (2) the prejudicial effect of testimony elicited by the trial court and the prosecutor's closing argument, relating to the use of microscopic hair comparisons in terms of statistical probabilities. We find that the evidence was sufficient to sustain the conviction but hold that there was prejudicial error in the overall reference to the defendant's guilt in terms of mathematical probabilities.

On January 18, 1978, at 12:10 P.M. two men robbed the North Shore State Bank in Duluth, Minnesota. Carl Massey, his sister Yvonne Massey, and John Murray were charged in the same indictment with the bank robbery. Yvonne Massey and John Murray plead guilty on the day of trial; only Carl Massey proceeded to trial. The Government's evidence included pictures recorded by an electronic surveillance camera of two men in the process of robbing the bank. The first man wore a ski mask with one center opening, surgical rubber gloves and a tan field jacket. The second man, wearing a ski mask with openings only for his eyes and mouth, leather gloves over cloth gloves, ragged jeans and a hooded parka, brandished two cocked guns during the course of the robbery. The two men fled the bank in a beige camper, which the police found abandoned a short distance from the bank. A police dog followed the scent of a track leading from the camper to a spot in a parking lot behind a hardware store, where the dog lost the scent. At approximately noon on the day of the robbery an employee and customer of the hardware store had observed a yellow Volkswagon with its motor running in the same parking lot. A young woman wearing a red scarf and wire rim glasses was observed in the driver's seat.

At approximately 12:45 P.M. on the day of the robbery, a witness, George Bowen, looked out his window and noticed a yellow Volkswagon parked in front of 1217 West Fifth Street, the residence of William and Lana Papberg. At about 1:10 P.M. he observed a man with blond hair, wearing a blue vest jacket, come out of the Papberg residence and begin rummaging through the car. At about 1:14 P.M. Sergeant Eli Miletich observed the yellow Volkswagon parked in front of the Papberg residence. The officer then went to a nearby phone, ran a license check and discovered that the Volkswagon was registered to Nancy Massey, mother of Carl Massey and Yvonne Massey. When he returned after running the license check, the Volkswagon was gone.

At approximately 2:15 P.M. the Volkswagon returned to the Papberg residence. Yvonne Massey was driving; John Murray was a passenger. The couple entered the Papberg residence and stayed for approximately 20 minutes.

When they returned Massey and Murray consented to a search of the car. In a suitcase belonging to John Murray the officer discovered a red notebook which contained a list of articles intended for use in criminal activity. The notebook also contained the following names and initials and accompanying percentage figures: "Jay, 100. Lana, 250. M.C., 10 percent. Y., 15 percent. C., 25 percent. G., 25 percent. J., 25 percent." Officers continued surveillance of the Papberg residence throughout the afternoon and into the evening.

The Government urges that the names and initials correspond to the first names of persons present at the Papberg residence on the night of the robbery: "Lana," Lana Papberg; "J," John Murray; "Y," Yvonne Massey; and "C," Carl Massey. The notebook also contained the following: Tuesday. Cars, 8:00 o'clock, 12:00 o'clock, 4:00 o'clock. One run through. Make all radios work. Gather all materials. Sacks for everything. Pick up M-1. Call Greg Monday. Look for cars 8, 12 and 4. One run through. A and E. Map cover. Stop at Jay's. 9. Folks. Scanner, walkie-talkie, stopwatch, clothes, mask, 357. Mac, lunch late, Lana, Elton, house, antenna, Nancy. Gloves. On another page, as described by a police officer, was:(A) drawing, a sketch of some kind with what could be a pot of gold with an arrow, leading to a vehicle, leading to another vehicle, leading to a third vehicle labeled "Safe and Clean", and another arrow proceeding to a house, and then an arrow proceeding to a, what could be a pot of gold. The notebook was admissible into evidence. However, the incriminating effect of such writings, at least as against Carl Massey, is questionable. Cf. United States v. Brown, 584 F.2d 252 (8th Cir. 1978).

At approximately 8:30 P.M. the Papberg residence was searched pursuant to a search warrant. Those present during the search included Lana Papberg, Nancy Massey, Yvonne Massey and Carl Massey, who was wearing a blue vest jacket. Among the many items seized in the basement were a ski mask with holes for the eyes and mouth, a red scarf, a pair of wire rim glasses and surgical rubber gloves. One of the bank's employees testified that the robber holding the guns wore a ski mask that covered everything but his eyes and mouth. There was also testimony that the red scarf and wire rim glasses were similar to the items worn by the woman sitting in the yellow Volkswagon on the day of the robbery. Also seized from the Papberg residence were two jackets, one of which an expert in photographic comparison analysis positively identified as the same jacket appearing in the surveillance photographs. The expert testified that the other jacket was similar to the one pictured by the surveillance camera but he was unable to make a positive identification.

F.B.I. Agent Robert Harvey, who participated in the search of the Papberg residence, found a blue ski mask with a center opening similar to the one described by a bank employee as belonging to one of the robbers. Five hairs found in the ski mask were microscopically compared with hair samples taken from Carl Massey's head. Agent James Hilverda, an expert in microscopic analysis, testified that Carl Massey's hair was similar to three of the five hairs found in the blue ski mask in all categories of microscopic comparison.

Two tellers at the bank testified that two days prior to the robbery, a young man with blond hair approached a teller's window and asked for a roll of quarters and some coin wrappers. Both tellers subsequent to the robbery identified the blond young man as Carl Massey.

In view of the jury verdict in favor of the Government, the evidence, although circumstantial, along with all reasonable inferences to be drawn therefrom, must be reviewed in the light most favorable to the Government. At oral argument the Government conceded, as we feel it must, that the observation of the defendant at the bank two days before the robbery, the notebook, the observation of a young man with blond hair rummaging through the Volkswagon shortly after the robbery and Massey's presence at the Papberg home at the time of the search are insufficient facts to sustain Massey's conviction.

Although the cumulative effect of this evidence points a strong finger of suspicion it is not in itself sufficient to prove that Massey was one of the bank robbers. The linchpin is established, however, when this evidence is weighed with the testimony of the Government's expert which established that three of five hair samples found in the blue ski mask (similar to the one used in the robbery) matched in all areas of microscopic comparison with known samples of defendant's hair. We therefore reject defendant's contention that the evidence was insufficient to support the conviction.

The evidentiary importance of the microscopic comparison of the hair samples, however, requires this court to scrutinize defendant's challenge to its foundation, the trial court's interpretation of it and its use by the prosecutor in closing argument. F.B.I. Agent Robert Harvey testified that five hairs were found in the blue ski mask. James Hilverda, an expert in microscopic analysis, then testified that Massey's hair was similar to three of the five hairs found in the blue ski mask in all categories of microscopic comparison. After so testifying the court interrogated Mr. Hilverda and the following colloquy occurred:

> THE COURT: How many people in the United States have hair that you couldn't distinguish? In other words, what's the likelihood that any one individual, his hair might meet all the characteristics of another individual?
>
> A. Well, first of all you could eliminate people from different races....
>
> You can say positively it came from a person of the Caucasian race or the Mongoloid race and the Negroid race, so basically exclude all those.
>
> You could exclude primarily those people who have mixed racial origin.
>
> So you can be positive on those.
>
> It's the type of thing hair evidence is the type that you can not positively say it comes from one individual to the exclusion of all others.
>
> And I would basically have to go with my experience in examining about probably in excess of 2,000 cases, that I have seen only on a couple of occasions I have seen hair from two different individuals I could not distinguish, that they were so similar, that some hairs within each sample were very close and I did not feel comfortable enough to work the case and to say that one hair came from one individual to the exclusion of the other.
>
> So I did not work the case in that instance.
>
> THE COURT: Two chances one chance in a thousand?
>
> A. Well that's very difficult you're putting it in probabilities because I do not take a hair from each case and compare it with hair from all the other cases.
>
> But that's the frequency with which I have seen it.
>
> The other thing, I could give you a study the Canadians have done a study where they have come up with one in 4,500, that a possibility that a hair which you have done or matched in the manner which I have set forth, there's a chance of one in 4,500 these hairs could have come from another individual.
>
> THE COURT: If you have one hair there, and then you could compare it with 4,500 others, and you might find one?
>
> A. You might find one that would show the same. That would be assuming that you took hairs from 4,500 other people of the same race. Of course, if you had more hairs than that, you would have multiplication factor at that point.

The defendant does not challenge the opinion evidence that certain hairs within the ski mask were microscopically identical to his hair. The fundamental challenge relates primarily to the prejudicial effect of the use of mathematical statistical probability as elicited and interpreted by the trial judge and argued by the prosecutor.

The actual testimony of the expert witness is somewhat confusing. It is difficult to understand just what tests he conducted in 2,000 cases because of lack of foundational detail to understand the type of experimentation or studies made. Yet the witness was candid in saying his own experience did not establish a statistical probability. Our reading of the record is simply that in his own experience the witness had microscopically examined 2,000 cases and in only one or two cases was he ever unable to make identification. There is no foundation to show the factual circumstances surrounding each of his examinations and certainly there is no statistical probability which could be drawn from his experience to show that there was only "one chance in a 1,000" that hair comparisons could be in error.

Concerning the witness' reference to the Canadian study there can be no doubt that the foundation for such testimony was insufficient. The witness testified that he did not know the nature and extent of the studies conducted from which the statistics were gathered. See Bogacki v. American Machine & Foundry Co., 417 F.2d 400, 408 (3d Cir. 1969).

At best, we conclude the court's colloquy with the witness concerning mathematical probabilities was speculative and confusing. The gravamen of the error came during the trial judge's comment construing the testimony in terms of the probability of making an erroneous identification through the comparison of hair samples.

However, prejudicial harm arises in the Government's closing summation, which exacerbated the trial judge's misunderstanding of the evidence. The prosecutor argued:

> I think there was some reference both to his own experience and the experience of a survey conducted by the Canadians as to the accuracy of the conclusions which one can draw from finding hair samples which are microscopically the same. I believe the witness said that out of 2,000 or 2,500 examinations only a handful I believe he said 3, but let's say 5 only a handful of instances occurred where he was unable to distinguish the hair from two different people. (This, of course, is not what the witness stated.) I believe that it was a handful out of 4,500 in the Canadian test.
>
> Now in order to convict the defendant, you must find him guilty beyond a reasonable doubt.

That is not beyond all doubt, but that is beyond a reasonable doubt.

> The Judge will tell you what that is. It is a fair doubt. It is a doubt that would make you hesitate in the more serious of your own affairs.
>
> A handful 3 to 5 out of 2,000. That's better than 99.44 percent; it's better than Ivory Soap, if you remember the commercial. It's very, very convincing.
>
> Now hair samples are not like fingerprints. It is not positive identification. There is a theoretical possibility (and it actually happened in the case of this examiner in 3 to 5 times out of say, 2,000) where the hairs of two different heads can look the same when you examine the whole range of their characteristics.
>
> However, it is infinitesimally rare, and when we talk about the range of proof which we can use in deciding questions for us, those kinds of percentages are higher than the percentage we use in any other area I can think of in terms of making a decision.
>
> I submit to you that if hair samples are found a known and an unknown and they are microscopically identical, that it is at the very least proof beyond a reasonable doubt that the unknown hair comes from the same head as the known hair.
>
> Now human hairs were found in this ski mask which I am holding, Government's Exhibit 31, which appears as the ski mask worn by the robber carrying the sack out of the bank and which was found in the same house as the fatigue jacket and the parka.
>
> The hair found in the ski mask was examined as to the 15 characteristics, and all 15 characteristics were microscopically the same as the known hair sample of the defendant Carl Massey.

> I argue to you that we have presented proof beyond a reasonable doubt that the clothing found in the Papberg residence was used in the robbery, and we have shown beyond a reasonable doubt that this ski mask was used by the defendant Carl Massey; and, while there is in fact other evidence which corroborates this, including the fact that he cased the bank, for example, We would argue to you that just the hair sample would be proof beyond a reasonable doubt because it is so convincing. (Emphasis added.)

By using such misleading mathematical odds the prosecutor "confuse(d) the probability of concurrence of the identifying marks with the probability of mistaken identification" of the bank robber. McCormick on Evidence s 204, at 487 (E. Cleary ed. 1972). In other words, the prosecutor has infused in the minds of the jury the confusion in identifying the hair with identifying the perpetrator of the crime. As stated in State v. Carlson, 267 N.W.2d 170 (Minn.1978), when reviewing similar opinion evidence:

Our concern over this evidence is ... with its potentially exaggerated impact upon the trier of fact. Testimony expressing opinions or conclusions in terms of statistical probabilities can make the uncertain seem all but proven, and suggest, by quantification, satisfaction of the requirement that guilt be established "beyond a reasonable doubt." Diligent cross-examination may in some cases minimize statistical manipulation and confine the scope of probability testimony. We are not convinced, however, that such rebuttal would dispel the psychological impact of the suggestion of mathematical precision, and we share the concern for "the substantial unfairness to a defendant which may result from ill conceived techniques with which the trier of fact is not technically equipped to cope."

Jurors being exposed to statistical probabilities "may find themselves 'surprised' at the strength of the inference of guilt flowing from the combination of mathematical and non-mathematical evidence. Indeed they may, and in a far deeper sense than with other equally obscure forms of expert testimony, for such testimony typically represents no more than an input into the trial process, whereas the proposed use of (statistical probability) methods changes the character of the trial process itself. When that change yields a 'surprisingly' strong inference of guilt in a particular case, it is by no means clear that, so long as one keeps one's numbers straight, 'this ... is no more than the evidence deserves.' Methods of proof that impose moral blame or authorize official sanctions on the basis of evidence that fails to penetrate or convince the untutored contemporary intuition threaten to make the legal system seem even more alien and inhuman than it already does to distressingly many." Tribe, Trial by Mathematics, 84 Harv.L.Rev. 1329, 1375–76 (1971) (footnotes omitted).

In view of the obvious importance of the hair comparison in conjunction with the weakness of the other circumstantial proof, we cannot say the error was harmless.

The judgment is reversed and remanded for a new trial.

BRIGHT, Circuit Judge, dissenting:

I respectfully dissent.

Upon examining the entire record, I am convinced that the Government provided sufficient evidence to sustain Massey's conviction for bank robbery. In this respect, I agree with the majority.

However, I cannot agree with the majority's conclusion that the testimony about statistical probability, as it relates to hair identification, constituted plain error.

The district court's interrogation of the FBI hair expert and the prosecutor's closing argument that its proof was better than the purity of Ivory soap improperly suggested that, as a matter of statistical probability, based upon undocumented studies referred to in the trial testimony, Carl Massey's guilt was established beyond a reasonable doubt.

Although the existence of suspect hair samples microscopically identical to known hair samples from Massey represents impressive evidence that such samples come from the same person, it does not constitute conclusive proof of guilt. To argue in terms of mathematical probability that such hair sample proof establishes Massey's guilt beyond a reasonable doubt tends to mislead the jury.

In this case, however, other evidence than the hair samples points to Massey as one of the two bank robbers. That other evidence by itself is probably insufficient to sustain a conviction. However, given the various coincidences in the evidence tending to link Massey to the crime, the jury could reasonably rule out other persons than Massey as the source of the crucial hair specimens found in the ski mask.

Plain error embodies the notion that an obvious mistake seriously affecting the fairness or integrity of a judicial proceeding has been committed. Courts exercise their power to notice plain error only in exceptional circumstances. United States v. Atkinson, 297 U.S. 157, 56 S.Ct. 391, 80 L.Ed. 555 (1936). We have described plain error as error "of such sufficient gravity as to substantially affect the rights of the defendant and prejudice him in the eyes of the jury to the extent that our failure to note (it) would perpetuate a miscarriage of justice." Chubet v. United States, 414 F.2d 1018, 1021 (8th Cir. 1969).

The errors alleged in this case may have unduly emphasized the importance of the hair samples, but, in light of all the evidence presented by the Government linking Massey to the crime, including the hair sample testimony, I do not deem the trial judge's interrogation or the prosecutor's closing argument as plain errors causing undue prejudice to Massey in the trial of this case.

I would affirm the conviction.

---

§ 4.57 The 50% Fallacy

Certain tests (for paternity in particular) often commence the testing process with a "prior probability assumption." Paternity testing, in particular, commences with a so-called "50% prior probability assumption." In other words, she says: "Yes," he says "No." Therefore, it is scientifically reasoned that absent other evidence, the probability of paternity is 50/50. From that perspective, a form of Bayes' Theorem is then employed for the purpose of modifying or "updating" the probability depending upon the number of other men who share the same genetic material.

This type of logic is not far afield from the Jack the Ripper analogy. Jack the Ripper was left handed. Approximately 8% of the population is left handed. I am left handed. Therefore, there is an 8% chance that I am Jack The Ripper.

§4.58 *The Picker Scam*

There are many ways in which probabilities can be manipulated by knowledgeable people. My favorite is the so-called picker scam. Football betters are the targets. The self-designated experts known as "pickers," claim, for a fee, to have a remarkable ability to pick the winning teams "by the spread." In truth, they are no better predictors than coin tossers. But they support their claims with a money back guarantee. Assume that I'm a picker and that I guarantee to pick the winner of the "big game" (by the Las Vegas spread). I sell you the name of the winning team the night before the game for a modest fee of $100.00. You then bet the farm on the game and retire on the winnings. I am so confident of my ability that I will return the money that you paid me if my pick turns out to be wrong. Sounds like a good deal, right? Take a moment and use your brain (something that the pickees lack). From the pickers' perspective, if I'm correct, I rake in $100.00 from several suckers. If I'm wrong (which I will be 50% of the time), I return the money and I break even (the poor loser who bet the farm, however, is sleeping under some bridge). In other words, as a picker, there are only two possibilities: I can only win or break even—not a bad set of bets.

Chapter 5

Statistics

"The man of the future is the man of statistics."

—Oliver Wendell Holmes

Whereas the subjects of algebra, geometry and physics reveal precise truths about the universe, statistics seems to deliver guesses, exaggerations and sometimes outright lies. It is a subject that is, nevertheless, indispensable to both science and law.

There are as many different statistical definitions, calculations and equations as there are ways to twist and distort their results. According to the United States Supreme Court, statistics come in an infinite variety. Their usefulness depends on all of the surrounding facts and circumstances. *International Broth. of Teamsters v. U.S.*, 431 U.S. 324, 97 S.Ct. 1843 (1977). And since they carry no presumption of accuracy, their credibility remains a question of fact for the jury. *State v. Ward*, 172 Conn. 163, 374 A.2d 168 (1976). Certain public records and vital statistics are admittedly entitled to special treatment as exemptions from the preclusive effects of the hearsay rule[1]; but here the term "statistics" more accurately describes mere record keeping as opposed to computation.

As a lawyer, be assured that if you do not know how statistical information is used and misused as a forensic weapon, you are certain to become its victim.[2] Unsupported lies are subject to attack and cross-examination. And, there are many evidentiary weapons to combat traditional hearsay, that most insidious form of evidence. Statistical information, however, is a special case of pure unadulterated multiple hearsay that has managed to work its way into the courtroom. It has a special aura of reliability about it, making it an even greater ally or threat, depending upon the skill of the user. It is a powerful weapon that can be used for both good[3] and evil.[4]

A statistical attack may be a heavy one involving a barrage of polls, experts and statisticians. Or it may be subtle. As a relatively new lawyer I can recall being asked to defend a young 17 year old accused of breaking and entering a residence in Lincoln Park, Michigan. At the arraignment the prosecuting attorney argued that the defendant should be held in jail pending the posting of a very large bond, alleging that: "Since the capture and jailing of the defendant, the crime wave of Lincoln Park has ended." Luckily, the judge stated

1. Rule 803(9) of the Federal Rules of Evidence provides a hearsay exception for: "Public Records of Vital Statistics: A record of a birth, death, or marriage, if reported to a public office in accordance with a legal duty."

2. "*Statistics are like alienists—They will testify for either side.*"—F. H. La Guardia.

3. See *Brown v. Board of Education*, 347 U.S. 483 (1954) wherein statistics were used to disprove the "Separate but equal" myth of *Plessy v. Fergusun*, 163 U.S. 537 (1896).

4. Disraeli stated that: "*There are three kinds of lies—lies, damned lies and statistics.*" The principle was well established prior to one of history's greatest misuse of statistical information; for many years, the tobacco industry, through its statisticians seemed to convince lawmakers that smoking did not cause lung cancer.

openly that he would totally disregard the allegation. To refute such hearsay would be difficult if not impossible, even with a team of my own statisticians. The defendant was released after posting a $25,000 bond. Lincoln Park, I am told, is still waiting for a crime wave to reignite.

To the non-statistician, the subject matter may seem complex, mystical and impenetrable. To the mathematician, however, the subject of Statistics is nothing more than a few basic common sense concepts derived from probability theory, mixed with some impressive symbols, drenched in a number of questionable definitions, and a massive amount of suspect conclusions. Do not be misled by the symbolic notation and complex definitions; the concepts are relatively simple.

Our discussions covering Statistics flows somewhat naturally from the subject of Probability (Chapter 4) which in turn, was derived from Set Theory (Chapter 3). The discipline known as Statistics is divided into two major topics: *Descriptive Statistics* and *Inferential Statistics.* The first describes the characteristics of a given set of elements. The second, as the name implies purports to base predictions or inferences upon the information gathered. Stated another way, the descriptive statistician merely summarizes the characteristics of a given set of numbers; the inferential statistician might then draw conclusions and inferences from the result.

§ 5.10 Descriptive Statistics

The Universe of descriptive statistics consists of sets and subsets of numbers. Any set that contains less than the Universe itself is a subset, which statisticians often refer to as **samples.** Once they have a set of numbers, as descriptive ("deductive") statisticians, they are concerned with summarizing various characteristics of the set. There are two primary types of descriptive devices: Measurements of Central Tendency (which might be thought of as measuring the similarities among the elements in the sample) and Measurements of Dispersion (which focus on the differences or aberrations).

§ 5.11 *Central Tendency*

The most common lay term used to describe central tendency is the "average." Most of us recognize this as the **mean**, which most accurately corresponds to our understanding of an average. But there are two other commonly accepted measures of "central tendency" that are also employed by those who might wish to convey a different form of "average." They are the **mode** and the **median.**

In order to examine the "central tendency" characteristics of a given set of numbers, it is common to arrange them in numerical order. Such an arrangement makes it easier to examine and to calculate the three primary measurements of central tendency: The **mode**, the **median** and the **mean.** There are others, to be sure.

As an example, we will use a hypothetical "*Save The American Bug*" organization (STAB) as our set of individuals to analyze. Their respective IQs will serve as our samples; in particular, we will assume that each person in the group has completed an objective intelligence (IQ) Test, with the following results:

1. Abe	150
2. Ike	145
3. John	125
4. Bill	125
5. Ronald	125
6. Newt	120
7. Al	100
8. Martin	100
9. George	70
10. Sarah	60

With our numerical arrangement alone, certain bits and pieces of information stand out. We notice, for example, that the list of IQs in the group extends from 60 to 150, yielding a **range** of 90. "Range," however, is a measure of dispersion, not central tendency and therefore will be dealt with in the subsection that follows.

We also notice that the most common score is 125 (the STAB people are not all morons), which non-coincidentally seems to fall in the mid-range of the data. If we were to construct a graphic representation of the data, we witness the rudimentary shape a bell curve (See *infra*, Figures 5-1 and 5-2). At the fringes of the data list are scores that appear infrequently, in this case only once for 150 and once for 60.

A. The Mode

The *mode* is nothing more than the number that occurs most frequently in a given set of data. Those proficient in second grade arithmetic might quickly determine that the mode for our STAB group is 125. There is only 1 mode in the above set, meaning that our group is "*unimodal*." If Newt had an IQ of 100 instead of 120, the group would be "bimodal," because "125" and "100" would share the top spot. A "multimodal" set means that there are more than 2 modes. Notice how impressive you can be from counting alone.

B. The Mean

The *mean* is the more commonly used than the mode when describing the central tendency of a set of numbers. The mean is nothing more than a simple "average." To understand it, we must advance to third grade arithmetic.

1. Abe	150
2. Ike	145
3. John	125
4. Bill	125
5. Ronald	125
6. Newt	120
7. Al	100
8. Martin	100
9. George	70
10. Sarah	60

$$1120 \div 10 = 112$$

Before proceeding further, you should take moment to appreciate how truly simple the "mean" is. It is nothing more than the average of all of the numbers in the set. There is nothing mysterious or complicated; we merely added up all of the numbers and divided by the total number of elements in the set (N). The problem: The simplicity of this definition, which is indeed a very important one, will not impress nor deceive anyone, except perhaps

George and Sarah. So, we need something more impressive. Greek letters always scare and impress people, especially jurors. Therefore, restate the above operation as follows:

$$\mu = \Sigma\, x_i / N$$

The funny looking "μ" is pronounced me-you; it represents the mean or average. Sure, we could use a simple "A" for average, but that would impress no one. If you really want to look and sound sharp you must master the pronunciation. Start to make a noise as if you were a cat and then immediately switch to a cow.

The "Σ" sign is where many non-mathematical individuals bite the dust. In truth, it means nothing more "Sum." Take more than a moment to study this common notation and the manner in which it is used below. Many intelligent people get lost in the notation, not the math.

1. Abe	150	$x_1 = 150$
2. Ike	145	$x_2 = 145$
3. John	125	$x_3 = 125$
4. Bill	125	$x_4 = 125$
5. Ronald	125	$x_5 = 125$
6. Newt	120	$x_6 = 120$
7. Al	100	$x_7 = 100$
8. Martin	100	$x_8 = 100$
9. George	70	$x_9 = 70$
10. Sarah	60	$x_{10} = 60$

x_i = Represents any one of the numbers in our set. It is nothing more than shorthand notation designed to "reduce" the complexity of more advanced operations. Because it would be clumsy to continue to use proper names in mathematical operations, we use the following more impressive notation.

If we add up the x_i elements, we obtain a sum, just as we did initially. Instead of the word "sum," we use the far more impressive symbol: Σ. As before, Σ = 1120. Or stated another way:

$$\Sigma x_i = 1120$$

We then merely divide by the total number of elements in the set (N), and give the result a fancy Greek letter accompanied by a neat animal sound and thereby reach the following result:

The mean IQ of The STAB group (μ) equals = The sum (Σ) of the elements (x_i) divided by the total number of elements (N) in the group. Or:

$$\mu = \Sigma\, x_i / N = 112$$

A so-called "modified mean" (a concept that is often frowned upon by experts and statisticians, but readily embraced by golfers) involves dropping the highest value and the lowest value prior to adding the numbers on the list; the sum would then be divided by N-2 instead of N.

C. The Median

The *median* is a simple concept of central tendency that seems to cause the most grief. Simply stated, it is the physical middle of the road. If you were to line up all of the numbers in order, the median would be the number in the middle. More specifically, the median is the number for which half of the numbers fall above and half fall below. Confusion seems to abound because there are several different definitions and methods for calculating this relatively simple concept.

If N (the number of elements in the set) is odd, and that odd element happens to appear in the middle of our list, there should be no problem identifying it as the median. Examine, for example, the following simple list of numbers from 1 to 9 (an odd number of elements). With "5" smack in the middle, there are exactly 4 elements above it and 4 below it. The number "5" clearly establishes the median:

1 2 3 4 **5** 6 7 8 9

If, however, N is even, such as in the case of our STAB group, then the Median position (M) is calculated as follows:

$$M = (N+1) / 2$$

In our example, there are 10 elements, an even number. The median would, therefore, lie somewhere between the scores of Ronald and Newt.

1. Abe	150
2. Ike	145
3. John	125
4. Bill	125
5. Ronald	125
☞ **Median**	
6. Newt	120
7. Al	100
8. Martin	100
9. George	70
10. Sarah	60

Using our simple equation, M = (N + 1) / 2, we arrive at he following:

Median Position = (10 + 1) /2 = 5.5 or halfway between Ronald and Newt.
Median = (120 + 125) / 2 = 122.5

It is important to note that the measures of central tendency (i.e., the mode, mean and median) have limited value where distributions are skewed as opposed to falling into a nice "bell shaped curve." The mode, for example, tells us little about the distribution; in the absence of a multi-modal set, identical scores might amount to nothing more than meaningless coincidence.

Notes and Problems

1. The defendant in an automobile negligence action has admitted liability. The 12 jurors are impaneled to determine the damage award for the plaintiff. They cannot agree. One juror believes that the plaintiff should receive next to nothing; another juror would like to provide an excessive award. The foreman has an idea. Each of the 12 jurors are to secretly write an amount on a slip of paper. The jury would agree in advance that foreman would then calculate the mean of the 12 numbers, which would constitute the jury's verdict. Is this a good idea? Where are the flaws with this methodology? With respect to so-called **quotient verdicts**, see *Huckle v. Kimble*, 172 Kan. 630, 243 P.2d 225 (1952), and by way of contrast, see *Shulz v. Chadwell*, 558 S.W. 123 (Ky.App., 1977).

2. In a controversial product liability case, the jury foreman asked for a secret ballot vote among the 12 jurors to help determine the amount that the jurors "in general" were willing to give to the severely injured plaintiff. Here are the results:

Juror 1: $12,000,000
Juror 2: $1,000,000
Juror 3: $20,000
Juror 4: $20,000
Juror 5: $20,000
Juror 6: $16,000
Juror 7: $8,000
Juror 8: $8,000
Juror 9: $5,000,000
Juror 10: $2,000
Juror 11: $0
Juror 12: $0

Calculate the mode, mean and the median. If limited to a selection of one of the three, which measure would you select? Why?

3. Where data forms a perfect bell curve, how will the median, mode and mean relate to one another?

4. Construct a set of numbers wherein the mean substantially exceeds the median. Then construct a set wherein the median substantially exceeds the mean. What, if anything, might you conclude about the reliability of both concepts?

§ *5.12 Dispersion*

As opposed to measures of central tendency, measures of dispersion are designed to quantify the differences or variations in the data. In other words, we know that George is probably an idiot, but measures of diversion might tell us just how stupid George really is as compared to the other members of STAB. As might be expected, the measurement of "differences" provides a tool for proving a wide variety of cases, most notably, those involving discrimination, products liability and toxic torts.

A. The Range

1. Abe 150
2. Ike 145
3. John 125
4. Bill 125
5. Ronald 125
6. Newt 120
7. Al 100
8. Martin 100
9. George 70
10. Sarah 60

The **Range** is the easiest measurement of dispersion to calculate. The top score (Abe the Brain) less the lowest score (Sarah the Imbecile) provides the Range. Or:

Range = High Score - Low Score. 150 - 60 = 90.

B. The Variance

So far our concepts have been somewhat intuitive. Now it is time to prepare for some definitions that are a bit more convoluted. Because they are commonly accepted and employed frequently, it is important that they are understood. The variance is a basic measure of dispersion. We are going to deal with another Greek letter: σ, which we will use to define **variance** as: σ^2 where:

$$\sigma^2 = \Sigma (x_i - \mu)^2 / N$$

There is no reason to be afraid, yet. Perhaps justifiably confused. You are probably asking yourself—*What kind of shmendrick creates a definition that is a square term?* I'm not speaking about a "normal" definition that is constructed of factors containing square terms; there are many of those. The area of a square (A), for example, equals one side squared. But who would start by defining A^2 prior to defining A? The answer to the question, lame though it may be, lies in the definition that follows. In the meanwhile, examine the above definition. Other than the strange manner in which we define our novel term, there are no new concepts that we don't already know. The variance (σ^2) is equal to the sum of numbers $(x_i - \mu)$ squared. Recall that μ = The mean. The result is then divided by the number of elements in the set.

The number of elements in the set $(x_i - \mu)$ looks fearful but is obtained by nothing more than simple arithmetic. There are a few things to keep in mind as your tackle our example. First, always work from the inside of the parentheses first. Second, never be intimidated by the statistical notation. The operations involve the simplest arithmetic combined with operations that you can perform on a basic calculator. Let's compute: $\sigma^2 = \Sigma (x_i - \mu)^2 / N$.

Starting with $(x_i - \mu)$, recall that we have already computed the value of μ (the mean), which is 112. Let's begin our computation with $(x_i - \mu)$. The notation implies that we are going subtract μ from each of our 10 elements. The starting point would, therefore, be: $(x_1 - 112)$. Since $x_1 = 150$, we only need to subtract: 150 - 112 = 38. Therefore:

1. Abe	150	$x_1 = 150$	$(x_1 - 112) = 150 - 112 = 38$
2. Ike	145	$x_2 = 145$	$(x_2 - 112) = 145 - 112 = 33$
3. John	125	$x_3 = 125$	$(x_3 - 112) = 125 - 112 = 13$
4. Bill	125	$x_4 = 125$	$(x_4 - 112) = 125 - 112 = 13$
5. Ronald	125	$x_5 = 125$	$(x_5 - 112) = 125 - 112 = 13$
6. Newt	120	$x_6 = 120$	$(x_6 - 112) = 120 - 112 = 8$
7. Al	100	$x_7 = 100$	$(x_7 - 112) = 100 - 112 = -12$
8. Martin	100	$x_8 = 100$	$(x_8 - 112) = 100 - 112 = -12$
9. George	70	$x_9 = 70$	$(x_9 - 112) = 70 - 112 = -42$
10. Sarah	60	$x_{10} = 60$	$(x_{10} - 112) = 60 - 112 = -52$

The next step involves squaring each $(x_i - 112)$ term **before** adding them all together. If you were to add them together first, you would obtain zero, which when squared will just yield another zero. This does not help us. By squaring the numbers first, the negatives become positives, so that they cannot cancel out the other positive numbers. Therefore:

1. Abe	150	$x_1 = 150$	$(x_1 - \mu) = 38$	$(x_1 - \mu)^2 = 38^2 = 1444$
2. Ike	145	$x_2 = 145$	$(x_2 - \mu) = 33$	$(x_2 - \mu)^2 = 33^2 = 1089$
3. John	125	$x_3 = 125$	$(x_3 - \mu) = 13$	$(x_3 - \mu)^2 = 13^2 = 169$
4. Bill	125	$x_4 = 125$	$(x_4 - \mu) = 13$	$(x_4 - \mu)^2 = 13^2 = 169$
5. Ronald	125	$x_5 = 125$	$(x_5 - \mu) = 13$	$(x_5 - \mu)^2 = 13^2 = 169$
6. Newt	120	$x_6 = 120$	$(x_6 - \mu) = 8$	$(x_6 - \mu)^2 = 8^2 = 64$
7. Al	100	$x_7 = 100$	$(x_7 - \mu) = -12$	$(x_7 - \mu)^2 = -12^2 = 144$

8. Martin	100	$x_8 = 100$	$(x_8 - \mu) = -12$	$(x_8 - \mu)^2 = -12^2 = 144$
9. George	70	$x_9 = 70$	$(x_9 - \mu) = -42$	$(x_9 - \mu)^2 = -42^2 = 1764$
10. Sarah	60	$x_{10} = 60$	$(x_{10} - \mu) = -52$	$(x_{10} - \mu)^2 = -52^2 = 2704$
				$\Sigma (x_i - \mu)^2 = 7{,}860$

$\sigma^2 = \Sigma (x_i - \mu)^2 / N = 7{,}860 / 10 = 786$

Therefore, the variance for our set of IQs = 786.

C. The Standard Deviation

The Standard Deviation is the most important and commonly used measure of dispersion. It is to dispersion as the mean is to central tendency. It is often said that the standard deviation is the counterpart of the mean. Mathematically, the standard deviation (σ) is nothing more than the square root of the variance (σ^2). The following definition of the standard deviation won't look quite as fearful if you always remember that once you have calculated the variance (using simple arithmetic), all that you are required to do to obtain the standard deviation is to hit the square root key on your calculator.

$$\sigma = \sqrt{\sigma^2} \quad = \quad \sqrt{\Sigma (x_i - \mu)^2 / N}$$

For STAB, the standard deviation for the set of IQs would be calculated as follows:

$$\sigma^2 = 786 \quad \Rightarrow \quad \sigma = 28.036$$

For data sets exhibiting a perfect normal bell-curve distribution, 68.27% of all of the elements would fall within 2 standard deviations, one on each side of the mean. And, 95.45% would fall within 4 standard deviations, and further, 99.73% would fall within 6 standard deviations. Figure 5-1 is a near perfect bell curve:

Figure 5-1:

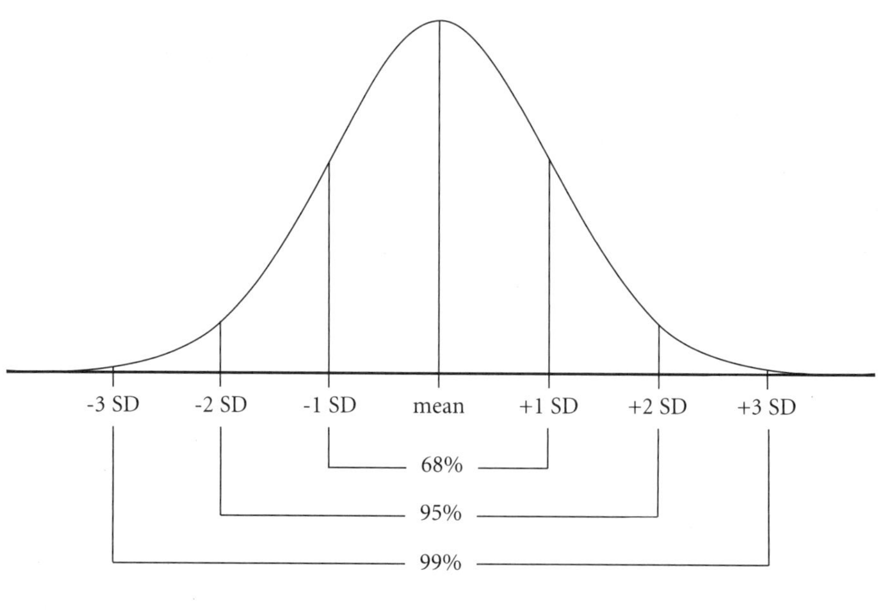

Notice that Figure 5-2 (like Figure 5-1) provides a bell curve. Notice also that all bell curves are not necessarily identical. Some may be perfectly symmetrical as is the case with both Figures 5-1 and 5-2. Others may be skewed.

Figure 5-2:

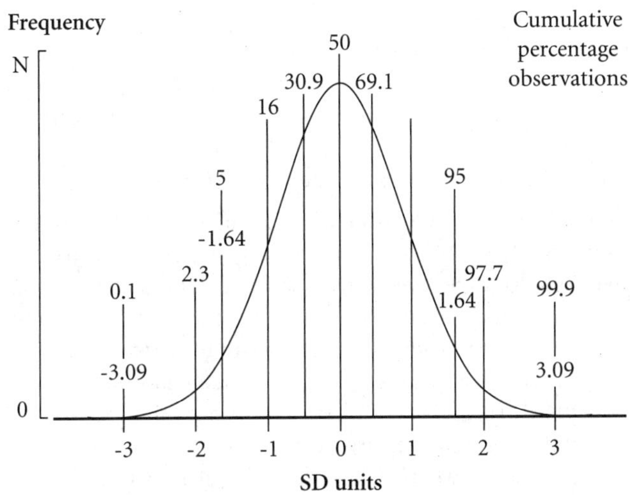

For want of a better term, the word "normal" describes an important concept for the legal profession, and for plaintiffs in particular. The standard deviation is nothing more than a sophisticated mathematical device to help us measure precisely how far-from-normal a given set of data might stray.

Assume that the average life expectancy for all men, women and children in the United States is 79 years. This does not mean that we expect everyone in country to die precisely at age 79; nor do we expect everyone to die within several years of age 79. On the contrary, for very large numbers we would expect a bell shaped curve similar to Figure 5-2. If the average age for a large city such as Detroit had a population with an average life expectancy of 75, we might not be surprised even though the number would be considered well below normal. If, however, the life expectancy were only 50, we might look for some causal relationship. In fact, we might say that there is a **statistical significance**. In turn, that statistical significance would be measurable in terms of standard deviations.

From a judicial perspective, *Equal Employment Opportunity Commission v. American Nat. Bank*, 652 F.2d 1176 (4th Cir., 1981) provided us with the following:

> The "standard deviation" is the measure of the predictable fluctuation in a random selection process. The difference between actual ("observed") numbers of the protected group in such a sample and the number that would be "expected" in a perfectly proportional process of selection from the appropriate pool can then be expressed in numbers of standard deviations. In turn, standard deviations can be expressed in terms of the mathematical probability that chance is the cause of the disparities (differences between "observed numbers and expected values") measured. As standard deviations increase numerically, the probability of chance as the cause of revealed underrepresentation of course diminishes. To the extent the probability of chance is shown to be quite small, the legal inference of discrimination based upon a rough legal assessment that disparities are manifestly "gross" or "substantial" is thus "scientifically" confirmed.

There is no universal agreement among the courts or statisticians as to the precise number of standard deviations that are significant in any particular matter; in other words, there is no bright line. But there are some pretty solid rules of thumb.

In a discrimination case, for example, the difference between the actual numbers of a protected group in a random sample and the number that would be expected in a perfectly proportional process of selection from the appropriate pool can be expressed in numbers of standard deviations, which, in turn, can be expressed in terms of a mathematical probability that "chance alone" is the cause of the disparities measured. As the standard deviations increase, the probability of pure "chance" as a cause of underrepresentation diminishes; and, to the extent that the probability of chance is shown to be quite small, the inference of discrimination becomes large.

According to *EEOC v. American National Bank*, 652 F.2d 1176 (4th Cir., 1981), the courts should be extremely cautious in drawing any conclusions from standard deviations in the range of **one to three**, but above that, analysis may perhaps safely be used to exclude chance as a hypothesis. Nevertheless, most courts and commentators have accepted the .05 level, a 1-in-20 probability (approximately two standard deviations) as being indicative of statistical significance. *Dalley v. Michigan Blue Cross/Blue Shield, Inc.*, 612 F.Supp. 1444 (E.D.Mich.1985), however, rejected the notion that **two standard deviations** should be equated with a prima facie case of discrimination under Title VII. The court thus refused to acknowledge a two-standard deviation litmus test.

In *Hazelwood School Dist. v. U.S.*, 433 U.S. 299, 97 S.Ct. 2736 (1977), the United States Supreme Court determined that where "gross statistical disparities" can be shown, they alone may, in a proper case, constitute prima facie proof of a pattern or practice of discrimination. This particular case, the difference between the observed and expected values was more than **six standard deviations**.

In *Castaneda v. Partida, infra*, the High Court in a grand jury composition discrimination case, declared that as a general rule for large samples, if the difference between the expected value and the observed number is greater than **two or three standard deviations**, then the hypothesis that selection occurred without regard to race would be suspect.

Castaneda v. Partida

The United States Supreme Court, 430 U.S. 482, 97 S.Ct. 1272 (1977)

Mr. Justice BLACKMUN delivered the opinion of the Court.

The sole issue presented in this case is whether the State of Texas, in the person of petitioner, the Sheriff of Hidalgo County, successfully rebutted respondent prisoner's prima facie showing of discrimination against Mexican-Americans in the state grand jury selection process. In his brief, petitioner, in claiming effective rebuttal, asserts: "This list (of the grand jurors that indicted respondent) indicates that 50 percent of the names appearing thereon were Spanish. The record indicates that 3 of the 5 jury commissioners, 5 of the grand jurors who returned the indictment, 7 of the petit jurors, the judge presiding at the trial, and the Sheriff who served notice on the grand jurors to appear had Spanish surnames."

I.

This Court on prior occasions has considered the workings of the Texas system of grand jury selection. See Hernandez v. Texas, 347 U.S. 475, 74 S.Ct. 667, 98 L.Ed. 866 (1954);

Cassell v. Texas, 339 U.S. 282, 70 S.Ct. 629, 94 L.Ed. 839 (1950); Akins v. Texas, 325 U.S. 398, 65 S.Ct. 1276, 89 L.Ed. 1692 (1945); Hill v. Texas, 316 U.S. 400, 62 S.Ct. 1159, 86 L.Ed. 1559 (1942); Smith v. Texas, 311 U.S. 128, 61 S.Ct. 164, 85 L.Ed. 84 (1940). Texas employs the 'key man' system, which relies on jury commissioners to select prospective grand jurors from the community at large. (The other principal state mode of juror selection is a random method similar to that used in the federal system.) The procedure begins with the state district judge's appointment of from three to five persons to serve as jury commissioners. Tex.Code Crim.Proc., Art. 19.01 (1966). (During the time period covered by this case, the statute was amended to omit the requirement that the commissioners be freeholders in the county. That change has no bearing on the issues before us.)

The commissioners then 'shall select not less than 15 nor more than 20 persons from the citizens of different portions of the county' to compose the list from which the actual grand jury will be drawn. Art. 19.06 (Supp.1976–1977). Prior to 1965, the law directed the commissioners to select 'sixteen men.' The legislature amended the statute that year to substitute the words 'twenty persons' for 'sixteen men.' In 1967, the law was amended again to provide the present range of from 15 to 20 persons. These changes in the number of persons required to be on the list account for the jump, from 16 to 20 in the grand jury list statistics.

When at least 12 of the persons on the list appear in court pursuant to summons, the district judge proceeds to 'test their qualifications.' Art. 19.21. The qualifications themselves are set out in Art. 19.08: A grand juror must be a citizen of Texas and of the county, be a qualified voter in the county, be 'of sound mind and good moral character,' be literate, have no prior felony conviction, and be under no pending indictment 'or other legal accusation for theft or of any felony.' Interrogation under oath is the method specified for testing the prospective juror's qualifications. Art. 19.22. The precise questions to be asked are set out in Art. 19.23, which, for the most part, tracks the language of Art. 19.08. After the court finds 12 jurors who meet the statutory qualifications, they are impaneled as the grand jury. Art. 19.26.

II.

Respondent, Rodrigo Partida, was indicted in March 1972 by the grand jury of the 92nd District Court of Hidalgo County for the crime of burglary of a private residence at night with intent to rape. Hidalgo is one of the border counties of southern Texas. After a trial before a petit jury, respondent was convicted and sentenced to eight years in the custody of the Texas Department of Corrections. He first raised his claim of discrimination in the grand jury selection process on a motion for new trial in the State District Court. (Petitioner abandoned the waiver point in his petition for certiorari.)

In support of his motion, respondent testified about the general existence of discrimination against Mexican-Americans in that area of Texas and introduced statistics from the 1970 census and the Hidalgo County grand jury records. The census figures show that in 1970, the population of Hidalgo County was 181,535. United States Bureau of the Census, 1970 Census of Population, Characteristics of the Population, vol. 1, pt. 45, s 1, Table 119, p. 914. Persons of Spanish language or Spanish surname totaled 143,611. Ibid., and id., Table 129, p. 1092. For our purposes, the terms 'Spanish-surnamed' and 'Mexican-American' are used as synonyms for the census designation 'Persons of Spanish Language or Spanish Surname.' Persons of Spanish language include both those whose mother tongue is Spanish and all other persons in families in which the head of the household or spouse reported Spanish as the mother tongue. Persons of Spanish surname, as the census uses that term, are determined by reference to a list of 8,000 Spanish surnames compiled by the Immigration and Naturalization Service. For Texas, social and economic characteristics are presented for persons of Spanish language combined with all other persons of Spanish surname in the census reports. U.S. Bureau of the Census, 1970 Census, Characteristics of the Population, vol. 1, pt. 45, s 2, App. B.

On the assumption that all the persons of Spanish language or Spanish surname were Mexican-Americans, these figures show that 79.1% of the county's population was Mexican-American. At oral argument, counsel for petitioner appears to have suggested that the presence of illegal aliens who have Spanish surnames might inflate the percentage of Mexican-Americans in the county's population. Tr. of Oral Arg. 10–12. We cannot agree that the presence of non-citizens makes any practical difference. Table 119 of the census breaks down the 181,535 people who composed the total county population into three groups: native of native parentage, native of foreign parentage, and foreign born. The only persons as to whom the assumption of noncitizenship would be logically sustainable are the foreign born. Even for them, it is probable that some were naturalized citizens.

Furthermore, only 22,845 persons were in the 'foreign born' category. If those persons are excluded from the population of the county, the total becomes 158,690. Assuming that every foreign-born person was counted as a Spanish-surnamed person (an assumption that favors the State), the total number of Mexican-Americans is reduced from 143,611 to 120,766. Using these adjusted figures, Mexican-Americans constitute 76.1% of the county's population, a figure only 3%, and thus negligibly, smaller than the one used throughout this litigation. For consistency, we shall continue to refer to the population figures for the entire county, particularly since the State has not shown why those figures are unreliable.

Respondent's data compiled from the Hidalgo County grand jury records from 1962 to 1972 showed that over that period, the average percentage of Spanish-surnamed grand jurors was 39%. The statistics for grand jury composition can be organized as follows:

Year	No. persons on grand jury list	Av. No. Spanish surnamed per list	Percentage Spanish surnamed
1962	16	6	37.5%
1963	16	5.75	35.9%
1964	16	4.75	29.7%
1965	16.2	5	30.9%
1966	20	7.5	37.5%
1967	20.25	7.25	35.8%
1968	20	6.6	33%
1969	20	10	50%
1970	20	8	40%
1971	20	9.4	47%
1972	20	10.5	52.5%

Of the 870 persons who were summoned to serve as grand jurors over the 11-year period, 339, or 39%, were Spanish surnamed. See table showing Hidalgo County grand jury panels from 1962 to 1972, App. 17–18.

In the 2½-year period during which the District Judge who impaneled the jury that indicted respondent was in charge, the average percentage was 45.5%. On the list from which the grand jury that indicted respondent was selected, 50% were Spanish surnamed. The last set of data that respondent introduced, again from the 1970 census, illustrated a number of ways in which Mexican-Americans tend to be underprivileged, including poverty-level incomes, less desirable jobs, substandard housing, and lower levels of education.

At oral argument, counsel for petitioner suggested that the data regarding educational background explained the discrepancy between the percentage of Mexican-Americans in the total population and the percentage on the grand jury lists. Tr. of Oral Arg. 8. For a variety of reasons, we cannot accept that suggestion.

First, under the Texas method of selecting grand jurors, qualifications are not tested until the persons on the list appear in the District Court. Prior to that time, assuming an unbiased selection procedure, persons of all educational characteristics should appear on the list. If the jury commissioners actually exercised some means of winnowing those who lacked the ability to read and write, it was incumbent on the State to call the commissioners and to have them explain how this was done. In the absence of any evidence in the record to this effect, we shall not assume that the only people excluded from grand jury service were the illiterate. Second, it is difficult to draw valid inferences from the raw census data, since the data are incomplete in some places and the definition of 'literacy' would undoubtedly be the subject of some dispute in any event. The State's failure to discuss the literacy problem at any point prior to oral argument compounds the difficulties. One gap in the data occurs with respect to the younger persons in the jury pool. The census reports for educational background cover only those who are 25 years of age and above. Yet the only age limitation on eligibility for grand jury service is qualification to vote. Tex.Code Crim.Proc., Art. 19.08 (Supp.1976–1977).

During the period to which the census figures apply, a person became qualified to vote at age 21. Tex.Elec.Code, Art. 5.01 (1967). (In 1975, Art. 5.01 was amended to give the franchise to all persons 18 and over. 1975 Tex.Gen.Laws, c. 682, s 3.) It is not improbable that the educational characteristics of persons in the younger age group would prove to be favorable to Mexican-Americans. Finally, even assuming that the statistics for persons age 25 and over are sufficiently representative to be useful, a significant discrepancy still exists between the number of Spanish-surnamed people and the level of representation on grand jury lists. Table 83 of the 1970 census shows that of a total of 80,049 persons in that age group, 13,205 have no schooling. (Data for McAllen-Pharr-Edinburg Standard Metropolitan Statistical Area. This SMSA is identical to Hidalgo County.) Table 97 shows that of the 55,949 Spanish-surnamed persons in the group, 12,817 have no schooling. This means that of the 24,100 persons of all other races and ethnic groups, 388 have no schooling.

Translated into percentages, 22.9% of the Spanish-surnamed persons have no schooling, and 1.6% of the others have no schooling. This means that 43,132 of the Spanish-surnamed persons have some schooling and 23,712 of the others have some schooling. The Spanish-surnamed persons thus represent 65% of the 66,844 with some schooling, and the others 35%. The 65% figure still creates a significant disparity when compared to the 39% representation on grand juries shown over the 11-year period involved here.

The suggestion is made in the dissenting opinion of The Chief Justice, post, at 1285–1286, that reliance on eligible population figures and allowance for literacy would defeat respondent's prima facie showing of discrimination. But the 65% to 39% disparity between Mexican-Americans over the age of 25 who have some schooling and Mexican-Americans represented on the grand jury venires takes both of The Chief Justice's concerns into account. Statistical analysis indicates that the discrepancy is significant. If one assumes that Mexican-Americans constitute only 65% of the jury pool, then a detailed calculation reveals that the likelihood that so substantial a discrepancy would occur by chance is less than 1 in 50.

We prefer not to rely on the 65% to 39% disparity, however, since there are so many implicit assumptions in this analysis, and we consider it inappropriate for us, as an appellate tribunal, to undertake this kind of inquiry without a record below in which those assumptions were tested. We rest, instead, on the fact that the record does not show any way by which the educational characteristics are taken into account in the compilation of the grand jury lists, since the procedure established by the State provides that literacy is tested only after the group of 20 are summoned.

The State offered no evidence at all either attacking respondent's allegations of discrimination or demonstrating that his statistics were unreliable in any way. The State District Court, nevertheless, denied the motion for a new trial.

On appeal, the Texas Court of Criminal Appeals affirmed the conviction. Partida v. State, 506 S.W.2d 209 (1974). Reaching the merits of the claim of grand jury discrimination, the court held that respondent had failed to make out a prima facie case. In the court's view, he should have shown how many of the females who served on the grand juries were Mexican-Americans married to men with Anglo-American surnames, how many Mexican-Americans were excused for reasons of age or health, or other legal reasons, and how many of those listed by the census would not have met the statutory qualifications of citizenship, literacy, sound mind, moral character, and lack of criminal record or accusation. Id., at 210–211. Quite beyond the uncertainties in the statistics, the court found it impossible to believe that discrimination could have been directed against a Mexican-American, in light of the many elective positions held by Mexican-Americans in the county and the substantial representation of Mexican-Americans on recent grand juries. The court noted that the foreman of the grand jury that indicted respondent was Mexican-American, and that 10 of the 20 summoned to serve had Spanish surnames. Seven of the 12 members of the petit jury that convicted him were Mexican-American. In addition, the state judge who presided over the trial was Mexican-American, as were a number of other elected officials in the county. In essence, the court refused to presume that Mexican-Americans would discriminate against their own kind.

After exhausting his state remedies, respondent filed his petition for habeas corpus in the Federal District Court, alleging a denial of due process and equal protection, guaranteed by the Fourteenth Amendment, because of gross under-representation of Mexican-Americans on the Hidalgo County grand juries. At a hearing at which the state transcript was introduced, petitioner presented the testimony of the state judge who selected the jury commissioners who had compiled the list from which respondent's grand jury was taken.

The judge first reviewed the State's grand jury selection process. In selecting the jury commissioners, the judge stated that he tried to appoint a greater number of Mexican-Americans than members of other ethnic groups. He testified that he instructed the commissioners about the qualifications of a grand juror and the exemptions provided by law. The record is silent, however, with regard to instructions dealing with the potential problem of discrimination directed against any identifiable group. The judge admitted that the actual results of the selection process had not produced grand jury lists that were 'representative of the ethnic balance in the community.'

The jury commissioners themselves, who were the only ones in a position to explain the apparent substantial underrepresentation of Mexican-Americans and to provide information on the actual operation of the selection process, were never called. The Federal District Judge observed, during the state judge's testimony, that the selection process for grand jurors in Hidalgo County typically resulted in a progressive reduction of the number of Mexican-Americans involved at each stage. See Alexander v. Louisiana, 405 U.S. 625, 92 S.Ct. 1221, 31 L.Ed.2d 536 (1972). For example, said the court, if 60% of the jury commissioners were Mexican-American, the jury panel might be only 55%, and the actual grand jury only 43%. The court speculated that the reason for this might be cultural. App. 84–85.

On the basis of the evidence before it, the court concluded that respondent had made out a 'bare prima facie case' of invidious discrimination with his proof of 'a long continued disproportion in the composition of the grand juries in Hidalgo County.' 384 F.Supp. 79, 90 (S.D.Tex.1974) (emphasis in original). Based on an examination of the reliability of the statistics offered by respondent, however, despite the lack of evidence in the record justifying such an inquiry, the court stated that the prima facie case was weak. The court believed that the census statistics did not reflect the true situation accurately, because of recent changes in the Hidalgo County area and the court's own impression of the demographic characteristics of the Mexican-American community. On the other hand, the court recognized that the Texas key-man system of grand jury selection was highly subjective, and was 'archaic and inefficient,' id., at 91, and that this was a factor arguing for less tolerance in the percentage differences. On balance, the court's doubts about the reliability of the statistics, coupled with its opinion that Mexican-Americans constituted a 'governing majority' in the county, caused it to conclude that the prima facie case was rebutted. The 'governing majority' theory distinguished respondent's case from all preceding cases involving similar disparities. On the basis of those findings, the court dismissed the petition.

The court suggested that the actual discrimination operating might be economic. The jury commissioners were from the higher socio-economic classes, and they tended to select prospective jurors from among their peers. Consequently, the number of Mexican-Americans was disproportionately low, since they were concentrated at the lower end of the economic scale. We find it unnecessary to decide whether a showing of simple economic discrimination would be enough to make out a prima facie case in the absence of other evidence, since that case is not before us. Cf. Thiel v. Southern Pacific Co., 328 U.S. 217, 66 S.Ct. 984, 90 L.Ed. 1181 (1946).

The United States Court of Appeals for the Fifth Circuit reversed. 524 F.2d 481 (1975). It agreed with the District Court that respondent had succeeded in making out a prima facie case. It found, however, that the State had failed to rebut that showing. The 'governing majority' theory contributed little to the State's case in the absence of specific proof to explain the disparity. In light of the State's abdication of its responsibility to introduce controverting evidence, the court held that respondent was entitled to prevail.

We granted certiorari to consider whether the existence of a 'governing majority' in itself can rebut a prima facie case of discrimination in grand jury selection, and, if not, whether the State otherwise met its burden of proof. 426 U.S. 934, 96 S.Ct. 2645, 49 L.Ed.2d 385 (1976).

III

A.

This Court has long recognized that 'it is a denial of the equal protection of the laws to try a defendant of a particular race or color under an indictment issued by a grand jury ... from which all persons of his race or color have, solely because of that race or color, been excluded by the State....' Hernandez v. Texas, 347 U.S., at 477, 74 S.Ct., at 670. See Alexander v. Louisiana, 405 U.S. 625, 628, 92 S.Ct. 1221, 1224, 31 L.Ed.2d 536 (1972); Carter v. Jury Comm'n, 396 U.S. 320, 330, 90 S.Ct. 518, 523, 24 L.Ed.2d 549

(1970). See also Peters v. Kiff, 407 U.S. 493, 497, 92 S.Ct. 2163, 2165, 33 L.Ed.2d 83 (1972) (plurality opinion); Id., at 507, 92 S.Ct., at 2170 (dissenting opinion).

While the earlier cases involved absolute exclusion of an identifiable group, later cases established the principle that substantial underrepresentation of the group constitutes a constitutional violation as well, if it results from purposeful discrimination. See Turner v. Fouche, 396 U.S. 346, 90 S.Ct. 532, 24 L.Ed.2d 567 (1970); Whitus v. Georgia, 385 U.S. 545, 552, 87 S.Ct. 643, 647, 17 L.Ed.2d 599 (1967); Swain v. Alabama, 380 U.S. 202, 85 S.Ct. 824, 13 L.Ed.2d 759 (1965); Cassell v. Texas, 339 U.S. 282, 70 S.Ct. 629, 94 L.Ed. 839 (1950). Recent cases have established the fact that an official act is not unconstitutional solely because it has a racially disproportionate impact. Washington v. Davis, 426 U.S. 229, 239, 96 S.Ct. 2040, 2047, 48 L.Ed.2d 597 (1976); see Arlington Heights v. Metropolitan Housing Dev. Corp., 429 U.S. 252, 264–265, 97 S.Ct. 555, 563, 50 L.Ed.2d 450 (1977). Nevertheless, as the Court recognized in Arlington Heights, '(s)ometimes a clear pattern, unexplainable on grounds other than race, emerges from the effect of the state action even when the governing legislation appears neutral on its face.' Id., at 266, 97 S.Ct., at 564. In Washington v. Davis, the application of these principles to the jury cases was considered:

'It is also clear from the cases dealing with racial discrimination in the selection of juries that the systematic exclusion of Negroes is itself such an 'unequal application of the law ... as to show intentional discrimination.'... A prima facie case of discriminatory purpose may be proved as well by the absence of Negroes on a particular jury combined with the failure of the jury commissioners to be informed of eligible Negro jurors in a community, ... or with racially non-neutral selection procedures.... With a prima facie case made out, 'the burden of proof shifts to the State to rebut the presumption of unconstitutional action by showing that permissible racially neutral selection criteria and procedures have produced the monochromatic result.' Alexander (v. Louisiana, 405 U.S.,) at 632, 92 S.Ct. 1221.' 426 U.S., at 241, 96 S.Ct., at 2048.

Thus, in order to show that an equal protection violation has occurred in the context of grand jury selection, the defendant must show that the procedure employed resulted in substantial underrepresentation of his race or of the identifiable group to which he belongs. The first step is to establish that the group is one that is a recognizable, distinct class, singled out for different treatment under the laws, as written or as applied. Hernandez v. Texas, 347 U.S., at 478–479, 74 S.Ct., at 670–671. Next, the degree of underrepresentation must be proved, by comparing the proportion of the group in the total population to the proportion called to serve as grand jurors, over a significant period of time. Id., at 480, 74 S.Ct., at 671. See Norris v. Alabama, 294 U.S. 587, 55 S.Ct. 579, 79 L.Ed. 1074 (1935).

This method of proof, sometimes called the 'rule of exclusion,' has been held to be available as a method of proving discrimination in jury selection against a delineated class. Hernandez v. Texas, 347 U.S., at 480, 74 S.Ct., at 671. Finally, as noted above, a selection procedure that is susceptible of abuse or is not racially neutral supports the presumption of discrimination raised by the statistical showing. Washington v. Davis, 426 U.S., at 241, 96 S.Ct., at 2048; Alexander v. Louisiana, 405 U.S., at 630, 92 S.Ct., at 1225. Once the defendant has shown substantial underrepresentation of his group, he has made out a prima facie case of discriminatory purpose, and the burden then shifts to the State to rebut that case. The idea behind the rule of exclusion is not at all complex. If a disparity is sufficiently large, then it is unlikely that it is due solely to chance or accident, and, in the absence of evidence to the contrary, one must conclude that racial or other class-related factors entered into the selection process. See Arlington Heights v. Metropolitan

Housing Dev. Corp., 429 U.S. 252, 266 n. 13, 97 S.Ct. 555, 564, 50 L.Ed.2d 450 (1977); Washington v. Davis, 426 U.S. 229, 241, 96 S.Ct. 2040, 2048, 48 L.Ed.2d 597 (1976); Eubanks v. Louisiana, 356 U.S., at 587, 78 S.Ct., at 973; Smith v. Texas, 311 U.S., at 131, 61 S.Ct., at 165.

B.

In this case, it is no longer open to dispute that Mexican-Americans are a clearly identifiable class. Cf. White v. Regester, 412 U.S. 755, 767, 93 S.Ct. 2332, 2340, 37 L.Ed.2d 314 (1973). The statistics introduced by respondent from the 1970 census illustrate disadvantages to which the group has been subject. Additionally, as in Alexander v. Louisiana, the selection procedure is not racially neutral with respect to Mexican-Americans; Spanish surnames are just as easily identifiable as race was from the questionnaires in Alexander or the notations and card colors in Avery v. Georgia, 345 U.S. 559, 73 S.Ct. 891, 97 L.Ed. 1244 (1953).

The dissenters argue that the subjectivity of the system cuts in favor of the State where those who control the selection process are members of the same class as the person claiming discrimination. The fact remains, however, that the class to which respondent belongs was substantially underrepresented on the grand jury lists of Hidalgo County. The dissenters' argument here is another aspect of the 'governing majority' theory; under the circumstances presented in this case, that theory does not dispel the presumption of purposeful discrimination created by the combined force of the statistical showing and the highly subjective method of selection.

The disparity proved by the 1970 census statistics showed that the population of the county was 79.1% Mexican-American, but that, over an 11-year period, only 39% of the persons summoned for grand jury service were Mexican-American. (Since the 1960 census did not compile separate statistics for Spanish-surnamed persons, it is impossible to ascertain whether the percentage of Mexican-Americans in the county changed appreciably over the period of time at issue. We therefore are forced to rely on the assumption that the 79.1% figure remained constant.) This difference of 40% is greater than that found significant in Turner v. Fouche, 396 U.S. 346, 90 S.Ct. 532, 24 L.Ed.2d 567 (1970) (60% Negroes in the general population, 37% on the grand jury lists). Since the State presented no evidence showing why the 11-year period was not reliable, we take it as the relevant base for comparison.

Statistical analysis of the grand jury lists during the 2½-year tenure of the State District Judge who selected the commissioners in respondent's case reveals that a significant disparity existed over this time period as well. Thus, the District Court's assumption that reference to a shorter time period would show that the prima facie case of discrimination could not be proved was unwarranted.

The mathematical disparities that have been accepted by this Court as adequate for a prima facie case have all been within the range presented here. For example, in Whitus v. Georgia, 385 U.S. 545, 87 S.Ct. 643, 17 L.Ed.2d 599 (1967), the number of Negroes listed on the tax digest amounted to 27.1% of the taxpayers, but only 9.1% of those on the grand jury venire. The disparity was held to be sufficient to make out a prima facie case of discrimination. See Sims v. Georgia, 389 U.S. 404, 88 S.Ct. 523, 19 L.Ed.2d 634 (1967) (24.4% of tax lists, 4.7% of grand jury lists); Jones v. Georgia, 389 U.S. 24, 88 S.Ct. 4, 19 L.Ed.2d 25 (1967) (19.7% of tax lists, 5% of jury list). We agree with the District Court and the Court of Appeals that the proof in this case was enough to establish a prima facie case of discrimination against the Mexican-Americans in the Hidalgo County grand jury selection.

If the jurors were drawn randomly from the general population, then the number of Mexican-Americans in the sample could be modeled by a binomial distribution. See Finkelstein, The Application of Statistical Decision Theory to the Jury Discrimination Cases, 80 Harv.L.Rev. 338, 353–356 (1966). See generally P. Hoel, Introduction to Mathematical Statistics 58–61, 79–86 (4th ed. 1971); F. Mosteller, R. Rourke, & G. Thomas, Probability with Statistical Applications 130–146, 270–291 (2d ed. 1970). Given that 79.1% of the population is Mexican-American, the expected number of Mexican-Americans among the 870 persons summoned to serve as grand jurors over the 11-year period is approximately 688. The observed number is 339. Of course, in any given drawing some fluctuation from the expected number is predicted. The important point, however, is that the statistical model shows that the results of a random drawing are likely to fall in the vicinity of the expected value. The measure of the predicted fluctuations from the expected value is the standard deviation, defined for the binomial distribution as the square root of the product of the total number in the sample (here 870) times the probability of selecting a Mexican-American (0.791) times the probability of selecting a non-Mexican-American (0.209). Id., at 213. Thus, in this case the standard deviation is approximately 12.

As a general rule for such large samples, if the difference between the expected value and the observed number is greater than two or three standard deviations, then the hypothesis that the jury drawing was random would be suspect to a social scientist. The 11-year data here reflect a difference between the expected and observed number of Mexican-Americans of approximately 29 standard deviations. A detailed calculation reveals that the likelihood that such a substantial departure from the expected value would occur by chance is less than 1 in 10,140. The data for the 2½-year period during which the State District Judge supervised the selection process similarly support the inference that the exclusion of Mexican-Americans did not occur by chance. Of 220 persons called to serve as grand jurors, only 100 were Mexican-Americans.

The expected Mexican-American representation is approximately 174 and the standard deviation, as calculated from the binomial model, is approximately six. The discrepancy between the expected and observed values is more than 12 standard deviations. Again, a detailed calculation shows that the likelihood of drawing not more than 100 Mexican-Americans by chance is negligible, being less than 1 in 1025.

Supporting this conclusion is the fact that the Texas system of selecting grand jurors is highly subjective. The facial constitutionality of the key-man system, of course, has been accepted by this Court. See, e.g., Carter v. Jury Comm'n, 396 U.S. 320, 90 S.Ct. 518, 24 L.Ed.2d 549 (1970); Akins v. Texas, 325 U.S. 398, 65 S.Ct. 1276, 89 L.Ed. 1692 (1945); Smith v. Texas, 311 U.S. 128, 61 S.Ct. 164, 85 L.Ed. 84 (1940). Nevertheless, the Court has noted that the system is susceptible of abuse as applied. See Hernandez v. Texas, 347 U.S., at 479, 74 S.Ct., at 671. Additionally, as noted, persons with Spanish surnames are readily identifiable. It has been said that random selection methods similar to the federal system would probably avoid most of the potential for abuse found in the key-man system.

The showing made by respondent therefore shifted the burden of proof to the State to dispel the inference of intentional discrimination. Inexplicably, the State introduced practically no evidence. The testimony of the State District Judge dealt principally with the selection of the jury commissioners and the instructions given to them. The commissioners themselves were not called to testify. A case such as Swain v. Alabama, 380 U.S., at 207, n. 4, 209, 85 S.Ct., at 828, 829, illustrates the potential usefulness of such testimony, when it sets out in detail the procedures followed by the commissioners. (This is not to say, of course, that a simple protestation from a commissioner that racial considerations played no part in the selection would be enough. This kind of testimony has been found insufficient on several occasions. E.g., Alexander v. Louisiana, 405 U.S., at 632, 92 S.Ct., at 1226; Hernandez v. Texas, 347 U.S. 475, 481, 74 S.Ct. 667, 671, 98 L.Ed. 866 (1954); Norris v. Alabama, 294 U.S. 587, 598, 55 S.Ct. 579, 583, 79 L.Ed 1074 (1935). Neither is the State entitled to rely on a presumption that the officials discharged their sworn duties to rebut the case of discrimination. Jones v. Georgia, 389 U.S. 24, 88 S.Ct. 4, 19 L.Ed.2d 25 (1967).) The opinion of the Texas Court of Criminal Appeals is particularly revealing as to the lack of rebuttal evidence in the record:

'How many of those listed in the census figures with Mexican-American names were not citizens of the state, but were so-called 'wet-backs' from the south side of the Rio Grande; how many were migrant workers and not residents of Hidalgo County; how many were illiterate and could not read and write; how many. were not of sound mind and good moral character, how many had been convicted of a felony or were under indictment or legal accusation for theft or a felony; none of these facts appear in the record.' 506 S.W.2d, at 211 (emphasis added).

In fact, the census figures showed that only a small part of the population reported for Hidalgo County was not native born. Without some testimony from the grand jury commissioners about the method by which they determined the other qualifications for grand jurors prior to the statutory time for testing qualifications, it is impossible to draw any inference about literacy, sound mind and moral character, and criminal record from the statistics about the population as a whole.

These are questions of disputed fact that present problems not amenable to resolution by an appellate court. We emphasize, however, that we are not saying that the statistical disparities proved here could never be explained in another case; we are simply saying that the State did not do so in this case. See Turner v. Fouche, 396 U.S., at 361, 90 S.Ct., at 540.

C.

In light of our holding that respondent proved a prima facie case of discrimination that was not rebutted by any of the evidence presently in the record, we have only to consider whether the District Court's 'governing majority' theory filled the evidentiary gap. In our view, it did not dispel the presumption of purposeful discrimination in the circumstances of this case. Because of the many facets of human motivation, it would be unwise to presume as a matter of law that human beings of one definable group will not discriminate against other members of their group. Indeed, even the dissent of Mr. Justice POWELL does not suggest that such a presumption would be appropriate. See Post, at 1290–1291, n. 6, 1291 n. 7. The problem is a complex one, about which widely differing views can be held, and, as such, it would be somewhat precipitate to take judicial notice of one view over another on the basis of a record as barren as this. This is not a case where a majority is practicing benevolent discrimination in favor of a traditionally disfavored minority, although that situation illustrates that motivations not immediately obvious might enter into discrimination against 'one's own kind.'

Furthermore, the relevance of a governing majority of elected officials to the grand jury selection process is questionable. The fact that certain elected officials are Mexican-American demonstrates nothing about the motivations and methods of the grand jury commissioners who select persons for grand jury lists. The only arguably relevant fact in this record on the issue is that three of the five jury commissioners in respondent's case were Mexican-American. Knowing only this, we would be forced to rely on the reasoning that we have rejected that human beings would not discriminate against their own kind in order to find that the presumption of purposeful discrimination was rebutted. Without the benefit of this simple behavioral presumption, discriminatory intent can be rebutted only with evidence in the record about the way in which the commissioners operated and their reasons for doing so. It was the State's burden to supply such evidence, once respondent established his prima facie case. The State's failure in this regard leaves unchallenged respondent's proof of purposeful discrimination.

Finally, even if a 'governing majority' theory has general applicability in cases of this kind, the inadequacy of the record in this case does not permit such an approach. Among the evidentiary deficiencies are the lack of any indication of how long the Mexican-Americans have enjoyed 'governing majority' status, the absence of information about the relative power inherent in the elective offices held by Mexican-Americans, and the uncertain relevance of the general political power to the specific issue in this case. Even for the most recent time period, when presumably the political power of Mexican-Americans was at its greatest, the discrepancy between the number of Mexican-Americans in the total population and the number on the grand jury lists was substantial. Thus, under the facts presented in this case, the 'governing majority' theory is not developed fully enough to satisfy the State's burden of rebuttal.

IV.

Rather than relying on an approach to the jury discrimination question that is as faintly defined as the 'governing majority' theory is on this record, we prefer to look at all the facts that bear on the issue, such as the statistical disparities, the method of selection, and any other relevant testimony as to the manner in which the selection process was implemented. Under this standard, the proof offered by respondent was sufficient to demonstrate a prima facie case of discrimination in grand jury selection. Since the State failed to rebut the presumption of purposeful discrimination by competent testimony, despite two opportunities to do so, we affirm the Court of Appeals' holding of a denial of equal protection of the law in the grand jury selection process in respondent's case.

It is so ordered.

Mr. Justice MARSHALL, concurring.

I join fully Mr. Justice BLACKMUN's sensitive opinion for the Court. I feel compelled to write separately, however, to express my profound disagreement with the views expressed by Mr. Justice POWELL in his dissent.

As my Brother POWELL observes, post, at 1287, there are three categories of evidence in this case that bear on the ultimate question whether respondent demonstrated by a preponderance of the evidence that the State had deliberately and systematically denied to members of (respondent's class) the right to participate as jurors in the administration of justice, post, at 1292, quoting Alexander v. Louisiana, 405 U.S. 625, 628–629, 92 S.Ct. 1221, 1224, 31 L.Ed.2d 536 (1972). First, there is the statistical evidence. That evidence reveals that for at least 10 years, Mexican-Americans have been grossly underrepresented on grand juries in Hidalgo County. As Mr. Justice BLACKMUN demonstrates, ante, at 1281 n. 17, it is all but impossible that this sizable disparity was produced by chance. The statistical evidence, then, at the very least supports an inference that Mexican-Americans were discriminated against in the choice of grand jurors.

Second, there is testimony concerning the grand jury selection system employed in this case. That testimony indicates that the commissioners who constructed the grand jury panels had ample opportunity to discriminate against Mexican-Americans, since the selection system is entirely discretionary and since Spanish-surnamed persons are readily identified. Indeed, for over 35 years this Court has recognized the potential for abuse inherent in the Texas grand jury selection plan. See Smith v. Texas, 311 U.S. 128, 130, 61 S.Ct. 164, 165, 85 L.Ed. 84 (1940); Hill v. Texas, 316 U.S. 400, 404, 62 S.Ct. 1159, 1161, 86 L.Ed. 1559 (1942); Cassell v. Texas, 339 U.S. 282, 289, 70 S.Ct. 629, 632, 94 L.Ed. 839 (1950); Hernandez v. Texas, 347 U.S. 475, 479, 74 S.Ct. 667, 671, 98 L.Ed. 839 (1954). Thus the testimony concerning the selection system, by itself, only buttresses the inference of purposeful discrimination suggested by the statistics.

In every other case of which I am aware where the evidence showed both statistical disparity and discretionary selection procedures, this Court has found that a prima facie case of discrimination was established, and has required the State to explain how ostensibly neutral selection procedures had produced such nonneutral results.

This line of cases begins with the decision almost a century ago in Neal v. Delaware, 103 U.S. 370, 26 L.Ed. 567 (1881), and extends to our recent decision in Alexander v. Louisiana. Yet my Brother POWELL would have us conclude that the evidence here was insufficient to establish purposeful discrimination, even though no explanation has been offered for the marked underrepresentation of Mexican-Americans on Hidalgo County grand juries.

The sole basis for Mr. Justice POWELL's conclusion lies in the third category of evidence presented: proof of 'the political dominance and control by the Mexican-American majority in Hidalgo County,' post, at 1287. Like the District Court, he appears to assume without any basis in the record that all Mexican-Americans, indeed all members of all minority groups have an 'inclination to assure fairness' to other members of their group. Post, at 1291. Although he concedes the possibility that minority group members will violate this 'inclination,' he apparently regards this possibility as more theoretical than real. Thus he would reject the inference of purposeful discrimination here absent any alternative explanation for the disparate results. I emphatically disagree.

In the first place, Mr. Justice POWELL's assumptions about human nature, plausible as they may sound, fly in the face of a great deal of social science theory and research. Social scientists agree that members of minority groups frequently respond to discrimination and prejudice by attempting to disassociate themselves from the group, even to the point of adopting the majority's negative attitudes towards the minority. Such behavior occurs with particular frequency among members of minority groups who have achieved some measure of economic or political success and thereby have gained some acceptability among the dominant group.

But even if my Brother POWELL's behavioral assumptions were more valid, I still could not agree to making them the foundation for a constitutional ruling. It seems to me that especially in reviewing claims of intentional discrimination, this Court has a solemn responsibility to avoid basing its decisions on broad generalizations concerning minority groups. If history has taught us anything, it is the danger of relying on such stereotypes.

The question for decision here is not how Mexican-Americans treat other Mexican-Americans, but how the particular grand jury commissioners in Hidalgo County acted. The only reliable way to answer that question, as we have said so many times, is for the State to produce testimony concerning the manner in which the selection process operated. Because the State failed to do so after respondent established a prima facie case of discrimination,

I join the Court's opinion affirming the Court of Appeals.

Mr. Chief Justice BURGER, with whom Mr. Justice POWELL and Mr. Justice REHNQUIST joined, dissented.

Notes and Problems

1. In *Castenada*, to what extent did the manner in which a citizen of Hidalgo County was categorized as "*Mexican-American*" affect the determination that 79.1% of the population of Hidalgo was Mexican-American? Was a three category classification scheme sufficient?

2. Essentially, Respondent Castenada complained that even though 79.1% of Hidalgo County is comprised of Mexican-Americans, only 39% of the grand jury panels consisted of persons who had Spanish surnames. Assuming these facts to be correct, why did the High Court find that it was statistically improbable that, absent discrimination, only 39%, of the population, randomly selected, would happen to be Spanish surnamed? Were those who assembled the grand jury required to select precisely 79.1% of the Mexican-American population? What "law" of probability theory would account for Court's dissatisfaction with explanations other than intentional discrimination?

3. With respect to the probability distribution, the majority opinion in *Castenada* found that:

> The 11-year data here reflect a difference between the expected and observed number of Mexican-Americans of approximately 29 standard deviations. A detailed calculation reveals that the likelihood that such a substantial departure from the expected value would occur by chance is less than 1 in 10,140.

Using the graph in Figure 5-2 (supra), attempt to estimate the position on the scale that would correspond to 29 standard deviations above the norm.

4. Assume that there is an opaque jar with 100 marbles; 20 of them are colored green and the remaining 80 (rounded up slightly from 79.1) are colored orange. If you were to reach into the jar and select one of them at random (without looking), what is the probability that you would select an orange one? Recall from Chapter 4:

P(Orange) = n (Orange) / (n (Green) + n (Orange))

P(Green) = 80 / 100 = .80 or 80%

There is, of course, no guarantee that any particular drawing would yield an orange marble, but as the selection process continued, one would expect that the percentage would approach 80% as opposed to 39%.

§ 5.20 Inferential Statistics

As we have seen, descriptive (or deductive) statistics focus on measurements of an existing universe of data wherein all of the numbers to be scrutinized are identifiable and present. For inferential statistics, however, our identifiable data is only a subset of a larger universe consisting of unknown data. An inferential statistician might typically take a set of data, analyze it using descriptive statistics and then apply principles of inferential statistics to predict, or draw conclusions from the data. The accuracy, integrity and reliability of those conclusions are themselves statistically measurable. There is no certainty, only increasing degrees of reliability depending upon the measurement methodology.

Statistical studies and information involving the jury selection process, jury verdicts, judicial rulings, settlements and case outcomes are all of great interest to the legal profession. One particular study revealed some fascinating information about the accuracy of predicting Supreme Court decisions. It seems that a crude mathematical model could out-predict the "legal experts" by a double digit margin of accuracy. See Theodore W. Ruger, Pauline T. Kim, Andrew D. Martin, & Kevin M. Quinn, "*The Supreme Court Forecasting Project: Legal and Political Science Approaches to Predicting Supreme Court Decision Making*," 104 Columbia L.Rev. 1150 (2004).

U.S. SUPREME COURT CASE OUTCOME FORECAST

	CORRECT	INCORRECT	TOTAL
MACHINE	75.0%	25.0%	100.0%
EXPERT	50.1%	49.9%	100.0%

The "Machine" employed a relatively simple model with six factors: The applicable circuit court from which the case originated, the type of issue involved, the nature of the petitioner, the nature of the respondent, whether the lower court ruling was conservative or liberal, and whether the petitioner was claiming that law or practice was unconstitutional. The above result, sometime referred to as the Martin-Quinn Score, surprised a lot people.

With respect to inferential statistics, attorneys must be particularly suspect. The equations and measurement devices carry impressive names.[5] None guarantee certainty, and none guarantee that the initial data was accumulated in an unbiased fashion.

Tagatz v. Marquette University

861 F.2d 1040 (7th Cir., Wis. 1988)

POSNER, Circuit Judge.

Dr. Glenn Tagatz, a professor in the Marquette University education school, brought suit against Marquette under Title VII of the Civil Rights Act of 1964, as amended, 42 U.S.C. §2000 et seq., and the Age Discrimination in Employment Act of 1967, as amended, 29 U.S.C. §626(c)(1). He charged that he had received smaller pay raises than colleagues who were either Catholic or under 40 years of age. Judge Warren, after a bench trial, entered judgment for Marquette, 681 F.Supp. 1344 (E.D.Wis.1988), and Tagatz appeals.

The case is remarkable because, for the first time ever so far as we know, the plaintiff testified as his own expert witness. (We have found a case where a party's employees testified as expert witnesses; the practice was criticized by the judge. See Rust Engineering Co. v. Lawrence Pumps, Inc., 401 F.Supp. 328, 334 (D.Mass.1975).) Dr. Tagatz, a specialist in statistical evidence in employment discrimination cases, prepared the statistical evidence on which his case rides, and was permitted to introduce the evidence as an expert witness. Rule 702 of the Federal Rules of Evidence, which governs the qualification of expert witnesses, is latitudinarian, and nothing in its language suggests that a party cannot qualify as an expert; nor did Marquette object to Dr. Tagatz's testifying as an ex-

5. How can you fail to be impressed by the *Poisson Distribution Formula, The Spearman Rank Correlation Coefficient, The Lagrangian Interpolation*, and *The Gamma Measure?*

pert witness. As Dr. Tagatz's counsel pointed out at argument, the fact that a party testifying as his own expert is not disinterested does not distinguish him from any other party who testifies in his own behalf; and hired experts, who generally are highly compensated—and by the party on whose behalf they are testifying—are not notably disinterested. There is a rule against employing expert witnesses on a contingent-fee basis (that is, against paying them more for their testimony if the party that hired them wins), and this rule might be thought to imply that a party—whose "reward" for testifying depends, of course, on the outcome of the suit—is not eligible to be an expert witness. But it is a rule of professional conduct rather than of admissibility of evidence. It is unethical for a lawyer to employ an expert witness on a contingent-fee basis, 3 Weinstein's Evidence ¶706[03] at pp. 706-23 to 706-24 (1987), but it does not follow that evidence obtained in violation of the rule is inadmissible. See United States v. Cervantes-Pacheco, 826 F.2d 310, 316 (5th Cir.1987) (en banc) (concurring opinion). The trier of fact should be able to discount for so obvious a conflict of interest. In any event, there was no objection to Dr. Tagatz's testifying as an expert witness, so we need not delve deeper into this intriguing subject.

Because the case was tried, the intricate issues concerning prima facie case that Dr. Tagatz presses on us are irrelevant. United States Postal Service v. Aikens, 460 U.S. 711, 715–16, 103 S.Ct. 1478, 1481–82, 75 L.Ed.2d 403 (1983); Morgan v. South Bend Community School Corp., 797 F.2d 471, 480 (7th Cir.1986); Washington v. Electrical Joint Apprenticeship & Training Comm., 845 F.2d 710, 713–14 (7th Cir.1988). The only question is whether Judge Warren's determination that Dr. Tagatz had not proved that he received smaller raises than he would have if he had been either Catholic or young was clearly erroneous. Although Dr. Tagatz attempts to present a disparate-impact claim, he has not identified any specific employment practice that he contends has a discriminatory effect. See Watson v. Fort Worth Bank & Trust, 487 U.S. 977, 108 S.Ct. 2777, 101 L.Ed.2d 827 (1988). The only issue, therefore, is whether he proved deliberate discrimination either against non-Catholic, or against middle-aged, faculty.

Marquette is a Jesuit institution. See Maguire v. Marquette University, 814 F.2d 1213, 1214 (7th Cir.1987). It gives preference to Catholic applicants for teaching jobs in fields where Catholic doctrine is important, such as theology and philosophy, as it is permitted to do by the exemption in Title VII for religious institutions. See 42 U.S.C. § 2000e-2(e)(2); cf. Pime v. Loyola University, 803 F.2d 351, 357–58 (7th Cir.1986) (concurring opinion). But its stated policy is not to discriminate against non-Catholics once they are hired; and whether because of this policy or otherwise, Marquette declined to plead the religious exemption as a defense to Dr. Tagatz's claim of religious discrimination.

Dr. Tagatz, who is 54 and an Episcopalian, received his doctorate in educational psychology from the University of Wisconsin and was hired by Marquette as an associate professor in 1968. In 1971 he was made chairman of the department of psychology in the education school but he was removed from that job in 1974. In 1976 he published a book that received a single, brutally unfavorable review, though the dean of the school of education said he rather liked it; since then Dr. Tagatz has published only one article in a professional journal. According to the Social Science Citation Index, Dr. Tagatz's book has never been cited in a scholarly publication; and none of his articles has been cited since 1982. Compare Weinstein v. University of Illinois, 811 F.2d 1091, 1093 n. 2 (7th Cir.1987).

Beginning in 1976 the dean placed annual raises on an explicitly meritocractic basis. Scholarly output (quality as well as quantity), teaching quality, and administrative and related service to the school or university are the criteria that the dean uses to determine how large a raise to give each member of the department. Between 1976 and 1981, Dr. Tagatz received average annual raises greater than some members of the department, but below the average. He brought this suit in 1982.

The evidence of discrimination is almost entirely statistical. It consists of a series of tables, prepared by Dr. Tagatz, that compare salary raises for Catholic and non-Catholic, over-40 and under-40 faculty members in the school of education. The samples are small, because the school of education has only a small faculty. In the table that Dr. Tagatz's counsel told us at argument contained the strongest evidence for his client, the salary raises between 1975 and 1985 of each of the 28 faculty members were compared with the faculty member's religion. The comparison showed that, on average, the Catholic members of the faculty received significantly larger annual raises than the non-Catholic ones. A similar table showed larger annual raises on a percentage basis for faculty members under 40 compared to those over 40. Marquette did not offer expert testimony, although Dr. Tagatz is incorrect to say that it offered no statistical evidence.

It presented a table showing that the most productive scholars in the school of education, as measured by number of publications, happened to be Catholics. (Dr. Tagatz's own conspicuous lack of scholarly productivity—attributed by the school's dean to Dr. Tagatz's having "gone on strike" after he was fired as department chairman—helped depress the average for the non-Catholic members of the department.) Another table presented by Marquette compared Dr. Tagatz's numerical teaching evaluations with the average for the school's faculty, and showed that in the six periods for which comparable data were available the average score was 57.02, but his score was only 53.07.

There was testimony that in addition to being a below-average teacher and at the bottom of the faculty in productivity, Dr. Tagatz was a poor citizen of the school—refusing an important committee assignment, failing to go through proper channels, enlisting students in his vendettas against other faculty members (the practice so memorably described in Mary McCarthy's novel The Groves of Academe), and generally behaving in trying ways. Against this there was testimony that Dr. Tagatz is a prominent member of the department and leads it in the number and size of research grants received.

Dr. Tagatz's main argument is that if the plaintiff in a Title VII case presents evidence that passes standard tests of statistical significance, the defendant must present a statistically significant rebuttal or lose. This argument misunderstands the significance of statistical significance, both generally and in this case. One of Dr. Tagatz's tables comparing salary raises for Catholic and for non-Catholic faculty in the school of education reveals a difference (favoring the Catholics) that is significant at the .0048 level. This means (speaking very crudely—see Kruskal, Tests of Significance, in 2 Int'l Encyclopedia of Statistics 944, 957 (1978)) that there is a probability of less than 5 in 1,000 that the difference is due to chance. But to infer as Dr. Tagatz would that there is a less than 5 in 1,000 chance (in the absence of contrary evidence) that the group that received the higher raises was favored because it was composed of Catholics is erroneous. All that the data show is that there is in all likelihood a real, not a spurious, difference between the means of the samples compared. The data do not show that the difference is due to a particular attribute (namely, being Catholic) which the members of the better-off group have and the members of the worse-off group lack. Correlation is not causation. Ste. Marie v. Eastern R.R. Ass'n, 650 F.2d 395, 400 (2d Cir.1981) (Friendly, J.). Members of each group may have other things in common, and one or more of those other things, rather than religion, may have caused the difference in mean salary increases. Marquette's evidence suggests that the better-off group is not only Catholic but also more productive, and the group's higher average annual salary raises may be due to its members' greater productivity rather than to their religion. Dr. Tagatz does not question the proposition that "scholarly achievement is clearly recognized as a nondiscriminatory reason for personnel action in academic settings." Anderson v. University of Northern Iowa, 779 F.2d 441, 443 (8th Cir.1985); see also Lieberman v. Gant, 630 F.2d 60, 66 (2d Cir.1980) (Friendly, J.).

All this is not to say that Dr. Tagatz's tables are no evidence. Cf. MacDissi v. Valmont Industries, Inc., 856 F.2d 1054, 1058 (8th Cir.1988). Correlations can be suggestive of causality; scientific induction means inferring causality from correlation. Having shown a correlation between religion and salary raises, Dr. Tagatz has shown that it is possible that notwithstanding Marquette's stated policy of treating its Catholic and non-Catholic faculty members alike, Marquette is discriminating against non-Catholics. But it is only a possibility. There are many other possibilities. One is that a Catholic university such as Marquette is more attractive to Catholic academics than to non-Catholic ones; this might explain why the Catholics in the school of education are more productive on average than the non-Catholics. Maybe it is a pure coincidence that the most productive members of the school of education are Catholic; for the sample on which both parties' tables are based is very small. Certainly in a case such as this where the plaintiff's statistical evidence is weak and equivocal, the defendant can rebut it by casual statistical evidence of its own, by nonstatistical evidence, or by a combination of the two types. Then it is up to the district judge to decide whether the plaintiff has carried his burden of persuasion.

Dr. Tagatz objects very strongly to Judge Warren's attempt to do a little statistical analysis of his own. One of Dr. Tagatz's tables compares the salary raises for the five Catholic and five non-Catholic faculty members who had been continuously employed since 1969, and shows that over this period the average salary raise for the Catholics was higher than for the non-Catholics (including Dr. Tagatz). All five Catholics were full professors at the end of the period, while only two of the non-Catholics were; and Judge Warren noted that if one compared the five Catholic full professors with the two non-Catholic full professors the latter group had done better. Dr. Tagatz points out correctly that a larger sample is more reliable than a smaller one, and complains that Judge Warren made an already very small sample significantly smaller by casting out three of its ten observations. "It is an unacceptable statistical procedure to turn a large sample into a small one by arbitrarily excluding observations." Washington v. Electrical Joint Apprenticeship & Training Comm., 845 F.2d at 713. And perhaps the exclusion of the three non-Catholic professors of lower rank was arbitrary. But Dr. Tagatz's failure to control for differences in rank, like his failure to control for differences in scholarly productivity and teaching evaluations, made his table essentially worthless—and such variables can be controlled for. See, e.g., Tuckman, Gapinski & Hagemann, Faculty Skills and the Salary Structure in Academe: A Market Perspective, 67 Am.Econ.Rev. 692 (1977). Judge Warren's reworking of the table was a means of demonstrating just how worthless the table was rather than an impermissible exercise in generating factual data on which the parties were given no opportunity to comment.

Judges, by the way, are not wallflowers or potted plants; and a district judge's effort to test the strength of a party's statistical evidence by determining how sensitive it is to the design of the sample—a standard method of evaluating statistical evidence—is rather to be commended than condemned. Compare Mister v. Illinois Central Gulf R.R., 832 F.2d 1427, 1432–33 (7th Cir.1987).

We have focused on the evidence of religious discrimination. The evidence of age discrimination was if anything less impressive. The discovery that young faculty members receive larger annual raises on a percentage basis is no evidence at all of age discrimination. As is well known, the professorial, like the Roman Catholic hierarchy, has only a few ranks. And unlike the Catholic hierarchy, the top rank—full professor—is normally reached relatively early in the academic's career (Dr. Tagatz became a full professor in 1970, when he was 36).

This rank structure implies, and one observes, that academics' salaries tend to rise rapidly in the early stages of their career and to reach a plateau when the academic becomes a full professor; and this regardless of age. The phenomenon of diminishing returns to years of experience is well documented in studies of academic salaries. Of course there are exceptions, but Dr. Tagatz's evidence was only of averages. It is unlikely—and there is no evidence—that the characteristic age profile of academic salaries reflects age discrimination. Among alternative explanations, tenure—and the job security that results from it—tends to take some of the edge off academic ambition. So we would expect tenured faculty to be somewhat less productive on average than nontenured faculty, and of course most nontenured faculty are under 40, and most faculty over 40 have tenure. A countervailing factor is that tenured faculty have passed a more careful screening process than untenured.

A comparison not between a faculty member before and after he got tenure but between tenured and (different) nontenured faculty members might therefore reveal that tenured faculty were more productive. Cutting back the other way is the possibility that new faculty are more mobile than old—another reason why one might expect the new (who generally are younger) to receive higher raises. A further consideration is that tenure itself is a form of compensation, a substitute for money; this may be another reason why salary increases are higher on a percentage basis for nontenured than for tenured faculty. Given plausible (we do not suggest compelling) explanations unrelated to age discrimination for the results of Dr. Tagatz's statistical comparisons, we can hardly say that Judge Warren committed a clear error in refusing to infer, from evidence unusually meager, that Marquette was discriminating against Dr. Tagatz because of his age.

Dr. Tagatz attempted to bolster his weak statistical evidence of religious and age discrimination with shards of testimonial evidence. There was testimony that the dean had suggested to one of the non-Catholic faculty members that she not wear her Masonic ring (the Catholic Church once regarded the Masons as dangerous heretics, and the Masons returned the compliment); that he told the faculty that Catholic doctrine should be taught wherever possible; and that the younger members of the faculty were more productive than the older ones. Judge Warren was not required to give controlling weight to such evidence, especially since Dr. Tagatz teaches statistics and there is no suggestion that the dean wanted statistics taught with a Catholic slant (e.g., regression of angels on the heads of pins).

Finally, Dr. Tagatz complains that Marquette discriminates in favor of Jesuit members of the faculty by giving them free housing and other (free) benefits. It is a nice question, not necessary to answer, whether discrimination in favor not of a religion but of a religious order is within the scope of Title VII. See Pime v. Loyola University, supra, 803 F.2d at 353–54 (concurring opinion). The discrimination in favor of Jesuits, if that is what it is, is as costly to the Catholic members of the faculty, except those who are Jesuits, as it is to non-Catholics such as Dr. Tagatz. In any event, he failed to prove discrimination. It is true that the Jesuits receive various benefits that the other faculty members do not receive. But this is not the whole picture. Jesuits do not receive their salaries. Instead the university pays those salaries to an association which maintains the residence where the Jesuits live. The association deducts the cost of feeding and housing the Jesuits and various other expenses and returns the balance to the university. The amount returned is about $4,000 per year per Jesuit. So while the Jesuits do get free housing, it is not at the expense of the university; and while they get health insurance and other fringe benefits without contributing (as other faculty members do) to the cost of those benefits, there is no evidence that these benefits cost the university more than the amount that it receives from the association. So far as appears, far from discriminating in favor of its Jesuit faculty members Marquette is obtaining their services at lower cost than the services of its other faculty members.

Despite his use of the expert of his choice, Dr. Tagatz presented a weak case which was properly dismissed. Compare Laborde v. Regents of University of California, 686 F.2d 715 (9th Cir.1982); Ottaviani v. State University of New York, 679 F.Supp. 288 (S.D.N.Y.1988).

AFFIRMED.

Notes and Problems

1. Clearly, police may not stop a motorist based on race or any other invidious classification; such conduct would violate the 14th Amendment. In *State v. Soto*, 324 N.J.Super. 66, 734 A.2d 350 (1996), there was unrebutted statistical evidence of disproportionate traffic stops against African—American motorists; those statistics established a de facto policy of targeting blacks for investigation and arrest, and thus established a selective enforcement policy that violated the equal protection and due process clauses and required the suppression of all evidence seized. Defendants based their claims of institutional racism primarily on statistics. During discovery, each side created a database of all stops and arrests by State Police members patrolling a Turnpike between exits 1 and 7A out of the Moorestown, New Jersey Station for thirty-five randomly selected days between April 1988 and May 1991; they used arrest reports, patrol charts, radio logs and traffic tickets. The databases were essentially the same. Both sides counted 3060 stops which the State found to include 1212 race identified stops (39.6%); the defense found 1146 (37.4%).

2. In *Stagi v. National R.R. Passenger Corp.*, 391 Fed.Appx. 133 (3rd Cir., 2010), employees commenced a putative class action suit against Amtrak, alleging that its promotion requirement for union employees had a disparate impact on female employees in violation of Title VII. A summary judgment for the employer was reversed. The suit apparently centered around Amtrak's "one-year blocking rule" whereby, a union member was required to occupy his or her current union position for at least one year to be eligible for a managerial promotion. A prima facie case of disparate impact discrimination, said the court, has two components. First, a plaintiff must identify "the specific employment practice that is challenged." *Watson v. Ft. Worth Bank & Trust*, 487 U.S. 977 (1988). Second, the plaintiff must show that the employment practice "causes a disparate impact on the basis of race, color, religion, sex, or national origin." 42 U.S.C. § 2000. For causation, the plaintiff must present "statistical evidence of a kind and degree sufficient to show that the practice in question has caused exclusion of applicants for jobs or promotions because of their membership in a protected group." Once the plaintiff establishes a prima facie case, the burden then shifts to the employer to show that the disputed employment practice is job-related with respect to the position in question and is consistent with business necessity. As the District Court noted, there is no "rigid mathematical formula" that courts can apply to determine whether plaintiffs have established a prima facie case. The Supreme Court has not provided any definitive guidance about when statistical evidence is sufficiently substantial; but a leading treatise notes that "The most widely used means of showing that an observed disparity is sufficiently substantial to satisfy the plaintiff's burden of proving adverse impact is to show that the disparity is sufficiently large that it is highly unlikely to have occurred at random." 1 B. Lindemann & P. Grossman, *Employment*

Discrimination Law 124 (4th ed.2007). This is typically done by the use of tests of statistical significance, which determine the probability of the observed disparity occurring by chance alone. The appellate court noted that there are two related concepts associated with statistical significance:

> Measures of probability levels and standard deviation. Probability levels (also called "p-values") are simply the probability that the observed disparity is random—the result of chance fluctuation or distribution. For example, a 0.05 probability level means that one would expect to see the observed disparity occur by chance only one time in twenty cases—there is only a five percent chance that the disparity is random. A standard deviation is a unit of measurement that allows statisticians to measure all types of disparities in common terms. In this context, the greater the number of standard deviations from the mean, the greater the likelihood that the observed result is not due to chance. To offer some sense of the relationship between these two measures, two standard deviations corresponds roughly to a probability level of 0.05; three standard deviations correspond to a probability level of 0.0027. See Lindemann & Grossman 126 n. 85 and accompanying text.

The court then cited *Castaneda* and other subsequent cases establishing the acceptance of a two-standard deviation measure, *Waisome v. Port Auth.*, 948 F.2d 1370 (2d Cir., 1991), also known as the "one chance in 20" rule, *Palmer v. Shultz*, 815 F.2d 84, (D.C.Cir., 1987), and sometimes termed a ".05 level of significance." The Equal Employment Opportunity Commission (EEOC) refers to it as the "80% rule." See also *Stout v. Potter*, 276 F.3d 1118 (9th Cir., 2002) applying yet another title: the "four-fifths rule" and calling it "rule of thumb."

A selection rate for any race, sex, or ethnic group which is less than four-fifths (4/5) (or eighty percent) of the rate for the group with the highest rate will generally be regarded by the Federal enforcement agencies as evidence of adverse impact, while a greater than four-fifths rate will generally not be so regarded.

3. In *Davis v. Califano*, 613 F.2d 957 (C.A.D.C., 1979), a Title VII employment discrimination suit, the plaintiff's statistical evidence indicated a substantial disparity between the percentage of female employees in the upper salary positions at an organization in which the plaintiff was an employee, as compared to the overall percentage of female employees in those organizational units; this was sufficient to make out prima facie case of discrimination. The appellate court ruled that only the minimum objective qualifications necessary for one to be eligible for promotion must be considered in the statistical data; not every conceivable factor relevant to a promotion decision must be included in order to make out a prima facie case. Should this holding affect the conduct and level of regression analysis required by the plaintiffs?

4. Reverse discrimination suits present additional challenges for statistical analysis, especially where the intent to discriminate is admitted by the defendant. With respect to university admissions policies, a valid desire for diversity among the student rank often conflicts with traditional admissions criteria. See *Davis v. Califano*, 613 F.2d 957 (C.A.D.C., 1979), and *Regents of Univ. of Cal. v. Bakke,* 438 U.S. 265, 98 S.Ct. 2733 (1978).

5. Plaintiffs in discrimination based lawsuits will often present some anecdotal evidence in addition to a statistical presentation. But where statistics are to be considered as the sole basis for inferring a discriminatory motive in an equal protection case alleging race discrimination, the disparity between the compared groups must be so large and significant that such disparity could not be caused by chance. How "large" is large? See *Santiago v. Miles*, 774 F.Supp. 775 (W.D.N.Y., 1991) citing *Castaneda*.

§ 5.21 *Correlation*

Correlation is a concept that we all recognize. It is a ubiquitous and an extremely important concept and one that is clearly recognized by the courts.

When two social or biological phenomena increase or decrease simultaneously and proportionally subject to identical external factors, they are said to correlate positively. Under the same circumstances, where one factor increases in the same proportion as the other decreases, the two are said to be negatively correlated.

A. The Coefficient of Correlation

The coefficient of correlation is an important number. Standing alone, it implies nothing unless it is predicated upon large numbers. The numerical value of the coefficient of correlation ranges between -1 and +1. For statistics involving sociological issues, perfect negative correlation (-1) and perfect positive correlation (+1) is rare. For many principles involving physics or mathematics, however, such results are not rare. For example, there is a perfect positive correlation between the radius of a circle and its area or its circumference. The coefficient of linear correlation is defined as follows:

$$r = \frac{n(\Sigma xy) - (\Sigma x)(\Sigma y)}{\sqrt{(n(\Sigma x^2) - (\Sigma x)^2)\,(n(\Sigma y^2) - (\Sigma y)^2)}}$$

Example 5-1: We suspect that for children between the ages of 4 and 18, as the child's age increases, so too does the number of hours spent playing video games. We do not know this to be true, and therefore, the conjecture is referred to as a **speculative hypothesis.**

For purposes of objectivity, the statistician might then typically refer to the **null hypothesis**, which is tantamount to an assumption that the speculative hypothesis is untrue. Stated differently, by statistical convention it might be stated that the speculated hypothesis is wrong (i.e., a null hypothesis) and that any perceived relationship between age and video game play is merely a matter of chance or coincidence. The statistician will then proceed to prove that the null hypothesis is incorrect and therefore nullified, thereby implying that an **alternative hypothesis** (i.e, the speculative hypothesis) is true.

It is universally accepted that a larger **sample population** yields greater accuracy than a smaller one. Nevertheless, in order to keep the calculations simple and manageable, we are going to consider just 5 children between the ages of 4 and 18:

Name	Age	Hours/Week
Dino	5	14
Desi	9	17
Donny	12	16
Denny	14	18
Bulgatti	17	23

Table 5-1

(x, y)	x	x^2	y	y^2	xy
(5, 14)	5	25	14	196	70
(9, 17)	9	81	17	289	153
(12, 16)	12	144	16	256	192
(14, 18)	14	196	18	324	252
(17, 23)	17	289	23	529	391
n = 5	$\Sigma x = 57$	$\Sigma x^2 = 735$	$\Sigma y = 88$	$\Sigma y^2 = 1{,}594$	$\Sigma xy = 1{,}058$

$$r = \frac{n(\Sigma xy) - (\Sigma x)(\Sigma y)}{\sqrt{(n(\Sigma x^2) - (\Sigma x)^2)\,(n(\Sigma y^2) - (\Sigma y)^2)}}$$

$$r = \frac{5(1058) - (57)(88)}{\sqrt{(5(735) - (57)^2)\,(5(1594) - (88)^2)}}$$

$$r = 0.883062706$$

Once the coefficient of correlation (often simply referred to as "correlation") has been calculated, there are many ways to interpret its significance depending upon the elements that are being studied. Table 5-2 provides some widely accepted rules of thumb.

Table 5-2

Correlation Coefficient	Correlation	Relationship
0–.20	Slight	Negligible
.20–.40	Low	Small
.40–.70	Moderate	Substantial
.70–.90	High	Marked
.90–1.00	Very High	Very Dependable

We start with two sets of data (i.e., list of numbers) and try to determine whether or not there is a relationship between them. Normally, we place one set on the X-axis and the other set on the Y-axis and then determine the pattern, if any. In many instances, there might not be any pattern implying that the correlation coefficient is zero.

If for, example, we were to consider a group of people and measure their IQs as compared to the number of digits in their home addresses, we would probably end with a zero correlation. On the other hand, if we were to consider a list of weights for cement blocks and compare that list to a list of the force exerted by those blocks, the correlation between the weights and the forces would be a perfect correlation of 1.0 The Newtonian laws of physics (in this case, Force = mass × acceleration) would guarantee it.

In most instances, the correlation coefficient will lies somewhere between -1.0 and +1.0 because there may be a relation between the sets of data that is less than perfect. If, for example, we were to compare a male's height with that of his father, we would expect a relationship that would be less than perfect.

Consider the following chart that seeks to compare the heights of fathers and their respective sons in order to determine whether or not there is any significant statistical correlation between the two:

Father	65	63	67	64	68	62	70	66	68	67	69	71
Son	68	66	68	65	69	66	68	65	71	67	68	70

Table 5-3

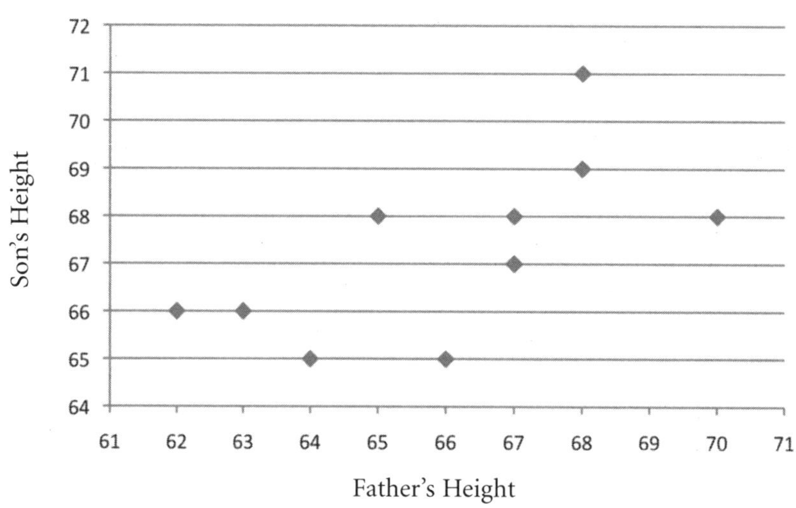

Notice that while the dots are not perfectly linear, the son's height, nevertheless trends upward as the father's height increases. Such a result tends to conform to our genetic expectations. We might predict from the chart alone, that there is a moderate to substantial correlation (.7027) between the height of the father and the height of the son. We might also conclude from the graph that sons tend to be taller than their fathers.

By contrast to the prior Table, consider the two sets of data presented in Table 5-4, which have no correlation to one another:

Table 5-4

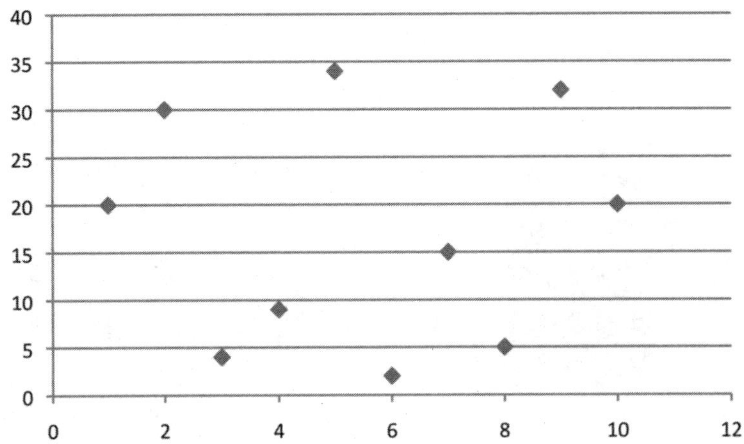

A negative coefficient of correlation signifies nothing more than a declining as opposed to an increasing pattern. It also signifies an inverse relationship. Table 5-5 might, for example, represent the number of hours during the day that a group of girls played with Barbie Dolls as their ages increased. The graph simply shows that as they grew older, their interest in Barbie decreased.

Table 5-5

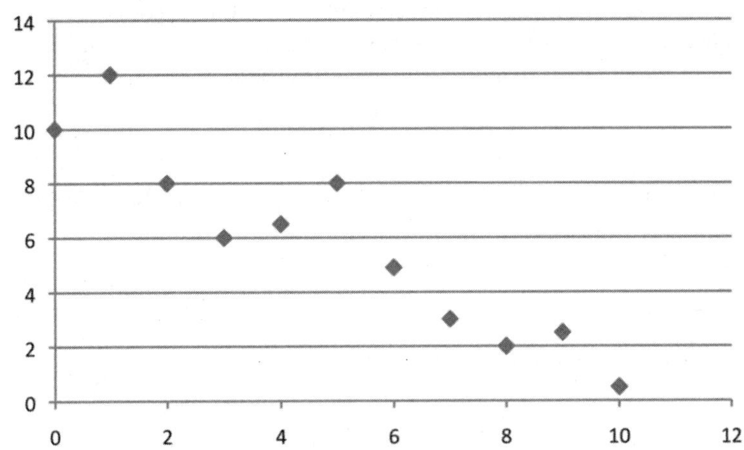

§ 5.22 Regression Analysis

Figure 5-3:

Not the regression we're referring to.

Figure 5-4:

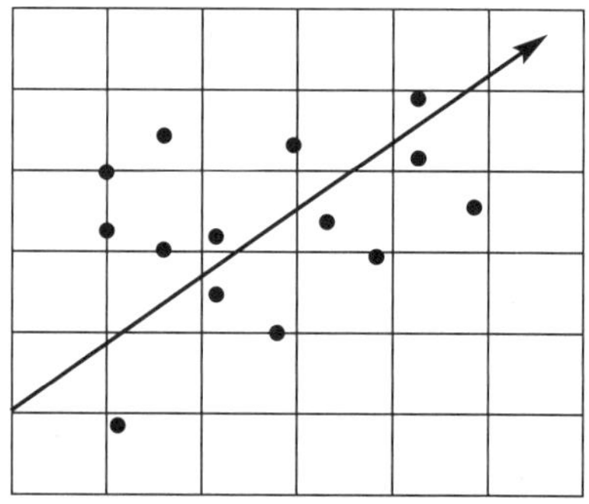

The regression we're referring to.

It's difficult to think of a more sophisticated name ("regression analysis") for drawing a simple line though a bunch of dots. Essentially that is what regression analysis is all about. But the process should not be trivialized. The primary purpose of regression analysis is to literally cut though a set of scattered data to derive a simple straight (or possibly curved) roadmap. As contrasted to the left figure of the lady with the bubble gum (referred to as a "scatterbrain"), our scattered data is sometimes referred to as a "scattergram."

The goal of the statistician is the determination of a straight line that connects all of the dots in the scattergram. Such a result would produce a 1.0 or -1.0 coefficient of correlation and suggest a strong causal connection between the X-axis and Y-axis data. The line should intercept the Y-axis in conjunction with the following "Point-Slope" equation that we customarily used in Analytic Geometry (b is sometimes referred to as the "Y-axis intercept"):

$$y = mx + b$$

Where a perfect line through all of the dots does not exist, there are several other potential curves to best satisfy the data:

$y = be^{mx}$ (Exponential)

$y = bx^{m}$ (Power)

$y = b + m (L_n x)$ (Logarithmic)

With our selected independent variable running across the X-axis, and our dependent variable along the Y-axis, a perfect correlation (+/- 1) will rarely occur. There may be other factors (and other independent variables) contributing the subject outcome. Regression analysis is designed to examine and then either confirm or eliminate the "other" potential independent variables. For example, in *Segar v. Smith*, 738 F.2d 1249 (D.C. Cir., 1984), the showing of salary regressions meeting or exceeding a .05 level of statistical significance supported an inference of Title VII discrimination with regard to salary, entry grade, evaluations and discipline.

Ottaviani v. State University of New York at New Paltz

875 F.2d 365, (2nd Cir., N.Y., 1989)

PIERCE, Circuit Judge

This is an appeal from a judgment of the United States District Court for the Southern District of New York, Kram, J., in which the court found in favor of defendants on all of the Title VII claims asserted by individual faculty members and a class of similarly situated plaintiffs, following a lengthy bench trial.

The district court ruled in favor of plaintiff-intervenor Harriet Klapper on her Equal Pay Act claim. That decision is not challenged on appeal by defendants, and appellants rely on the favorable ruling only insofar as it supports their class-wide claims of discrimination.

The decision of the district court is published in a thorough and lengthy opinion at 679 F.Supp. 288 (S.D.N.Y.1988), familiarity with which is assumed herein. Appellants contend the district court erred in its decision and principally attack the district court's treatment of the evidence presented in support of their Title VII claims. For the reasons that follow, we affirm.

BACKGROUND

This complicated Title VII suit was commenced by and on behalf of full-time, academic rank female faculty members at the State University of New York ("SUNY") at New Paltz ("the University") who were employed in the University's Division of Liberal Arts and Sciences at any time between academic years 1973 and 1984. The plaintiffs alleged that between 1973 and 1984, the University discriminated against female members of its faculty on the basis of gender in three separate categories: (1) placement in initial faculty rank at the University, (2) promotion into higher rank, and (3) salary. Judge Kram conducted a bench trial which extended over nine months on all of the plaintiffs' claims, and both parties presented extensive evidence to the court. For the sake of brevity, we will discuss only so much of the proceedings below as is relevant to our discussion of the key issues raised on appeal.

During the trial, the district court basically considered two types of evidence—objective statistical evidence and extensive "anecdotal" evidence. The statistical evidence presented by both sides consisted primarily of data produced by means of various "multiple regression analyses." Depending upon the party presenting the statistical evidence, the data was intended to either demonstrate or rebut the plaintiffs' claim of a pattern of ongoing discrimination against women within the University in all three of the contested categories.

A. The Statistical Evidence

Multiple regression analysis is a statistical tool commonly used by social scientists to determine the influence that various independent, predetermined factors (so-called "independent variables") have on an observed phenomenon (the so-called "dependent variable"). See Eastland v. Tennessee Valley Auth. 704 F.2d 613, 621 (11th Cir.1983), cert. denied, 465 U.S. 1066, 104 S.Ct. 1415, 79 L.Ed.2d 741 (1984); Fisher, Multiple Regression in Legal Proceedings, 80 Colum.L.Rev. 702, 702, 705–06 (1980).

In disparate treatment cases involving claims of gender discrimination, plaintiffs typically use multiple regression analysis to isolate the influence of gender on employment decisions relating to a particular job or job benefit, such as salary. See, e.g., Sobel v. Yeshiva Univ., 839 F.2d 18, 21–22 (2d Cir.1988); EEOC v. Sears, Roebuck & Co., 839

F.2d 302, 324–25 & n. 22 (7th Cir.1988); Palmer v. Schultz, 815 F.2d 84, 90–91 (D.C.Cir.1987).

The first step in such a regression analysis is to specify all of the possible "legitimate" (i.e., nondiscriminatory) factors that are likely to significantly affect the dependent variable and which could account for disparities in the treatment of male and female employees. See Sobel, 839 F.2d at 20–21; Segar v. Smith, 738 F.2d 1249, 1261 (D.C.Cir.1984), cert. denied, 471 U.S. 1115, 105 S.Ct. 2357, 86 L.Ed.2d 258 (1985); Fisher, supra, at 713–14. By identifying those legitimate criteria that affect the decision making process, individual plaintiffs can make predictions about what job or job benefits similarly situated employees should ideally receive, and then can measure the difference between the predicted treatment and the actual treatment of those employees. If there is a disparity between the predicted and actual outcomes for female employees, plaintiffs in a disparate treatment case can argue that the net "residual" difference represents the unlawful effect of discriminatory animus on the allocation of jobs or job benefits. See Palmer, 815 F.2d at 90–91; D. Baldus & J. Cole, Statistical Proof of Discrimination § 3.2, at 94 (1980); id. § 8.02[1], at 245–46. Another way in which statisticians can measure the influence of gender on a particular employment decision is by using gender as one of the independent variables in a regression analysis. For each independent variable in a multiple regression analysis, the statistician calculates a coefficient, which is a measure of the effect that the variable has on the dependent variable being examined. If the regression coefficient for gender is sufficiently large, then it is probative of the impact that gender has on the employment decision at issue. D. Baldus & J. Cole, supra, §§ 8.01 to 8.02[1], at 240–45; see, e.g., Segar, 738 F.2d at 1261–62.

In this case, the parties' statistical experts each determined what factors they thought were relevant to the setting of salaries and rank at the University, and used those factors as independent variables in their multiple regression analyses. By accounting for all of the "legitimate" factors that could affect salary and rank in general, the plaintiffs hoped to prove that there was a net "residual" difference or disparity between the predicted and actual salaries and rank of female faculty members that could only be attributed to ongoing gender discrimination within the University. Conversely, the defendants sought to attribute observed disparities in the pay and rank of male versus female faculty members to "legitimate" factors such as unequal job qualifications.

1. Plaintiffs' Proof of Salary Discrimination

a. Plaintiffs' Main Salary Study

The plaintiffs' main salary study was contained in Trial Exhibit 882 and purported to demonstrate the difference in salaries between male and female faculty members at New Paltz. According to the plaintiffs' statistical expert, Dr. Mary Gray, women actually earned from $1,036 to $2,277 less than their predicted salaries in each year of the class period. The defendants challenged these findings on several grounds, but principally attacked the plaintiffs' study for its failure to include certain independent variables which the defendants claimed were influential in the setting of faculty salaries at the University.

The plaintiffs' main salary study incorporated the following independent variables: (1) number of years of full time teaching experience prior to hire at New Paltz; (2) number of years' teaching experience in academic rank at New Paltz; (3) possession of a doctorate degree; (4) number of years since obtaining the doctorate degree; (5) number of publications; (6) other experience prior to hire at New Paltz; and (7) years of full-time high school teaching experience. The plaintiffs' statistical expert, however, did not include academic rank variables in her main salary study such as prior rank, current rank, and years in current rank. Although Dr. Gray conceded that these three factors may influence salary decisions, she maintained that academic rank itself was subject to discrimination at New Paltz, and that the use of rank variables would therefore be inappropriate.

In connection with this assertion, the plaintiffs attempted to demonstrate that female faculty members were placed in lower academic ranks at New Paltz than their male counterparts, and promoted more slowly into higher academic ranks than their male counterparts, solely because of their gender. (There are four types of "academic rank" at New Paltz: (1) professor, (2) associate professor, (3) assistant professor, and (4) instructor. Faculty members in one of these academic ranks either hold tenure or are on a "tenure track.") The defendants' statistical expert, Dr. Judith Stoikov, responded by attempting to prove that rank at New Paltz was not discriminatory. After considering all the evidence as to rank, the district court rejected plaintiffs' proof as "unpersuasive," and concluded that plaintiffs had "failed to prove that rank at New Paltz was discriminatory." Ottaviani, 679 F.Supp. at 306.

The district court's rejection of plaintiffs' claims as to discrimination in rank at New Paltz had two important consequences for the plaintiffs' case. First, the court's ruling eliminated two of the contested categories of discrimination at New Paltz, and left the salary discrimination claim as plaintiffs' only remaining Title VII claim. Second, and equally important from the plaintiffs' perspective, the court's ruling "validated" academic rank as one of the legitimate factors to consider in accounting for salary disparities between male and female faculty members. Since the court considered the academic rank of faculty members to be a legitimate influencing factor on faculty salaries at New Paltz, and since the plaintiffs' main salary study failed to include academic rank variables, the court found the plaintiffs' principal study to be fundamentally flawed and less probative of discrimination than it otherwise might have been.

b. Plaintiffs' Other Salary Studies

Apart from their main salary study, the plaintiffs had also performed salary regressions which did include rank variables. Since these other studies did include what the court considered to be most of the relevant legitimate factors which could influence salary at New Paltz, the court accordingly looked primarily to these studies to determine whether the plaintiffs had made out a prima facie case of gender discrimination.

After considering and weighing all the evidence presented, the district court reached certain conclusions with respect to both the plaintiffs' and the defendants' statistical evidence. While the district judge found some of the plaintiffs' statistical evidence "persuasive," she thought that it was insufficient to establish a prima facie case of gender discrimination. On the other hand, the district judge did not believe that defendants' statistical evidence was sufficient to rebut the plaintiffs' discrimination claims altogether.

Since the judge found the statistical evidence to be inconclusive one way or the other, she ruled that whether or not the plaintiffs could prevail on their discrimination claims would depend upon whether the totality of the evidence adduced at trial supported a finding of discrimination. Accordingly, the district judge next considered whether the extensive anecdotal evidence proffered by plaintiffs supported their claims of discrimination.

B. The Anecdotal Evidence

The anecdotal evidence at trial consisted of various narrative descriptions of events at the University which the plaintiffs contended illustrated or proved that the University had discriminated against its female faculty members. Specifically, the plaintiff class members sought to establish that: (1) the University did not have a viable affirmative action program; (2) New Paltz's methods for identifying and correcting existing salary inequities from 1973 to 1984 were either flawed or non-existent; (3) the University either retrenched or eliminated faculty positions to the detriment of its female faculty members; and (4) the University demonstrated a disdain for women's issues through its handling of the Women's Studies Program at New Paltz. Eleven witnesses also testified about individual instances of alleged salary discrimination at New Paltz, which the plaintiffs contended were illustrative of the administration's policies toward women as a whole. The plaintiffs presented testimonial evidence of discrimination against: three New Paltz faculty members who were neither class members nor individual plaintiffs, four individual members of the class who had not brought individual Title VII claims, and four members of the class who had asserted individual Title VII claims as well.

On rebuttal, the defendants sought to negate the plaintiffs' claims through the specific testimony of University administrators and faculty members, and other types of anecdotal evidence. The defendants contended that such evidence demonstrated that there were nondiscriminatory reasons for all of the actions taken by the University during the period in question which negatively affected its female faculty members, and that none of the employment practices at issue were motivated by discriminatory animus.

After reviewing the anecdotal evidence, the district judge held that the plaintiffs had not proven their Title VII claims against the University. Although she found that the anecdotal evidence supported an inference of prima facie discrimination in a few of the individual class members' cases, in each of those cases she either accepted the defendants' explanations for the pay disparities, or found that the isolated incidents of discrimination were insufficient to support the class' claim of a pattern or practice of gender discrimination. Accordingly, the district court entered judgment in favor of defendants on all of the Title VII claims.

On appeal, the appellants contend that the district court erred in its treatment and analysis of the evidence in several key respects, and that as a result, the court's finding of no discrimination was erroneous. First, appellants challenge the district court's determination that the statistical evidence was inconclusive. Appellants contend that the statistical evidence adduced at trial was more than sufficient to establish a prima facie case of gender discrimination as a matter of law.

Moreover, they also contend that the district judge's decision to allow allegedly "tainted" variables such as "rank" to be used in the multiple regression analyses minimized the overall impact of defendants' alleged discriminatory treatment of female faculty members, and resulted in weaker statistical proof. Appellants also take issue with the district court's rejection of the proffered anecdotal evidence of discrimination. Finally, appellants contend that the district court erroneously excluded or ignored evidence of pre-Title VII discrimination, in contravention of the Supreme Court's decision in Bazemore v. Friday, 478 U.S. 385, 106 S.Ct. 3000, 92 L.Ed.2d 315 (1986).

For the reasons that follow, we hold that Judge Kram did not clearly err in finding in favor of the defendants, and we affirm the decision of the district court.

DISCUSSION

We begin by noting that the district court correctly stated the familiar legal standards to be applied in Title VII cases. Since the plaintiff class herein had raised a "disparate treatment" claim under Title VII, the claimants bore the burden of not only establishing discriminatory intent on the part of SUNY administrators, but that "unlawful discrimination [was] a regular procedure or policy followed by [the University]." International Bhd. of Teamsters v. United States, 431 U.S. 324, 360, 97 S.Ct. 1843, 1867, 52 L.Ed.2d 396 (1977); see Coser v. Moore, 739 F.2d 746, 749 (2d Cir.1984) ("In order to prevail on their claim of a pattern and practice of discrimination, plaintiffs had to show by a preponderance of the evidence that [the defendant]'s 'standard operating procedure—the regular rather than the unusual practice' is to discriminate on the basis of sex.") (quoting Teamsters, 431 U.S. at 336, 97 S.Ct. at 1855).

The Supreme Court has established a three-step process for evaluating disparate treatment claims brought pursuant to Title VII. McDonnell Douglas Corp. v. Green, 411 U.S. 792, 802–04, 93 S.Ct. 1817, 1824–25, 36 L.Ed.2d 668 (1973); Woodbury v. New York City Transit Auth., 832 F.2d 764, 769 (2d Cir.1987). First, the plaintiffs bear the burden of establishing a prima facie case of discrimination by a preponderance of the evidence, McDonnell Douglas Corp., 411 U.S. at 802, 93 S.Ct. at 1824, which gives rise to a rebuttable presumption of unlawful discrimination, Texas Dep't of Community Affairs v. Burdine, 450 U.S. 248, 254 & n. 7, 101 S.Ct. 1089, 1094 & n. 7, 67 L.Ed.2d 207 (1981). Once they have done so, the burden of production shifts to the defendants to "articulate some legitimate, nondiscriminatory reason" for the challenged employment practice. McDonnell Douglas Corp., 411 U.S. at 802, 93 S.Ct. at 1824. In cases where the plaintiff has relied on statistical evidence to establish a prima facie case of discrimination, the defendants may also attempt to undermine the plaintiffs' prima facie case by attacking the validity of that statistical evidence, or by introducing statistical evidence of their own showing that the challenged practice did not result in disparate treatment. See Berger v. Iron Workers Reinforced Rodmen Local 201, 843 F.2d 1395, 1412 (D.C.Cir.1988) (citing Teamsters, 431 U.S. at 360, 97 S.Ct. at 1867); Coble v. Hot Springs School Dist. No. 6, 682 F.2d 721, 730 (8th Cir.1982). If the defendants meet this burden, the plaintiffs must then show either that the defendants' statistical proof is inadequate, or that the defendants' explanation for the challenged practice is merely a pretext for discrimination. See Zahorik v. Cornell Univ., 729 F.2d 85, 92 (2d Cir.1984).

Even though the burden of production shifts to the defendants during the second stage of the process, the ultimate burden of persuasion rests always with the plaintiffs to prove their claims of discrimination. See United States Postal Serv. Bd. of Governors v. Aikens, 460 U.S. 711, 716, 103 S.Ct. 1478, 1482, 75 L.Ed.2d 403 (1983); Burdine, 450 U.S. at 253, 101 S.Ct. at 1093. As stated earlier, appellants contend herein that they met their burden of proof, and that the evidence at trial clearly established a pattern and practice of discrimination by University administrators with respect to faculty rank and salary. We will consider plaintiffs' principal claims of error seriatim, beginning with the district court's treatment of the statistical evidence adduced at trial.

A. Significance of Plaintiffs' Statistical Evidence

At trial, the plaintiffs herein contended that the statistical evidence alone was sufficient to establish a prima facie case of discrimination. According to plaintiffs, female faculty members were clearly treated less favorably than their male counterparts, and that unfavorable, disparate treatment was due solely to gender bias. The district court, however, found that the plaintiffs' statistical evidence was not "statistically significant" enough to establish a prima facie case of discrimination. For the reasons that follow, we conclude that the district court did not clearly err in ruling that the plaintiffs' proffered statistical evidence was not dispositive of their Title VII claims.

As discussed earlier, plaintiffs in a disparate treatment case frequently rely on statistical evidence to establish that there is a disparity between the predicted and actual treatment of employees who are members of a disadvantaged group, and to argue that such disparities exist because of an unlawful bias directed against those employees. Not all disparities, however, are probative of discrimination. Before a deviation from a predicted outcome can be considered probative, the deviation must be "statistically significant."

Statistical significance is a measure of the probability that a disparity is simply due to chance, rather than any other identifiable factor. See Segar, 738 F.2d at 1282; see also Castaneda v. Partida, 430 U.S. 482, 496 n. 17, 97 S.Ct. 1272, 1281 n. 17, 51 L.Ed.2d 498 (1977); Mister v. Illinois Cent. Gulf R. Co., 832 F.2d 1427, 1430–31 (7th Cir.1987), cert. denied, 485 U.S. 1035, 108 S.Ct. 1597, 99 L.Ed.2d 911 (1988). Because random deviations from the norm can always occur, see Frazier v. Consolidated Rail Corp., 851 F.2d 1447, 1451–52 (D.C.Cir.1988); Palmer, 815 F.2d at 91; Mister, 832 F.2d at 1430–31; Fisher, supra, at 705, statisticians do not consider slight disparities between predicted and actual results to be statistically significant. See Eubanks v. Pickens-Bond Constr. Co., 635 F.2d 1341, 1347–48 (8th Cir.1980); Fisher, supra, at 706. As the disparity between predicted and actual results becomes greater, however, it becomes less likely that the deviation is a random fluctuation. When the probability that a disparity is due to chance sinks to a certain threshold level, statisticians can then infer from the statistical evidence, albeit indirectly, that the deviation is attributable to some other cause unrelated to mere chance. See D. Baldus & J. Cole, supra, § 9.42, at 191–93 (Supp.1987); Berger, 843 F.2d at 1412.

One unit of measurement used to express the probability that an observed result is merely a random deviation from a predicted result is the "standard deviation." D. Baldus & J. Cole, supra, § 2.224, at 63; id. § 9.03, at 294–95; see also Berger, 843 F.2d at 1412; Coates v. Johnson & Johnson, 756 F.2d 524, 536 n. 11 (7th Cir.1985). The standard deviation "is a measure of spread, dispersion or variability of a group of numbers." D. Baldus & J. Cole, supra, at 359.

Generally, the fewer the number of standard deviations that separate an observed from a predicted result, the more likely it is that any observed disparity between predicted and actual results is not really a "disparity" at all but rather a random fluctuation. Conversely, "[t]he greater the number of standard deviations, the less likely it is that chance is the cause of any difference between the expected and observed results." Coates, 756 F.2d at 536 n. 11; see Woodbury, 832 F.2d at 770; Segar, 738 F.2d at 1282. A finding of two standard deviations corresponds approximately to a one in twenty, or five percent, chance that a disparity is merely a random deviation from the norm, and most social scientists accept two standard deviations as a threshold level of "statistical significance." See Castaneda, 430 U.S. at 496 n. 17, 97 S.Ct. at 1281 n. 17; Melani v. Board of Higher Educ., 561 F.Supp. 769, 774 (S.D.N.Y.1983); Cooper v. University of Tex. at Dallas, 482 F.Supp. 187, 194 (N.D.Tex.1979) (citing N. Nie, C. Hull, J. Jenkins, K. Steinbrenner & D. Bent, Statistical Package for the Social Sciences 222 (2d ed. 1975)), aff'd, 648 F.2d 1039 (5th Cir.1981). When the results of a statistical analysis yield levels of statistical significance at or below the 0.05 level, chance explanations for a disparity become suspect, and most statisticians will begin to question the assumptions underlying their predictions.

Cognizant of the important role that statistics play in disparate treatment cases, the Supreme Court has held that "[w]here gross statistical disparities can be shown, they alone may in a proper case constitute prima facie proof of a pattern or practice of discrimination." Hazelwood School Dist. v. United States, 433 U.S. 299, 307–08, 97 S.Ct. 2736, 2741–42, 53 L.Ed.2d 768 (1977). The threshold question in disparate treatment cases, then, is: "[A]t what point is the disparity in selection rates ... sufficiently large, or the probability that chance was the cause sufficiently low, for the numbers alone to establish a legitimate inference of discrimination [?]" Palmer, 815 F.2d at 92 (emphasis added); see Frazier, 851 F.2d at 1452. In answer to this question, "most courts follow the conventions of social science which set 0.05 as the level of significance below which chance explanations become suspect." D. Baldus & J. Cole, supra, § 9.02, at 291; see, e.g., Castenada, 430 U.S. at 496 n. 17, 97 S.Ct. at 1281 n. 17; Frazier, 851 F.2d at 1452; Berger, 843 F.2d at 1412; Segar, 738 F.2d at 1262. The existence of a 0.05 level of statistical significance indicates that it is fairly unlikely that an observed disparity is due to chance, and it can provide indirect support for the proposition that disparate results are intentional rather than random. The commentators are careful to point out, however, that no matter how great the number of standard deviations is, statistical tests can never entirely rule out the possibility that chance caused the disparity. See Palmer, 815 F.2d at 91; D. Baldus & J. Cole, supra, § 9.42, at 191–93 (Supp.1987). By no means, however, is a five percent probability of chance (or approximately two standard deviations) considered an "exact legal threshold." Palmer, 815 F.2d at 92. In the present case, the three salary studies which the district court considered most probative of a pattern or practice of discrimination produced a range of standard deviations between approximately one and five, and of the total thirty-three standard deviation measures cited, twenty-four exceeded two standard deviations. Significantly, however, nine of the measures cited fell below two standard deviations. Also, the negative residuals associated with being female were not significant in every year of the liability period.

As discussed supra, two standard deviations corresponds roughly to a 1 in 20 chance that the outcome is a random fluctuation. Three standard deviations corresponds to approximately a 1 in 384 chance of randomness. Finally, a range of four to five standard deviations corresponds to a probability range of 1 chance in 15,786 to 1 chance in 1,742,160.

Given the range of standard deviations associated with their salary regressions, the plaintiffs contended that the statistical evidence clearly gave rise to a presumption of discrimination. As discussed earlier herein, however, although the district judge found the studies to be "persuasive," she nevertheless held that these levels of "statistical significance" alone were "not sufficiently high to support a prima facie claim of salary discrimination." Ottaviani, 679 F.Supp. at 309.

On appeal, appellants argue inter alia that, as a matter of law, a finding of two standard deviations should be equated with a prima facie case of discrimination. According to appellants, the district court therefore erred in finding that they had not met their burden of establishing a prima facie case. In support of this argument, appellants point out that several courts have accepted two standard deviations as prima facie proof of discrimination. See, e.g., Berger, 843 F.2d at 1412 ("if the likelihood that a fluctuation from expected results occurred by chance is five percent or less, a statistically significant difference is proved, and a prima facie case of discrimination is established") (citing Segar, 738 F.2d at 1282–83); Eldredge v. Carpenters 46 N. Cal. Counties JATC, 833 F.2d 1334, 1340 n. 8 (9th Cir.1987) (. 045 level of statistical significance (approximately two standard deviations or 1 chance in 22) sufficient to give rise to an inference that discriminatory system rather than chance is responsible for women's lower admission rates to apprenticeship program), cert. denied, 487 U.S. 1210, 108 S.Ct. 2857, 101 L.Ed.2d 894 (1988); Dalley v. Michigan Blue Cross/Blue Shield, Inc., 612 F.Supp. 1444, 1451 n. 18 (E.D.Mich.1985) ("Most courts and commentators have accepted the .05 level," or 1 in 20 probability, as indicative of statistical significance). While appellants' argument that a finding of two standard deviations should be equated with a prima facie case of discrimination under Title VII is not without initial appeal, we are constrained to reject such a formal "litmus" test for assessing the legitimacy of Title VII claims.

It is certainly true that a finding of two to three standard deviations can be highly probative of discriminatory treatment. See Segar, 738 F.2d at 1282–83. As tempting as it might be to announce a black letter rule of law, however, recent Supreme Court pronouncements instruct that there simply is no minimum threshold level of statistical significance which mandates a finding that Title VII plaintiffs have made out a prima facie case. See, e.g., Watson v. Fort Worth Bank & Trust, 487 U.S. 977, 108 S.Ct. 2777, 2789 n. 3, 101 L.Ed.2d 827 (1988) ("We have emphasized the useful role that statistical methods can have in Title VII cases, but we have not suggested that any particular number of 'standard deviations' can determine whether a plaintiff has made out a prima facie case in the complex area of employment discrimination."); see also Palmer, 815 F.2d at 92 (noting that Supreme Court has not established "an exact legal threshold at which statistical evidence, standing alone, establishes an inference of discrimination"); Coser v. Moore, 739 F.2d at 754 n. 3 (a significance level of 5% probability of chance "has no talismanic importance"); EEOC v. American Nat'l Bank, 652 F.2d 1176, 1192 (4th Cir.1981) ("courts of law should be extremely cautious in drawing any conclusions from standard deviations in the range of one to three"), cert. denied, 459 U.S. 923, 103 S.Ct. 235, 74 L.Ed.2d 186 (1982); D. Baldus & J. Cole, supra, § 9.4, at 188–89 (Supp.1987) (courts should use tests of statistical significance only as "an aid to interpretation" and not as a "rule of law"). Accordingly, in accordance with Supreme Court pronouncements, we must reject appellants' suggestion that this court announce a rule of law with respect to what level of statistical significance automatically gives rise to a rebuttable presumption of discrimination.

Moreover, as a practical matter, the issue of whether or not two or more standard deviations establish a prima facie case of discrimination is not pivotal in this case. It is well-established that "once a Title VII case has been 'fully tried on the merits,' the question whether the plaintiff has established a prima facie case 'is no longer relevant.'" Mitchell v. Baldrige, 759 F.2d 80, 83 (D.C.Cir.1985) (quoting Aikens, 460 U.S. at 714–15, 103 S.Ct. at 1481–82). As the Supreme Court stated in Bazemore v. Friday, "the only issue to be decided at that point is whether the plaintiffs have actually proved discrimination." 478 U.S. at 398, 106 S.Ct. at 3008 (emphasis added). Despite appellants' efforts to deemphasize on appeal what transpired at trial, the district court herein actually proceeded as though the plaintiffs had made out a prima facie case of discrimination, and it cannot be gainsaid that the plaintiffs' claims were fully litigated on the merits.

At the close of the plaintiffs' case-in-chief, the defendants moved to dismiss the action specifically on the grounds that plaintiffs had not met their burden of establishing a prima facie case. The court denied the motion without explanation, however, and directed the defendants to proceed with their case. Later, at the close of defendants' rebuttal case, the court also allowed the plaintiffs to respond to the defendants' rebuttal evidence. Accordingly, the question remaining for the district court at the close of the entire trial was whether the plaintiffs should ultimately prevail on their Title VII claims. See Bazemore, 478 U.S. at 398, 106 S.Ct. at 3008. Since the plaintiffs fully litigated those claims, any arguments on appeal directed at whether or not plaintiffs established a prima facie case are arguably misleading and misplaced. See EEOC v. Sears, Roebuck & Co., 839 F.2d at 312 n. 9.

The net import of Judge Kram's rulings regarding the significance of plaintiffs' statistical evidence is that she found the evidence to be "persuasive" but not dispositive. Contrary to appellants' assertions, it is clear from the district judge's rulings that she did not simply ignore the statistical evidence of discrimination presented by plaintiffs. The court found this evidence sufficient to cause her to deny the defendants' motion to dismiss at the end of plaintiffs' case, and to accept rebuttal evidence from the defendants. On rebuttal, however, the defendants were able to successfully undermine the plaintiffs' case by attacking the validity of the plaintiffs' statistical evidence, and by introducing statistical evidence of their own to negate the inference of discrimination that had been raised. Cf. Berger, 843 F.2d at 1416 ("Mere conjecture or general assertions of inadequacies in the opponent's statistical case, without demonstrating their effect on the results, will not suffice."); Sobel, 839 F.2d at 34 (same).

Specifically, the defendants criticized the plaintiffs' most probative studies for excluding one factor which they claimed exerted a "highly significant positive influence ... on current salary," namely, whether a faculty member had held a prior, full-time administrative position at SUNY New Paltz before returning to full-time teaching. The defendants also criticized these studies because the salary regressions were "fitted" only to male faculty members, i.e., they used independent variables that were derived only from the male population. The district court noted in its opinion that a "males only regression" based exclusively on values existing only in the male population might have tended to overestimate the predicted salaries of certain female faculty members, because it might not have taken into account legitimate factors existing solely in the female population which could have affected the rate of pay for women teachers at the University. See Ottaviani, 679 F.Supp. at 307. If the predicted salary for a female faculty member was overestimated, this type of regression arguably would have overestimated the discrepancies between male and female salaries at the University.

Finally, the defendants criticized these studies because they inappropriately aggregated Instructors and Assistant Professors into a single "rank." The defendants pointed out at trial that when the two ranks were combined into a single rank, the predicted salary of a female Instructor would essentially be based on the higher salary of an Assistant Professor, and hence the net residual difference between the predicted and actual salaries of a female Instructor would be overstated. Apart from these criticisms of plaintiffs' statistical evidence, the defendants also offered persuasive anecdotal evidence to negate the plaintiffs' claims of discriminatory animus. After considering all of the evidence presented, both statistical and anecdotal, the district court simply found that plaintiffs had failed to preponderate on their claims.

Recent Supreme Court precedent has made it clear that this court can reverse such a factual determination "only if it is clearly erroneous in light of all the evidence in the record or if it rests on legal error." Palmer, 815 F.2d at 101 (citing, inter alia, Bazemore, 478 U.S. 385, 106 S.Ct. 3000). Especially in cases where statistical evidence is involved, "great deference is due the district court's determination of whether the resultant numbers are sufficiently probative of the ultimate fact in issue." EEOC v. Sears, Roebuck & Co., 839 F.2d at 310 (citation omitted); see Griffin v. Board of Regents of Regency Univ., 795 F.2d 1281, 1289–90 (7th Cir.1986). As the Supreme Court cautioned in the Teamsters case, "statistics are not irrefutable; they come in infinite variety and, like any other kind of evidence, they may be rebutted. In short, their usefulness depends on all of the surrounding facts and circumstances." 431 U.S. at 340, 97 S.Ct. at 1857. The district judge herein gave due consideration to all of the evidence presented, and after reviewing the record, we do not perceive a convincing basis for finding her interpretation of that evidence to be clearly erroneous. Accordingly, we affirm her rulings with respect to the statistical evidence presented.

B. Use of Rank Variables

In conjunction with their attack on the district court's assessment of the sufficiency of plaintiffs' statistical evidence, appellants also challenge the district court's determination that "rank" was an appropriate factor to consider in assessing pay disparities between male and female faculty members. According to appellants, if the court had rejected the rank variables and considered only those salary studies which excluded rank, then the number of standard deviations associated with their findings of discrimination would have been much greater, and their statistical proof would have been even more probative.

Although we recognize that the use of rank variables in testing for salary discrimination against women faculty members is not universally accepted, see Finkelstein, The Judicial Reception of Multiple Regression Studies in Race and Sex Discrimination Cases, 80 Colum.L.Rev. 737, 741–42 (1980); D. Baldus & J. Cole, supra, § 8.23, at 113–14 (Supp.1987), in Sobel v. Yeshiva University, this court specifically upheld the use of rank variables in a multiple regression analysis, stating that rank could be used as a legitimate factor in explaining pay disparities so long as rank itself was clearly not tainted by discrimination, 839 F.2d at 35. As the plaintiffs' statistical expert, Dr. Mary Gray, explained in her own report: "In a bias-free system, one could use rank as a measure of productivity since the review process for promotion or hire should evaluate teaching, scholarship and service." (Emphasis added). See D. Baldus & J. Cole, supra, § 8.2, at 114.

The question to be resolved, then, in cases involving the use of academic rank factors, is whether rank is tainted by discrimination at the particular institution charged with violating Title VII. Although appellants reiterate on appeal their claim that rank at New Paltz was tainted, it is clear that the district judge accepted and considered evidence from the parties on both sides of this issue, and that she rejected the plaintiffs' contentions on this point.

At trial, the plaintiffs failed to adduce any significant statistical evidence of discrimination as to rank. As the district court stated in its opinion, the plaintiffs' studies of rank, rank at hire, and waiting time for promotion "were mere compilations of data" which neither accounted for important factors relevant to assignment of rank and promotion, "nor demonstrated that observed differences were statistically significant." The defendants, on the other hand, offered persuasive objective evidence to demonstrate that there was no discrimination in either placement into initial rank or promotion at New Paltz between 1973 and 1984, and the district court chose to credit the defendants' evidence. Upon review of the record, we cannot state that the court's rulings in this regard were clearly erroneous. Accordingly, the district court's decision to focus primarily on those studies which included rank as an essential independent variable was not improper, and appellants' contentions to the contrary must be rejected. See Presseisen v. Swarthmore College, 442 F.Supp. 593, 614, 619 (E.D.Pa.1977) (inclusion of rank variable appropriate when evidence showed no discrimination with respect to hiring and promotion), aff'd mem., 582 F.2d 1275 (3d Cir.1978); see also EEOC v. Sears, Roebuck & Co., 839 F.2d at 327 (court's decision to focus generally on those regression analyses which did not omit "major factors" was proper); Rossini v. Ogilvy & Mather, Inc., 798 F.2d 590, 603–04 (2d Cir.1986) (trial court's reliance on studies which incorporated controversial variables not clearly erroneous, where court's decision came after "extensive testimony from experts on both sides of the issue").

C. Anecdotal Evidence

Appellants also contend on appeal that the district court did not give sufficient weight to the anecdotal evidence adduced at trial, and that the court should have rejected the explanations proffered by University administrators to explain pay and rank inequities as "pretextual." Our review of the anecdotal evidence, however, is limited to ascertaining whether the district judge committed clear error in making her findings. See Anderson v. City of Bessemer, 470 U.S. 564, 573, 105 S.Ct. 1504, 1511, 84 L.Ed.2d

518 (1985); Pullman-Standard v. Swint, 456 U.S. 273 (1982). It is not the function of this court to reweigh the evidence anew, particularly when findings by a district court are based on in-court credibility determinations. Anderson, 470 U.S. at 575, 105 S.Ct. at 1512. Rather, under the clearly erroneous standard, we may only reject findings by the trial court when we are left with the "definite and firm conviction that a mistake has been committed.

In this case, the district court found that the defendants had successfully rebutted the plaintiffs' anecdotal proof, and that, in any event, the anecdotal evidence on its face was too limited to prove class-wide discrimination. After reviewing the entire record, we do not think that the court's decision to credit the testimony of the defendants rather than that of the plaintiffs was clearly erroneous. Since the district court's "account of the evidence is plausible in light of the record viewed in its entirety," we may not overturn the findings of the court even if we might "have weighed the evidence differently," had we been sitting as the trier of fact. Anderson, 470 U.S. at 573–74, 105 S.Ct. at 1510–11. Accordingly, we affirm the findings of the district court with respect to the anecdotal evidence presented.

D. Bazemore Claim

In Bazemore v. Friday, 478 U.S. 385, 106 S.Ct. 3000, 92 L.Ed.2d 315 (1986), the Supreme Court held that employers have an obligation to eradicate employment discrimination that began prior to the effective date of Title VII (1972), if the discrimination continues into the post-1972 liability period. Id. at 397, 106 S.Ct. at 3007. The Supreme Court also stated that statistical evidence of pre-Act discrimination can be probative of ongoing, post-Act discrimination. Id. at 402, 106 S.Ct. at 3010.

On appeal, appellants contend that the district court erroneously excluded evidence of pre-Act discrimination in violation of the Supreme Court's dictates in Bazemore. In particular, appellants claim that the district judge improperly excluded Exhibit 990, which purported to document statistically significant evidence of discrimination as to initial faculty rank. This claim is without merit, however. At trial, the defendants objected to the admission of Exhibit 990 not because it was offered to prove pre-Act discrimination, but because it was unreliable and incomplete. While the weakness of statistical evidence should not ordinarily preclude its admission, see Bazemore, 478 U.S. at 400, 106 S.Ct. at 3009, the Supreme Court has recognized that some statistical evidence may be so unreliable as to be irrelevant, see id. at 400 n. 10, 106 S.Ct. at 3009 n. 10; see also Penk v. Oregon State Bd. of Higher Educ., 816 F.2d 458, 465 (9th Cir.) ("Bazemore ... does not give blanket approval to the introduction of all evidence derived from multiple regression analyses."), cert. denied, ___ U.S. ___, 108 S.Ct. 158, 98 L.Ed.2d 113 (1987). Apparently the district judge herein thought that to be the case with respect to this particular exhibit, because she sustained the defendants' objection to its admission on the grounds that it was irrelevant and unduly confusing. Upon review of the record, we do not find the district court's decision to exclude the study to be clearly erroneous, and therefore we affirm the evidentiary ruling.

Moreover, we note that appellant's reliance on this court's decision in Sobel v. Yeshiva University as support for their more generalized, Bazemore-type claims is misplaced. In Sobel, the plaintiffs introduced evidence specifically designed to prove that women were discriminated against prior to the effective date of Title VII, and argued that "Yeshiva had a legal obligation to equalize women's salaries immediately upon application of Title VII to universities." 839 F.2d at 27. In the present case, even though the Supreme Court handed down its decision in Bazemore the same month that plaintiffs' trial was commenced, the plaintiffs did not introduce any statistical evidence of substance to prove that there was discrimination at New Paltz prior to the effective date of Title VII. Instead, nearly all of the plaintiffs' studies focused on the class liability period, which covered the years 1973 to 1984. This is in marked contrast to Sobel and Bazemore, wherein the plaintiffs offered direct, independent proof of pre-Act discrimination. See Bazemore, 478 U.S. at 401, 106 S.Ct. at 3009; Sobel, 839 F.2d at 30. Accordingly, we find appellants' arguments on this point generally to be without merit.

CONCLUSION

In sum, the burden of persuasion was on the plaintiffs to prove by a preponderance of the evidence that there was a pattern or practice of discrimination at SUNY New Paltz, and they failed to meet that burden. We have considered all of the arguments presented on appeal, and find them to be without merit. For the reasons stated above, the judgment of the district court is affirmed.

§ 5.30 Misuse of Statistical Theory

I can remember one cynical judge who would always, after unsuccessfully attempting to force a settlement on the parties, announce the start of trial by banging his gavel and saying: "Let the perjury begin." He had a variety of names for expert witnesses that cannot now be repeated. As obscene as those titles may have been, they couldn't come close to words that he used to refer to statisticians. The judge was a cynic for sure, but he also "coincidentally" happened to be a mathematician.

Bending statistical information has cost many lives on several separate occasions. In 2011, pharmaceutical giant Merck & Company, Inc. agreed to pay 950 million dollars in fines as it pled guilty to criminal charges over the marketing and sales of the painkiller Vioxx. Based upon its own statistical trials, the makers of the popular pain killer knew that it dramatically increased cardiac events and was responsible for many heart attacks and deaths. In a twisted comparison to Bayer's pain killer Aleve, Merck managed to mask some remarkable numbers indicating that Vioxx users were 400 times more likely to suffer heart attacks than Aleve users. To make its own numbers more palatable, Merck claimed that Aleve reduced the risk of cardiac events by 80%, a claim which was totally baseless.

If a product's harmful statistics happen to be generated by a competitor or other third party, a different approach might be employed. Instead of pulling a profitable product off of the market, the pharmaceutical company might elect to find a hired gun from among companies that specialize in discrediting statistical studies, regardless of their accuracy. Such was the case for Phenylpropanolamine, more popularly known as Accutrim or Dexatrim (and other brand names). The weight loss drug, not only suppressed appetites, it also caused bleeding in the brain and strokes. Despite a dramatic statistical link to the problems, the manufacturer managed to delay its removal from market for close to three decades.

Coverups and statistical manipulation, unfortunately, have not been limited to the pharmaceutical industry. There are many others; but the *Oscars* for the promulgation of statistical deception clearly belongs to the tobacco and asbestos industries.

Unfortunately, there are many ways in which statistics can be mis-used or mis-applied. They have only scratched the surface. National advertisers and their agencies are particularly adept at skewing and twisting statistical information. The number of misleading commercials appearing on television within a year alone is, no doubt, countless. The most common misuses can be grouped as follows:

§ 5.31 *Correlation v. Causation*

Of the many deadly, capital sins committed by statisticians, one of the most common involves mistaking correlation for causation. If, for example we were to conduct a study that compared the number of shark attacks on surfers with the temperature of the water, we might find a very strong correlation between the temperature of the water and the number of shark attacks. We might, therefore, conclude that the higher the temperature, the more aggressive the sharks become. From that, we might then conclude that warm water *causes* sharks to become aggressive. This conclusion would totally ignore the perfectly logical explanation that cold weather kept more surfers out of the water; therefore, there were fewer surfers in the water to get attacked.

Mistaking causation for correlation occurs more frequently than most would assume. For example, studies show that people who purchase flood insurance are more likely to have their homes flooded than people who do not; that is because people who live on flood planes or risky areas are more likely to purchase flood insurance. Purchasing flood insurance alone does not signal Mother Nature to bring on the rain. Taller children spell better than shorter children; however, being tall doesn't make you a better speller—Taller children are normally older. This next one is questionable: Studies show that married people are happier than single people. Does that necessarily mean that you will be happier if you get married? (Actually, married people don't know that they're happier without being told so by their nagging spouses).

On the other hand, it would be inaccurate and foolish to always assume that a strong correlation is merely coincidental. Such foolishness can prove deadly. Arguably, it is no less dishonest to ignore a correlation than to overstate it. The tobacco companies managed to convince the courts for years that there was no causation with respect to cigarette smoking and lung cancer. To best suit their interests, the expert witnesses for the tobacco companies managed to redefine "causation" claiming that it requires absolute "equivalence." Consider the following re-enactment:

Q: (Court) Is there a correlation between smoking and lung cancer.

A: (Tobacco Expert) Yes. But that does not mean that smoking causes lung cancer.

Q: But is there a connection?

A: I'm not sure what you mean by connection.

Q: Does smoking cause lung cancer?

A: In a scientific sense?

Q: Yes.

A: No. It does not.

Q: How do you arrive at that?

A: What do you mean?

Q: Everything I've ever read says that you are wrong.

A: In a scientific sense, there are people who smoked all of their lives and never contracted lung cancer. Likewise, there are many people who never smoked who contracted lung cancer.

Q: So, you are saying that out of a statistical sample of 1,000 people who smoke, if 1 person does not contract lung cancer, then that would disprove any causation. Is that what you are saying?

A: Yes. In the world of science, one counterexample is all that you need to disprove a hypothesis.

§ 5.32 *Semantic Gymnastics*

For this treatise, effort has been exerted to seek out the most widely accepted definitions, terms and equations. That has not been easy. Pick up any two books on the subject of *Statistics*, put them side by side and see if it is possible to derive a set of common terms and definitions, even for some of the most fundamental concepts. Notice how many different definitions there are for basic concepts such as "correlation" or "margin of error."

Is the language of statistics intentionally confusing? Or is it simply a matter of tailoring one's definitions to suit the particular hypothesis that one is attempting to prove, in much the same manner that the tobacco company experts redefined causation in the preceding subsection.

One of the most mis-used terms is the "average," which some will use to describe a mode or median (as opposed to a mean) as if the three terms were equivalent. We know from § 5.11 (*supra*) that they are far from equivalent.

§ 5.33 *Biased Sampling*

It is no coincidence that political candidates fill their so-called town hall meetings and rallies with members of their own party, never with a representative cross section of the voting constituency. Polls can be notoriously biased. It is common for a candidate to purchase or subscribe to a poll that shows him or her to be leading. Consider a questionnaire with the following:

> Please check the box of the political party with whom you consider yourself to be affiliated:
>
> ❐ A. Liberal Pinko Democrat
>
> ❐ B. Right Wing Republican
>
> ❐ C. Loyal American Independent
>
> Perhaps the most famous pictorial comment on the accuracy of pollsters is the infamous photo of presidential election winner Harry Truman holding up a newspaper headed "Dewey Defeats Truman."

The sampling process itself is subject to error. No one can ever be absolutely certain whether or not any given sample is representative of the subject population.

The concept of scientific sampling seemed to take hold following the defeat of Alfred Landon by Franklin Roosevelt in the election of 1936; it was the single most lopsided presidential election wherein Roosevelt carried every state except New Hampshire and Vermont. A highly respected magazine, The Literary Digest (founded in 1890) issued sample ballots to 10,000,000 people, whose names and addresses were obtained from telephone books and various lists. From approximately 2,000,000 returned post cards, the magazine confidently predicted that Republican presidential candidate Alfred Landon would win 57 percent of the popular vote and 370 electoral votes. Meanwhile, a relative newcomer named George Gallup only polled 50,000 for the infamous 1936 election, but did so "scientifically," by carefully selecting a *sample* designed to be representative of all of the nation's voters. He predicted that Roosevelt would win with 55.7 %, of the vote; Roosevelt received 60.8% of the tally, creating a deviation of -5 points.

The 1936 election changed polling forever. Except for the election of 1992 (Predicted to receive 49% of the vote, Clinton only received 43%), The Gallup Pole's presidential election predictions have always witnessed lower margins of error than the 1936 election. Gallup managed to predict some percentages close to dead-on (leaving a near-zero deviation): Richard Nixon in 1968 (43%), George McGovern in 1972 (38%), Ronald Regan in 1984 (59%), George Bush in 1992 (37%), Robert Dole in 1996 (41%), and George W. Bush in 2000 (48%). Gallup's poorest presidential prognostication in 1936 predicted that Alfred Landon would receive 44% of the vote when he actually only garnered 37%. But even Gallup's worst was a masterpiece compared to The Literary Digest's prediction that Alf would win the election with 57% of the vote.

In re Estate of Marcos Human Rights Litigation

910 F.Supp. 1460 (D.Hawai'I, 1995)

REAL, District Judge.

I. BACKGROUND

Victims of torture, summary execution and disappearance filed suits for damages, in the form of a class action as well as individual direct actions, against the Estate of the former President of the Philippines, Ferdinand E. Marcos (MARCOS), for human rights violations. Specifically, the violations are alleged to have occurred during the period in which MARCOS, as President of the Philippines, declared martial law, from September 21, 1972 to February 25, 1986.

In 1986 MARCOS fled the Philippines and arrived in the State of Hawaii. MARCOS was a resident of Hawaii at the time he was served with the complaints that are the subject of this litigation but he died during the pendency of these actions. The Estate of Ferdinand E. Marcos (the ESTATE) has been substituted in MARCOS' place; his widow, Imelda Marcos, and his son, Ferdinand E. Marcos, Jr., have appeared before this Court as representatives of the ESTATE.

The action was tried in the three phases: (1) liability, (2) exemplary damages, and (3) compensatory damages, over a nine year period—from 1986 to 1995. In the compensatory damages phase, Phase III, this Court allowed the jury to consider the damages to a random sample of plaintiffs as representative of the injuries suffered by those in the three subclasses; i.e. (1) plaintiffs who were tortured; (2) the families of those individuals who were the subjects of summary execution; and (3) the families of those who disappeared as the result of the actions of MARCOS. Pragmatically, the jury could not hear testimony of nearly 10,000 plaintiffs in this action within any practicable and reasonable time, to do justice to the class members. The individual plaintiffs who opted out of the certified class action each presented his or her individual claim for compensatory damages to the jury in a separate part of the Trial.

This opinion addresses the compensatory damages phase of the trial. The Court deals here with the propriety of the use of inferential statistics to ascertain the damages suffered by each of the 9,541 class members.

II. MARCOS REGIME

MARCOS was elected President of the Philippines in 1965 and was re-elected in 1969. The Philippine Constitution of 1935, still in effect in 1972, was similar to the United States Constitution, in that it limited election of the President to two four-year terms. Thus, MARCOS would have had to leave the office of the Presidency by the end of 1973, but he did not. On September 21, 1972 MARCOS imposed martial law on all of the Philippines through Proclamation 1081, which suspended the Constitution, in order to keep himself in office. The stated purpose for the imposition of martial law, as expressed in Proclamation 1081, was: "to maintain law and order throughout the Philippines, prevent or suppress all forms of lawless violence as well as any act of insurrection or rebellion and to enforce obedience to all the laws and decrees, orders and regulations promulgated by me personally or upon my direction."

At the time martial law was declared, a Constitutional Convention, elected by the people, had been meeting and was near completion of proposed revisions to the 1935 Constitution. On orders from MARCOS, some delegates to the Convention were arrested and placed under detention while others went into hiding or left the country leaving the revisions uncompleted. Without allowing for ratification of the new Constitution by a plebiscite, on January 17, 1973, MARCOS ordered ratification of a revised Constitution, tailor-made for his maintenance of power. With those actions MARCOS planted the seeds for what grew into a virtual dictatorship in the Philippines.

The new Constitution nullified the term limits for the President and provided that MARCOS could function as President, using his own judgment, for as long as necessary. Until he convened a new legislative body, MARCOS also had sole authority to rule in the Philippines.

Proclamation 1081 not only declared martial law, but also set the stage for what plaintiffs alleged, and the jury found, to be acts of torture, summary execution, disappearance, arbitrary detention, and numerous other atrocities for which the jury found MARCOS personally responsible. MARCOS gradually increased his own power to such an extent that there were no limits to his orders of the human rights violations suffered by plaintiffs in this action. MARCOS promulgated General Order No. 1 which stated he was the Commander-in-Chief of the Armed Forces of the Philippines. The order also stated that MARCOS was to govern the nation and direct the operation of the entire Government, including all its agencies and instrumentalities. By General Orders 2 and 2–A, signed by MARCOS immediately after proclaiming martial law, MARCOS authorized the arrest, by the military, of a long list of dissidents. By General Order 3, MARCOS maintained, as captive, the executive and judicial branches of all political entities in the Philippines until otherwise ordered by himself personally.

Immediately after the declaration of martial law the issuance of General Orders 1, 2, 2A, 3 and 3A caused arrests of persons accused of subversion, apparently because of their real or apparent opposition to the MARCOS government. These arrests were made pursuant to orders issued by the Secretary of Defense Juan Ponce Enrile ("ENRILE"), or MARCOS himself.

The arrest orders were means for detention of each of the representatives of the plaintiff class as well as each of the individual plaintiffs. During those detentions the plaintiffs experienced human rights violations including, but not limited to the following: 1. Beatings while blindfolded by punching, kicking and hitting with the butts of rifles; 2. The "telephone" where a detainee's ears were clapped simultaneously, producing a ringing sound in the head; 3. Insertion of bullets between the fingers of a detainee and squeezing the hand; 4. The "wet submarine", where a detainee's head was submerged in a toilet bowl full of excrement; 5. The "water cure", where a cloth was placed over the detainee's mouth and nose, and water poured over it producing a drowning sensation; 6. The "dry submarine", where a plastic bag was placed over the detainee's head producing suffocation; 7. Use of a detainee's hands for putting out lighted cigarettes; 8. Use of flat-irons on the soles of a detainee's feet; 9. Forcing a detainee while wet and naked to sit before an air conditioner often while sitting on a block of ice; 10. Injection of a clear substance into the body a detainee believed to be truth serum; 11. Stripping, sexually molesting and raping female detainees; one male plaintiff testified he was threatened with rape; 12. Electric shock where one electrode is attached to the genitals of males or the breast of females and another electrode to some other part of the body, usually a finger, and electrical energy produced from a military field telephone is sent through the body; 13. Russian roulette; and 14. Solitary confinement while hand-cuffed or tied to a bed.

All of these forms of torture were used during "tactical interrogation," attempting to elicit information from detainees concerning opposition to the MARCOS government. The more the detainees resisted, whether purposefully or out of lack of knowledge, the more serious the torture used. Eventually, MARCOS, his family and others loyal to him fled to Hawaii in February of 1986. One month later, a number of lawsuits were filed, including those that are the subject of this case.

III. CLASS ACTION

On September 22, 1992, in the liability phase of the trial, the jury found defendants liable to 10,059 plaintiffs, for the acts of torture, summary execution and disappearance. On February 23, 1994 the jury awarded plaintiffs $1.2 billion in exemplary damages.

In the compensatory damages phase, the class action plaintiffs presented their case to the jury by using damages sustained by a random sample of plaintiffs as representative of damages suffered by the entire class. After reviewing the deposition of 137 claimants and hearing the live testimony of several class members who could come to Court, the Special Master presented a report to the jury recommending the damages suffered by the 137 claimants, to give the jury a statistically valid representation of damages suffered by the entire class. On January 20, 1995, the jury reconvened and after hearing several representatives of the class and the testimony of the Special Master found the defendant liable to the class for over $766 million in compensatory damages, with individual plaintiff's awards ranging from $150,000 to $700,000.

The Court held that damages of 137 of the claimants, presented to the jury in the form of a report presented by the Special Master, was representative of damages sustained by the entire class, and introduction of such report did not offend due process. Furthermore, the fact that defendants did not have the opportunity to cross-examine all class plaintiffs, because only the testimony of 137 claimants was presented in the report, did not violate defendant's right to a jury trial under the Seventh Amendment of the United States Constitution. Defendant was given the opportunity to depose any of the randomly chosen 137 class members whose testimony was the subject of the Special Master's Report. They also had the opportunity to depose any of the 10,059 class members. This opinion will address judgment as to the class plaintiffs only.

IV. DISCUSSION

A. Issues Presented

All threshold issues in this case have been previously resolved by the Ninth Circuit. At this time there are two issues before this Court. The primary question is whether the use by this Court of a random sample of plaintiffs, as representative of the injuries suffered by others in the class, violates defendant's due process rights. The second question is whether use of the random sample violates the defendant's Seventh Amendment right to a jury trial.

The Ninth Circuit concluded, "Marcos' acts of torture, execution and disappearance were clearly acts outside his authority as President ... Marcos' acts were not taken within any official mandate and were therefore not the acts of an agency or instrumentality of a foreign state.

B. Random Sampling

1. Introduction

The ESTATE asserts random sampling is inappropriate for this case, and each claim should be individually tried. This Court holds otherwise. The use of aggregate procedures, with the help of an expert in the field of inferential statistics, for the purpose of determining class compensatory damages is proper.

James Dannemiller, an expert in the field of inferential statistics and survey sampling for twenty five years, assisted in this case. He has testified as an expert in those areas in both state and federal courts. Mr. Dannemiller formulated a plan so that only 137 randomly selected claims, of the 9,541 claims found to be valid, would have to be examined in order to achieve a 95% statistical confidence level that all claims would fall within the ambit of the 137 randomly selected claims. (A 95% confidence level certainly meets any due process or confrontation claim made by the defendant.)

Mr. Dannemiller testified that inferential statistics is a recognized science which uses mathematical equations to infer the probability of events occurring or not occurring. One branch of that science is the sampling theory, which deals with the selection of sample sizes sufficient to produce results that can be applied to a larger population from which the sample was selected with a specified probability of error. The formula Dannemiller used in this case is a well-known statistical tool that is found in Leslie Kish, Survey Sampling 53 (New York, John Wiley and Sons 1962) (KISH FORMULA).

Mr. Dannemiller testified under the KISH FORMULA, 137 randomly selected valid claims examined from a larger population of 9,541 validly submitted claims by class members would produce a 95% confidence level. The Court then considered the details of deposing 137 randomly selected claimants.

This Court appointed a Special Master to facilitate the taking of depositions of 137 randomly selected plaintiffs. The Special Master's appointment had a three-fold purpose: first, he supervised the taking of the 137 depositions in the Philippines; second, he served as a court-appointed expert on damages, under Federal Rule of Evidence 706, to review the deposition transcripts along with the claim forms; finally, he made recommendations on compensatory damages for the 137 claimants as well as the remaining class members to the jury.

Using the compensatory damages determined for the 137 claimants using the depositions the Special Master then reviewed all 9,541 valid claims and determined his recommendation to the jury of the aggregate damages suffered by each subclass category.

The distribution of funds to individual class members was left for later determination by the Court.

The depositions which the Special Master oversaw were noticed and taken in accordance with the Federal Rules of Civil Procedure. Although having notice of the depositions of the 137 class member sample and the names of the individual class members, the ESTATE chose not to participate and did not appear at any of the depositions, which were taken during October and November of 1994. Nor did the ESTATE choose to depose any of the 9,541 class members to test the procedure employed by the Court, or to acquire evidence to refute the fairness to the defendant of this random selection process using inferential statistical methodology.

The Special Master was directed by this Court to review the depositions for the following three elements: (1) whether the abuse claimed fell within one of the three definitions, with which the Court charged the jury at the liability phase of the trial; (2) whether the Philippine military or paramilitary was involved in such abuse; and (3) whether the abuse occurred during the period of September 1972 through February 1986. The claims of all the class members were filed with the Court and examined by the Special Master. Each claim was made under oath. After considering the deposition of the 137 claimants and the claims filed by each of the class members, the Special Master prepared the attached report.

Of the 137 randomly sampled claims, 67 were torture victims, 52 were execution victims and 18 were disappearance victims. (Both the execution victim claims and the disappearance claims are treated as wrongful death claims for the purpose of determining damages.) Based upon the depositions of each of the 137 randomly selected class member's claims and review of all the claims of the remaining class members, the Special Master recommended damages under Philippine, International, and American law, for each of the three categories of claims. During the Special Master's testimony, the Court advised the jury that they, in determining damages, could accept, modify or reject the recommendations of the Special Master. The jury was also instructed that they could, independently, on the basis of the depositions of the 137 randomly chosen class members, make their own judgment as to the individual damages of the 137 claimants and the aggregate damages suffered by the class. Copies of the Special Master's and Court–Appointed Expert's Report and Addendum thereto were supplied to each member of the jury. After five days of deliberations, the jury returned a verdict of over $766 million, approximately $1 million less than the Special Master had recommended.

In his report and testimony, the Special Master made damage determinations for torture victims by ranking each claim from 1–5, with 5 representing the worst abuses and suffering. The torture claims were evaluated based upon Judge Real's decision in Trajano v. Imee Marcos–Manotoc, aff'd, In re: Estate of Ferdinand E. Marcos Litigation, 978 F.2d 493 (9th Cir.1992), cert. denied, 508 U.S. 972, 113 S.Ct. 2960, 125 L.Ed.2d 661 (1993), as part of this matter, and the following considerations: (1) physical torture, including what methods were used and/or abuses were suffered; (2) mental abuse, including fright and anguish; (3) amount of time torture lasted; (4) length of detention, if any; (5) physical and/or mental injuries; (6) victim's age; and (7) actual losses, including medical bills. Although each claim of torture could have been but were not totally unique, as the Court Appointed Expert on damages, the Special Master, was able to determine that there were sufficient similarities within a rating category to recommend a standard damage amount to each victim within that grouping.

For summary execution and disappearance claims, whether there was any torture prior to a victim's death or disappearance weighed into the damages recommended for these two categories. Applying Philippine law, loss of earnings in torture claims was also factored into each of the Special Master's recommendations. After the aggregate lost earnings were computed, the result was converted into American dollars by dividing the figure by twenty-four, which was the approximate exchange rate, as of December 1994, between U.S. dollars and Philippine pesos.

Because there were discrepancies between some transcripts stating income earned in gross, some in net, and some giving no amount, serving as the Court Appointed Expert on damages, the Special Master, recommended that it was necessary to place a cap upon lost earnings. $120,000 was the maximum a claimant could receive.

When a witness did not state the amount of income earned by a summary execution or disappearance victim, an average for the victim's occupation was utilized. For example, when the victim was a farmer, the average earnings for one harvesting the same crop on the same amount of land was used. If a person stated the victim's income in terms of per-harvest, and if there were three harvests per year, for example, then that victim's earnings would be multiplied by three. For any victim who did not work, there was no award given for lost earnings. Defendant made no objection nor did defendant cross-examine the Special Master application of the Special Master's damages calculations.

For computing the total amount of damages for summary execution and disappearance victims, depending on the individual facts, there were different variables which went into the equation: (1) torture prior to death or disappearance; (2) the actual killing or disappearance; (3) the victim's family's mental anguish; and (4) lost earnings, calculated in the above described manner.

2. Precedent

The utilization of random sampling was fully examined in Cimino v. Raymark Industries, Inc., 751 F.Supp. 649 (E.D.Tex.1990), an asbestos class action. There, Chief Judge Robert M. Parker of the Eastern District of Texas allowed the use of inferential statistics in determining the amount of damages to award each class plaintiff. The 2,298 class members were divided into five categories based on the plaintiff's individual claims. The Court then selected a random sampling from each disease category created by the asbestos and the damages attributed to each sample category was then presented to the jury. Each plaintiff who was a member of the random sampling was awarded his individual verdict and the average verdict for each disease category constituted the damage awarded for each non-sampled class member.

As in Cimino, here class members were divided into three categories: torture, summary execution and disappearance—based upon a plaintiff's claims. The court next selected a random sampling from the population of 9,541 plaintiffs. Each plaintiff who was in the random sampling category and was found to have a valid claim was awarded his or her individual verdict. The average verdict for each category was the amount awarded to the class members who were not in the sample class.

3. Due Process

In a case such as this one, where there are 9,541 class members, most of whom live in other areas of the world, a balancing of interests must occur to obtain justice to the parties. A due process analysis must weigh defendant's claim to the right to trial in each individual case against judicial economy and manageability by use of a valid statistical procedure. The Court in Cimino, rejected defendants argument that they were entitled to a one-on-one trial for each of the 2,298 cases. The Court held due process is not necessarily limited to the traditional sense as argued by defendants, "but should also encompass the impact on plaintiffs and even the obvious societal interests involved." Id. at 666. The Court in that case was concerned that a one-on-one trial for each case, assuming the Court could close thirty cases a month, would take six and one-half years.

This Court was moved by the same concerns as Chief Judge Parker in Cimino. Here, individual trials for each of the 9,541 plaintiffs would take decades. Most of that time would be wasted since the nature of the injuries would be similar, if not identical, the testimony would be largely duplicative. Utilizing the procedure employed by the Court the injuries could be accurately categorized, and the source of the injuries would be identical.

This Court believes individual testimony from each of the plaintiffs, i.e., testimony of all 9,541 plaintiffs, could well have been repetitive. Although the facts in this case are not identical to those in Cimino, the damages here are much more objective, i.e., the torture, summary execution and disappearance come from the same source with the same objective. Inferential statistics with random sampling produces an acceptable due process solution to the troublesome area of mass tort litigation.

The issue remains whether this Court's use of inferential statistics in using aggregate procedures, denied defendant's their constitutional due process right to a one-on-one trial. This Court believes, "the aggregate trial is, in some vital respects, superior to the individual trial" and does not violate the substantive or procedural due process rights of either the plaintiffs or the defendant.

This Court finds persuasive the analysis of Professors Saks and Blanck in their discussion that aggregate trials do not violate due process. In Mathews v. Eldridge, 424 U.S. 319, 96 S.Ct. 893, 47 L.Ed.2d 18 (1976) the Supreme Court identified three factors to be balanced by the judge in determining what kind of process is due. The three factors to be balanced are: (1) the private interest affected, (2) the risk of erroneous deprivation of the interest through the procedures used, and (3) the government's interest, including all fiscal and administrative burdens that the additional procedure would require.

The use of random samples in aggregate procedures meets every requirement of the Mathews standard. Under the first prong of Mathews, since liability was established in Phase I of the trial, the private interest affected is the harm done to plaintiffs by MARCOS and what damages defendant would be required to pay plaintiffs for their injuries. It is probable the judgment against the ESTATE, had this Court allowed one-on-one trials, would be significantly more than the judgment the jury returned in the aggregate procedure. Nor were amounts awarded nearly as high as the amounts awarded in other "bipolar" human rights litigation. See, e.g., Filartiga v. Pena–Irala, 577 F.Supp. 860 (E.D.N.Y.1984) (over $10 million for a single summary execution); see Estate of Ferdinand E. Marcos Litigation, 978 F.2d 493 (9th Cir.1992) (over $4 million for a single summary execution), aff'g Trajano v. Imee Marcos–Manotoc. It is therefore evident that one-on-one trials would produce substantially higher verdicts than those returned in the aggregate.

The second prong of the Mathews test involves the erroneous deprivation of defendants rights through the procedures used, i.e., aggregated trials. With aggregate trials, the only issue decided is damages; liability is determined subject to objections and appeals; and liability is tried before the jury. Thus there is little risk of erroneous deprivation of defendants interest through the procedures used. In this case there can be no erroneous deprivation of the ESTATE's interest, which is to pay as little in damages as possible. There is no proof that the ESTATE would pay less had damages been determined on a bipolar basis. In fact, as stated above it appears that had such a procedure been utilized, each claim would have brought a higher judgment against the ESTATE.

Finally, aggregate trials are consistent with the third prong of the Mathews test. Clearly it cannot be questioned that a one-on-one trial is more burdensome for the Court than an aggregate trial. The costs involved in conducting bipolar trials with 9,541 plaintiffs in this case would substantially surpass the costs of an aggregate trial which lasted only about one and one-half weeks. The judicial and administrative time and costs of holding bipolar trials would also have been virtually, if not absolutely, prohibitive.

Lastly, because class members are mostly impecunious, the cost of bringing them to the forum or even taking their depositions would have prevented their claims from ever being determined. To deny relief to persons who have suffered significant damage simply based on the number of persons a single tortfeasor has hurt is unconscionable. Moreover the whole jurisprudence of class action treatment of numerous claims supports the conclusion that the ESTATE has suffered no due process violation.

4. Seventh Amendment

The issue here is whether the use of random samples, in an aggregate trial, violates the Seventh Amendment right to a jury trial. The Seventh Amendment "was designed to preserve the basic institution of jury trial in only its most fundamental elements, not the great mass of procedural forms and details, varying even then so widely among common-law jurisdictions." Galloway v. United States, 319 U.S. 372, 392, 63 S.Ct. 1077, 1088, 87 L.Ed. 1458 (1943). The Seventh Amendment provides no formula for the procedures to be used in a trial by jury. Rather, it is the rules of evidence and procedure that impact jury trials. Pragmatic application of these rules, consistent with justice, is all that is necessary for the presentation of the facts necessary for a jury determination. To claim otherwise certainly raises form over substance to a new level in today's jurisprudential world.

Here, the jury did determine the facts of the case, as the substance of the action was presented to the jury. There would be no benefit to either side in having the entire class testify given the repetition in the claims. Rule 23 of the Federal Rule of Civil Procedure does not mandate the presence of each member of the class. Therefore, by choosing a random sample of 137 claimants in an aggregate trial, neither side was deprived of even the form of their right to a jury trial.

C. Federal Common Law

Because plaintiffs in this action are citizens of the Philippines who are complaining of human rights abuses which occurred in that country, this case arises under two statutes—the Alien Tort Statute, 28 U.S.C. § 1350, and the newly enacted Tort Victim Protection Act ("TVPA") of 1991 (codified in the note to 28 U.S.C. § 1350). The Alien Tort Statute does not address damages; the TVPA only provides that an individual who abuses others under color of law of a foreign country is subject to liability for damages. Virtually all of the nations of the world, including the United States and the Philippines, are in agreement that human rights victims should have enforceable rights to fair and adequate compensation. Therefore, the issue of damages is one of federal law.

Because Congress in the TVPA offered no methodology as to how damages should be determined, federal courts are free to and should create federal common law to provide justice for any injury contemplated by the Alien Tort Statute and the TVPA or treaties dealing with the protection of human rights. See Textile Workers Union of America v. Lincoln Mills of Alabama, 353 U.S. 448, 457, 77 S.Ct. 912, 918, 1 L.Ed.2d 972 (1957) ("Some [problems] will lack express statutory sanction but will be solved by looking at the policy of the legislation and fashioning a remedy that will effectuate that policy."); Banco Nacional de Cuba v. Sabbatino, 376 U.S. 398, 425, 84 S.Ct. 923, 939, 11 L.Ed.2d 804 (1964) ("… an issue concerned with a basic choice regarding the competence and function of the Judiciary and the National Executive in ordering our relationships with other members of the international community must be treated exclusively as an aspect of federal law."); National Audubon Society v. Dept. of Water, 858 F.2d 1409 (9th Cir.1988); First National City Bank v. Banco Para, 462 U.S. 611, 103 S.Ct. 2591, 77 L.Ed.2d 46 (1983). Therefore, the aggregation of compensatory damage claims is appropriate under federal common law for human rights claims.

V. CONCLUSION

The use of an aggregate procedure for determining compensatory damages, under the procedures followed in this litigation, was neither a violation of the parties' due process rights nor their right to a jury trial under the Seventh Amendment. The aggregation of compensatory damage claims vindicates important federal and international policies, permits justice to be done without unduly clogging the court system, and was shown to be fair to the defendant.

Judgment shall be entered for plaintiffs.

§ 5.34 *Extraneous Variables*

A legitimate pollster walks house-to-house, during normal business hours, with a set of perfectly unbiased questions for a sample of a given community. If no one answers the door bell, the pollster continues to the next home. Innocently and unwittingly, the pollster has introduced an unintended extraneous variable by excluding those individuals who were not at home or who failed to answer the door bell. That variable could be quite significant since it might involve the elimination of all persons who are at work. There are insidious ways in which the polling process itself could introduce unintended variables.

One particularly important variable is the polling process itself and the propensity of individuals to provide answers designed to achieve a personal purpose or agenda. Human purposes come in infinite varieties to choose from. A desire to provide answers that will satisfy the pollster is but one relatively innocent motive that can throw results askew. But perhaps the most interesting problem is the simple fact that the act of questioning a subject may, in and of itself, change the result. A fundamental precept of quantum mechanics is Heisenberg's famous *Uncertainty Principle*, which holds that the mere act of measurement must always alter or change the entity that is being measured.

In the 1947 film *Magic Town*, a comedy directed by William A. Wellman, starring James Stewart and Jane Wyman, our hero played by Jimmy, discovered a pollster's dream come true—a perfect town with a population that represented a perfect sample of the entire United States. A pollster could conduct polls on virtually every subject with 100% accuracy, until the word leaked out to the people of the town who then set up mini-poll booths on every corner, selling their personal opinions to anyone interested. As you might guess, as soon as this happened, the accuracy of the samples faded into the main street mainstream.

§ 5.35 *The Least Integer Principle*

Look at the people around you. Someone is smarter (i.e., has a higher IQ) than everyone else in the room. Sure, you may be in a room filled with imbeciles, morons and reality show participants. Nevertheless, as a sound principle of mathematics, there is a smartest, a tallest, a smallest, etc.; everything capable of quantitative measurement can be lined up from greatest to least. It's that simple. So when a scientific survey once found that Old Gold Cigarettes contained the least amount of harmful tar and nicotine as compared to all of the other brand names, it manufacturers were quick to seize upon the information and send it directly to Madison Avenue. The resulting commercials, of course, forgot to mention that Old Golds still contained the requisite amount of harmful substances to cause cancer and emphysema. Instead, it was billed as the least harmful. This is akin to the manner in which the Tony awards are given to the "least bad" new shows on Broadway. Winning a Tony, therefore, does not assure that the winning show is worth seeing. The same of course can be said about the Emmy Awards and the Oscars.

§ 5.36 Small Sample Scams

There's a hockey or soccer game tomorrow night. I'll bet you that the sheet of paper that I'm holding in my hand (which I'm willing to place in a sealed envelope) has the exact score of tomorrow night's game. How do I do it? Simple. The sheet has the following numbers written on it: 1-0, 2-0, 3-0, 4-0, 5-0, 2-1 ... etc. How many possible combinations are likely? If I write out enough numbers, one set is certain to be correct. The "small sample statistical scam" is similar. For example, assume that I want to conduct a survey or poll among 120 soda drinking consumers to whether they prefer Coca Cola or Pepsi Cola. I could simply poll 120 consumers and ask each drinker which drink he or she preferred; that would be the honest way to do it. But a more effective way to guarantee a desired outcome would be to break the group up into 10 sets of 12 drinkers and poll each set separately. The different sets are bound to produce different results; therefore, I merely select the set with the most desirable results and discard the rest. Then I can claim that "10 out of 12 prefer ________." (Fill in your favorite colored sugar-water.)

As the preceding discussion suggests, it is critical that thorough discovery be conducted whenever the opposition proposes to introduce or use *any* statistical surveys. Consider the following interrogatory as a simple first round discovery opener:[6]

> With respect to the subject statistical study that you propose to introduce as evidence, or use, during the course of the trial, please set forth the following information specifically and in general:
>
> A. Please identify all documents (electronic included), notes, memoranda, records and writings that either directly or indirectly pertain to the subject statistical study.
>
> B. Please identify all documents (electronic included), notes, memoranda, records and writings that either directly or indirectly pertain to all statistical studies that relate to the subject statistical study or were conducted in proximity to the subject statistical study.
>
> C. Please identify all documents (electronic included), notes, memoranda, records and writings that either directly or indirectly pertain to every statistical study of which you are aware that relates to the subject matter of the subject statistical study.[7]

---

6. For discovery strategy in general, see Lipson, *Guerrilla Discovery*, James Publishing Co. (2003, Updated Annually).

7. It would be advisable to reword "subject matter" so as to specifically identify the particular subject of the study.

D. Please provide all of the details as to the manner in which the survey of the subject statistical study was conducted, including but not limited to your sample selection procedures and the methods by which your data was extrapolated.

E. Please identify and provide all documents, ballots, survey forms and writings that were used, either directly or indirectly to conduct your survey.

F. Did you conduct any surveys or obtain any samples that have not been identified in your prior responses or your responses to any prior discovery requests? If so, for each such survey or sample, please provide detailed information.

G. What, if any, conclusions do you or your experts intend to present? For each such conclusion, please set forth each hypothesis upon which the conclusion was predicated, detailing the manner in which you used your statistical data to arrive at your conclusion. Set forth details and calculations pertaining to error analysis, margin of error, and regression analysis for each such conclusion.

Another weapon against the divide-and-conquer technique described in the hypothetical Coke test involves the so-called Rule of Completeness, embodied in Rule 106 of the Federal Rules of Evidence and its state counterparts. The Rule prevents the enemy from cherry-picking and choosing among the most favorable pieces of an evidentiary puzzle:

> If a party introduces all or part of a writing or recorded statement, an adverse party may require the introduction, at that time, of any other part—or any other writing or recorded statement—that in fairness ought to be considered at the same time.

Notice that the Rule permits the adverse party to introduce any "other part," not necessarily the entirety. For the example stated above, however, it would seem that "fairness" would dictate the introduction of the remaining 9 sets of survey results.

§ 5.37 *Sample Switching*

One of the most common techniques used to evaluate the performance of political leaders is the so-called "sample switch." The discrepancies inherent in comparing administrations are inevitable. The times are different, the global and domestic circumstances are also different. Some switches are innocent; others are insidious.

If, for example, I were to compare sample salaries before and after the election of a particular governor, at least we might all be aware of uncontrollable factors such as inflation and economic downturn. If, on the other hand, I were to use a small town sample for my "before" and big city sample for my "after," in order to make it appear that the newly elected governor caused an increase in salaries and earnings, that would clearly be dishonest.

§ 5.38 *Detached Logic*

"Three out of every four doctors prescribe Sneezies Cold Medicine." There is no cure for the common cold and yet commercials would commonly imply otherwise without saying so. Statistical mis-statements, such as this example falsely permit the claimant to deduce that A implies C without proving B.

§ 5.39 *Mathematical Manipulation*

Uncovering statistical manipulation could be a full time job for many people. It would be akin to providing fixes for computer viruses; the list is never-ending. Here are a few examples:

1. Use of decimals carried out to several places to provide a false sense of precision and accuracy.
2. Discounting "coincidence" might be unjustifiable. Frequent variations within the first standard deviation are not only common, they are inevitable.
3. Failing to perform an adequate regression analysis can lead to false implications and misleading correlation. Closely related to this problem is a failure to include relevant independent variables.
4. Margins of error may be improperly presented or accounted for.

With the countless number of methods, definitions, equations and descriptions that statistics engenders, confusion and manipulation are inevitable.

Notes and Problems

1. In *THOIP v. Walt Disney Co.*, 690 F.Supp.2d 218 (S.D.N.Y., 2010), the New York District Court provided a list of factors for assessing the validity and reliability of surveys conducted by experts; in particular, the court should consider whether:

   1. The proper universe was examined and the representative sample was drawn from that universe;
   2. The survey's methodology and execution were in accordance with generally accepted standards of objective procedure and statistics in the field of such surveys;
   3. The questions were leading or suggestive;
   4. The data gathered were accurately reported; and
   5. The persons conducting the survey were recognized experts.

2. In *Brettelle v. Life Ins. Co. of North America*, 691 F.Supp.2d 1249 (S.D.Cal., 2010), the California District Court refused to take judicial notice of statistics from the National Highway Traffic Safety Administration (NHTSA). According to an ERISA plan administrator's interpretation, 27% of motorcycle fatalities involved operators with blood alcohol content levels greater than 0.08%. This was used to demonstrate that the death of the participant's husband (who operated a motorcycle at an excessive speed while moderately intoxicated) was substantially likely to occur, and thus was not covered by a death benefits policy. The statistics, according to the court, shed no light on the correlation between driving a motorcycle while under the influence and being involved in a serious injury accident; instead, the stats indicated that sober motorcycle operators were more than twice as likely to die in an accident than alcohol-impaired operators!

§ 5.40 Challenging Statistical Testimony

Statistical evidence can present itself in many different forms. It is capable of assuming all of the existing species of evidence (Testimonial, Documentary, Real and Demonstrative). As became evident from some of the preceding cases, it may also present itself during closing arguments. The most common presentation, however, will come from an expert witness.

When testifying as an expert, statisticians are subject to the same "gatekeeper" restrictions that confront other experts. See *Daubert v. Dow*, 509 U.S. 579 (1993), and *Kumho Tire Company, Ltd. v. Carmichael*, 526 U.S. 137 (1999). Rule 702 of the *Federal Rules of Evidence* permits an expert to present an opinion. The rule states that: "A witness who is qualified as an expert by knowledge, skill, experience, training, or education may testify in the form of an opinion or otherwise if: (a) the expert's scientific, technical, or other specialized knowledge will help the trier of fact to understand the evidence or to determine a fact in issue; (b) the testimony is based on sufficient facts or data; (c) the testimony is the product of reliable principles and methods; and (d) the expert has reliably applied the principles and methods to the facts of the case." The checklist presented in Rule 702 should provide some guidance for the cross-examination of the alleged expert. All discovery rules both state and federal require that detailed information pertaining to the expert witness's qualifications must be provided in advance. Therefore, for an enemy statistician, dwell only upon those elements that are lacking.

1. Question the witness about his or her training in the fields of statistics; inquire about education, advanced degrees (Mathematics in particular), employment, number of years in practice, published books, articles and contributions to other works.

2. What equation or formula did the expert employ to generate the numbers presented as evidence? (Require the expert to explain in detail, every variable and the reason for its selection.)

3. What constants or numbers were inserted into the equation or formula? Where did the numbers come from? Were they reliable?

4. Upon what evidence or authority does the expert predicate his or her claim that the selected equation or formula is valid?

5. Are there other alternative equations or formulae that could have been employed? If so, then why weren't they?

6. What assumptions are used for the particular formula or equation? Are they objective or subjective?

7. Has a thorough regression analysis been conducted? In that regard, what are all of the potential independent variables that may have effected the subject results.

8. Did the expert, during the course of his or her analysis or presentation employ any of the deceptive practices mentioned in the preceding sections or cases?

9. Knowing that the threshold requirement for any expert witness testimony is simplification and clarification, was the expert's presentation or explanation too complex?

10. Did the expert exceed his or her expertise, draw any unwarranted inferences, improperly state conclusions or otherwise invade the province of the judge or jury?

11. What statistical information, reports or studies were conducted but not used or reported by the expert during the opposition's case in chief.

§ 5.41 Computerized Hijacking

Never accept an expert's unchallenged explanation that: "*The statistics were generated by a computer, and are, therefore, accurate or reliable*," regardless of the degree of acceptance in the relevant scientific community. Be acutely aware of the distinction between a computer operation that merely runs arithmetic calculations and one that makes statistical decisions.[8]

In a relatively benign, uncelebrated case, *Allstate Ins. Co. v. Palumbo*, 109 Conn.App. 731, 952 A.2d 1235 (2008), Allstate Insurance Company filed an equitable subrogation action against a homeowner's fiancé, seeking to recover reimbursement after the fiancé, who had lived at the house with the homeowner, negligently installed water heating elements which caused a fire that damaged the house and personal property. The court permitted the use of computer programs used to "assist" experts in arriving at property damage estimates, finding that the results were properly authenticated and admissible. The first expert prepared an estimate regarding structural damage and testified as to the value of the structural damage; the second expert created a list of the items of personal property that had been destroyed in the fire and testified about the process that she followed to value the items that were destroyed. The computer program "assisted" the experts regarding the structural damage estimate and the typewritten list. Moreover, reliability must be the watchword in determining the admissibility of computer generated evidence, and in order to lay a proper foundation for computer generated evidence, there must be testimony by a person with some degree of computer expertise, who has sufficient knowledge to be examined and cross-examined about the functioning of the computer. In addition to the reliability of the evidence itself (which must be established to admit computer generated evidence) is the reliability of the procedures involved; defense counsel must have the opportunity to cross-examine the witness as to the methods used.

Notes and Problems

1. On more than one occasion, we have selected the tobacco companies for our discussions of statistical deception. This selection was not based on randomness. See *U.S. v. Philip Morris USA, Inc.*, 449 F.Supp.2d 1 (D.D.C., 2006): declaring:

> Since the 1950s, Defendants have researched and recognized, decades before the scientific community did, that nicotine is an addictive drug, that cigarette manufacturers are in the drug business, and that cigarettes are drug delivery devices. The physiological impact of nicotine explains in large part why people use tobacco products and find it so difficult to stop using them. Moreover, Defendants have sought to exploit the addictive quality of smoking and nicotine for decades in order to develop new products and increase sales. Notwithstanding the understanding and acceptance of each Defendant that smoking and nicotine are addictive, Defendants have publicly denied and distorted the truth as to the addictive nature of their products for several decades. Defendants have publicly denied that

8. See Ashley S. Lipson, *Art of Advocacy—Demonstrative Evidence* (LexisNexis, 2011) at § 16.05[3].

nicotine is addictive, have suppressed research showing its addictiveness, and have repeatedly used misleading statistics as to the number of smokers who have quit voluntarily and without professional help.

2. Not all institutions that employ misleading statistics are evil. Even our most revered companies can be lured by temptation. With respect to publishers' circulation statistics, see *Penthouse Intern., Ltd. v. Playboy Enterprises, Inc.*, 663 F.2d 371 (2nd Cir., 1981).

3. Particular attention must be directed at the population that is the subject of the sample and its relevance to the issue under consideration. Consider *Lanigan v. Huron Valley Hosp., Inc.*, 282 Mich.App. 558, 766 N.W.2d 896 (2009), involving a medical malpractice action based upon an alleged failure by physicians to make a timely diagnosis which would have permitted the plaintiff to undergo heart bypass surgery as opposed to a heart transplant. The defense argued that because of the heart transplant, the plaintiff, in fact ended up with a "better result" claiming that heart patients statistically survive longer than bypass patients. The court observed:

> Dr. Belen argued in his brief in support of the motion for summary disposition that plaintiff's opportunity to survive actually increased as a result of the heart transplant because patients receiving heart transplants have a 65 percent chance of surviving 10 years, whereas patients suffering from cardiogenic shock have a fatality rate of 70 percent. However, the deposition testimony of plaintiff's expert, Dr. Daniel Wohlgelernter, indicates that the 70 percent mortality rate applies to "all-comers," which would include individuals in exceedingly poor health, unlike plaintiff. Just as a defendant must take a frail plaintiff as he finds him, so must a defendant take a strong and healthy plaintiff as he finds him. In this case, the 70 percent mortality rate was not limited to those similarly situated to the plaintiff. Thus, this general statistical evidence was without selective application to plaintiff. Without some connection to the plaintiff, the statistical evidence was, in the context offered and standing alone, only marginally relevant and was an improper basis upon which to grant summary disposition.

4. Statistical surveys are often created, conducted and employed solely for litigation purposes, as was the case in *Marcos*. Such surveys may be indispensable to trademark infringement actions, where the controversies most often center around confusion in the minds of the public with respect to products and brands. Regardless of the type of litigation, however, the accuracy and validity of the survey is always subject to scrutiny and attack by the opposition. Consider *Stender v. Lucky Stores, Inc.*, 803 F.Supp. 259 (N.D.Cal., 1992), a sex discrimination action; ruling in favor of the plaintiff, the court chastised the defendant's "job interest survey" with a litany of disrespect, finding that it failed to survey terminated employees, failed to poll resigned employees, used a stratified sample, exaggerated differences between men's and women's work interests, failed to present concrete rather than abstract choices to respondents, employed poor methods for standardizing preference scores and basically used as its base a "feeder pool" for management's position.

5. This chapter ends with a piece of statistical information that is of potential vital interest to all litigators. According to an article appearing in the "Science and Technology" section of the April 16, 2011 issue of the *Economist*, titled "The Science of Justice—I Think It's Time We Broke For Lunch," a statistical study seemed to indicate that the sympathy of judges, when deciding whether or not to grant parole to incarcerated prisoners, decreased throughout the morning as the judges became hungrier, reaching a low point just before lunch. According to the study, compassion was restored after lunch.

Chapter 6

Graphs and Diagrams

"Seeing is deceiving."

—Anonymous

Combine the captivating visual effects of demonstrative evidence with a persuasive presentation of mathematical or scientific information and you have a fairly effective method for convincing a jury of almost any position:[1]

> "The superiority of visual images over oral descriptions cannot be disputed. Studies have shown that visual images have a greater impact than verbal presentations; also these studies indicate that a person given a visual demonstration as opposed to an oral description is likely to retain information for a longer period of time (Citations omitted)."

Charts and graphs, in particular, can be extremely effective devices for presenting and summarizing numerical data. In a courtroom setting, numbers seem to convey an atmosphere of truth and reliability. It, therefore, should come as no surprise that battles over their admissibility, appearence and interpretation are no less fierce than the fights involving their statistical counterparts. See, for example, *Rewis v. U.S.*, 503 F 2d 1202 (5th Cir., 1974).

§ 6.10 General Admissibility

The courts tend to lump together charts, diagram, drawings, graphs, summaries and PowerPoint demonstrations when considering their admissibility into evidence. From this indelicate grouping, a set of general rules accompanied by a critical distinction has emerged. The distinction centers around whether the demonstrative exhibit is to be formally admitted as evidence or used "for demonstrative purposes only." If the exhibit is to be formally admitted, the rules of evidence apply with all of their foundation requirements, not to mention the hurdles of the hearsay rule. If, on the other hand, an exhibit is to be used only for demonstrative or pedagogical purposes, there is a simple requirement—that it clarify or simplify a relevant issue for the trier of fact. In either event, the exhibit is going to be viewed by the jury and should be predicated upon previously admitted proper evidence.

As a general rule, charts, graphs and diagrams are admissible or usable as demonstrative evidence if they assist the trier of fact by clarifying or simplifying issues, claims or defenses that are relevant to the proceeding.[2] As long as the exhibit explains the un-

1. See Ashley S. Lipson, *Art of Advocacy—Demonstrative Evidence* (LexisNexis, 2011).
2. Id., § 8.05.

derlying testimony or evidence and does not confuse the jury, it should be admissible as evidence or allowable for demonstrative purposes only.[3]

Charts that are misleading or substantially prejudicial can be excluded altogether.[4] In addition, charts that purport to present scientific information by unqualified individuals will not be permitted.[5]

The subject matter of this chapter is split into two divisions: Graphs (which happen to represent pure numeric data) and Diagrams, which, for our purposes include all of the other miscellaneous visual presentations used to convey physical, scientific or financial data.

§ 6.20 Graphs

Much of our focus is directed toward the manner in which graphs are used and abused as tools of persuasion and deception. The chapter is not intended as a course in graph construction. Microsoft's *Excel* and Adobe's *Quattro Pro* adequately fulfill that function.

Lying with statistics is closely akin to lying with charts and graphs. But whereas Chapter 5 uncovered methods for the manipulation of statistical data, this chapter will focus more upon a manipulation of the visual images that might be derived or generated from statistical data and other physical evidence.

Terra Securities Asa Konkursbo v. Citigroup, Inc.

740 F.Supp.2d 441, S.D.N.Y., 2010.

VICTOR MARRERO, District Judge.

Plaintiffs Terra Securities ASA Konkursbo ("Terra") and seven Norwegian municipalities — Bremanger, Hattfjelldal, Hemnes, Kvinesdal, Narvik, Rana and Vik (the "Municipalities") — filed an amended complaint in this action, dated March 15, 2010 (the "Terra Complaint"), naming as defendants Citigroup, Inc., Citigroup Global Markets, Inc., Citigroup Alternative Investments LLC, and Citigroup Financial Products, Inc. (collectively, "Defendants"). Terra and the Municipalities assert securities fraud claims under § 10(b) of the Securities Exchange Act of 1934, 15 U.S.C. § 78, Rule 10b–5 promulgated thereunder, 17 C.F.R. § 240.10b–5, and § 20(a) of the Exchange Act, 15 U.S.C. § 78t(a), as well as common law fraud and negligent misrepresentation claims.

On April 20, 2010, Banca Carige S.P.A. — Cassa Di Risparmio Di Genova E Imperia, Carige Vita Nuova S.P.A., and Carige Assicurazioni S.P.A. (collectively, "Banca Carige", together with Terra and the Municipalities, "Plaintiffs"), filed a complaint (the "Banca Carige Complaint") asserting nearly identical securities and common law fraud claims against Defendants arising out of underlying events and operative facts substantially similar to those asserted in the Terra Complaint.

Defendants now move to dismiss the Terra and Banca Carige Complaints pursuant to Federal Rule of Civil Procedure 12(b)(6), asserting that Plaintiffs fail to state a claim upon which relief can be granted. For the reasons stated below, Defendants' motion to dismiss

---

3. *State v. Daly*, 278 Neb. 903, 775 N.W.2d 47 (2009); *Zemke v. Zemke*, 860 P.2d 756 (N.M., 1993); *Dunn v. Moss*, 598 N.Y.S.2d 350 (App.Div., 1993).

4. See *Rule* 403 of the *Federal Rules of Evidence*.

5. *Barmeyer v. Montana Power Co.*, 657 P.594 (Mont., 1983).

the Terra Complaint is GRANTED in part and DENIED in part, and the motion to dismiss the Banca Carige Complaint is GRANTED.

I. BACKGROUND

A. PROCEDURAL HISTORY

Terra and the Municipalities filed this action in August 2009. In October 2009, Defendants moved to dismiss the original complaint on subject matter jurisdiction and forum non conveniens grounds. By Decision and Order dated February 16, 2010, the Court denied Defendants' motion. Thereafter, Terra and the Municipalities filed the amended Terra Complaint, dated March 15, 2010. As noted above, the Banca Carige Complaint was filed in April 2010, alleging claims substantially similar to those brought in the Terra Complaint. By Order dated May 4, 2010, the Court consolidated the two actions, and Defendants now move to dismiss both the Terra and Banca Carige Complaints for failure to state a claim under Rule 12(b)(6).

B. FACTUAL ALLEGATIONS

The operative facts and events underlying the Terra Complaint are largely set forth in the February 2010 Decision, familiarity with which is assumed. Here, the Court will briefly review additional facts relevant to this motion to dismiss, as well as the factual background and allegations derived from the Banca Carige Complaint.

1. Plaintiffs' Fund–Linked Investments

In May and June of 2007, Defendants sold over $115 million in securities to the Municipalities through Terra, a Norwegian securities firm. The securities constituted fund-linked notes ("FLNs") linked to the Citi Tender Option Bond Fund (the "Citi TOB Fund"). The FLNs were arranged by Defendants, issued by Banque AIG and Starling Finance, PLC, and purchased by Terra "for the benefit of the investing Municipalities."

Similarly, in or around January 2007, Defendants sold 10 million euros worth of FLNs linked to Defendants' Offshore Tender Option Bond Fund (the "Offshore TOB Fund," together with the Citi TOB Fund, the "Funds") to Banca Carige, and entered into a fund-linked "Total Return Swap" (the "TRS") agreement with Banca Carige.

2. Marketing Materials

In or about April 2007, Defendants began marketing the FLNs to the Municipalities through Terra. In May and June of 2007, Terra entered into distribution agreements with Defendants, which governed the terms of their FLN distribution, and mandated distribution of marketing materials, including the presentation (the "Presentation") that Plaintiffs allege contained material misstatements and omissions of fact. On two separate occasions, Defendants allegedly provided the Presentation to Terra with full knowledge and intention that it would be transmitted to the Municipalities and/or would serve as the basis for Terra's advice to the Municipalities with respect to their investment in the FLNs. Similarly, Banca Carige alleges that Defendants provided it with the Presentation along with other materials marketing the Offshore TOB Fund beginning in or around November 2006.

The Presentation marketed the Funds by describing the Funds' investment strategy, detailing their structure, and purporting to demonstrate the historical performance of municipal yields hedged with interest rate swap agreements. The Presentation described the Funds' investment strategy as an arbitrage opportunity for investors, whereby the Funds take advantage of the relative steepness of the long-term, nontaxable, municipal curve as against a taxable London Interbank Offered Rate ("LIBOR") curve (the "Hedging Strategy"). The Funds purported to hedge against a drop in municipal bond values with LIBOR swap agreements that traded a fixed interest rate for a floating rate, and Defendants represented that the net amount long-term municipal bonds pay over the cost of short-term LIBOR loans (the "Arbitrage") was consistent over time.

Both the Terra and Banca Carige Complaints rely primarily on allegations that Defendants materially misrepresented the Hedging Strategy by portraying the correlation between long-term municipal bond rates and LIBOR swap rates as "virtually perfect, with a factor of almost .97 out of a possible 1."

Specifically, Plaintiffs rely on a graph contained in the Presentation entitled "Correlation Between Municipal and LIBOR rates" (the "Graph"), which purports to represent a regression analysis of the taxable and non-taxable rates over the last thirty years. Defendants represented that the Arbitrage was the result of "market inefficiency due to investor preference for shorter term municipal maturities, the risk of changes in tax law, and the lack of any short market on municipal bonds."

Plaintiffs allege that the correlation presented in the Graph was "blatantly false and misleading." The Graph incorrectly compared levels of interest rates and not rates of change, and thereby generated the misrepresentative, near-perfect correlation. Plaintiffs also condemn statistical flaws that led to the misrepresentative Graph, including an unacceptable standard error.

By contrast to the Graph's representation, Plaintiffs allege that if more accurate correlation analyses for the hedging strategy were employed, the result would have been a significantly lower correlation factor—as low as .562 or .27 out of a possible 1, depending on the municipal benchmark used. As a result, Plaintiffs allege that Defendants misrepresented the risk associated with the FLNs and the TRS, and that they relied on these misrepresentations to their financial detriment.

3. Failure to Disclose Credit and Liquidity Risk

Plaintiffs allege that the Presentation materially omitted to disclose two additional risk factors: (1) credit risk—the risk that municipal borrowers may become less creditworthy; and (2) liquidity risk—the risk that investors may become increasingly concerned about the liquidity of their long-term municipal bonds. Plaintiffs assert that the risk that insurers to the municipal debt offerings would falter could not be effectively hedged with interest rate swap agreements and thus should have been disclosed in the Presentation.

4. Financial Loss

Six weeks after the Municipalities' purchase of the FLNs, Defendants' representations regarding the high correlation between municipal and LIBOR rates were proven "demonstrably false, and those rates began to dramatically diverge." As the rates diverged, the Citi TOB Fund's assets, and thus correspondingly the Municipalities' FLNs, lost value. Municipal bond yields increased, causing the value of long-term municipal bonds to fall, but the Funds' taxable LIBOR swap agreements did not offset the non-taxable losses. Instead, "as investors worried about credit and liquidity risk," yields fell on taxable instruments, and thus both assets held by the Citi TOB Fund fell in value.

Ultimately, at the request of Defendants and Terra, the Municipalities sold their FLNs in late 2007 and early 2008. Combined, the Municipalities lost approximately $90 million, and these investment losses also allegedly resulted in Terra's financial ruin. Following the precipitous decline of the FLNs, Norway's Financial Supervisory Agency launched an investigation into Terra, which forced Terra to cease operations and declare bankruptcy.

For its part, Banca Carige alleges damages in excess of $47 million resulting from Defendants' alleged fraud. In or about the spring of 2008, after their fund-linked investments had suffered losses, Defendants contacted Banca Carige and assured them that the Offshore TOB Fund was stabilizing. Later that year, in December 2008, Banca Carige retained an Italian law firm to evaluate their fund-linked investments, an evaluation which ultimately prompted the instant action.

Plaintiffs allege that where they lost, Defendants gained: the fund-linked investments were set up to provide insurance for Defendants' investments in the Funds, as well as hefty advisory and dealing fees.

II. LEGAL STANDARD

In assessing a motion to dismiss under Rule 12(b)(6), dismissal of a complaint is appropriate if the plaintiff has failed to offer factual allegations sufficient to render the asserted claim plausible on its face. See Ashcroft v. Iqbal, — U.S. ——, 129 S.Ct. 1937, 1949, 173 L.Ed.2d 868 (2009) (quoting Bell Atl. Corp. v. Twombly, 550 U.S. 544, 570, 127 S.Ct. 1955, 167 L.Ed.2d 929 (2007)).

III. DISCUSSION

A. FEDERAL SECURITIES FRAUD

The parties do not dispute that the FLNs that Plaintiffs purchased were listed on European stock exchanges and the TRS was sold in Europe. Accordingly, the Court grants Defendants' motion to dismiss Plaintiffs' federal securities fraud claims under §§ 10(b) and 20(a).

B. COMMON LAW FRAUD

In addition to their federal securities fraud claims, Plaintiffs bring common law fraud claims arising out of the same allegedly fraudulent conduct. To state a claim for common law fraud under New York law, a plaintiff must allege: "(1) a material representation or omission of fact; (2) made with knowledge of its falsity; (3) with an intent to defraud; and (4) reasonable reliance on the part of the plaintiff, (5) that causes damage to the plaintiff." Haggerty v. Ciarelli & Dempsey, 374 Fed.Appx. 92, 94 (2d Cir.2010). Claims of common law fraud must satisfy the requirements of Rule 9(b) of the Federal Rules of Civil Procedure (" Rule 9(b)").

Defendants argue that Plaintiffs fail to plead two elements of their common law fraud claims: (1) reasonable reliance upon the alleged material misstatements, and (2) that the reliance caused their alleged injuries.

1. Reasonable Reliance

New York law requires that plaintiffs alleging common law fraud establish reasonable reliance on a material misrepresentation. A plaintiff must demonstrate that there was some basis for it to have relied on the alleged misstatement or omission. To determine, on a motion to dismiss, whether a plaintiff has alleged reasonable reliance, a court may "consider the entire context of the transaction, including ... the sophistication of the parties, and the content of any agreements between them." Emergent Capital Inv. Mgmt., LLC v. Stonepath Group, Inc., 343 F.3d 189, 195 (2d Cir.2003).

a. Sophisticated Investors

The reasonableness of a claim of reliance must be considered in light of the plaintiff's sophistication. "It is well established that where sophisticated businessmen engaged in major transactions enjoy access to critical information but fail to take advantage of that access, New York courts are particularly disinclined to entertain claims of justifiable reliance." Lazard Freres & Co. v. Protective Life Ins. Co., 108 F.3d 1531 (2d Cir.1997). Sophisticated investors must "investigate the information available to them with the care and prudence expected from people blessed with full access to information." Hirsch v. du Pont, 553 F.2d 750 (2d Cir.1977); see also Banque Franco v. Christophides, 106 F.3d 22 (2nd Cir.1997) (finding that analysis of reasonable reliance in "fraud cases under New York law has taken account of the degree to which the truth was accessible to the defrauded person").

Defendants argue that, as sophisticated investors, Terra and Banca Carige had a duty to investigate and inquire regarding any purported misrepresentations, and any reliance on their part was thus unreasonable as a matter of law. Plaintiffs counter that reasonableness of reliance is a question of fact, generally not properly resolved on a motion to dismiss. Even if properly considered here, Plaintiffs argue that Defendants' misrepresentations and omissions were not discoverable from "available information" provided to them prior to their investments. As an initial matter, the Court notes that the reasonableness of reliance is properly considered at the motion to dismiss stage. The Court thus turns to consideration of whether Terra and Banca Carige are sophisticated financial entities that, as a matter of law, could not have reasonably relied on the alleged misrepresentations. (The parties agree that the Municipalities are not sophisticated investors for the purposes of this inquiry.) Representations by a plaintiff that it had "knowledge and experience in financial and business matters and that it could readily evaluate the risks of the transaction" can indicate sophistication for the purposes of this analysis. In their respective Complaints and agreements relevant to this dispute, Terra and Banca Carige acknowledge their own sophistication as investors.

In addition, as noted above, Terra entered into Distribution Agreements with Citigroup to market and distribute the FLNs at issue here. The Distribution Agreements provide, in pertinent part: "Terra has sufficient knowledge, experience, and professional advice to make its own evaluation of the merits and risks of a transaction of this type," and "The Distributor represents and warrants ... that ... it is experienced in distributing structured products and understands the risks inherent in them...." Likewise, Banca Carige, in the TRS agreement, avowed that "[i]t is capable of evaluating and understanding (on its own behalf or through independent professional advice), and understands and accepts, the terms, conditions, and risks of [the transaction.]."

Terra and Banca Carige assert that the information underlying the alleged misrepresentations was within the peculiar knowledge of Defendants and unavailable to them. However, Plaintiffs mis-characterize what constitutes available information for the purposes of the present inquiry. The availability of information in this context is not whether the requisite material was made available to Plaintiffs by Defendants. Rather, available in this context denotes accessible—would the information necessary to unmask the alleged fraud have been accessible to the sophisticated party through minimal diligence. See Christophides, 106 F.3d at 27 ("[A]nalysis of justifiable reliance in fraud cases under New York law has taken account of the degree to which the truth was accessible to the defrauded person."); Emergent Capital, 165 F.Supp.2d at 623 ("[T]he sophisticated investor such as Emergent must show that he or she has made an independent inquiry into all available information.").

While Plaintiffs correctly articulate the applicable standard—a heightened degree of diligence is required of sophisticated plaintiffs only where red flags or other circumstances existed that made reliance unreasonable—they misapply it to the facts at issue here. The Court finds that Terra and Banca Carige knew or should have known that they were in a position to acquire additional information regarding the alleged misrepresentations. See Ashland Inc. v. Morgan Stanley, 700 F.Supp.2d 453, 469 (2d Cir.2010) ("Where the investor knows that he or she is in a position to acquire additional information, but does not inquire, the Second Circuit has found that the duty to exercise minimal diligence renders the investor's reliance unreasonable.")

First, the full context of the Presentation should have flagged for Terra and Banca Carige that an independent appraisal of the Graph and the Hedging Strategy was warranted. The Presentation provides an explicit risk disclosure (the "Risk Disclosure") addressing the Hedging Strategy and the Graph: "[T]he Fund may experience substantial volatility due to dissensions in the relationship of the municipal bond investments and the hedging instruments." Plaintiffs contend that this cautionary language about the possible future divergence of the taxable and non-taxable rates does not inoculate the plausible marketing power of a misrepresentation about historical performance of those rates. To an untrained eye this may hold true, but the Graph and Risk Disclosure, when viewed in tandem by a sophisticated appraiser, should have resulted in a dissonant picture that looked decisively like a red flag.

The Court finds that Terra and Banca Carige's respective Complaints also reveal additional hints of misstatement that render their alleged reliance unreasonable. Among the reasons Terra and Banca Carige set forth to allege that the Graph fraudulently misrepresented the Hedging Strategy are that Defendants "failed to set forth a standard error for its statistical analysis," and relied on municipal bond data that did not correspond, in significant part, with the class of municipal bonds in the Funds. Both of these flaws are readily apparent from the Presentation: there is no standard error indicated for the Graph, and the data source, which Plaintiffs now maintain fraudulently misrepresented the Funds' bond investments, is indicated below the Graph. Viewing Plaintiffs in light of their sophistication as investors, the Court finds that Terra and Banca Carige cannot now in hindsight rely on these statistical flaws and characterize them as misrepresentations when they were sufficiently on notice of them and their potential implications before investing in the FLNs. Terra and Banca Carige knew or should have known that they were in a position to acquire additional information regarding the statistical analysis portrayed in the Graph, as well as the precise nature of the bond investments in the Funds.

Finally, neither Banca Carige nor Terra allege that the statistical information underpinning the allegedly misleading Graph, or the existence of any credit or liquidity risk, was inaccessible or non-public information. The main premise of Plaintiffs' lawsuit—that an alternative representation of the relationship between taxable and non-taxable rates would not have been misleading—itself demonstrates that the information was public and readily available to them before their entering the securities transactions at issue. Indeed, Terra and Banca Carige now rely on "market research" to claim that Defendants' representations in the Graph were false and misleading. While it is true that the reliance requirement is not "designed to shield perpetrators of fraud by forcing investors to conduct exhaustive research every time they invest money," Alexander v. Evans, No. 88 Civ. 5309 (S.D.N.Y.1993), it is also true that this doctrine generally applies to ordinary purchasers of securities, not to sophisticated investors, and especially not to financial entities like Terra and Banca Carige that hold themselves out to the public as experienced investment managers, bankers and brokers, and do conduct exhaustive research in connection with the services they provide to their customers.

Here, the Court is not persuaded that Banca Carige and Terra had no duty to conduct an independent appraisal of a statistical representation based on publicly-available interest rates. See Lazard Freres, 108 F.3d at 1542 ("Where the representation relates to matters that are not peculiarly within the other party's knowledge and both parties have available the means of ascertaining the truth, New York courts have held that the complaining party should have discovered the facts and that any reliance under such circumstances therefore would be unjustifiable"); Ashland, 700 F.Supp.2d at 469–71 (finding that complaint of sophisticated investors provided no indication that information about student loan auction rate securities market was so closely kept so as to render the information inaccessible for the purposes of alleging reasonable reliance); cf. Abu Dhabi Commercial Bank, 651 F.Supp.2d at 181 (holding that plaintiffs adequately pleaded reasonable reliance where rating agencies had access to nonpublic information that even sophisticated investors could not obtain).

Accordingly, the Court finds that Terra and Banca Carige's reliance on the Presentation to be unreasonable as a matter of law and grants Defendants' motion to dismiss Terra and Banca Carige's common law fraud claims. Terra and Banca Carige further argue that any sophistication they may have as a Norwegian securities firm and Italian bank does not translate to sophistication with respect to the United States municipal bond market, and thus that they were not under any obligation to access information that may have exposed Defendants' alleged misrepresentations. In essence, Plaintiffs argue that sophistication in this context is territorially-limited and issue-specific. However, the duty so circumscribed misstates the applicable standard. While the sophistication of an investor in a specific financial sector may be relevant, Plaintiffs point to no authority that would shield a foreign financial firm from the duty to conduct their own research and analysis of United States-based investments and publicly-available interest rates. In any event, as noted above, Terra and Banca Carige explicitly represented that they had the sophistication and know-how to conduct their own independent appraisal of the FLN investments. As such, the Court finds that their claimed reliance on the Graph and other marketing misrepresentations to be unreasonable as a matter of law.

b. Risk Disclosures

The Court next turns to consideration of whether the Municipalities have adequately alleged reliance. The Municipalities, like Terra and Banca Carige, assert that Defendants fraudulently misrepresented the risk of a divergence between the municipal non-taxable and taxable rates, and that they reasonably relied on this misrepresentation in deciding to invest in the FLNs. Defendants contest the reasonableness of the Municipalities' alleged reliance, arguing that they disclaimed reliance, and that extensive risk disclosures in the Presentation bespoke caution. A securities fraud claim must be dismissed as a matter of law where cautionary language expressly warns of or directly relates to the risk that brought about plaintiffs' loss, and thus where no reasonable investor could have been misled about the risk when he invested.

c. Particularity

Defendants argue that the Municipalities have failed to plead their surviving fraud claim with the particularity required by Rule 9(b). Specifically, Defendants argue that Plaintiffs have failed to plead direct reliance on the alleged misrepresentations by failing to identify specific municipal employees who relied on the Presentation, and by acknowledging that Terra may have served as an intermediary to the FLN transactions. The Court is not persuaded. Rule 9(b) does not require Plaintiffs to allege direct communication between the Municipalities and Defendants. See Securities Investor Protection Corp. v. BDO Seidman, LLP, 222 F.3d 63, 77 (2d Cir.2000) ("[S]o long as the plaintiff receives the substance of a defendant's misstatements, he or she may establish reliance on that information even if the defendant did not communicate with the plaintiff directly.").

The Court finds that the Municipalities have pled with sufficient particularity the circumstances under which they reasonably relied on the alleged misstatements, and thus have adequately pled reliance under Rule 9(b).

2. Causation

The causation element of common law fraud requires that Plaintiffs adequately plead a causal connection between Defendants' alleged misrepresentations and Plaintiffs' alleged losses. See Emergent Capital, 343 F.3d at 197. Plaintiffs must show that their damages were foreseeable and were caused by the materialization of the concealed risk. See Lentell v. Merrill Lynch, 396 F.3d 161, 173 (2d Cir.2005); Castellano v. Young & Rubicam, Inc., 257 F.3d 171, 186 (2d Cir.2001).

The Municipalities assert that their investment losses were caused by a risk concealed by the Presentation and its enclosed Graph, namely the likelihood of a divergence of the taxable and non-taxable rates. Defendants argue that the Municipalities have failed to plausibly distinguish losses attributable to the alleged misrepresentations from those resulting from a market-wide credit crisis, and from those losses resulting from Terra's own intervening, wrongful conduct. The Court is not persuaded and finds that Plaintiffs have plausibly alleged that the divergence of the non-taxable and taxable rates, as well as the materialization of other omitted risks, caused their financial losses.

In cases such as this, where multiple explanations could plausibly surface as the proximate cause of alleged losses, the chain of causation is generally "a matter of proof at trial and not to be decided on a Rule 12(b)(6) motion to dismiss." Emergent Capital, 343 F.3d at 197. Among other reasons, to find differently would effectively preclude most, if not all, fraud claims which materialize during a time of economic crisis. Here, the Municipalities do not merely allege that an otherwise intact investment vehicle fell victim, like so many other investments, to the credit and liquidity risk symptomatic of the market collapse. Rather, Plaintiffs allege that Defendants' Hedging Strategy, based on a purportedly near-perfect correlation between taxable and non-taxable rates, was a lemon from the get-go, unable to traverse difficult market terrain despite contrary warranties in the Presentation. Simply put, the Municipalities allege that: the Presentation misrepresented the historical volatility of Defendants' Hedging Strategy in times of market instability, the market destabilized, the taxable and nontaxable rates diverged, and the Municipalities suffered losses.

At the motion to dismiss stage, the Municipalities need not demonstrate that Defendants' alleged misrepresentations caused all or even most of their losses. Rather, they need allege only "facts that would allow a factfinder to ascribe some rough proportion of the whole loss to [the defendant's alleged] misstatements." Lattanzio v. Deloitte & Touche LLP, 476 F.3d 147, 158 (2d Cir.2007). The Municipalities plausibly assert that if the correlation between taxable and non-taxable rates had remained as near-perfect as represented in the Graph, some or all of their losses would not have occurred. Accordingly, the Court finds that the Municipalities have adequately pled causation and denies Defendants' motion to dismiss the Municipalities' common law fraud claim.

C. NEGLIGENT MISREPRESENTATION

Terra and the Municipalities allege that the material misrepresentations underpinning their fraud claims also amount to negligent misrepresentation. To allege a claim of negligent misrepresentation, a plaintiff must plead that "(1) the defendant had a duty, as a result of a special relationship, to give correct information; (2) the defendant made a false representation that he or she should have known was incorrect; (3) the defendant knew that the plaintiff desired the information supplied in the representation for a serious purpose; (4) the plaintiff intended to rely and act upon it; and (5) the plaintiff reasonably relied on it to his or detriment." Pension Comm. of the Univ. of Montreal Pension Plan v. Banc of Am. Sec., LLC, 446 F.Supp.2d 163, 198 (S.D.N.Y.2006).

For the reasons described above, Terra has failed to plausibly allege reasonable reliance on Defendants' alleged misrepresentations. The heightened duty placed on sophisticated investors applies with equal force to the reliance prong of negligent misrepresentation. See Century Pac., Inc. v. Hilton Hotels Corp., 354 Fed.Appx. 496, 498 (2d Cir.2009) (finding reliance by sophisticated party unreasonable for the purposes of a negligent misrepresentation claim). Accordingly, the Court grants Defendants' motion to dismiss Terra's negligent misrepresentation claim.

The Court, however, finds that the Municipalities plausibly allege a negligent misrepresentation claim. Although it does not appear from the face of the Terra Complaint that Defendants were in contractual privity with the Municipalities, Plaintiffs have alleged sufficient facts to support a reasonable inference that Defendants nevertheless owed them a duty of care. To show that a defendant not in privity with a plaintiff nevertheless owes a duty to give that plaintiff accurate information, the plaintiff must show that there is "(1) an awareness by the maker of the statement that it is to be used for a particular purpose; (2) reliance by a known party on the statement in furtherance of that purpose; and (3) some conduct by the maker of the statement linking it to the relying party and evincing its understanding of that reliance." Credit Alliance Corp. v. Arthur Andersen & Co., 65 N.Y.2d 536, 493 N.Y.S.2d 435, 483 N.E.2d 110, 118 (1985).

In addition to the allegations of reliance described above, the Municipalities allege that Defendants insisted that Terra provide them the Presentation for the express purpose of marketing the FLN investments. The Terra Complaint thus plausibly alleges (1) that Defendants were aware that the Presentation was to be used for the purpose of marketing FLN investments; and (2) linking conduct by Defendants evincing their understanding that the Municipalities would rely on the Presentation to evaluate the prospect of FLN investments.

Accordingly, the Court denies Defendants' motion to dismiss the Municipalities' negligent misrepresentation claim.

IV. LEAVE TO REPLEAD

For the reasons discussed above, the Court has determined that Terra and Banca Carige cannot sustain plausible fraud or negligent misrepresentation claims based on Defendants' alleged misrepresentations in the Presentation. It is therefore possible that a repleading would be futile if it were based upon the same type of alleged misrepresentations or omissions. Nonetheless, the Court will grant leave to replead upon a request by Terra or Banca Carige showing that such a repleading would not be futile. Plaintiffs may submit any such request within ten days of this Order.

§ 6.21 *Selective Deception*

A simple single **line graph** consists of a set of ordered pairs of numbers that are placed along an X-axis and Y-axis forming points on a grid that are joined by lines or curves (Figure 6-3). The presentation sounds objective enough. Such a graph can present a straightforward and unbiased function (Figure 6-1) or it can serve to persuade (Figure 6-2), or deceive; the thin "*lines*" among these purposes are not always clear.

Figure 6-1:

Figure 6-2:

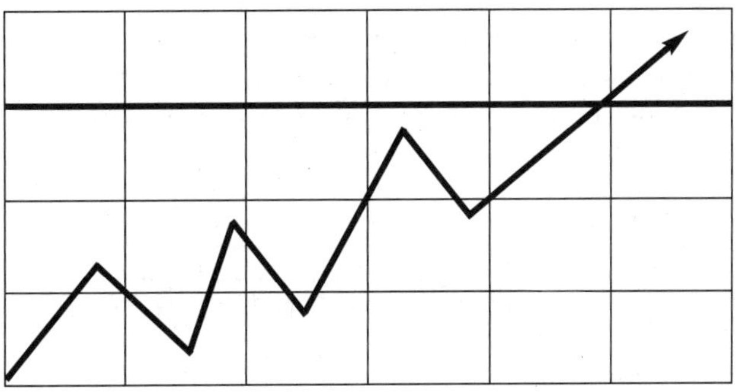

In detecting the most common tricks and deceptions, the practitioner must inspect not *only* that which is presented, but that which is missing. Consider the following:

Example:

Stock prices go up and down. In fact, you can rest assured that statistical information relating to *any* economic trend, issue or dollar-related data will most likely fluctuate over time. In the midst of that fluctuation, an event will occur. Some grieving plaintiff or a group of them will lose some money, some health or some body parts, then either rightly or wrongfully, point the finger at some unlucky defendant. The plaintiff will claim that the subject event resulted from some illegal, negligent, improper or unethical act or breach by the defendant which interrupted the normal course and flow of whatever it was that had been previously peacefully flowing. This malicious, willful, wanton or negligent misconduct may come in the form a tortious act, a Ford Pinto gas tank, a Gray water heater, a Toyota accelerator, the breach of a contract, the infringement of a trademark or copyright, or it may have resulted from the monopolization of an entire industry. The following chart might represent the plaintiff's income, business profits or any quantifiable measure relating to the incident (which allegedly occurred in Month 10). The next four graphs, despite their apparent differences, are all predicated upon the *same identical* data:

Figure 6-3:

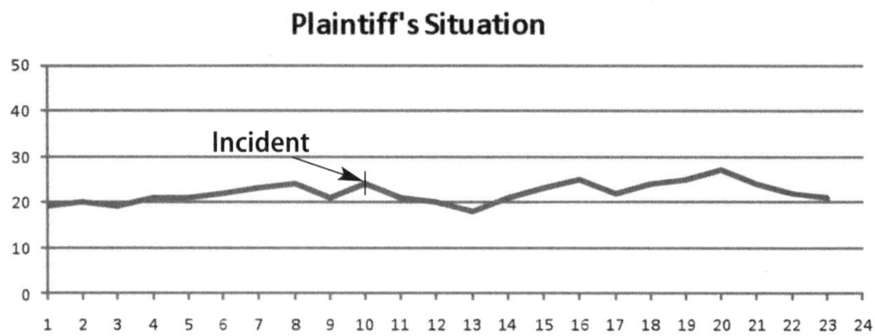

Figure 6-3 provides a relatively benign picture of the Plaintiff's economic circumstances. The fluctuation in profit, gross receipts, business activity (or whatever else the Y-axis may represent) seems restrained and moderate. The Month-10 incident or accident does not, at least from the graphic data, appear to have had a significant impact.

In order to provide a more dramatic "damages" picture for the jury, there is no need to falsify data or distort the numbers. In fact, the only thing that we need to do is adjust the scale and the period covered, thereby yielding an entirely different picture:

Figure 6-4:

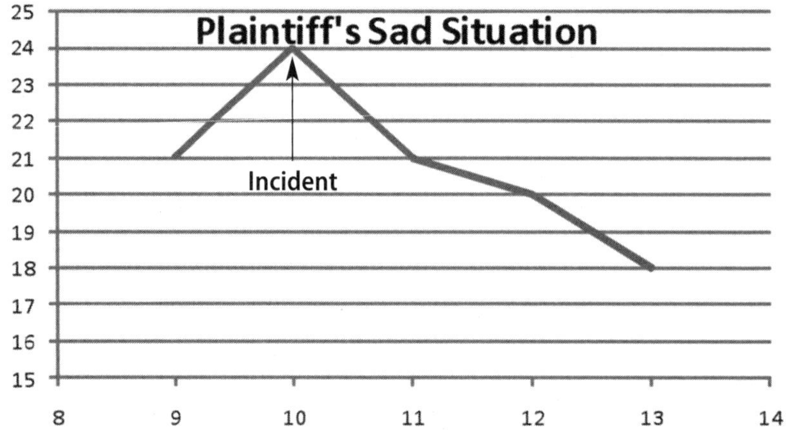

The underlying data for Figure 6-4 is identical to that of Figure 6-3. The only changes involve cherry picking the data and altering the scales. Notice how Figure 6-4 managed to chop off the profitable period following Month 13, while it simultaneously stretched out the three month downturn following the subject incident. Deriving this particular image from Figure 6-3 was not difficult. In fact, Microsoft's Excel will do your dirty work for you by automatically adjusting the **domain** (length of the X-axis) and **range** (length of the Y-axis), once you have specified the portion of data that you wish to use for your graph.

The good news is that you are far from powerless when it comes to confronting graphical manipulation. For the "damages-related" expert witness, who introduced Figure 6-4 to support his claim that the plaintiff suffered egregiously following the willful and licentious acts of the defendant, it's time to have some fun.

Hypothetical Cross-Examination of Expert

The following portion of cross-examination presupposes the introduction of Figure 6-4 into evidence or use by the expert for demonstrative purposes only.

Q. [By Defense Counsel]: You testified on direct examination that Figure 6-4 represents the alleged downturn that the Plaintiff suffered following the subject incident. Is that correct?

A. [By Expert Witness]: Yes.

Q. Where, or from whom, did you receive the data that you used to make up the graph?

A. I didn't *make it up*; I carefully constructed it based upon the data that was provided to me. The plaintiff and his agents supplied me with the information that I used.

Q. During the process of *this construction*, can you tell us why the plaintiff found it necessary to hide information or otherwise keep it from you?

[Plaintiff's Counsel]: Objection! Counsel has no basis whatsoever for making such an accusation, nor does he have the right to speculate as to what my client was thinking.

Q. Let me rephrase the question. Why then did your graph omit important data extending beyond the subject incident?

A. I don't know what you're talking about.

Q. Let me rephrase the question once again. Why then did your graph omit important information?

A. What information?

Q. Let me show you Defendant's Exhibit (Figure 6-3) and ask whether or not it represents *all* of the data that you had access to over the 24 month period represented in the chart?[6]

A. Figure 6-3 appears to represent a 24 month period.

Q. [Displaying both charts to the jury]. But yours conveniently omits 20 of those 24 months. Which chart, *Sir*, presents a more complete and correct picture for the jury?

A. It's not a question of correct or complete, it's a....

Q. [Interrupting]. I'll repeat my question. Which chart, Sir, presents a more complete and correct picture for the jury?

---

6. Springing the chart on the witness for the first time during the course of trial is dangerous and not recommended. A claim that Figure 6-3 is being used "for impeachment purposes only" may fool or persuade some judges. The preferred course, however, would have been for the defendant to provide the chart to the enemy during early discovery disclosures to assure admissibility. See generally, Lipson, *Guerrilla Discovery*, James Publishing Company (2002–2011, updated annually).

A. As I tried to explain. It's not a question of correct or complete, it's a....

[Plaintiff's Counsel]: Objection! Counsel is arguing with the witness. He can't complete his answer.

Q. I apologize. Please complete your answer. We all want to hear it.

A. As I tried to explain. It's not a question of correct or complete, it's a matter of which period we were focusing on.

Q. Is there something particularly relevant about the 4th month after the incident, but not the 6th month?

A. No.

Q. Is there something particularly relevant about the 4th month after the incident, but not the 7th month or the 8th month or the 9th month or the 10th month ...?

[Plaintiff's Counsel]: We get the point.

Q. No further questions.

[Save the expert's disembowelment for closing arguments]

Figure 6-5 provides a relatively pretty picture from the perspective of the defense. As is the case with Figure 6-4, it too delivers a very selected, cherry-picked portion of the entire picture.

Figure 6-5:

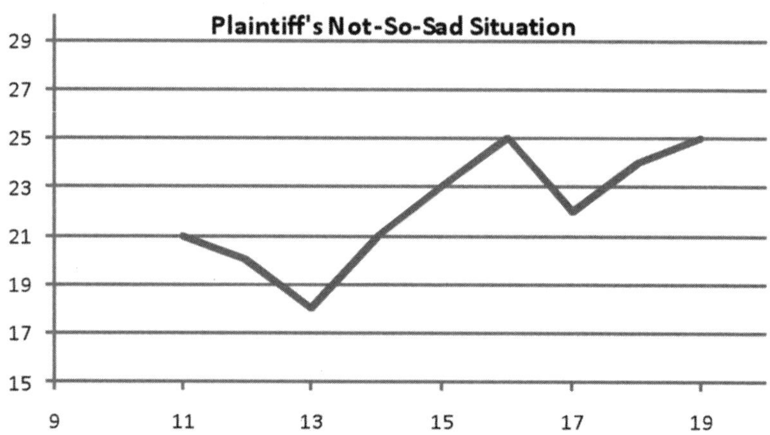

Particular attention should be directed at the scaling along the Y-axis. There is nothing sacred about the space between the numbers. Nor can one claim deception merely because larger or smaller spaces were selected. One should, however, be extremely suspicious where the scaling is dropped altogether, as is the case with Figure 6-6.

Figure 6-6:

Scale manipulation of simple two dimensional graphs becomes even more effective when shades or fills are used. Compare these next two graphs, which like the preceding graphs, are based upon identical data.

Figure 6-7:

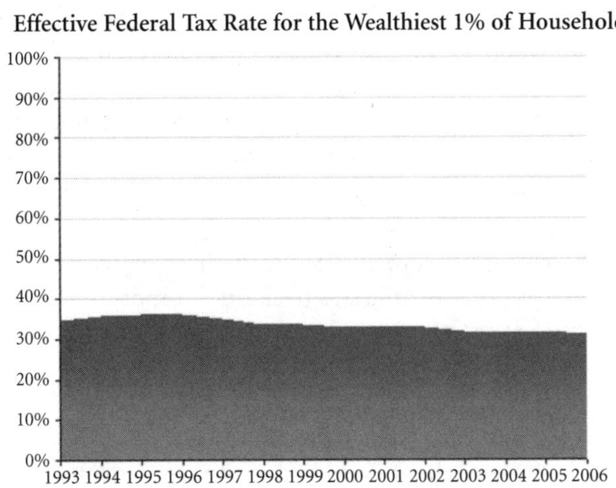

Figure 6-8:

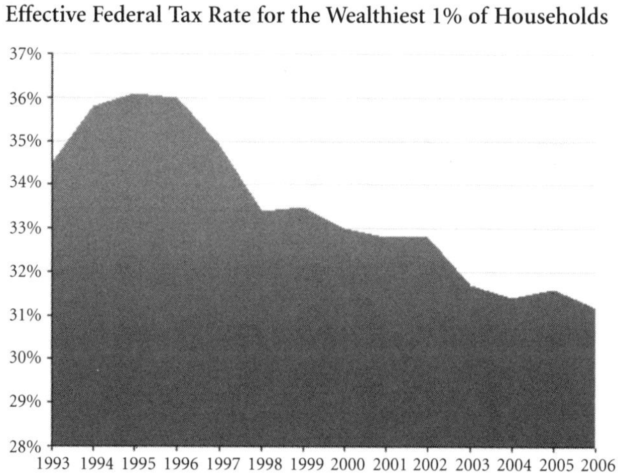

Pfizer, Inc. v. Merial Ltd.

1 Trades P 72,927 (S.D.N.Y., 2000)

MARTIN, District J.

Plaintiff Pfizer, Inc. brings this action against Merial Limited for false advertising under the Lanham Act as well false advertising, deceptive practices, and unfair competition under state law. Both companies are in the business of making flea and tick control products for dogs and cats. Pfizer claims that Merial's recent television advertising and direct marketing brochures sent to veterinarians in support of its Frontline product are false and misleading and moves to enjoin the ads and brochures while this case is pending. While Pfizer has established that a Merial television ad and one of Merial's brochures are false and misleading, it has failed to establish that it will be irreparably harmed if an injunction does not issue. Therefore, the motion is denied.

Until recently tick and flea infestations on pets have been treated with powders, sprays, and flea collars containing pesticides. In the last few years several pharmaceutical companies have developed "spot-on" products that involve application of a small amount of liquid to a spot on the animal's back once a month. Pfizer's "Revolution" and Merial's "Frontline" are spot-on products. Both are available only with a veterinary prescription, and both have been the subject of extensive marketing campaigns nationwide.

Under well-settled law in the Second Circuit, a plaintiff is entitled to a preliminary injunction upon a showing (1) of either a likelihood of success on the merits or of sufficiently serious questions going to the merits to make them fair ground for litigation and a balance of hardships decidedly favoring the plaintiff; and (2) that it will suffer irreparable harm. See Warner–Lambert Co. v. Northside Dev. Corp., 86 F.3d 3, 6 (2d Cir.1996); Castrol, Inc. v. Ouaker State Corp., 977 F.2d 57, 62 (2d Cir.1992).

Section 43(a)(2) of the Lanham Act prohibits "false or misleading representations of fact" in advertising that misrepresent the nature, characteristics or qualities of the advertiser's or another person's products. To succeed on the merits of a false advertising claim, "a plaintiff must demonstrate that 'an advertisement is either literally false or that the advertisement, though literally true, is likely to mislead and confuse consumers." ' See Castrol, 977 F.2d at 62 (quoting McNeil–P.C.C. v. Bristol Myers Squib Co., 938 F.2d 1544, 1549 (2d Cir.1991)). When the claim in a comparative ad is literally false, a court may enjoin the use of the claim without reference to the ad's impact on consumers. See id.

Pfizer primarily takes issue with a graph produced by Merial that reported the results of a comparative efficacy test between Frontline and Revolution. The graph appeared in at least one veterinary journal and was included in a brochure that Merial distributed directly to veterinarians. The graph appears under the headline, "FRONTLINE BEATS REVOLUTION vs. fleas" which is followed by smaller text reading "FRONTLINE works FASTER, performs BETTER & lasts LONGER against Fleas. In a recent controlled, head-to-head study, FRONTLINE outperformed REVOLUTION in key aspects of flea control." The graph itself is labeled "Comparative Efficacy of FRONTLINE Top Spot TM and REVOLUTION Against Fleas on Dogs."

The graph's vertical axis indicates percent efficacy of each product in killing fleas and the graph's horizontal axis indicates time over a period of 30 days with the labels "Day 1," "Day 9," "Day 16," "Day 23," and "Day 30" running across the bottom. The graph contains data lines representing the purported efficacy of each product over the 30–day period. Below the graph small print text explains that the test involved 30 dogs in three groups — 10 untreated controls, 10 treated with Frontline, and 10 treated with Revolution. It further explains that the dogs were treated on Day 0 and re-infested with fleas on Day 8, Day 15, Day 22, and Day 29 and that flea counts were taken on Day 1, Day 9, Day 16, Day 23, and Day 30.

The data line in the graph depicts dramatic differences between the two products. For Frontline the line begins at what looks like 98% efficacy on Day 1 then rises to 100% on Day 9 and continues there through Day 30. The Revolution line begins at around 73% on Day 1 climbs sharply to nearly 100% for Day 9 falls a bit to 98% for Day 16 then plunges to 75% on Day 23 and a mere 50% on Day 30.

The problem with this graph is that it depicts continuous lines of efficacy for what were really five individual flea counts taken on the dogs in the study. Merial counted fleas 24 hours after the initial treatment and 24 hours after each of the five flea re-infestations. It then represented those flea counts as dots on a graph — so far, so good. The problem arises when those five dots are connected by straight lines. The lines appear to represent the efficacy of the products at all points between the five flea counts. For example, half way between Day 23 and Day 30 on what would be Day 27 the line shows Revolution's efficacy as approximately 65%. But as Pfizer points out and Merial admits, if flea counts are taken 36 hours after re-infestation (instead of 24) Revolution's efficacy returns to nearly 100%. Therefore, the data line at many points throughout the graph is patently false.

Ordinarily, Pfizer would be entitled to a preliminary injunction against the use of the graph. But perhaps realizing the weakness of its position, defendant Merial ceased using the graph shortly after the initiation of this lawsuit. It has begun using a new graph to represent the same study in its advertising and has represented to the Court that it will never use the old graph again. Therefore, Pfizer's motion for a preliminary injunction as to the original graph is moot. See Consumers Union of U.S. v. General Sign Corp., 724 F.2d 1044, 1052 (2d Cir.1983); American Express v. Mastercard, 776 F.Supp. 787, 791 (S.D.N.Y.1991). Furthermore, Pfizer cannot show a likelihood of irreparable harm caused by a graph that Merial is no longer using.

Pfizer argues that the Court should enjoin Merial's use of its new graph because it is also false and misleading, but Pfizer has not shown a sufficient likelihood of success on the merits as to the new graph. The new graph appears under the same headline and introductory text. However, the graph is now labeled "Comparative Efficacy of FRONTLINE Top Spot TM and REVOLUTION Against Fleas and Flea re-infestations on Dogs at 24 hours." Instead of data lines the graph contains five pairs of bars representing Frontline and Revolution on each of the five days that flea counts were taken. As in the first graph, the differences between the products appear dramatic, but the graph clearly represents the comparative efficacy only at the five times that flea counts were taken and makes no representation as to what happened between those five times.

According to Pfizer, the new graph is still false and misleading because the overall message of the ad is that Pfizer's Revolution is not effective and that Merial's Frontline is overwhelmingly more effective. This message is false, Pfizer contends, because comparative efficacy looks completely different when based on a "real world" study or, as Merial admits, on a study that takes flea counts at 36 hours after infestation. Essentially, Pfizer contends that Revolution performs as well as Frontline on aspects of efficacy other than those tested in Merial's study. However true this contention may be, it does not render Merial's graph false or misleading.

The message that the graph and the ad as a whole convey is relatively clear: a study demonstrated that Frontline is more effective than Revolution against fleas within 24 hours of treatment and re-infestation, and that the difference becomes more acute later in the month. Basically, Pfizer's arguments regarding the new graph amount to an expression of its wish that Merial would have used a different study to test the relative efficacy of the two products.

The products may be comparable when flea counts are taken at 36 hours after re-infestation, but there is nothing inherently false or misleading about choosing to take counts at 24 hours. While Pfizer claims that a 12 hour difference in killing speed means nothing in the grand scheme of 30–day pest control, it is up to vets and pet owners to decide how important it is to them that Frontline kills fleas a bit faster than Revolution. Pfizer is free to conduct its own studies based on "real world" situations or based on flea counts at 36 hours after re-infestation, circulate the findings to vets or the public, and explain why its studies are more relevant to overall efficacy.

Furthermore, Merial has targeted the ad at veterinarians—sophisticated readers who are more likely to understand the limitations of Merial's study. See Sandoz Pharmaceuticals Corp, v. Richardson–Vicks, Inc., 902 F.2d 222, 230 (3d Cir.1990) (rejecting claim that bar chart used in advertising distributed to doctors was false or misleading and noting that a target audience's special knowledge is highly relevant to any claim that the audience would be misled).

Pfizer asks the court for a broad injunction similar to the one issued in Castrol v. Quaker State, but in that case the court found that the defendant's tests did not substantiate its advertising claims. See Castrol, 977 F.2d at 61–62. Pfizer has made no claim here that Merial falsified its flea counts or otherwise improperly conducted its tests. The test results were valid and support Merial's graph; Pfizer just wishes Merial had used a different test. Because Pfizer is unlikely to prevail on its claim that the new graph is false or misleading, the Court will not enjoin Merial's use of the graph in its advertising.

The next issue is whether Merial's current television advertising for Frontline should be enjoined. Pfizer initially challenged four television spots, but Merial has discontinued three of them and has represented that it has no intention of using them again. The only ad that Merial is currently running is a 30–second spot titled "Answers" which features an actress playing a veterinarian. The ad depicts the "vet" in a white lab coat examining dogs and cats in her office and at one point holding up a Frontline brochure. In relevant part, she says:

I think people are definitely confused about which flea and tick protection to use for their dogs and cats. Yeah, there's others out there, but the fact is FRONTLINE is still the best. That's because FRONTLINE kills 100% of fleas within a day. And only FRONTLINE kills all major ticks, including the one that can carry Lyme disease.

The bone of contention here is the statement that "only FRONTLINE kills all major ticks, including the one that can carry Lyme disease ." This statement is literally false because another product, "Defend Exspot," also kills all major ticks including the one that carries Lyme disease.

Merial argues that the statement when taken in the context of the ad as a whole is not false or misleading. Apparently, Defend Exspot kills all major types of ticks but also has a tendency to kill cats. Therefore, vets can only prescribe it for use on dogs. Merial argues that because the actress in their ad initially says, "I think people are definitely confused about which flea and tick protection to use for their dogs and cats," the overall message of the ad is that "only Frontline kills all major ticks, including the one that carries Lyme disease for dogs and cats." But that is not what the ad says. The ad says that "only Frontline kills all major ticks," and the larger context of the ad does not change the literal falsity of that statement. For a cat owner who views the ad the effect of the false statement is not particularly misleading; Frontline is the only product available that kills all major ticks and can be used on the viewer's cat. But for the dog owner who views the same ad the effect is unequivocally both false and misleading; there is another product available that kills all major ticks and can be used on the viewer's dog.

The only problem with the preliminary injunction motion as to the TV ad is that Pfizer has not sufficiently shown a danger of irreparable harm if the motion is not granted. Pfizer argues that because the ad in question is literally false, irreparable harm should be presumed. However, the rule Pfizer refers to has only been applied in cases involving literally false comparative advertising involving the plaintiff's product. See e.g., McNeil–P.C.C., Inc. v. Bristol–Myers Squibb Co., 938 F.2d 1544, 1545 (2d Cir.1991); McNeilab, Inc. v. American Home Products Corp, 848 F.2d 34, 35 (2d Cir.1988). Moreover, the Second Circuit has stated explicitly that there is no presumption of irreparable harm on a Lanham Act preliminary injunction motion. See Coca–Cola Co. v. Tropicana Products, Inc., 690 F.2d 312, 316 (2d Cir.1982); Johnson & Johnson v. Carter–Wallace, Inc., 631 F.2d 186, 190 (2d Cir.1980). While the party seeking an injunction does not have to produce specific evidence of actual sales losses, it must come forward with some evidence to show that it is likely that defendant's advertising has caused or will cause such losses. See Johnson & Johnson, 631 F.2d at 190.

In this case, Pfizer's Revolution is not mentioned in Merial's Frontline ad, and the ad's claims are absolutely true as to Revolution. Unlike Frontline, Revolution does not kill all major ticks. The only real harm Frontline's false statement will cause is to the makers of Defend Exspot-the other product that kills all major ticks. At argument, Pfizer's counsel cited the "halo effect" caused by the Frontline commercials as a harm to Pfizer. In other words, statements that a product is "the best" or "the only one" tend to raise the product's esteem in consumers' minds, making them more likely to buy the product whether or not they need the touted quality. In this instance, the logic is that even people whose pets are not afflicted by ticks will be more likely to buy Frontline over Revolution because of the "halo" created by the claim that Frontline is the only one that kills all major ticks.

But this potential injury to Pfizer is speculative at best. Logically, it is safe to say that the false statement in the Frontline ad will only have an substantial impact on consumers who are looking for a product that kills all major ticks. Even assuming that these consumers purchase Frontline because of the false statement, there is absolutely no harm to Pfizer because Revolution does not kill all major ticks. Pfizer has not produced any evidence to counter this reasoning or support its "halo effect" argument raised for the first time at oral argument. On this scant record, the Court does not find a sufficient likelihood of irreparable harm to support a preliminary injunction.

Pfizer's motion for a preliminary injunction is denied.

SO ORDERED.

§ 6.22 *Slanted Perspectives*

Line graphs are not the only means available to distort one's perception of numerical data. On the contrary, virtually any type of graph, drawing or image can be used to deform otherwise reliable or objective numerical data.

I'll bet you never knew that the number 36 was more than double the number 30. Only two types of professionals could reach such a ridiculous conclusion—statisticians and bookies. Two different methods of deception are work in Figure 6-9. First, there is a small range of data (.36 - .26 = .10) that commences at .26 instead of zero. Next, the perspective is designed to greatly enlarge those objects that are closest to the viewer. The resulting deception is very dramatic when you consider that the actual numerical difference between the largest and the smallest objects is only .06.

Figure 6-9:

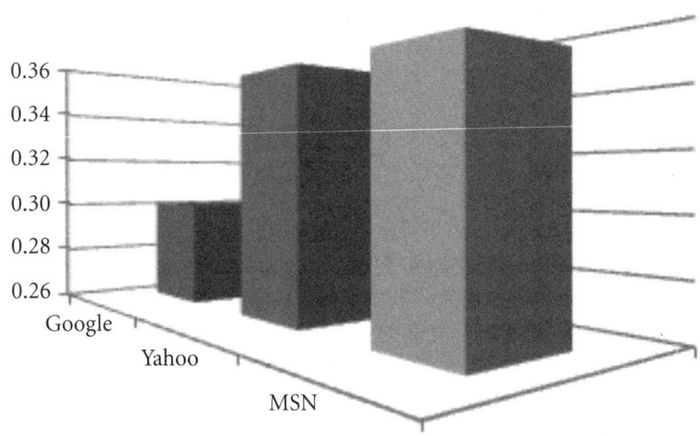

Figure 6-10 provides an additional illustration of the power of scale manipulation alone. There is no need to distort the perspective or create the optical illusion that we witnessed in Figure 6-9.

If we accept the figures provided in the next chart, unemployment was higher in the Clinton years than in the Bush years. (For openers, one might be curious as to which "Bush" the chart is referring to, since Clinton was both preceded and succeeded by one.) In any event notice that the difference in unemployment rates is minuscule:

5.2052 - 5.1852 = 0.02.

Taking an incremental approach:

$$0.02 / 5.2052 = 0.0038.$$

Thus, if the graph represented the incremental drop in unemployment witnessed by the Bush presidency, your eyes would not be able to detect any difference (which would be less than 1%). Yet, the difference in area (an attribute that is recognized as having the greatest visual significance) is nearly 300%.

Figure 6-10:

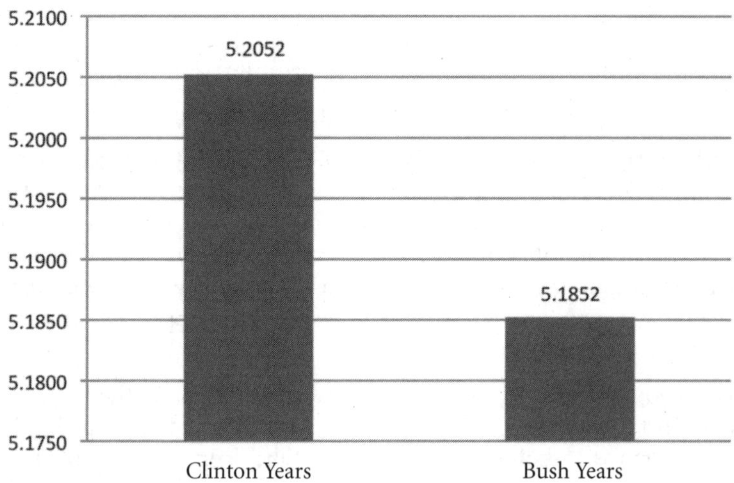

Another instrument of deception involves the increase or decrease of a particular "rate." I am not referring to the increase or decrease in actual numbers, but rather the change in the *rate* of change. Consider for example the following raw data, which represents the number of annual murders for a hypothetical city with a population hypothetically fixed at 100,000:

2007.......................................500 People Murdered

2008.......................................650 People Murdered

2009.......................................750 People Murdered

2010.......................................840 People Murdered

2011.......................................920 People Murdered

2012.......................................960 People Murdered

2013.......................................990 People Murdered

From the above data, it is fair to state that the city has a relatively high murder rate; most civilized societies would consider it to be clearly unacceptable for every year listed. For the calendar year 2013 nearly 1% of the entire population has been murdered. Also, from the data, it is clear that the number of murders per 100,000 has risen every year for six straight years. And while it would be correct to state that the murder rate has increased every year, it would also be correct (though potentially deceptive) to say that:

> "The annual increase in the murder rate has *declined* every year for each of the six years."

This particular type of statement, though true, is often misinterpreted and taken to mean that the murder rate itself has declined. It hasn't. Even though the murder rate has increased, the *amount* of the increase has gone down every year. The increase from 2007 to 2008 was 650—500 = 150. But the increase from 2008 to 2009 was only 100; and the increase from 2009 to 1010 was only 90. In fact, the magnitude of the increase continues to decline though 2013, which delivered an increase over 2012 of only 30. Therefore, the amount of the increase has decreased every year. Some statisticians will go so far as to provide a declining bar graph representing the diminishing increases. The public (i.e., those who are lucky enough to survive) will, of course, be easily fooled into believing that the actual murder rate has been diminishing, when in fact both the number of murders and the murder rate have been increasing.

Haller v. AstraZeneca Pharmaceuticals LP

598 F.Supp.2d 1271 (M.D.Fla., 2009)

Seroquel, a prescription drug manufactured and sold by AstraZeneca Pharmaceuticals, was first approved for use in the United States by the Food and Drug Administration (FDA) in 1997. The drug, known chemically as quetiapine fumarate, is still on the market today as one of a number of medications known as second-generation, or "atypical," antipsychotics. Although initially indicated to treat the "manifestations of psychotic disorders," including schizophrenia, Seroquel has since been approved for various uses in patients suffering from bipolar disorder.

Several years ago, people who had been prescribed Seroquel by their doctors began filing lawsuits against AstraZeneca, primarily alleging that they had developed diabetes and other related disorders as a result of their ingestion of the drug. In July 2006, the Judicial Panel on Multidistrict Litigation transferred ninety-two actions involving alleged injuries resulting from the use of Seroquel to this Court for consolidated and coordinated pretrial proceedings. Since that time, this consolidated action has grown to include the personal injury claims of several thousand people from across the nation, including the plaintiff in the present case, David Haller.

> [Patient Haller sued AstraZeneca, the manufacturer of the anti-psychotic drug asserting claims for strict product liability, negligence, fraud, and civil conspiracy. He alleged that his weight gain and diabetes was caused by taking anti-psychotic medication. The manufacturer moved for summary judgment. A major factor in the granting of the motion was unreliable methodology and analysis employed by Haller's expert witness, Dr. Tulloch, whose testimony was excluded.]

When confronted with his deposition testimony in which he admitted that Haller's weight gain while he was on Seroquel did not cause him to have diabetes, Dr. Tulloch agreed that he had said that, but added: "I can't say that in the current setting. I have shown you how much he gained weight when he was exposed to Seroquel, with all the other issues that were present. And we have shown that while his weight was high, with all the other medications that had been present, this man developed diabetes following a period of pancreatitis. So if [sic] that context, I might have been thinking of it in a different context. But I think we have shared the data today, and now we have a nice simple graph [Exhibit 19] to show the weights in relationship to the diagnosis."

Another glaring example is Exhibit 19, the weight vs. time graph Dr. Tulloch relied on at the Daubert hearing. As a visual representation of Haller's weight gain and loss over time, Exhibit 19 is seriously flawed, for at least two reasons. First, the graph is grossly misleading because the horizontal date axis is not scaled. For example, the distance on the chart between 1–22–97 and 8–26–00 (a 3 [ONE HALF] year interval) is the same as between 6–23–04 and 6–24–04 (a 1–day interval). This results in an artificially steep slope at some points on the chart. Second, the graph is comprised of two stapled-together segments, and the lines signifying weight increments on the vertical axis are misaligned from the first to the second segment.

The Court is at a loss to conceive how a graph so visually misleading could be considered an aid to the Court. Further, this demonstrates that the reliability problems associated with Dr. Tulloch's methodology were not confined to the early stages of his involvement in the case, but instead continued all the way through the Daubert hearing.

Beyond the problems associated with the way in which Dr. Tulloch reached his opinions, the stark fact is that the grounds for his causation opinion have been a veritable moving target. And what is most troubling is that the underpinnings of his opinions have changed in direct response to AstraZeneca's motion practice.

The Court thus concludes that Dr. Tulloch has employed a flawed methodology in reaching his specific medical causation opinions. Moreover, those opinions are impermissibly conjectural and speculative; they require too great a leap of faith. For these reasons, Dr. Tulloch's proposed expert testimony fails both the reliability and relevance prongs of Daubert. The predominant lesson of Daubert is that federal courts are not simply to take the expert's "word for it." Because Dr. Tulloch's specific causation opinions in this case amount to no more than the ipse dixit of an expert witness, they must be excluded.

Motion granted.

§ *6.23 Tasteless Pies*

The Pie Chart provides one of the simplest and least confusing methods for presenting a percentage-distribution or breakdown. Like the graph, the visual image of the pie chart causes the viewer to make certain presumptions. First, we are taught to presume that the entire pie represents 100% or "all" of something (which is why Figure 6-11 should immediately raise an eyebrow). Second, and most importantly, the size or magnitude of each slice is supposed to be in proportion to its percentage of the "whole" (Figure 6-12).

Figure 6-11:

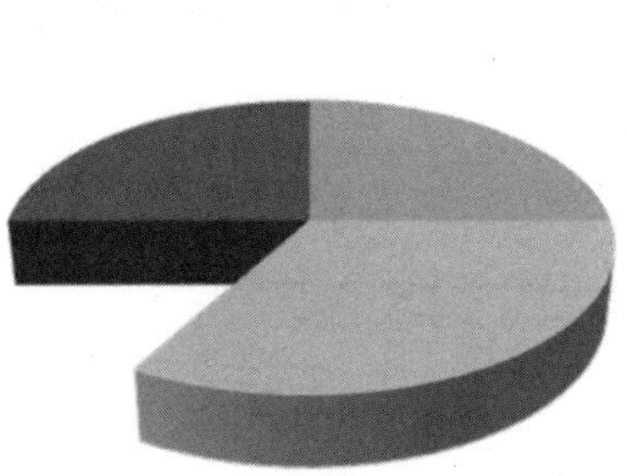

Figure 6-12:

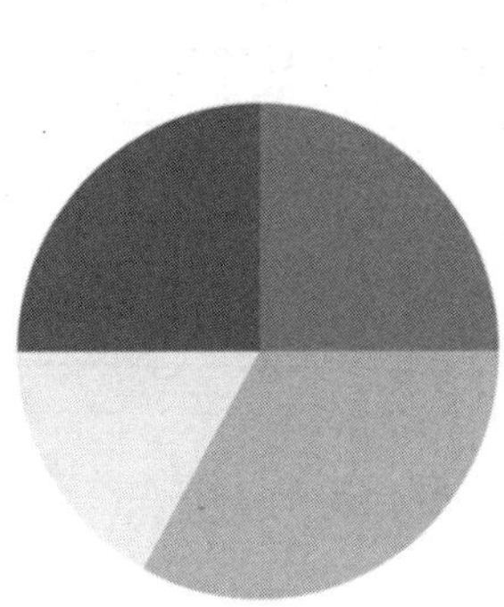

A common, subtle method of increasing the "magnitude" and appearance of the slice that one wishes to accentuate involves the use of a simple three-dimensional pie, with the favored slice in front. Notice that even though the front slice (Figure 6-13) represents no more than 1/3 of the pie, it occupies nearly 1/2 of the area of the total image. You may be quick to understand this; jurors, however, may not be quite as astute.

Figure 6-13:

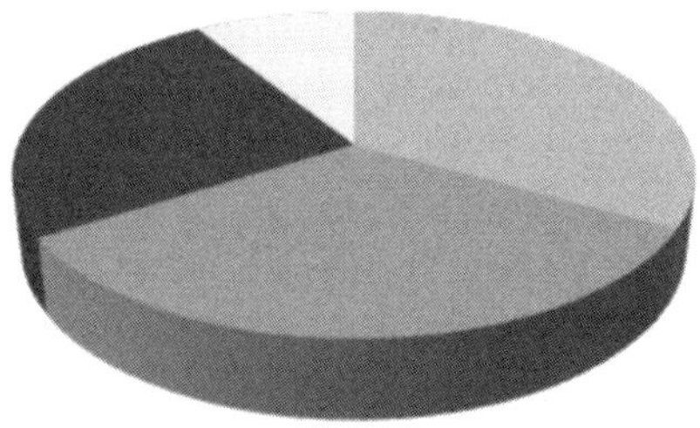

With today's software, a truly creative and deceptive advocate can accomplish some rather remarkable artwork.

Figure 6-14:

Figures 6-14 and 6-15 are masterpieces of deception. (Escher and Dali would both be green with envy.) In Figure 6-14, notice how the tallest slice does not represent more than 20% of the entire pie, and yet its area and prominence comprise far more. Notice also the manner in which that slice obfuscates and diminishes the "equal" slice to its immediate left, not to mention the manner in which it purports to dominate the entire landscape.

Figure 6-15:

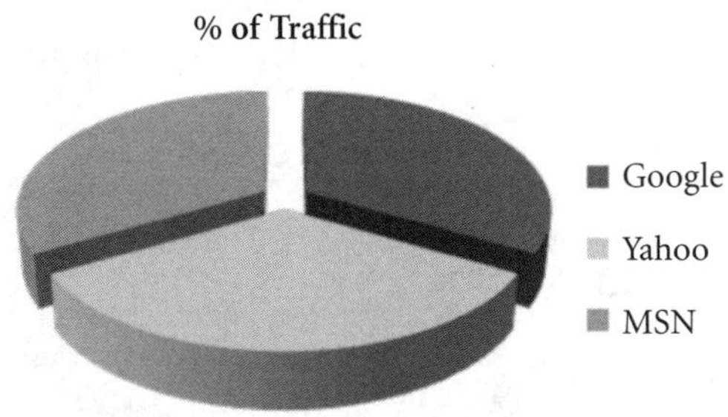

Here is an epicurean's ecstacy of visual deception. Quickly determine the size of the slice nearest the viewer (i.e., the front slice). At first glance it seems as if it occupies more than 1/2 of the entire pie and yet, in truth, all three slices are equal.

In addition to the endless visual imagery available for the presentation of mathematical and statistical data, labeling also provides an excellent opportunity for deception and obfuscation. In a simple pie chart, for example, a slice might have its magnitude labeled using several different numerical schemes. Figure 6-16 represents the most common.

Consider a random pie that represents 1,000 objects or elements. A slice of that pie representing 200 of those 1,000 items or amounts could be represented in any one or more of the following methods (The possibilities for deception and confusion are endless especially if multiple formats are **mixed** together in a single chart):

FORMATS:

1. 200
2. 1/5
3. 200/1000
4. 20%
5. 1:5
6. 0.200
7. 0.20

Any one or more of the above formats might legitimately be employed. Selection may ultimately depend upon which one best communicates the desired information to the jury. It might be even more impressive, however, to select the decimal format and then use 5 or 6 decimal places to falsely convey a sense of extreme precision that does not exist. 19.99987% might be far more impressive than just plain 20%.

Figure 6-16:

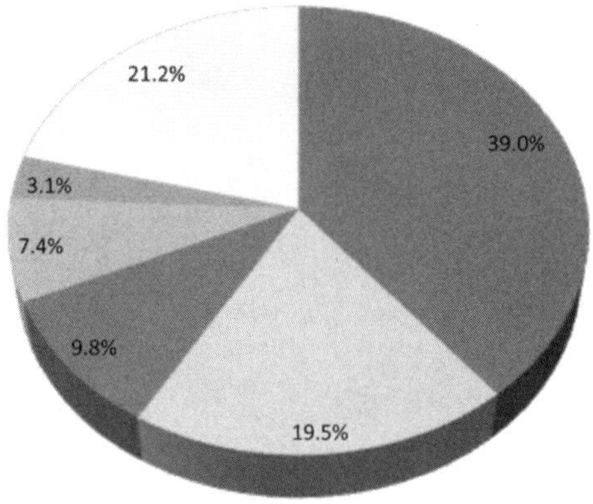

There are several theories as to which slice is in the most important position, depending upon the manner in which the eye moves. Some suggest that the slice representing the 1:00 to 3:00 o'clock position would be most favored because regardless of whether we are clock watchers or scientists, our initial attention seems to be directed to that area.

Notes and Problems

1. *Model Imperial Supply Co. Inc. v. Westwind Cosmetics, Inc.*, 829 F. Supp. 35 (E.D. N.Y., 1993) presented a battle over counterfeit men's fragrances. The court found that the gas chromatography tests conducted by the prevailing party were more reliable than the tests conducted by the opponent for several reasons including the fact that the winner's graphs displayed approximately 100 peaks for each mixture while the competing test produced graphs with approximately eleven peaks for each sample. How would you analogize this finding with Figures 6-3 versus 6-4? Can you draw a comparison of both sets of graphs with manner in which high resolution images (i.e., images with greater numbers of pixels) provide a "clearer" picture than low resolution images?

2. What degree of accuracy should be required and demonstrated before a machine (in the form of a graph generator) should be permitted to provide decisive evidence for the court to rely upon? See Rule 702 of the Federal Rules of Evidence; see also § 1.22.

§ 6.24 *Cheaters' Checklist*

It is not possible to catalogue all of the many ways that graphs can be used to persuade, mislead and lie. And even if cataloguing were possible, like viruses on the Internet, new bugs would come faster than the methods for catching them.

We have attempted to identify those techniques and methods that are promulgating the vast majority of deceptive graphs and charts. Here is a checklist to supplement the conduct that we are trying to expose:[7]

---

7. Assistance acknowledged from: Lipson, *Art of Advocacy—Demonstrative Evidence*, LexisNexis (updated annually), and Jones, *How to Lie with Charts*, Author's Choice Press, 2000 and Google.

❒ 1. One of the oldest tricks in the book is to use carefully skewed data points for a graph or chart to show a trend or support an argument (See *supra*, Figures 6-3 et seq.).

❒ 2. In a three-dimensional pie chart, the item located at the lowest point on the chart always appears to be disproportionately larger than its size warrants, due to its three dimensional perspective.

❒ 3. Using several different pie charts of different sizes always confuses a jury even if the data's proportions are correct.

❒ 4. Using two-dimensional graphs that contain more than a single line or curve is per se confusing.

❒ 5. Turning a conventional bar chart on its side causes confusion and side steps the impact that would otherwise be obtained if it were vertical.

❒ 6. The term "average" is often misused to denote "mean," "median," or "mode," whichever term best suits the purpose of the person offering the statistics. The three functions, of course, can differ significantly, depending upon the structure and distribution of the underlying data.

❒ 7. For two-dimensional charts, data on the X-axis that is totally unrelated to data on the Y-axis, may nevertheless, appear to be related, regardless of the shape of the resulting curves or lines. One assumes that such data is related by the mere fact that a two-dimensional axis is used.

❒ 8. All scales are not necessarily linear. Using charts predicated upon exponential or logarithmic functions can smooth out the rough lines and give the appearance of a trend that does not exist. To avoid highlighting steep exponential trends, sometimes semi-log charts are used.

❒ 9. When viewing a map or diagram, one is accustomed to assuming that "up" is North, "down" is South, "left" is East and "right" is West. Switching or reversing these items confuses the viewer.

❒ 10. Where there is a natural order, whether it is geographical or otherwise, rearranging objects out of their natural order can cause confusion. For example: "West, East, Central" or "First, Third, Second."

❒ 11. Selecting the "wrong" type of chart causes confusion and obscures information that might otherwise be derived from a proper presentation. For example, using a pie chart to demonstrate a trend or a bar chart to show relative proportions would certainly cause confusion.

❒ 12. Colors and contrast play an important role in the manner in which viewers attach significance to data. For instance, lightly colored data only stands out if it is placed against a dark background. As a corollary, light data against a light background or dark data against a dark background tends to be obscured or diminished in appearance.

❒ 13. Areas are often the key to deception. As we have seen in prior examples, the viewer's attention is directed toward the images as opposed to the numbers (After all, that is the purpose of presenting a graphic image instead of a stream of numbers).

❒ 14. Labels can cause confusion in a variety of ways. Confusing color codes or naming conventions along the X or Y axis are common.

❒ 15. Sometimes the title or text that precedes a chart may not be particularly appropriate or may otherwise seem "out of place," thereby causing additional confusion.

In the last analysis, there are many ways to deceive and many ways to combat deception. The key to being forearmed, of course is to be forewarned. Lying in wait with a slicing cross-examination may be the most aggressive defense. But there are others. For example, Rule 403 of the *Federal Rules of Evidence* (and most state counterparts) provide:

> The court may exclude relevant evidence if its probative value is substantially outweighed by a danger of one or more of the following: unfair prejudice, confusing the issues, misleading the jury, undue delay, wasting time, or needlessly presenting cumulative evidence.

It is important to note that Rule 403 is available to attack and exclude evidence even though that evidence is otherwise probative and relevant.

---

§ 6.30 Diagrams

The purpose of a diagram is to convey some physical, numerical or financial information. It will normally involve numbers, distances or other quantitative data. The only feature distinguishing the diagram from the graph is the rigid multiple-axis structure of the latter.

Stenger v. Germanos

265 Ill.App.3d 942, 639 N.E.2d 179 (1994)

Justice SCARIANO delivered the opinion of the court:

On March 23, 1989, plaintiff's decedent, Frank E. Stenger, was struck by an automobile being operated by defendant Gregg Germanos as Stenger crossed Arlington Heights Road near its intersection with Frederick Avenue in the village of Arlington Heights. He later died as a result of the injuries sustained in the collision. Plaintiff John Stenger, who was decedent's brother, filed a two count complaint in the circuit court grounded in the Wrongful Death and Survival Acts, and the action proceeded to trial before a jury.

The jury found damages in the amount of $306,000 and allocated 51% of the fault to defendant and 49% to decedent resulting in a judgment amount of $150,000. After the court rejected his post trial motion, defendant filed a timely notice of appeal....

Next, defendant complains about a scale diagram prepared by Newman and offered into evidence. He also takes issue with the introduction of photos Newman took of the scene, which he claims depicted decedent's remains after the accident, and thus, by their gruesomeness, inflamed the jury's passion.

As to the former, it is well settled that the admissibility of diagrams is addressed to the sound discretion of the trial court, and a decision to admit such evidence will not be disturbed unless its exercise is clearly abused. There is nothing in the record before us to indicate that the diagram at issue did anything other than to assist the jury in visualizing the scene of the accident, a permissible and, in fact, paradigmatic use of demonstrative evidence. Hargrove v. Neuner (1985), 138 Ill.App.3d 811, 818, 93 Ill.Dec. 145, 151, 485 N.E.2d 1355, 1361 ("The diagrams undoubtedly gave the jury a clearer picture of the events in question so that they could more intelligently consider the testimony of the witnesses").

* * *

Because we find that defendant received a trial which fairly fixed his fault and liability, the judgment of the circuit court is affirmed.

Affirmed.

Libby v. Hill

687 S.W.2d 264 (Mo.App., 1985)

LOWENSTEIN, Presiding Judge.

This case involved a wrongful death suit arising out of a motorcycle/school bus collision. Appellant, Jean Libby, is the widow of the motorcycle driver, Richard Libby. The jury found for the defendants, Brooks, the bus driver, and Hill, Brooks' employer, finding the decedent either failed to keep a careful lookout, drove at excessive speed, or failed to operate his motorcycle on the right half of the roadway.

The accident occurred in an early morning fog when the bus was making a left hand turn off a Highway 24 onto a gravel road near Lexington. The motorcycle, going the same direction as the bus, hit it broadside, just to the front of the rear wheel on the drivers side. Libby's claim was that the bus had overshot its turn and had backed up on the pavement and so was at a more perpendicular angle than it should have been. Libby also claimed the bus lights were not on so as to give the decedent ample time to brake, and that the bus had not made a left turn signal.

Two of Libby's points on appeal concern diagrams that were kept out of evidence. One was Plaintiff's Exhibit 2B, a tracing of a drawing of the accident scene made by the investigating Highway Patrolman. Libby contends the diagram was critical for proving the angle of the bus. However, the trial court sustained the defendants' objection to its admission on the grounds that the diagram was not drawn to scale, nor based on measurements, thus they were only an approximation and speculative and not competent evidence.

Libby cites a number of cases, such as Martin v. Sloan, 377 S.W.2d 252 (Mo.1964), for the proposition that whether a drawing is to scale goes to the weight of evidence not its admissibility. But that is not the only problem here. The officer admitted on direct examination that he had not measured the angle of the bus in relationship to the highway. Thus, there was no guarantee of accuracy of the drawing regarding the very aspect on which Libby planned to rely. Trial judges are afforded wide discretion on permitting a witness to utilize diagrams in explaining his testimony. The court in Cantrell v. Superior Loan Corp., 603 S.W.2d 627, 643 (Mo.App.1980), said that "when the use of such articles would not assist the jury to better understand the testimony of the witness, the court, may in its discretion refuse to admit an exhibit." See also Campbell v. Manhattan & Bronx Surface Transit Operating Authority, 81 A.D.2d 529, 438 N.Y.S.2d 87, 88 (1981), where a police officer's diagram was properly excluded because he was not a witness to the accident and had taken no measurements and made no independent investigation.

While admission of this diagram would also have properly been within the trial judge's discretion, this court's scope of review is limited by Rule 84.13(b) to whether the alleged error materially affected the merits of the action. As this court said in Davis v. Mathews, 649 S.W.2d 256, 259 (Mo.App.1983), "error in the admission of evidence is not grounds for reversal if it does not prejudice the complaining party or adversely affect the jury in reaching its verdict." Green v. Stanfill, 641 S.W.2d 490, 492 (Mo.App.1982).

In this case Libby was able to thoroughly examine the police officer about his observations. Furthermore, the theory that the bus had backed up was put before the jury through the testimony of a witness who spoke with the bus driver shortly after the accident. It cannot be said the absence of this diagram materially affected Libby's case. The trial court's ruling to reject the diagram which might have aided the jury in understanding what happened, did not however, amount to an abuse of discretion. O'Laughlin v. Barstow, 654 S.W.2d 95, 97 (Mo.App.1983).

The second diagram, Plaintiff's Exhibit 115, that was kept out of evidence was a rough sketch by defendant Hill, drawn at the scene the day of the collision. Libby wanted to use this diagram to impeach Hill's testimony about the angle of the bus. But as with the police diagram, no measurements were taken of the angle—in fact the sketch was not made on his own observation of the bus, but on hearsay of others.

The drawing was made merely to orient Hill with what had happened, and involved nothing more than speculation. For the same reasons as the first point discussed, the trial court did not abuse its discretion in refusing to admit Hill's sketch.

In sum, the plaintiff's theory of the bus missing the turn and backing up to make it, was not kept from the jury by the non-receipt of the two diagrams. The trooper and two witnesses testified as to the location and angle of the bus just after the accident. The trooper's exhibit probably should have been allowed in evidence for whatever its worth. That it wasn't does not amount to error materially affecting the merits of the action and causing reversal. Rule 84.13(b). Both points on the diagrams are denied

. . . .

All concur.

Notes and Problems

1. In *Libby*, if the accuracy of a chart goes to the weight of the evidence as opposed to its admissibility, then why didn't the court permit Libby's proposed drawings to be presented to the jury? For admissibility of an admittedly rough drawing, is it necessary that every angle be measured?

2. In *Libby*, wouldn't the rough sketch drawn by defendant Hill constitute an admission? Should a statement made by a defendant that lacks an intelligent basis be admitted against that defendant nevertheless? What if defendant Hill were called to the witness stand and asked to provide a simple rough sketch for demonstrative purposes only?

3. Can you distinguish *Stenger* and *Libby* or do the different holdings simply boil down to a matter of court discretion?

Ladeburg v. Ray

508 N.W.2d 694 (Iowa, 1993)

TERNUS, Justice.

Plaintiff, Helen Ladeburg, was struck by a semitrailer driven by defendant Ronald Thomas Ray and owned by defendants Rayloc, a division of Genuine Parts Company, and Genuine Parts Company. Plaintiff filed this negligence action which ultimately resulted in a jury verdict for the defendants.

Plaintiff argues that the district court erred when it (1) allowed defendants to use computer-generated evidence which had not been timely disclosed, (2) submitted instructions on plaintiff's comparative fault in failing to maintain a proper lookout, and (3) refused to grant a new trial on the basis that the court incorrectly submitted the issue of defendants' fault to the jury. We affirm.

I. Computer-Generated Diagrams.

Plaintiff filed her petition at law and request for jury trial in February 1991. Although the court administrator had set deadlines for discovery and designation of experts, the parties informally agreed to extend the deadlines for their own convenience. As a consequence, plaintiff did not learn that the defendants' expert intended to use computer-generated diagrams of the accident scene until she took the expert's deposition five days prior to trial on May 1, 1992. Plaintiff then filed a motion in limine requesting that the court exclude the computer-generated diagrams because the defendants had given plaintiff insufficient notice of their intent to use such evidence. The district court denied plaintiff's motion.

Trial began on May 6, 1992. During the trial plaintiff objected to defendants' use of the diagrams. The district court overruled this objection and the computer-generated documents were admitted into evidence. Plaintiff argues that the district court should have excluded the diagrams. We review for an abuse of discretion. Preferred Mktg. Assocs. Co. v. Hawkeye Nat'l Life Ins. Co., 452 N.W.2d 389, 393 (Iowa 1990); White v. Citizens Nat'l Bank, 262 N.W.2d 812, 816 (Iowa 1978).

The evidence to which plaintiff objects consists of diagrams of the accident scene produced by defendants' expert, an engineer. These diagrams were drawn by a computer. The expert testified that he used the computer as a drafting tool. The computer did not determine where the lines, geometric patterns, and letters should be placed on the drawings. Rather, the expert made those calculations and decisions, entered that data into the computer, and the computer "drew" the diagrams.

Plaintiff contends that the use of the diagrams should have been prohibited for two reasons. First, plaintiff argues defendants did not comply with the discovery requirements of Iowa Rule of Civil Procedure 125(c) requiring disclosure of expert information at least thirty days before trial.

Secondly, plaintiff contends a party should be required to give advance notice of an intent to use computer-generated documents. She claims jurors are unduly persuaded by computer-generated evidence. Advance notice would give the opposing party an adequate opportunity to rebut or impeach such evidence.

Plaintiff argues that in the absence of advance notice, the prejudicial effect of one party's use of computer evidence outweighs its probative value. Plaintiff asserts that such evidence is inadmissible under these circumstances. See Iowa R.Evid. 403. See generally

Randolph A. Bain and Cynthia King, Guidelines for the Admissibility of Evidence Generated by Computer for Purposes of Litigation, 15 U.C.Davis L.Rev. 951, 961–62 (1982).

Turning first to plaintiff's complaint based on rule 125, we note that defendants did not reveal their expert and produce the diagrams within the time constraints of rule 125 or in accordance with the deadlines set by the court administrator. However, the record is clear that the parties informally agreed to proceed with discovery beyond the deadlines imposed by the rules and the scheduling order. Plaintiff cannot now complain that defendants' late disclosure of the computer-drawn diagrams violated rule 125(c). Cf. Marine Am. State Bank v. Lincoln, 433 N.W.2d 709, 713 (Iowa 1988) (defendants could not seek sanctions under Iowa Rule of Civil Procedure 134(b) where they filed neither a motion to compel nor a motion for sanctions prior to the late production of documents); Provenzano v. Wetrich, McKeown & Haas, P.C., 481 N.W.2d 536, 540 (Iowa App.1991) (plaintiff waived right to object to late designation of one of defendants' experts because of parties' agreement to designate an additional expert beyond the time limitations imposed by Iowa Code section 668.11 (1989)).

Nor do we believe that the trial court abused its discretion in concluding that the admission of these diagrams did not unfairly prejudice the plaintiff. The diagrams were merely mechanical drawings made by a computer and the expert who prepared them was available for cross-examination. Under these circumstances, the trial court was well within its discretion in admitting the drawings into evidence. Cf. Perma Research & Dev. v. Singer Co., 542 F.2d 111, 115 (2d Cir.), cert. denied, 429 U.S. 987, 97 S.Ct. 507, 50 L.Ed.2d 598 (1976) (not abuse of discretion to admit expert testimony concerning results of computer simulation).

II. Instructions on Plaintiff's Comparative Fault.

Plaintiff argues that the trial court erred in submitting instructions on her comparative fault to the jury. Error in giving an instruction does not require reversal unless the error is prejudicial. Gore v. Smith, 464 N.W.2d 865, 868 (Iowa 1991). In Gore, we held that any alleged error in a damage instruction could not have prejudiced the plaintiff because the plaintiff had failed to establish the liability of the defendants. Id.

Likewise, in the present case, plaintiff could not have been prejudiced by the instructions on her comparative fault. This case was submitted to the jury on special interrogatories. The jury found that defendants were not at fault in response to the first question on the verdict form. Consequently, the jury did not answer the interrogatories concerning plaintiff's fault. Therefore, because there could be no prejudice to plaintiff, we need not address her objection to the submission of instructions on her comparative fault.

III. Motion for New Trial.

Plaintiff asserts that the trial court erred in failing to grant her motion for a new trial. She claims that the facts supported a determination that the defendants were at fault as a matter of law. Because the jury answered a special interrogatory finding that the defendants were not at fault, plaintiff reasons that the jury verdict failed to administer substantial justice.

Our standard of review of a denial of a motion for new trial depends upon the grounds for new trial asserted in the motion and ruled upon by the court. Julian v. City of Cedar Rapids, 271 N.W.2d 707, 708–09 (Iowa 1978). If the motion and ruling are based on a discretionary ground, the trial court's decision is reviewed on appeal for an abuse of discretion. E.g., Mays v. C. Mac Chambers Co., 490 N.W.2d 800, 803 (Iowa 1992) (motion for new trial based on Iowa Rule of Civil Procedure 244(b), attorney misconduct); Shepherd Components, Inc. v.

Brice Petrides-Donohue & Assocs., Inc., 473 N.W.2d 612, 618 (Iowa 1991) (motion for new trial based on rule 244(f), sufficiency of the evidence). However, if the motion for new trial and ruling are based on a claim that the trial court erred on an issue of law, then the trial court's decision stands or falls on the correctness of its ruling on the legal question. Julian, 271 N.W.2d at 709 (motion for new trial based on rule 244(h), error in instructions).

In the present case, plaintiff asserted in her motion for new trial that "the Court should have properly instructed the jury that the Defendant was negligent as a matter of law." Plaintiff's counsel acknowledged at oral argument that plaintiff's motion for new trial was based on rule 244(h) which permits the court to grant a new trial if there is an error of law in the proceedings. Thus, this case comes within the rule that the trial court's refusal to grant a new trial is not a matter of discretion and will be reviewed by the appellate court for the correctness of the trial court's decision on the legal question.

Before we consider whether the court erred in failing to instruct the jury that the defendants were negligent as a matter of law, we must determine whether this alleged error was properly preserved for review. We have previously held that objections first raised in a motion for new trial will not be considered on appeal because such objections must be made before jury arguments. Julian, 271 N.W.2d at 709; Dutcher v. Lewis, 221 N.W.2d 755, 758–59 (Iowa 1974); Iowa R.Civ.P. 196.

Here the plaintiff did not object to the court's instructions and the verdict forms submitting the question of the defendants' negligence to the jury. Not until plaintiff filed her motion for new trial did she claim that the court should have instructed the jury that the defendants were negligent as a matter of law. We hold error, if any, in the court's instructions was waived.

IV. Summary.

In summary, we hold that the trial court did not abuse its discretion in admitting the computer drawings of defendants' expert into evidence. Second, any error in the court's instructions on plaintiff's comparative fault did not prejudice the plaintiff. Third, the trial court did not err in denying plaintiff's motion for new trial.

AFFIRMED.

Notes and Problems

1. Do you agree with the *Ladeburg* decision? Does the court's acknowledgment that the material "diagrams were drawn by a computer" trouble you?

2. Can you distinguish between computer software programs that act merely as a drawing tool and those that present conclusions?

3. There once was a time when computer-generated output, like other new forms of evidence, required a rigorous foundation in order to establish its authenticity and trustworthiness before being admitted during trial. Now the computer, having invaded every aspect of our lives, is no longer new; and its reliability is rarely questioned. With minimal foundation requirements trial courts now freely admit computerized simulations, reconstructions, digital graphics, animations, and stored business records. And well they should, in most instances. See generally, Lipson, Art of Advocacy—Demonstrative Evidence, Lexis/Nexis (2012).

Chapter 7

Classical Physics

> *"Physicists are trained to investigate a problem before arriving at a decision. Lawyers, advertisers and others are trained to do exactly the opposite: to seek data to confirm a determination that has already been made."*
>
> —Robert Crease

Sometimes referred to as "Newtonian Physics," classical physics takes us from the teachings of Aristotle (384–322 B.C.) through the 19th Century, during which Sir Isaac Newton and James Maxwell provided mathematical unification for many of the most significant laws and concepts of basic physics. Most of basic Newtonian physics is predicated upon a simple definition: $F = ma$. Force equals mass times acceleration; it's that simple. The term "*Newtonian Physics*" (often referred to as "mechanical physics") is commonly used to distinguish it from the dramatic scientific developments that occurred throughout the early 1900s when Einstein published his universe-shattering concepts; those are dealt with in the chapter that follows.[1] Unlike Einstein's Universe, with Newton, there was consistency everywhere and obedience to his three famous laws:

§ 7.10 Newton's Laws of Motion

1. **First Law:** An object that is at rest or in a state of uniform motion, shall remain at rest or in uniform motion unless it is acted upon by some external force.
2. **Second Law:** If a force acts on an object, it will cause an acceleration of that object.
3. **Third Law:** For every action, there is an equal and opposite reaction.

The First Law is not entirely intuitive. It suggests that a body in motion will continue to move forever absent some outside force or interference. The law makes more sense when you consider that friction is the force that slows objects down. In space, an object could travel indefinitely. The *Law of Conservation of Momentum* is an extremely significant principle that flows from Newton's First Law. It states that: The total momentum of a system remains constant unless it is acted upon by some external force. From this principle, we may derive a corollary: An object moving in a straight linear motion (i.e., a straight line) at a constant velocity will continue to move in a straight line at that velocity until it is acted upon by some outside force.[2]

---

1. The same Sir Isaac Newton who is responsible for developing differential calculus, laid the groundwork for almost everything that we now know about mechanical physics.

2. Conservation might be illustrated by a bouncing ball. Were it not for outside forces, a dropped ball would bounce forever. Each bounce would reach the same height as the bounce before it. An article in the *American Journal of Physics* defined "perfectly happy balls" and "completely unhappy balls."

From a forensic perspective, energy and momentum become very significant. An automobile crashing into another vehicle or person may transfer the momentum and energy with catastrophic results. Consider the mismatch between the small automobile as it collides with a train or a large truck.

The Second Law is embodied in the equation F = ma, which is more akin to a definition than a law. Force is defined in terms of an acceleration applied to an object. Do not confuse speed (or velocity) with acceleration. Acceleration is defined as a *change* in speed or velocity. And, as the equation implies, an applied force will always be directly proportional to acceleration and mass.

It is important to note that in the world of physics, the term "acceleration" refers to any change in movement or velocity. "Speeding up" or increasing velocity is most familiar. But a change in direction or a slowing down would also constitute an acceleration. Physicists might refer to a negative acceleration but they rarely use the term "deceleration."

Newton's Third Law is best expressed by the *recoil* from the gun, that occurs from the warning shot that the officer might fire into the head of a fleeing suspect.

If we were to translate the three laws into legalese, we might restate Newton's First Law by saying: Any body (i.e., thing) standing still or in motion, will stay still or keep going unless something or somebody or something butts in. Restating the Second Law: The force of the push equals the size of the thing that you're pushing times the increase of the speed of the movement. The Third Law is somewhat self-explanatory: For every action, there is an equal and opposite price to pay.

Newton's three simple laws take on many shapes and forms. For example, the universally accepted *Fundamental Laws of Accident Reconstruction* are squarely predicated on them. Compare these to the formal statement of Newton's law, *supra*:[3]

A. A body at rest tends to remain at rest. A body will remain still unless moved by some force.

B. A body in motion tends to remain in motion in a straight line at a uniform velocity. The body will move at a constant speed unless it is acted upon by an unbalanced force.

C. If an unbalanced force acts on a moving body, the body accelerates in the direction of the force; that acceleration is directly proportional to the unbalanced force and inversely proportional to the mass of the body.

D. For every action there is an equal and opposite reaction.

To better understand the three laws, physics employs some simple building blocks, familiar to us all. They are also three in number: **mass (m)**, **distance (d)** and **time (t)**. It might be noted that while both *mass* and *weight* are related to size, they are not equivalent concepts.

Distance and time are relatively simple, straightforward concepts. In Newton's world, they do *not* vary under any circumstances or conditions. Time is measured in seconds. As is the case with mass and distance, in the world of pure Newtonian Physics, time won't bend, twist or keep you young. Those exotic contortions will have to wait until the next chapter.

A perfectly happy ball bounces to its original height when dropped on a massive, rigid surface. A completely unhappy ball does not bounce at all. Rod Cross, "*The Coefficient of Restitution for the Collisions of Happy Balls, Unhappy Balls and Tennis Balls*," 68 Am.J.Phy. 1025 (2000).

3. See Baker, *Open Forum: Scientific Reconstruction of an Automobile Accident*, 25 Ins. Counsel J. 438 (1958) and Zurmuhlen, *Automobile Accident Reconstruction*, 18 Trial Law. Q 49 (Winter 1987).

§ 7.11 *Measuring Up*

Table 7-1 provides a unique arrangement. The chart is based on the notion that many of our most important physics concepts are constructed from powers of three concepts. The quickly summarized concepts below could fill volumes for those who have the time and inclination to study them. We lawyers do not.

Notice how each concept builds upon the one that precedes it. Each is constructed with nothing more than the three basic items used for Newton's laws: **mass (m)**, **distance (d)** and **time (t)**. Notice also how each definition employs increasing exponents and nothing else. Recall that *any* number raised to the zero power is always equal to 1; therefore, a component with a zero exponent means that the component is not a factor. Mass, for example, does not effect velocity or acceleration.

Table 7-1 [Unit Definitions]

I. VELOCITY:	V =	m^0	·	d / t
II. ACCELERATION:	A =	m^0	·	d / t^2
III. MOMENTUM:	ρ =	m	·	d / t
IV. FORCE:	Newton =	m	·	d / t^2
V. ENERGY:	Joule =	m	·	d^2 / t^2
VI. POWER:	Watt =	m	·	d^2 / t^3

Throughout our discussions, it is very important to keep certain ubiquitous natural forces in mind. Because they are "forces," they are measured in Newtons (md / t^2). *Gravity* is a force. *Friction* is also a force. According to the *Law of Conservation of Momentum,* without friction, we could give an object such as an automobile a push and it would continue on its way without any need for gasoline. Without gravitational pull from planets, an object shot into space could travel forever or until it collided with some extraterrestrial entity.

§ 7.12 *The Gravity of the Situation*

Size matters with respect to mass and its interaction with human affairs. This is, therefore, a good time to distinguish weight from mass, particularly since, as you may recall from § 2.23, the United States and a very few other backward countries have failed to adopt the metric system.

Weight refers to the mass of an object when subjected to a particular gravitational pull. In other words, a rock of a given mass might weigh one pound on Earth but that same mass would have a far greater "weight" on Jupiter which, because of that planet's size, would exert a much greater gravitational pull. Therefore, in the world of Newtonian Physics, mass is a concept that does not vary with movement nor location.

The metric "Gram" is a measure of mass. One gram of matter on the Earth will equal one gram of matter anywhere in the universe. Weight, however, depends upon two factors: mass and gravitational force. The equation relating mass to earthly weight is a fundamental one: Weight (Newtons) = Mass (Kilograms) × Gravity (m/s^2). The value of gravity on Earth at sea level is a constant 9.8 meters per second squared (9.8 m/sec^2). Therefore: W = m × 9.8 m/sec^2.

Example 7-1: How much does a 4,000 gram block of chocolate weigh on Earth?

Solution: 4,000 grams = 4 Kilograms
4 Kg × 9.8 m/sec^2 = 39.2 Newtons

Example 7-2: The force of gravity on Jupiter would be many times that of Earth because of its great size. By the same token, the force of gravity on the moon is much less than that of Earth; to be exact it is equal to 1.67 m/sec^2. How much does a 4,000 gram block of chocolate weigh on the moon? What is the difference in weight between Earth and moon?

Solution: 4,000 grams = 4 Kilograms
4 Kg × 1.625 m/sec^2 = 6.5 Newtons

The difference in weight between Earth and moon = 39.2 N - 6.5 N = 32.7 N

Stated otherwise: On the moon, you would weigh 6.5 / 39.2 = 16.6% of what you weigh on Earth.

Example 7-3: For your Jenny Craig weigh-in, which planet would yield the best results as you step on the scale? Uranus or Mars?

Solution: Mars would help keep the scale down because that planet has far less mass than Uranus. Appearance-wise, it wouldn't matter. Per the Law of the Conservation of Energy, it's a matter of caloric intake v. calories burned; i.e., "diet and exercise."

In a free falling elevator, the contents (i.e., people) would be traveling at 32 ft/sec^2 and could suffer some rather unfortunate consequences if the building's basement floor were used to bring the runaway elevator to a stop. Not every elevator malfunction, however, involves an abrupt ending.

Gavoni v. Dobbs House, Inc.

164 F.3d 1071 (7th Cir., Ill., 1999)

CUDAHY, Circuit Judge.

The plaintiffs, Barbara Gavoni, Angela Rosendale and Lela Jordan, worked together as manicurists and hair dressers at a salon in La Crosse, Wisconsin and attended a cosmetology convention at a Chicago-area hotel owned and operated by the defendant, Dobbs House. On the morning of March 29, 1993, the plaintiffs and another convention attendee boarded an elevator on the eighth floor hoping and expecting to go down to the lobby.

Their hopes were dashed—literally and figuratively. The elevator went up to the eleventh floor, stopped briefly, descended to the lower lobby at an uneven rate and abruptly stopped. When the doors did not open, Jordan pressed the alarm button and tried calling for help on the emergency telephone. The riders eventually made enough noise to alert the hotel staff. A maintenance employee, Edward Johnstone, arrived within minutes and rescued the riders. Johnstone testified that when he opened the elevator doors he found "three young ladies standing there laughing and joking around."

He also claimed that each held a drinking glass with liquid in it. The plaintiffs admitted that Jordan made a sarcastic comment when asked if they wanted to be rescued but denied both that they were jocular and that they had drinks in hand. The plaintiffs told their rescuers that they were not injured, sat down for several minutes and then proceeded to the convention. That afternoon, Jordan and Rosendale competed in a fingernail decoration competition while Gavoni returned to their room. The plaintiffs later dined together and attended a cocktail party and hair show. Each had several alcoholic drinks; Jordan and Rosendale were dancing.

The next day, on the bus back to La Crosse the plaintiffs discussed the elevator incident and decided to go to the emergency room together that same night. Each complained of neck, shoulder and upper back pain. Over the next several years, each of the plaintiffs continued to experience pain and discomfort.

The plaintiffs brought this diversity suit against Dobbs House and the elevator manufacturer, Westinghouse. Pursuant to Fed.R.Civ.P. 68, Dobbs House made an offer of $10,000 "to be divided among all three plaintiffs, with costs then accrued." The plaintiffs rejected the offer. Prior to trial, the plaintiffs settled their claims against Westinghouse for a total of $105,000: Gavoni received $17,850; Rosendale got $44,100; and Jordan received $43,050.

The case against Dobbs House proceeded to trial. Dobbs House never denied that the elevator had malfunctioned; experts from both sides testified that the incident was likely caused by a faulty electrical connection. Instead, Dobbs House argued that the plaintiffs had inflated their injuries and consequent damages, making a federal case out of a minor accident.

The plaintiffs presented testimony from a single doctor that the elevator incident caused each plaintiff a variety of ailments, from sore knees to cracked teeth to chronic back pain. Dobbs House presented expert testimony contradicting this conclusion. At closing, the plaintiffs sought $825,000 in damages: $230,000 for Gavoni, $320,000 for Rosendale and $275,000 for Jordan. The jury found against Dobbs House on liability but awarded the plaintiffs a relatively paltry $6500—$2000 for Gavoni, $2000 for Rosendale and $2500 for Jordan.

Following the verdict, the plaintiffs moved for costs as the prevailing party under Fed.R.Civ.P. 54(d). They also moved under Fed.R.Civ.P. 59 for a new trial on various grounds. Dobbs House, for its part, moved for costs pursuant to Fed.R.Civ.P. 68. The court denied all three motions.

The plaintiffs raise five separate issues on appeal. Their first four complaints were originally included in their Rule 59 motion, and the plaintiffs do little, if anything, to develop these arguments on appeal. Although we might be free to ignore these undeveloped arguments, cf. Indurante v. Local 705, Int'l Bhd. of Teamsters, AFL-CIO, 160 F.3d 364, 366–67 (7th Cir.1998) (citing cases), we will nonetheless briefly consider them. The plaintiffs also appeal the denial of their Rule 54(d) motion for costs and the defendant cross-appeals from the denial of its Rule 68 motion for costs.

The plaintiffs first allege that the district court erred by allowing the jury to view one side (the left side) of an expert witness's chart. Dobbs House's elevator expert, John Donnelly, referred to a chart which compared the G-forces of a normally functioning elevator (the right side of the chart) with the G-forces of everyday human activity (the left side of the chart). Donnelly had prepared only the right side, but was familiar with the left.

After a prolonged argument at side bar and a foundation voir dire by the judge and defense counsel, the district court ruled that the chart was:

> "expert data that is utilized in the field and, therefore, for the limited purpose of his testimony I will permit the inquiry."

We will not disturb a district court ruling on expert testimony unless it is manifestly erroneous, see Deimer v. Cincinnati Sub-Zero Prod., Inc., 58 F.3d 341, 344 (7th Cir.1995), and there is no such error here. The district court made an extensive inquiry into the foundation for the expert's testimony and admitted it subject to being later stricken. Further, and directly relevant to their present argument, the plaintiffs did not ask that the left side of the chart (not prepared by the witness) be covered or redacted. Defense counsel later displayed the entire chart during closing argument and the plaintiffs now argue that this was error. Failure to demand redaction or that the chart be covered waives these arguments on appeal. See Miksis v. Howard, 106 F.3d 754, 761 (7th Cir.1997); Holmes v. Elgin, Joliet & Eastern Ry. Co., 18 F.3d 1393, 1398 (7th Cir.1994).

[The other issues raised on appeal are omitted]

The district court did not err in any of its evidentiary decisions or its post-trial rulings on the parties' motions for costs. The case is therefore Affirmed. FLAUM, Circuit Judge, dissented in part.

Notes and Problems

1. In *Gavoni*, if the malfunctioning elevator were permitted to free fall for 4 stories (assume that each story of the building was 12' in height), what speed would the passengers have attained before hitting the bottom? Might the litigious passengers have fared better in their lawsuit had they been lying on the floor of the elevator, with spilled drinks soaking their twisting, writhing bodies? Did the plaintiffs' counsel correctly deal with the opposition's presentation of the subject chart? What steps should have been taken? See generally, Chapter 6.

2. If you want to witness a runaway elevator scene with a terrifying application of Newton's First Law (and one that is far more exciting than the Dobbs House elevator), rent the motion picture, *Damien-Omen II*, directed by Don Taylor, starring William Holden and Lee Grant (1978).

3. Thinking of weight (typically measured by "static" ounces and pounds) in terms of "force" (measured in Newtons) might seem confusing, particularly since we tend to think of pounds as describing nothing more than motionless "dead weight" instead of a more dynamic force. But if you pretend, for the moment, that you are lying flat on your back as someone places a heavy cement block on your stomach, you would quickly appreciate the fact that the block continues over time to exert a very real measurable, if not painful force that does not discontinue until the block is removed.

4. Equating *motion* with *weight* is a bit trickier but far more interesting. That is precisely how Einstein developed his universally accepted General Theory of Relativity (later discussed in Chapter 8) which encompasses the manner in which both time and space are bent and stretched by the mass of nearby objects.

If the malfunctioning elevator in the *Gavoni* case had free fallen for 11 floors, during that ride, the litigious comedians inside would have experienced complete weightlessness with respect to one another and with respect to the elevator in which they were riding. On the other hand a truly weightless person in outer space would experience "weight" if he were placed inside of a rapidly ascending elevator. In fact, reasoned Einstein, there is absolutely no difference whatsoever between an accelerating body in space and a static body subjected to the simple pull of gravity while standing on Earth. Therefore, the physicist deduced that gravity itself must be nothing more than the force of motion that results from bends in space-time.

5. For measuring the gravitational attraction of two bodies to one another, Newton's Universal Law of Gravity still prevails. It's the law that governs the motion of all of the

planets throughout the Universe: F = **MmG /d²** where F = The force of gravity, M and m equal the masses of the larger and smaller bodies respectively, G = The gravitational constant and d = The distance between the two bodies.

6. There are four basic forces in the universe: (1) The strong nuclear force (which holds together the nucleus of the atom), (2) the weak nuclear force (involving radioactive decay), (3) electromagnetism and (4) gravity. Gravity is, by far, the weakest of the four forces.

§ *7.13 Moving Objects*

A discussion of motion best begins by clarifying an important distinction between speed and velocity. The false notion that the two concepts mean the same thing has caused many a lawyer sleepless nights.

Consider this: You're absolutely positive about the speed of that unsympathetic train—the one that scattered remnants of your client about two counties. And all of the witnesses, not to mention your math, say that the train must have been doing at least 95 miles per hour. Yet, based on the very same data and testimony, the defending railroad expert testifies that the velocity of the train was only 45 miles per hour during its entire journey. Your figures are correct; but so are those of the expert. How so?

If you speak in terms of speed, you are dividing the total distance traveled by the time required to traverse that distance. If, however, the expert is speaking in terms of velocity, he or she may also divide the distance by the required time, but unlike your measurements, the expert's distance will be measured point-to-point. In other words, velocity (a vector) is concerned with distance "as the crow flies," an unrealistic concern for the twisting, turning curves of ground transportation.

Consider, for example, two separate routes that stretch from Los Angeles to Ontario, California in the Inland Empire. Interstate 10 is straight as an arrow. Highway 60 joins the same two cities, but with its twists and turns, the route is longer. Two automobiles start the journey from Los Angeles to Ontario at the same time. Both reach their destinations at the same time. Their speeds or rates for the trip would, of course, be different. But their velocities would be the same because they both traveled between the same two points at the same time. By analogy, it would be correct to say that even though the average speed for the winner of the Indy 500 was just under 200 miles per hour; nevertheless, its velocity was zero. That's because the car ended up right where it started from, at the start/finish line. With that in mind, consider cross-examination of the Railroad's expert by a plaintiff's attorney:

Q. (By Physics-literate lawyer) What did you say about the train's speed?

A. (Expert) I said that its velocity was 45 miles per hour.

Q. Velocity measures the total distance divided by time; does it not?

A. Yes.

Q. But the only possible relevance to this case is the train's ability to stop and its force upon impact; isn't that correct?

A. Yes.

Q. And both of those factors involve the train's speed immediately before impact; don't they?

A. Yes.

Q. So the speed of the train hours before the collision would not have any bearing on those factors would it?

A. That's true.

Q. So you could have completely and accurately provided all of the relevant factors by giving the mass of the train and its speed immediately prior to impact; isn't that true?

A. Yes.

Q. Well then, who told you to work in those numbers about velocity?

To tie the preceding principles together, let's consider the lawyers' favorite moving object, an engine propelled automobile operated by an inebriated alcoholic driver:

VELOCITY: The alcoholic gets into his automobile and begins to drive. The first element to consider would be his velocity (V). If he is traveling at the rate of 1 meter in 1 second, his velocity, V = 1 m/s. Mass is not a relevant consideration at this juncture. In our simplified chart (§7.11), m was assigned a zero exponent. Therefore, mass = $m^0 = 1$.

ACCELERATION: A driver, drunk or sober, would be required to accelerate his automobile from a zero velocity. Acceleration is a concept that is not directly regulated by law enforcement statutes. A by-product of heavily increased acceleration such as squealing tires or street racing may, however, be punishable.

MOMENTUM: Momentum is the first definition on our list that involves mass as a factor. Two vehicles exhibiting the same velocity or acceleration would exhibit different momentum, depending upon their respective masses. Where the defendant happens to be the owner or driver of a large truck or a train, the target plaintiff would certainly prefer to speak in terms of momentum as opposed to velocity, especially if the large vehicle happened to be traveling rapidly (high velocity) at the time of the confrontation.

FORCE (from Newton's Second Law) equals Mass × Acceleration. Force is the thing that causes injuries, dents, not to mention hints, allegations and lawsuits.[4] In particular, it is the **impact** or stopping force that causes the grief. Or, to put it more euphemistically, it is not the speed that kills you, but rather the stopping.

ENERGY comes in two flavors: Potential Energy and Kinetic Energy. **Potential energy** is energy that is stored in a body or system by virtue of its position or status. If, for example, you haul a bowling ball to the top of the Empire State Building, that ball, due to its position, would possess a great deal of potential energy. **Kinetic energy**, on the other hand, is energy in motion. The automobile being driven by our inebriated drunk, presents a good example of kinetic energy. Kinetic bears the following relation to the other measurements on Table 7-1:

$$E = \tfrac{1}{2} mv^2$$

WORK, sometimes referred to as "mechanical work," is the amount of energy transferred by a force acting through a distance. It is equal to force multiplied times distance:

---

4. Even though "mass" is the correct measurement device, in the United States (and three other backward countries), "weight" may be used instead when describing force.

$$W = F \cdot d$$

You may notice from Table 7-1 that if we multiplied force times distance, we would essentially obtain "Energy."

$$F = m \cdot d / t^2$$

$$(d)\ F = m \cdot (d)\ d / t^2 = m \cdot d^2 / t^2 = E$$

$$E = dF$$

Why then do we need this concept called "work"? There is a reason for splitting energy up into the two separate components of force and distance, particularly as we study the tree that our drunk has selected for impact. But first, let's consider a very important relationship between kinetic energy and force:

$$E_{Kinetic} = \frac{1}{2} mv^2$$

Next, by way of simple substitution, let's calculate the "Impact Force that is going to be required to bring our drunk to an unceremonious halt (We will employ a "negative" value since we are seeking the force that will counteract the existing kinetic energy as the vehicle comes to an abrupt halt). Trees generally do not move.

$$E = -\frac{1}{2} mv^2$$

$$dF = -\frac{1}{2} mv^2$$

$$\text{Therefore: } F_{Impact} = -mv^2 / 2d$$

IMPACT may simplistically be thought of as the force encountered by a sudden stop. Notice that in the above equation, the force of impact, represented by F_{Impact} assumes a negative value. This makes sense when you view the force of impact as "opposing" the velocity (a positive value) of the moving mass.

The d factor is a very important key to understanding the concept of *impact.* The greater the distance (d) that the vehicle "gives" upon impact, the less the impact force. The term: d represents the distance that the object travels after the initial point of collision.

Those who have both water-skied and rollarbladed, are personally acquainted with the concept of impact. In both sports you might fall with the same force, but the impacts will be significantly different. That is because the distance traveled *after* you hit the water is significantly greater than the distance you travel after kissing the concrete while rollerblading. With water, there is "give." Concrete on the other hand does not move. This accounts for the tragic results that inevitably occur when snow skiers collide with trees. Simply stated, trees do not move. Thus, for trees, the factor d is extremely small, thereby making the impact very large.

Because damage, destruction and injury are common by-products of impact, a great deal of human energy and research has been devoted to its reduction. Automobile air-bags, for example, are specifically designed and engineered to increase the "d" factor by only a couple of feet. Yet, that short distance will often measure the difference between death and a complete lack of injury. The process of "cushioning" is a science unto itself. See *Pullman Inc. v. ACF Industries Inc.*, 269 F.Supp. 279 (D.C.N.Y., 1967), a patent infringement suit; consistent with the preceding definitions, the *Pullman* court discussed cushions and their efficiency, noting that: "A cushion is commonly characterized by its force and its travel. In engineering terms, the 'work' it does—or the energy it absorbs—when impacted is the product obtained by multiplying its average force by the distance over which it travels.".

Scientifically, all quantities should be calculated using metric quantities; however, as previously indicated, the United States has yet to embrace it. So let's continue to examine the fate of our drunk, using the same principles of physics, but at the same time switching our units of measurement from grams to pounds so the jurors can fully appreciate what the numbers mean.

Assume that the weight of the automobile is 3200 pounds. Assume further that the distance that automobile moved after impact was only 6" or ½ foot. Assume further that our drunk hit the tree going 30 miles per hour.

Recall that:

Weight = mass x 9.8 m/sec^2. For Americans, this means that

Weight = mass x 32 feet/sec^2 which solving the equation for mass, yields:

mass = Weight / (32 feet/sec^2)

Therefore, mass = 3200 pounds/ (32 feet/sec^2) = 100 pounds/feet/sec^2

Also 30 miles per hour = 44 feet/sec

Since $F_{Impact} = - mv^2 / 2d \quad \Rightarrow$

$$F_{Impact} = - \frac{(100)\,(44)^2}{2\,(.5)} = 193{,}600 \text{ Foot Pounds}$$

Knowing that 193,600 pounds equals 98.6 tons, using our knowledge of physics, we might come to the conclusion that it doesn't pay to drive drunk.

Algeo v. Pittsburgh Rys. Co.

202 Pa.Super. 548, 198 A.2d 415 (1964)

WOODSIDE, Judge.

This is an appeal from judgments entered in trespass actions against the Pittsburgh Railways Company after the denial of a motion for a new trial by the lower court. The judgments were entered on jury verdicts for the plaintiffs in two suits, joined for trial, which arose out of an intersection collision between an automobile and a streetcar.

The appellant contends as its sole ground for reversal that the verdicts were against the weight of the evidence.

The accident occurred at an intersection when the plaintiff-driver, Marando, attempted to make a right hand turn into the street being traveled by the streetcar which approached the intersection from the plaintiffs left. The intersection had no traffic controls. The street on which the streetcar was proceeding had two tracks. Automobiles were parked along both sides of the street, leaving insufficient space for traffic to pass between the streetcar tracks and the parked automobiles. The street on which the automobile had been proceeding was narrow.

Two of the plaintiffs, the driver and his passenger, Algeo, testified that the former stopped at the curb line prior to entering the intersection, stopped again while making the right hand turn and prior to entering upon the streetcar tracks, and was also stopped when the streetcar struck the automobile. The plaintiffs also testified that the operator of the streetcar accelerated his speed as he approached the intersection and made no effort to stop. This testimony was corroborated by two disinterested eye witnesses.

The operator of the streetcar testified that as he reached the intersection he saw the automobile coming from his right, that he immediately tried to stop, and that the automobile struck the right side of the streetcar at the front doors on an angle.

As a result of the collision the streetcar was derailed, the front set of wheels leaving the track to its left. The front of the streetcar crossed another set of tracks and struck a parked automobile. The automobile driven by Marando, in the words of the operator of the streetcar, was 'hanging right on the side of the streetcar' and 'rode right along'.

It is the contention of the appellant that 'the movement of the two vehicles after the collision up until the time they came to a complete stop show unmistakably and incontrovertibly that the theory of the plaintiffs is absolutely untenable'.

The appellant rests his contention upon the testimony of a physicist whom he called as an expert witness. The physicist, an Associate Head of the Department of Physics at Carnegie Tech. testified that the speed of the automobile at the moment of collision had to be at least twice that of the streetcar.

He explained to the jury Newton's Laws of Motion and pointed out that the momentum of the streetcar (weight multiplied by its velocity) and the momentum of the automobile determined the direction taken by the vehicle after the collision. Knowing the weight of the vehicles and the direction the vehicles took after the accident, he said he could determine the relative speed of the two vehicles. Using this method of calculation, he concluded that the automobile was traveling twice the speed of the streetcar. He further said that in the light of the direction taken by the vehicles after the collision it was 'impossible' for the accident to have occurred in the manner described by the plaintiffs and their witnesses.

'Impossible' results from vehicular collisions are well known to investigators of accidents. The reason seemingly impossible results occur is because nobody can determine accurately all of the factors which are operating at the moment of collision. Accidents do not happen with the simplicity used in a laboratory to demonstrate the rules of motion. In this case there are an unlimited number of unknown factors that may have contributed to the derailing of the streetcar and influenced its subsequent direction. For example, the real cause of derailment may have been a part of the automobile or streetcar, possibly no larger than a bolt, falling on the track. The cobblestones which the wheels struck upon leaving the track, the direction in which the streetcar was swinging at the instant at which it left the track, the grade of the street and any number of other unknown factors could have influenced the direction taken by the vehicles. Furthermore, when one vehicle strikes another at an angle there is a tendency of the striking vehicle to follow the angle of the struck vehicle.

Whether accurate or not, the testimony of the professor concerning the application of the laws of physics to the facts was opinion evidence which the jury could disregard. Ray, to Use of Miller v. Philadelphia, 344 Pa. 439, 25 A.2d 145 (1942); Gaita v. Pamula, 385 Pa. 171, 174, 175, 122 A.2d 63 (1956). Furthermore, the professor assumed that the automobile struck the streetcar. The weight of the evidence was to the contrary. An examination of the photographs picturing the damage to the automobile and the streetcar indicate that the front of the automobile was practically undamaged and that the blow was taken by it on the side. It is quite possible that after the collision the vehicle became hooked to the streetcar between the automobile's front and back wheels and 'rode right along' with the streetcar as the motorman testified.

This case does not come within the rule of incontrovertible physical facts. Ropele v. Stewart, 185 Pa.Super. 522, 528, 529, 137 A.2d 895, 898 (1958). Whether the accident happened in the precise manner contended by the plaintiffs does not control liability. The jury may determine from all the evidence what occurred and having thus reconstructed the accident may then determine whether the defendant was liable. As we view it, the value of the expert's testimony to the jury was limited to a consideration by it of the recognized laws of physics, which probably were already known by the jury in a practical sense, if not in the technical sense. It was within the province of the jury, and not the witness, to apply the rules of physics as well as the rules of law to the evidence.

This case was for the jury, and we believe that a verdict against the defendant is in accord with the weight of the evidence. The court below did not abuse its discretion in refusing the defendant's motion for a new trial.

Judgments affirmed.

Winterberg v. Thomas

126 Colo. 60, 246 P.2d 1058 (1952)

MOORE, Justice.

We will herein refer to the parties as they appeared in the trial court, where plaintiff in error was defendant and defendant in error was plaintiff.

The complaint was based upon a collision between a motorcycle ridden by plaintiff and an automobile driven by defendant. The accident occurred at the intersection of Grant street and Forest Drive in the City and County of Denver on May 12, 1950. Plaintiff sought to recover damages for personal injuries sustained in said collision, and alleged that defendant's negligence was the proximate cause of the accident and resulting injuries. Defendant denied that negligence attributed to him by plaintiff, and alleged that the injuries and damage sustained by plaintiff were proximately caused or contributed to by plaintiff's own carelessness.

Trial was to the court without a jury and resulted in judgment for plaintiff in the sum of $7,500, together with interest and costs. Defendant seeks reversal of the judgment by writ of error.

It is important in considering the issues to visualize the physical circumstances at the scene of the accident. Grant street is a one-way street going south. Forest Drive is a two-way narrow street on the south side of Cherry creek running northwesterly and southeasterly. Speer Boulevard is on the north side of Cherry creek paralleling Forest Drive. At the intersection of Speer Boulevard and Grant street, traffic is controlled by traffic lights. Similar traffic controls govern the movement of traffic at the intersection of Forest Drive and Grant street.

That portion of the evidence concerning which there was no dispute, was to the effect that plaintiff was riding a motorcycle in a northwesterly direction on Forest Drive and defendant was driving his automobile south on Grant street. At the time of the collision plaintiff's motorbike was across the center line of Grant street in the intersection and on the right, or north, side of the middle line of Forest Drive. The accident occurred at approximately 6:15 A.M. and plaintiff did not regain consciousness until approximately 8½ hours thereafter. He remained in the hospital for 12 days, and serious permanent injuries were sustained by him as a result of the collision. There was no dispute concerning the sequence in the operation of the traffic control signals at the intersections involved in the case.

The specification of points upon which plaintiff in error relies chiefly for reversal of the judgment is that the court erred in finding that defendant ran through a red light and that plaintiff was proceeding through the intersection on the green light. It is asserted that the overwhelming weight of the oral testimony and the undisputed physical facts establish exactly the contrary and therefore plaintiff was guilty of contributory negligence which precluded recovery.

In order to make clear the evidence upon which this argument is based, we set forth the substance of plaintiff's testimony concerning the accident. He testified that he drove his vehicle north on Logan street from Ellsworth street to Forest Drive, a distance of several blocks. Logan street is a one-way street for northbound traffic and runs parallel to Grant street one block to the east.

The intersection of Logan street and Forest Drive is controlled by traffic signals similar to those at the intersection of Grant street and Forest Drive. Upon arriving at the intersection of Logan street and Forest Drive he brought his vehicle to a stop and waited for the light to turn green for northbound traffic across said intersection. When this light turned green he made a left turn on Forest Drive and proceeded in a northwesterly direction.

Q. And you were going about fifteen miles an hour, as you call it, down Forest Drive?

A. Yes, sir, down about half way.

Q. I believe you testified when you were about half way down you looked up and saw both of these lights green at the intersection of Grant, is that right?

A. Yes, sir, they were green.

....

Q. And that is where you were when you saw these two green lights down on Grant?

A. Yes, sir, I was about in the middle of the block.

Q. Then you slowed down to about seven miles an hour?

A. About seven miles an hour, yes, sir.

Q. And did you watch both of those lights?

A. Yes, sir, they were still green.

The above quoted testimony was given upon cross-examination. His statement upon direct examination was that the lights were green as he entered the intersection. He further said:

> Well, I went down Ellsworth to Logan Street and then down Logan Street to Forest Drive, and then I hit Forest Drive. I was going down Forest Drive here and I got down about half a block here; I looked up and both lights were green, this light and this one here, and then I looked through the trees there and across the bridge I could see some cars approaching the bridge, which was coming down this way, and after I got up about ten yards I seen a car parked waiting for the light, about here; and then in the meantime, back here about fifty yards, I slowed up to about seven miles an hour and then I went into the intersection and something struck me suddenly, hit me on the right leg and the rear rump and knocked me out and I never knew any more.

One other witness testified to the effect that at the time of the collision plaintiff had the green light, but the evidence of this witness was highly incredible and the trial court in its findings eliminated from consideration the testimony of that witness. Relatives of the plaintiff testified that at the Denver General Hospital, a few days after the accident defendant appeared to offer blood for a transfusion to plaintiff and that he then stated, in explaining how the accident happened, that he had run through a red light.

The making of any such statement was vigorously denied by defendant. He stated at the scene of the accident, to the officers who arrived almost immediately, that, 'The light was green and I started across and when the bicycle came to my left and ran the red light, I hit it.' The findings of the trial court include the following statement:

> In this case, the evidence is in hopeless conflict, but the court has followed it very carefully, and even disregarding the testimony of Mrs. Finnicum and disregarding the testimony of the admissions made by the defendant the court, nevertheless, feels that the defendant was guilty of negligence in three particulars....

The court found that defendant was guilty of driving faster than the speed limit of 25 miles per hour; that he ran through a red light; and that he was guilty of reckless driving.

On behalf of defendant three eye-witnesses to the collision were called. Two of them, Mr. Rupp and Mr. Patterson, were riding in a car following that which defendant was driving. They were employees of the Gates Rubber Company and were on their way to work. They testified positively that defendant's automobile entered the intersection of Grant street and Forest Drive with the green light and that the light did not change at any time prior to the accident. Witness Eichman testified that he was driving his car south on Grant street approximately 20 to 30 feet behind defendant's car. He, too, testified positively that defendant's car entered the intersection with the green light. These witnesses were not acquainted with either of the parties to the action. Mr. Bishop, an attorney who investigated the accident shortly after it happened, testified that he called upon plaintiff at the hospital three or four days after the accident and that plaintiff stated that he did not know how fast he was going or what color the light was when he entered the intersection. Expert medical testimony from Dr. Robert G. Packard, Dr. Harry C. Hughes and Dr. Samuel P. Newman was to the effect that it would be reasonable to suppose that plaintiff suffered an obliteration of memory for about 30 seconds prior to the accident, and that where an injured person is knocked unconscious for a substantial period of time such a condition may reasonably be expected.

Witness Rudden, the city engineer who had charge of designing, building and timing the electric traffic controls, explained in minute detail the sequence of the traffic lights at the intersections involved. From his testimony it is clear that a person intending to travel on Forest Drive from Logan to Grant streets, who starts from Logan when the light turns green for northbound traffic at Logan and Forest Drive, would have 39 seconds of red light burning for northwesterly traffic at the intersection of Forest Drive and Grant street. It is undisputed that at the instant the green light came on for northbound traffic on Logan street across Forest Drive, the light turned red at Forest Drive and Grant street for northwesterly traffic on Forest Drive. This undisputed testimony definitely established that plaintiff, going north on Logan street and continuing his journey on Forest Drive when the light on Logan street turned from red to green, could not possibly have crossed the intersection of Forest Drive and Grant street within 39 seconds without running through a red light. Simple mathematics will demonstrate conclusively that at 15 miles per hour one would travel 22 feet per second. Plaintiff testified that when he reached a point half way down the block, a distance of approximately 200 feet, the traffic light turned green at Forest Drive and Grant street.

Simple mathematics will demonstrate conclusively that he reached this halfway point, if he traveled as he claimed at the rate of 15 miles per hour, in 9 or 10 seconds. There still would be a full 30 seconds of red light, and even if he traveled the last 200 feet at only 7 miles per hour he could not possibly consume the full 30 seconds in traveling the remaining 200 feet.

If, as plaintiff claims, the first half of the block was traveled at the rate of 15 miles per hour and the second half of the block at the rate of 7 miles per hour, the average would be about 11 miles per hour for the entire block. Simple mathematics will demonstrate that the entire distance from Logan to Grant street—407 feet—would be covered in approximately 25 seconds of time, leaving approximately 14 additional seconds of time during which the red light would be against traffic in the direction traveled by plaintiff.

The crux of the contention of defendant is that the foregoing admitted facts demonstrate conclusively that plaintiff entered the intersection against the red light and that this conclusion is inescapable, notwithstanding the statement of plaintiff that the lights were green.

Question to be Determined

Do the undisputed physical facts in the instant case, when subjected to the application of simple mathematics, demonstrate beyond question that the findings of the court are based upon assertions which cannot possibly be true, by reason of which, as a matter of law, the judgment should be reversed? The question is answered in the affirmative. It is true, as a general rule, that a party to an action is entitled to have his case submitted to the jury, or the court, as the trier of the facts, upon his own interested testimony although contradicted by disinterested witnesses, and usually a judgment rendered under such circumstances will not be nullified by an appellate court. However, where, as in the instant case, the testimony of a party not only is opposed to that of three disinterested eyewitnesses, but is shown to be untrue by the incontrovertible physical facts, a different situation is presented.

A court cannot accept as true that which the indisputable evidence demonstrates is false. Lessig v. Reading Transit & Light Co., 270 Pa. 299, 113 A. 381, 382. The applicable rule was quoted with approval in that case from Elliott on Evidence, section 39, as follows:

> Even though it may not be authorized to weigh evidence and pass upon the facts, it may, and should, so use its judicial knowledge as to bring about justice. Thus, there are often undisputed physical facts clearly shown in evidence, and by applying to them a well-known law of nature, of mathematics, or the like, it is demonstrated beyond controversy that the verdict or finding is based upon what is untrue and cannot be true. In such case, it is very generally held that the appellate court should take judicial notice of the law of nature or mathematics or quality of matter, or whatever it may be that rules the case and apply it as a trial court should have done.

From the nature of the question under discussion individual cases are of but limited value as precedents since no two cases present identical physical facts leading to the same conclusion. Every case must rest upon its own factual basis. It is firmly established in the law, however, that appellate courts, upon review of a judgment, will not hesitate to take judicial notice of the unquestioned laws of mathematics, and where the trial court has failed to recognize them and has entered judgment supported by some evidence which is 'so far at variance with the undisputed physical facts as to demonstrate that the evidence is false and the verdict unjust,' the judgment will be set aside. Sheppard v. Wichita Ice & Cold Storage Co., 82 Kan. 509, 108 P. 819, 28 L.R.A., N.S., 648.

We have recognized the applicable rule and have reversed judgments because of the force thereof in the following cases: Westerkamp v. Chicago B. & Q. Ry. Co., 41 Colo. 290, 92 P. 687; McLennon v. Whitney-Steen Co., 63 Colo. 568, 167 P. 771; Brice v. Miller, 121 Colo. 552, 218 P.2d 746. We quote from the Westerkamp case as follows [41 Colo. 290, 92 P. 689]:

> Where a physical situation renders the right of a matter clearly beyond all reasonable controversy, there is no conflict to be solved by a jury, because no just verdict can be rendered contrary to all reasonable probabilities. In such circumstances, the testimony of a witness to that which is physically impossible must be rejected, and a court will treat as unsaid by a witness that which, in the very nature of things, could not be as said. A trial court should not submit to a jury the determination of a fact about which, from all the testimony, there can be no dispute.

We are satisfied that by a simple mathematical calculation the statement of plaintiff that he went through the intersection of Forest Drive and Grant street on a green light is shown conclusively to be false. The findings and judgment of the trial court are without support in the evidence.

Accordingly the judgment is reversed and cause remanded with directions to dismiss the action.

STONE, J., dissents.

State v. Holzapfel

2010 WL 2502216 (Ohio App. 10 Dist., 2010).

CONNOR, J.

Plaintiff-appellant, State of Ohio ("the State"), filed a motion for leave to appeal The State seeks leave to appeal a judgment from the Franklin County Municipal Court granting defendant-appellee, Frank J. Holzapfel's ("defendant"), motion for a new trial. Because we find no abuse of discretion, and because the State has failed to sufficiently demonstrate a probability that its claimed errors did in fact occur, we deny the State's motion for leave to appeal.

On December 29, 2008, defendant was traveling in his vehicle heading eastbound on West Broad Street. As he was making a left-hand turn onto North Terrace Avenue, he collided with a motorcycle driven by Glenn Stepp. The passenger on the motorcycle, Pamela Stepp, was thrown from the motorcycle, thereby sustaining injuries from which she later died. Following an investigation, defendant was charged with vehicular manslaughter and failure to yield the right of way on a left turn. Defendant proceeded to a jury trial on the vehicular manslaughter charge and was convicted on September 25, 2009. Defendant was also found guilty of failure to yield the right-of-way on a left turn by the trial court.

In October 2009, defendant filed a timely motion for new trial in municipal court pursuant to Crim.R. 33, alleging an irregularity in the proceedings, as well as newly discovered evidence. Defendant alleged that Detective Cornute, a Columbus Police Detective with the accident investigation unit who conducted the investigation at issue and testified at the trial on behalf of the State, had made a mathematical error in calculating the speed of the motorcycle. As a result, defendant alleged this error warranted a new trial.

Under Crim.R. 33(A), a new trial may be granted upon the motion of the defendant for causes materially affecting his substantial rights, including: (1) Irregularity in the proceedings, or in any order or ruling of the court, or abuse of discretion by the court, because of which the defendant was prevented from having a fair trial or ... (6) When new evidence material to the defense is discovered, which the defendant could not with reasonable diligence have discovered and produced at the trial.

Following a hearing in which Detective Cornute, as well as defendant's expert, Gregory DuBois, testified, the trial court issued a written decision granting defendant's motion for a new trial on both offenses. The trial court determined that Detective Cornute's testimony regarding the estimated speed of the motorcycle, which was based upon an erroneous mathematical calculation, was an irregularity in the proceedings. The trial court also determined the range of the speed as properly calculated was newly discovered evidence. The State then filed the instant motion for leave to appeal, alleging the trial court abused its discretion in granting defendant's motion for new trial.

Although R.C. 2505.03 generally provides that every final order or judgment may be reviewed on appeal, appeals by the State in criminal proceedings are specifically governed by R.C. 2945 .67(A). Pursuant to R.C. 2945.67(A), the State may appeal as of right an order which: (1) grants a motion to dismiss all or any part of an indictment, complaint, or information; (2) grants a motion to suppress evidence; (3) grants a motion for the return of seized property; and (4) grants post-conviction relief. The statute further provides that, with the exception of final verdicts, the State may appeal any other decision in a criminal case by leave of the court of appeals.

Turning to the substantive issues in this matter, the State contends the trial court lacked discretion to grant defendant a new trial based upon the facts of this case. Specifically, the

State argues the trial court abused its discretion in granting a new trial by finding that: (1) the State's expert testimony regarding the estimated speed of the motorcycle was erroneous and constituted an irregularity in the proceedings; (2) the range of speed as calculated by defendant's expert constituted newly discovered evidence that was not readily discoverable; and (3) defendant was prevented from having a fair trial as a result of the "new" speed calculations of his expert.

In order to grant a new trial based upon a claim of newly discovered evidence, it must be demonstrated that the newly discovered evidence: (1) discloses a strong probability that it will change the result if a new trial is granted, (2) has been discovered since the trial, (3) is such as could not in the exercise of due diligence have been discovered before the trial, (4) is material to the issues, (5) is not merely cumulative to former evidence, and (6) does not merely impeach or contradict the former evidence.

The State submits that Detective Cornute's trial testimony did not constitute an irregularity in the proceedings. The State also asserts there is currently no "new" evidence which could not have been discovered with reasonable diligence and produced at trial, since defendant had received Detective Cornute's work product and report prior to the trial. In addition, the State argues there is not a strong probability that the allegedly "new" evidence would change the result of the trial. The State further argues that the evidence alleged as being "new" is merely cumulative to evidence formerly presented at trial and also merely impeaches or contradicts the evidence presented at trial.

The critical factual issue in this case appears to rest on whether the motorcycle was proceeding lawfully at the time of the accident, since defendant, while making a left-hand turn, had a duty to yield the right-of-way to oncoming traffic that was proceeding lawfully. If the motorcycle was not proceeding lawfully at the time of the accident, then its driver and passenger forfeited the right-of-way, and defendant was not under a duty to yield the right-of-way and thus, would not be guilty of the offenses upon which he was originally convicted. The opposite analysis is also true, so if the motorcycle was proceeding lawfully, then defendant was under a duty to yield the right-of-way and would be guilty of the offenses. Thus, any testimony regarding the estimated speed of the motorcycle is critically important.

The posted speed limit in the area where the accident occurred was 35 miles per hour. At trial, Detective Cornute testified that the estimated speed of the motorcycle at the time of impact was 27.9 miles per hour. This determination was made using a mathematical formula known as the "vault formula."

Detective Cornute used an angle of departure (also known as trajectory) of zero degrees and 45 degrees. While he did not testify at trial as to the specific estimated speed "range," at the hearing, he indicated that, had he testified to that "range" of speed (under the improper calculation) it would have been 17.9 to 27.9 miles per hour. (It is apparent that this speed "range" was contained within his work product and his report, which were provided to defendant's counsel and expert.) However, at the hearing on the motion for new trial, Detective Cornute acknowledged that his trial testimony was mistakenly based upon a computational error. This error occurred in his speed computation using the zero degree angle of departure and resulted in an incorrect speed of 17.9 miles per hour.

At the motion hearing, Detective Cornute testified that since the trial, he had not recalculated the speed of the motorcycle using a zero degree angle of departure. Instead, he had corrected his calculation and used angles of departure of 18 and 45 degrees. Although he had testified at trial that he did not know the exact trajectory the motorcycle rider had traveled in reaching her final resting place, Detective Cornute now testified that the angle had to be a minimum of 18 degrees, and thus his previous use of a zero degree angle of departure was incorrect. This minimum angle was due to the fact that the motorcycle rider did not or would not have been able to go through the motorcycle driver in front of her, the motorcycle, or the car in order to reach her final resting place and thus would have "vaulted" over them at an angle of 18 degrees or more. Under this new calculation, Detective Cornute indicated that the estimated range of speed would be 27.9 to 34 miles per hour. Despite this new "range," Detective Cornute testified at the hearing that this new information would not alter his trial testimony estimating the motorcycle's speed to be 27.9 miles per hour, based upon other evidence obtained from the accident.

At the motion hearing, defendant's expert, Gregory DuBois, testified that Detective Cornute made a mathematical error in his original calculation. Mr. DuBois also took exception with Detective Cornute's testimony that it was scientifically impossible for the departure angle to be less than 18 degrees. He testified that he believed the zero and 45 degree angles of departure should be used in the formula, rather than the 18 and 45 degree angles of departure Detective Cornute was now using. While he generally disputed that the vault formula was the proper formula to apply in these circumstances, Mr. DuBois testified that, at a zero degree angle of departure, the maximum speed of the motorcycle would have been 86.6 miles per hour.

In reviewing the trial court's decision to grant defendant's motion for a new trial under an abuse of discretion standard, and keeping in mind that we cannot substitute our judgment for that of the trial court, we cannot say that the trial court acted unreasonably, arbitrarily, or unconscionably in granting defendant a new trial on the basis that the computational error subsequently resulted in the discovery of new evidence. Furthermore, although the State submits that defendant did not meet the six factors required under the Petro test in order to receive a new trial on the grounds of newly discovered evidence, we disagree.

In addressing the Petro requirements, the State first argues the evidence discovered was not "new" and that it was either known prior to trial or could have been discovered prior to trial if the defense had exercised due diligence. "Newly discovered evidence" is "evidence of facts in existence at the time of trial of which the party seeking a new trial was justifiably ignorant." Campbell v. Am. Foreign S .S. Corp. (C.A.2, 1941), 116 F.2d 926, 928.

The trial court determined that the range of speed as properly calculated was newly discovered evidence, finding that Detective Cornute's testimony (and likely Mr. DuBois' testimony), as based upon the corrected mathematical calculation, constituted new evidence not known during the trial, despite the fact that it was not new "physical" evidence upon which to base the calculation. Additionally, as to the issue of whether this was evidence which could have been discovered prior to trial if the defense had exercised due diligence, the trial court determined that it was not readily discoverable before trial, due to the complexity of the applicable mathematical formula and the likelihood that it was not understood by the average lay person or attorney.

Black's Law Dictionary defines due diligence as "[t]he diligence reasonably expected from, and ordinarily exercised by, a person who seeks to satisfy a legal requirement or to discharge an obligation." Black's Law Dictionary (9th ed.2004). Furthermore, in Love, the First District Court of Appeals found that "[d]ue diligence has a somewhat accordion-like definition, as where one court might find expansively that a defendant acted with due diligence, while another court might restrictively deem that the defendant had not."

In the instant case, although the defense was provided with Detective Cornute's work product regarding his calculations, it cannot be disputed that the formula in question involves complicated computations and variables. Even Detective Cornute himself was unaware of the mistaken calculation or the incorrectly computed speed range until after the trial. Given this, we cannot say that the trial court abused its discretion in determining that this error could not have been discovered before trial even in the exercise of due diligence. In addition, we note there is nothing to indicate that defendant withheld his knowledge about the incorrect calculation with the intent to use it in reserve as grounds for a new trial.

The State also argues that the new evidence is merely cumulative to the former evidence and merely impeaches or contradicts that former evidence. However, it is apparent that the trial court did not find this to be the case. Although the new evidence would most certainly contradict the former evidence, the new evidence, as noted by the trial court, shows that Detective Cornute's testimony regarding the estimated speed of the motorcycle is based upon a rationale that is very different from the rationale used to reach his conclusion provided at trial. Furthermore, as the trial court noted, given this new rationale, the questioning of Detective Cornute upon cross-examination, as well as Mr. DuBois' questioning on direct, would most certainly be different and would elicit different testimony and different evidence to be considered by the jury. Moreover, this new evidence provides specific areas of conflict between the two experts and their analysis of the accident and thus is not "merely cumulative" nor "merely" contradictory.

Finally, the new evidence is clearly material to the issues. The trial court noted that the issue of the speed of the motorcycle was "the key factor" in the jury's determination of whether the motorcycle was being operated in a lawful manner. (December 10, 2009 decision and entry granting new trial, at 2.) Additionally, this determination is essential to establishing defendant's guilt.

Given the analysis as set forth above, and taking into account that (1) the posted speed limit in the area was 35 miles per hour; (2) Detective Cornute's new speed estimate under the vault formula ranges from 27.9 to 34 miles per hour and only estimates speed at the time of impact, and not in the seconds prior to impact; (3) an independent witness testified at trial that she was traveling at a speed of 35 miles per hour and that the motorcycle passed her; and (4) a determination that Mr. DuBois' new testimony under the revised calculation (but using the zero and 45 degree angles of departure) would establish an estimated speed range of 27.9 to 86.6 miles per hour, the new evidence discloses a strong probability that it will change the result if a new trial is granted.

Therefore, we find the trial court did not abuse its discretion in ordering a new trial on the basis of newly discovered evidence. We also find the State failed to sufficiently demonstrate a probability that its claimed errors on this ground did in fact occur. Because we have determined there was no abuse of discretion in ordering a new trial on this ground, we need not address the State's argument regarding the trial court's finding of an irregularity in the proceedings. Accordingly, the State's motion for leave to appeal is denied.

Motion denied.

SADLER, J., dissenting. Because I believe that the State sufficiently demonstrated a probability that its claimed errors involving the granting of a new trial did in fact occur, I respectfully dissent and would grant the State leave to appeal.

Addison v. Moss

122 N.C.App. 569, 471 S.E.2d 89 (1996)

JOHN, Judge.

Defendants appeal judgment in the amount of $3,500, plus attorney's fees of $6,000. Defendants' primary contention is that the trial court erred by excluding certain testimony from their accident reconstruction expert. We agree.

Pertinent procedural and background information is as follows: As plaintiff drove her automobile in a northerly direction on Highway 301 in Wilson County on the afternoon of 24 July 1992, one or two bundles of empty burlap tobacco sheets fell from the bed of a truck traveling in front of her. Plaintiff testified that she hit the sheets, lost control of her vehicle, swerved to her left into the median, and executed a 360 degree turn within the median before coming to a stop.

Plaintiff subsequently filed suit 3 December 1992 against James R. Moss (Moss), the driver of the truck, and Howard A. Tyson (Tyson), the truck's owner, alleging injuries as a result of the collision. Defendants' answer denied Moss was negligent in the operation of the truck, and further asserted plaintiff was contributorily negligent in "fail[ing] to keep a proper lookout," "fail[ing] to keep her vehicle under proper control," and "fail[ing] to reduce speed upon approaching a special hazard." Following a jury trial and the judgment in favor of plaintiff, defendants timely appealed to this Court.

Defendants first contend the trial court erroneously precluded testimony from David S. Brown (Brown), an accident reconstruction expert, regarding experiments he performed to determine how tobacco bundles behave after falling from the back of a truck. The purport of Brown's experiments, according to defendants, was to illustrate the law of inertia, the physical principle that objects in motion tend to stay in motion. Defendants wished Brown to testify that, in conformity with the law of inertia, when the bundles fell from the back of the truck in the accident at issue, they kept moving forward in a northerly direction—significantly, away from plaintiff's automobile—for some distance before the friction of the road brought them to a stop. Brown's experiments attempted to determine just how far forward the bundles traveled after falling from defendants' truck. Defendants contend this information would have been extremely relevant at trial because the greater the distance plaintiff was able to travel before reaching the bundles, the greater likelihood she was contributorily negligent in failing to apply her brakes and stop before striking them.

Initially, we note a court may take judicial notice, whether requested or not and at any stage of the proceeding, of facts capable of accurate and ready determination by resort to sources whose accuracy cannot be reasonably questioned. N.C.G.S. §8C-1, Rule 201 (1992). Inertia is reliably defined as "a property of matter by which it remains at rest or in uniform motion in the same straight line unless acted upon by some external force." Webster's Third New International Dictionary 1156 (1966). We therefore take judicial notice that the bundles were subject to the property of inertia when they fell from the back of Tyson's truck. Concerning the admission of experimental evidence, our Supreme Court has held:

> Experimental evidence is competent when the experiment is carried out under circumstances substantially similar to those existing at the time of the occurrence in question and tends to shed light on it. It is not required that the conditions be precisely similar, the want of exact similarity going to the weight of the evidence with the jury.

Exact reproduction of the original occurrence is not required, particularly when an expert is available to explain relevant differences between conditions of the experiment and the original occurrence and their possible effects on results. Short v. General Motors Corp., 70 N.C.App. 454, 455, 320 S.E.2d 19, 20, disc. review denied, 312 N.C. 623, 323 S.E.2d 924 (1984). Whether substantial similarity exists is a question of law reviewable as any other question of law by the appellate courts, id.; however, a trial court's ruling on this issue will normally be upheld unless found to be "too wide of the mark." State v. Jones, 287 N.C. 84, 91, 214 S.E.2d 24, 29 (1975) (citation omitted).

On voir dire, Brown testified he met with defendants near Tyson's farm in Nash County and performed experiments by dropping bundles of tobacco sheets off the back of a truck and measuring the distances they traveled after falling. Plaintiff's counsel interposed the following objection:

> [Brown's] presentation is flawed because he testified that he did tests in Nash County trying to see what reaction a bundle of sheets would have upon being pushed out of a truck. And he testified that he did not duplicate that on Highway 301, and of course I would hope that he didn't go out there and do it on Highway 301. But it is well known that with regard to experiments that the circumstances basically have to be the same. You can't take results out of Nash county where we don't know the terrain and that kind of thing, and try to transfer that to Wilson County.

Brown thereafter explained the similarities between the conditions of the experiment and those of the actual accident. He testified that the bed of the truck used to conduct the experiment was the same distance from the ground as the trailer involved in the accident, that the truck traveled during the experiment at the same speed traveled by Moss at the time of the accident, that the bundles he used duplicated those which had originally fallen from the truck in shape and size, and that the terrain of the road involved was the same as that of Highway 301. Brown further stated:

Every effort was made to have everything the same in Nash County that it would be in Wilson County, based on everything I had to work with as to the terrain. Terrain, of course, is not a matter, it's simply a roadway, because all of my tests were done in the roadway. There's not a question of whether it was a ditch or whether it sloped this way or that, even off of the edge of the pavement. Everything was as near the same as it could possibly be done, which is what you do in a test.... I would say that just literally hundreds and thousands of tests that have been done by the standards writers and the other technical organizations such as [the Society of Automotive Engineers] to establish how the different pavement surfaces will perform not only with tires but with other things that hit them or gouge them or slide on them or anything else, would indicate to me that unequivocally that the difference between what we did in terms of the tests and what actually happened is identical.

Upon thorough review of the record, we conclude the trial court was "too wide of the mark" in sustaining plaintiff's objection grounded upon the experiments having been conducted in Nash County while the accident took place in Wilson County. The evidence adequately established that the conditions under which the experiments were performed in Nash County were "substantially similar," Brown, 280 N.C. at 597, 187 S.E.2d at 91, to those in neighboring Wilson County. Further, plaintiff's assertions notwithstanding, Brown's testimony regarding the similarity between the relevant roads in Nash and Wilson counties is not inherently incredible simply because he visited the accident site at 301 after performing the experiments. All evidence showed that the road at the accident site on Highway 301 was unremarkable—flat, straight, and made of asphalt—and therefore easily capable of replication. Indeed, plaintiff's counsel essentially admitted at trial that defendants would have been unable to reenact the accident on Highway 301 itself due to heavy traffic. See State v. Wright, 52 N.C.App. 166, 174, 278 S.E.2d 579, 586, disc. review denied, 303 N.C. 319 (1981) (not reasonable or possible to perform test under precise conditions existing when collision occurred). The record reflects that Brown's experiments were conducted upon a similar, acceptable, and safer alternative road.

Plaintiff's counsel also relied at trial upon Brown's testimony that "you wouldn't expect the bundle to fall the same way every time" to argue that the measurements taken by Brown were a poor reflection of the actual behavior of the bundles at the time of the accident. However, Brown thereafter explained that in his experiment the bundles tended to either roll or slide once they hit the pavement and that "because of the difference between rolling and sliding, you would get some variation in the total distance it would go before it stopped." Brown further indicated that he performed the experiment a sufficient number of times to develop a range of values for the distance traveled by the bundles, that this range of distances was narrow, and that in fairness to plaintiff he utilized the shortest distance recorded, fifty feet, in developing a chart of the accident for demonstration to the jury. Earlier, Brown also testified he was able, with separate physics calculations, to corroborate as accurate the distances observed during his experiment.

Finally, it appears that any discrepancies between Brown's experiments and the original accident could have been brought out on cross-examination and that Brown had the ability to explain how certain differences might have affected the data he gathered. See Short, 70 N.C.App. at 455, 320 S.E.2d at 20. It would then have been for the jury to determine the weight and credibility of Brown's testimony.

We next consider whether the exclusion of Brown's testimony was prejudicial to defendants. See N.C.G.S. § 1A-1, Rule 61 (error in exclusion of evidence must amount to denial of substantial right to warrant new trial). Absent the excluded evidence, the jury at trial lacked a basis for considering the role played by inertia in the accident at issue. Without the expert testimony and the option of determining its weight and credibility, the jury was left to its own devices and may well have assumed the bundle or bundles would naturally remain in the spot where they landed after falling, or, more damaging to defendants, that the bundles in falling from the truck would have rolled towards plaintiff's oncoming vehicle, thereby reducing the amount of time within which she might have avoided striking them. We therefore hold the exclusion of Brown's testimony constituted error prejudicial to defendants and grant them a new trial. We decline to address defendants' remaining assignments of error as they may not occur upon retrial.

New trial.

§ 7.14 *There Shall Be Friction*

Friction is a force. Unlike a nuclear explosion or the acceleration of a vehicle, it is a more insidious force that seemingly lies in wait until it acts vigorously to oppose some form of movement. Were it not for friction, we could nearly achieve perpetual motion. Objects, such as automobiles could nearly travel forever were it not for friction. There are many different forms of friction, ranging from intergalactic to subatomic. We are most interested in the simple mechanical form of friction.

Technically, friction might be considered as the force required to move two sliding surfaces over one another. It is reduced once the motion has started. At that juncture, we move from static to kinetic friction.

To understand friction, we must first understand the so-called normal force. The normal force acts on any object that touches a surface (either directly or indirectly). Law books sitting on a table would exert a constant normal force (F) on the table equal to the mass of the books times the gravitational constant:

$$F_{Normal} = - mg$$

At the same time, the table would be exerting an upward force on the books. The two forces (i.e., the force of the books and the force of the table). The two forces are equal and offsetting because the books are static. If the downward force of the books exceeded the force that the upward force that the table was able exert, the books would simply crash through the table and end up on the floor.

Example 7-4: A normal person weighing 70 Kilograms is standing on a normal floor in a normal law library. How much normal force would he be exerting on the floor?

Solution: $F_{Normal} = - mg$
$m = 70$ kg
$g = - 9.8$ m/sec^2

(Physicists like to consider the downward pull of gravity as being "-" negative).

$$F_{Normal} = - mg = - (70 \text{ kg})(-9.8 \text{ m/s}^2) = 686 \text{ N (Newtons)}$$

Friction can be thought of as a force that opposes the movement of an object across a surface. In the preceding problem, for example, assume that our normal fellow (who weighs 70 kg) neglected to pay his bookie, who sent Guido to pay a visit on the normal fellow. Assume that Guido knocked our normal guy unconscious and proceeded to drag his body across the floor. The force of friction would be working against Guido. Because the force of friction is, indeed a force (as is the case with gravity and the normal force), it is also measured in Newtons:

$$F_{Friction} = \mu \, F_{Normal}$$

The term "μ" is a constant referred to as the "coefficient of friction." The **coefficient of friction** is basically a measure of roughness between two opposing surfaces, one of which is to be dragged across the other. A large coefficient of friction implies a rough surface or surfaces. A low coefficient of friction implies slipperiness. A zero coefficient of friction implies that there is no friction whatsoever between the two surfaces, a condition which is non-existent on this planet.

Anyone who has ever driven an automobile during a Midwest winter as fresh snow was beginning to fall on a slippery street knows the meaning of a near-zero coefficient of friction. Brakes are useless, and slowing down to near zero miles per hour doesn't prevent collisions. Here are some coefficients:

Table 7-2 [Coefficients of Friction]

Surface 1	Surface 2	μ
Rubber	Dry Asphalt	1.000
Rubber	Wet Asphalt	0.950
Rubber	Ice	0.005
Steel	Ice	0.010
Steel	Steel	0.180
Book	Wooden Floor	0.100

Example 7-5: A law book weighing 500 grams is dragged across the wooden floor. What is the force of friction acting against the movement of the dragger?

Solution: $F_{Normal} = - mg = - (0.5) \times (-9.8) = 4.9N$
$F_{Friction} = \mu \, F_{Normal}$
$F_{Friction} = \mu \, (4.9N) = (0.100) \times 4.9N = 0.49N$

People v. Herman

City Magistrates' Court of New York, Traffic Court, Borough of Queens, 174 Misc. 235, 20 N.Y.S.2d 149 (1940)

HORN, C. M.

The defendant is charged with the offense of operating a motor vehicle on a public highway in the borough of Queens, city of New York, at a rate of speed in excess of thirty-five miles per hour in violation of section 14 of article 3 of the Traffic Regulations of the City of New York, promulgated pursuant to section 435 of the New York City Charter, which regulation establishes the speed limits for automobiles operated within the city of New York.

This complaint was ordered by the court as a result of evidence presented before the magistrate at a criminal homicide hearing.

On December 1, 1939, at about six-thirty A.M., the defendant, while operating a Ford coach on Roosevelt avenue near Junction avenue in the county of Queens (following an accident), applied the brakes of his automobile, which locked and caused continuous tire skid marks from all four wheels for a distance of over sixty four feet and one inch in length upon a dry asphalt pavement.

About twenty minutes after the accident occurred, members of the homicide squad of the police department, in the course of their duties, arrived upon the scene, and after the skid marks were identified by the defendant as having been caused by the tires of his automobile, measured their length. Only those skid marks which were firm and continuous were taken into consideration. The testimony shows that the brakes were applied sooner, but since the first skid marks were intermittent and not firm and continuous, measurements were only taken from the start of the firm, continuous skid marks. Measuring from that point to the rear of the vehicle and then deducting its length, a measurement of sixty-four feet, one inch was determined. The four wheel brakes were then duly examined and being found to be in good working condition the same automobile was driven by one of the officers over the same stretch of roadway and when the speedometer recorded about thirty miles per hour the brakes were suddenly and firmly applied, causing similar skid marks of forty-three feet three inches in length.

The only evidence in dispute is with respect to the length of the skid marks made by defendant's car. A lay witness, one Roper, who passed by following the accident, approximated them to be between thirty-five and forty feet long. In view of the fact that the police officers, who are specially trained for such investigations, made actual measurements, it is obvious that their figure is the correct one.

When questioned at the scene of the accident the defendant stated to a police officer that he had been driving his car at a speed of about thirty to thirty-five miles per hour at the time he applied his brakes. Outside of this admission by the defendant, who did not testify upon the trial, and the testimony of Mr. Roper, no evidence was offered other than the testimony of the two members of the homicide squad who, after being qualified as experts, estimated the speed of the automobile on the basis of a mathematical formula.

While evidence of the speed of an automobile determined by the application of this formula has been used as evidence in both civil and criminal cases, none has come to the court's attention where it was used as a basis for prosecuting a defendant for speeding. Usually the evidence relied upon in such cases is the testimony of police officers who, after following a defendant in a motor cycle or in an automobile, have 'clocked' on the speedometer of their vehicle the speed at which the defendant is charged with having operated his automobile. This was so because no other means was devised for measuring speed.

The science of police work has made rapid strides within the last decade. Among other things, the homicide squad of the New York police department, in its desire to determine the cause of many accidents and in an endeavor to minimize them, caused an intensive study to be made of the relationship of skid marks to that of the speed of an automobile in order to arrive at a formula whereby, based on certain known factors, the speed of an automobile, at the time of the commencement of the skid, might be determined. Lieutenant Andrews and Detective Grant in collaboration with Dr. F. C. Stanley, professor of physics and chemistry and technical research director for one of the large brake companies, and Mr. Tilden, professor of mechanical engineering, and also associated with one of the large brake companies, studied the problem for about two years. As a result of their study, which included a series of scientific tests and compilation of statistics, they determined that by the application of the formula $s = v^2/30 \times f$, the speed of an automobile can be determined. The formula is used by the New York police department, other agencies and brake companies. In interpreting this formula:

s = distance of skid

v = velocity (speed) of vehicle

30 = gravitational constant in miles per hour (transform feet per second to miles/hour)

f = coefficient of friction

A skid mark is caused by the burning of rubber from the tire as a result of the friction between the surface of the tire and the surface of the pavement (with brakes locked). This friction varies according to the holding power of the road surface. From thousands of tests made by the use of a drawbar and a dynometer, the coefficient of friction of pavements of streets throughout New York city has been determined. The coefficient of friction of asphalt pavement was determined to be .78. Others vary below and above, up to .95, which is the coefficient of brushed concrete.

The formula applied by the People has a valid basis in the science of physics and mathematics and is more clearly explained as follows: It is a fundamental truth that if a man travels 20 miles in four hours, his average velocity (speed) would be 5 miles per hour; in terms of a formula this would be

average velocity = distance/time or

distance = average velocity × time or

(1) $S = v_a \times t$

Acceleration is defined as change in speed per unit of time. When the speed is increasing, the acceleration is positive, and when the speed is decreasing, the acceleration is negative (or deceleration). To illustrate, the acceleration of a falling body is 32.2 feet per second; i.e., a body will fall at a rate of speed of 32.2 feet per second at the end of the first second, at a rate of speed of 64.4 feet at the end of the second second, etc. From this it will be observed that if the acceleration is constant the speed acquired will be directly proportional to the time. Take our example of the falling body, the speed (velocity) at any particular point in time will be the product of the number of seconds multiplied by 32.2 (the force of gravity—g);

v—1st second = 1 × 32.2 or 32.2 ft./sec.

v—2nd second = 2 × 32.2 or 64.4 ft./sec.

v—3rd second = 3 × 32.2 or 96.6 ft./sec. etc.

Writing this in terms of a formula, we have

final velocity = acceleration × time or

$v_f = at$ or

(2) $t = v_f / a$

Any elementary physics book will tell us that the average velocity of a body moving with uniform acceleration equals one-half the sum of its initial and final velocity. Applying this to a falling body whose acceleration is accepted as fundamental, we would have a body starting from rest or a speed of zero to a velocity of 32.2 feet per second at the end of the first second. The distance covered, 16.1 feet in the first second, is elementary. The average speed or velocity would, therefore, be 16.1 feet per second. Written mathematically this would be:

average velocity = ½ (initial velocity, plus, final velocity) or substituting

average velocity = ½ (0 plus 32.2)

average velocity = 16.1 or

(3) $v_a = \frac{1}{2} \times v_f$

when the initial velocity is zero. But substituting the value of v_f as found in formula (2), we have average velocity equals one-half acceleration times the time,

(4) $v_a = \frac{1}{2} at$

But formula (1) shows that distance traversed equals average speed (velocity) multiplied by time. Therefore, substituting the values in formula (4) for its equivalent in formula (1) we have

s (distance) = $(\frac{1}{2} at) \times t$ or

(5) $s = \frac{1}{2} at^2$

Reading this formula we may say that where the acceleration is constant, the distance traversed from rest varies as the square of the time. In formula (2) we have t equals v_f/a. In formula (5) we have s = ½ at^2, now substituting algebraically the value of t, we have

$s = ½\ a \times (v_a/a)^2$

$s = ½\ a \times v_f^2/v_a^2$ or

$s = v_f^2 / 2a$ transposing the figures we have

(6) $v_f^2 = 2$ or $as = v_f^2/2$

therefore, if a body starts from rest and acceleration is constant, the velocity varies as the square root of the distance traversed.

Physics has established that every body in motion possesses kinetic energy; that is, it will do a certain amount of work against a resisting force before it stops moving. The fundamental equation for the work done by any force acting through a distance is

(7) Work = $f \times s$

Whenever a force is acting on a body, the body has an acceleration in the direction in which the force acts and the acceleration is proportional to the force, that is:

(8) $f_1 / f_2 = a_1 / a_2$

One way to get a force to act on a body is to let the body fall. In this illustration the force acting is known, for it is the weight (w) of the body. The acceleration is also known for it is gravity (g) which is 32.2 feet per second, per second. So that the weight of the body and its acceleration when falling can always be used as two of the numbers in the proportion just mentioned (8), that is,

(9) $f/w = a/g$ or $f = wa/g$

substituting formula (9) in formula (7) we have

(10) $fs = (wa)/g$ or $fs = w/g \times as$

But the term as has been determined in formula (6) to be equal to $v_f^2/2$, substituting this valuation of as in formula (10), we have

$fs = w/g \times v_f^2/2$ or

(11) $fs = w \times v_f^2/2\ g$

transposing

(12) $s = w \times v_f^2/2 \times gf$

The coefficient of friction is determined by the pull or force of friction that is necessary to move a given weight over a given surface, or

coefficient of friction = force of friction/total normal pressure (weight), or

force of friction (f) = coefficient of friction × weight, or

(13) f = coefficient of f × w; substituting this valuation of f in formula (12) we have

$s = w \times v_f^2/2g \times$ coefficient of f × w

with weight canceling weight this resolves itself into

(14) $s = v_f^2/2\ g \times$ coefficient of f

This formula represents velocity in feet per second. In order to change it to miles per hour, we make use of simple arithmetic, and

$s = (5280/3600)^2 / 64.4 \times$ coefficient of f, or

(15) $s = v_f^2 /30 \times$ coefficient of f (Black & Davis, *Practical Physics*, Macmillan Co. 1924.)

Applying the formula and using the known facts, we have

64' 1" or $64 = v_f^2 / 30 \times .78$

$v_f = 38.6$ miles per hour.

The witnesses Andrews and Grant were cross-examined at length both by counsel for defendant and the court as to various factors which might show that the formula was erroneous. They also testified that after having arrived at the formula they have experimented with it for a period of two and a half years and have found it substantially accurate. From all the evidence the court is satisfied beyond a reasonable doubt that the defendant was driving an automobile in excess of twenty-five miles per hour.

The defendant further argues, assuming the formula and computations to be correct, there still has been no violation proved, for the People have failed to show:

(1) That the speed limit at the particular point was twenty-five miles per hour.

(2) That the defendant was proceeding at a speed in excess of twenty-five miles per hour for at least one-quarter of a mile.

Section 14 of Article 3 of Chapter 14 of the Traffic Regulations of the City of New York provides:

1. A driver of any vehicle shall not drive at a rate of speed faster than twenty-five miles per hour, except where signs authorized by proper authority are posted indicating a greater or lesser speed.

2. No driver shall drive at a rate of speed faster than ten miles per hour when turning corner.

There being no evidence before the court indicating that signs, authorized by proper authority, were posted indicating a greater or lesser speed, the general prohibition of a speed faster than twenty-five miles per hour must apply.

With respect to the second point raised, the defendant refers to the Vehicle and Traffic Law of the State of New York, paragraph b of subdivision 1 of section 56, which provides: "A rate of speed, by a motor vehicle other than a motor truck, in excess of forty miles an hour for a distance of one quarter of a mile, shall be presumptive evidence of driving at a rate of speed which is not careful and prudent."

This contention of the defendant is erroneous, for the two statutes are entirely dissimilar. The regulation under which the defendant is being prosecuted is one which contains a specific prohibition against the operation of an automobile in excess of twenty-five miles per hour. The Vehicle and Traffic Law section of which the defendant has quoted a part also provides under paragraph a of subdivision 1 as follows:

"Every person operating a motor vehicle upon a public highway shall drive such vehicle in a careful and prudent manner and at a rate of speed so as not to endanger the property of another or the life or limb of any person."

It is to be noted that the State statute does not prohibit driving at a speed in excess of a fixed rate of speed, but merely seeks to prohibit driving at a speed which is dangerous to the property or person of another. In that connection paragraph b merely creates the presumption that driving at a rate of speed in excess of forty miles per hour for a distance of one-quarter of a mile is a violation of the section. Its effect is merely to shift the burden of proof upon the defendant to show that, though exceeding forty miles per hour for a distance of a quarter of a mile, he was not under the circumstances driving in a careless or imprudent manner. In other words, speed alone is not the proper test of careless driving under the statute. (People v. Carrie, 122 Misc. 753; People v. Mellen, 104 id. 355.)

Furthermore, Section 54 of the Vehicle and Traffic Law provides: "... and provided further that nothing in this chapter contained shall impair the validity or effect of any ordinances regulating the speed of motor vehicles ... heretofore or hereafter made, adopted or prescribed pursuant to law in any city of the first class."

Hence, it is clearly apparent that the Legislature reserved to cities of the first class the power to regulate the speed of motor vehicles operated upon its streets.

Since Section 435 of the New York City Charter gives the police department the power and duty to regulate, direct, control and restrict the movement of vehicular traffic, for the facilitation of traffic and the convenience of the public as well as the proper protection of human life and health, and the commissioner the power to make rules and regulations to accomplish that purpose, and since the provisions of section 14 of article 3 of chapter 14 of the Traffic Regulations of the City of New York are specific, evidence of speed in excess of twenty-five miles per hour, even though for a distance of less than one-quarter of a mile, would prove a violation of the regulation.

Under the circumstances, the defendant is found guilty, but due to the co-operation of the defendant and his counsel in the preparation and submission of this cause which is in the nature of a test case, sentence is suspended.

Notes and Problems

1. **Pre-impact Velocity:** As indicated in *Herman*, you can determine with some degree of accuracy the pre-crash velocity (V) by using the coefficient of friction and the length of the skid mark (s); however, as indicated in Note 2, *infra*, employing the Drag Factor (D), as opposed to the unmodified coefficient of friction, would result in greater accuracy:

$$s = \frac{V^2}{30\ D}$$

Of the many equations used for calculating the pre-impact velocity, another commonly accepted alternative includes the acceleration factor for the force of gravity g (32 ft. / second^2):

$$V^2 = 2\ g\ f\ s$$

(As a side note on *Herman*, consider the derivation of the constant "30." Was the court correct?)

2. **Post-impact Velocity:** Acknowledging that pre-impact velocity will normally be of primary concern (particularly in criminal matters) to accident reconstructionists, nevertheless, the action and movement of two colliding vehicles following an accident may also be of major significance. There are many equations that have been accepted by experts and the courts. Here is a generally accepted equation using as factors: V (Post-im-

pact velocity), v_f (Velocity after application of acceleration), D (Drag factor), acceleration of gravity (g) and d (distance traveled after impact):

$$V^2 = (v_f^2) + (2D)(gd)$$

The Drag Factor is a modification or adjustment to the Coefficient of Friction (See Table 7-2). Using the Coefficient without such an adjustment presumes that upon braking, all four wheels will lock and therefore scrape against the surface of a level road. Under these conditions, the Drag Factor would equal the Coefficient. If, however, at least one of the wheels fails to lock or if the road is wet or at an incline, the Drag Factor would be different.

3. **A Word of Caution!** The factors discussed above are by no means exhaustive. There are many other physical factors both mathematical and descriptive that should be taken into account by the accident reconstruction expert. See Badger, "*Reconstruction of Traffic Accidents*," 9 Am. Jur. 3d Proof of Facts 115 (1990); Schorr, "*The Techniques of Traffic Accident Reconstruction*," 27 The Practical Law. 77 (1981); an Caufield, "*Cross-Examination of Accident Reconstruction Expert by Defense Attorney to Refute Opinion of Speeding and Inattentiveness Based on Formulas*," in R. Habush, *Art of Advocacy: Cross Examination of Non-Medical Experts*, 41-20 (1991).

4. Accidents on curved roads are particularly tricky. While rounding a curve, the centrifugal force tends to cause the automobile to fly off the road in a straight line, while the counteracting centripetal force (friction) tends to keep the car on the curved road. A loss of control on the curve or a crash also creates marks on the road, but they are not the same marks that we typically encounter in a linear collision. They are measured differently and called by different names, such as yaw marks, scrub marks or sideslip. The following equation is commonly used to determine the maximum speed at which a vehicle can safely make a turn (Critical Speed):

$$V = \sqrt{g\,f\,R}$$

where V = Critical Speed, g = Acceleration attributable to gravity (32.2 feet/Second^2), f = Drag factor and R = Radius of the curve.

5. The yaw marks referred to in the preceding note have their own special significance in determining the material factors for accident reconstruction experts. One such factor involves the radius of the marks, which can be calculated according to the following equation:

$$R = \frac{C^2}{8\,m} + \frac{m}{2} + \frac{T}{2}$$

where R = The radius of the yaw mark, C = Chord of the arc created by the yaw mark, T = axle width and m = The middle ordinate (the distance from the bisector of the cord to the yaw mark).

6. Another factor in accident equation is the angle of the "grade" or rise of a road or highway. Not all streets are flat. A simple one-step formula for determining the stopping distance of an automobile includes grade as a factor:

$$D = \frac{S^2}{30\,(f + G)}$$

where D = The stopping distance in feet, S = Speed in miles/hour, f = The Coefficient of Friction and G = The rise in vertical feet per horizontal foot (tantamount to a sine function).

Hoffman v. Oakley

184 N.C.App. 677, 647 S.E.2d 117 (2007)

GEER, Judge.

Plaintiff Paula Ann Hoffman and her daughter, third-party defendant Catherine Michelle Hoffman (the "Hoffmans"), appeal from a judgment in favor of defendants/third-party plaintiffs, Shawn Cherri Oakley and David Reade Oakley, entered in accordance with a jury verdict, concluding that Catherine Michelle Hoffman had been contributorily negligent in an automobile collision. The primary issue on appeal is whether the trial court erred by admitting the testimony of the defendant/third-party plaintiffs' accident reconstruction expert, which, the Hoffmans contend, constituted improper expert testimony regarding the speed Catherine was traveling.

It has long been the law, in North Carolina, that an expert witness may not testify regarding the speed of a vehicle unless he or she personally observed the vehicle. See 2 Kenneth S. Broun, Brandis & Broun on North Carolina Evidence § 183, at 37–38 n. 166 (6th ed.2004) (urging elimination of limitations on accident reconstruction expert testimony). Although our legislature has recently amended Rule 702 to overturn this doctrine, the amendment applies only to "offenses" committed on or after 1 December 2006. 2006 N.C. Sess. Laws 253, sec. 33. Since the automobile collision in this case occurred on 13 March 2003, we must apply the former law. Nevertheless, we hold that the expert's testimony did not amount to an opinion on Catherine Hoffman's speed, but rather was the type of testimony admissible even under the previously existing law.

Facts

At approximately 7:30 a.m. on 13 March 2003, Catherine Hoffman was driving her mother's 1996 Honda Civic on Brooks Avenue in Raleigh, North Carolina. As Catherine approached the defendant Oakleys' home on Brooks Avenue, Shawn Oakley was backing David Oakley's mini-van out of their driveway when the two cars collided.

On 28 April 2003, Paula Hoffman filed suit against the Oakleys, in Wake County District Court, for losses resulting from the property damage to her car. She alleged that Shawn Oakley had been negligent in backing the mini-van out of her driveway and had caused the collision. The Oakleys filed an answer denying the relevant allegations of Paula Hoffman's complaint and, subsequently, filed an amended answer and a third-party complaint against Catherine Hoffman. The Oakleys' third-party complaint alleged that Catherine's negligence had been the sole cause of the collision or, alternatively, that her contributory negligence precluded her mother's recovery.

The case was tried before a jury on 8 and 9 August 2005 in Wake County District Court, with the parties stipulating that any negligence by Catherine Hoffman was to be imputed to Paula Hoffman. After hearing testimony from the Hoffmans, Shawn Oakley, the police officers who arrived on the scene after the collision, and an expert in accident reconstruction, the jury determined that although Paula Hoffman's vehicle was damaged by Shawn Oakley's negligence, Catherine Hoffman—and, therefore, Paula Hoffman—was contributorily negligent. Accordingly, the trial court entered judgment ordering that the Hoffmans recover nothing from the Oakleys.

The Hoffmans' subsequent motions for a new trial or judgment notwithstanding the verdict were denied, and the trial court awarded the Oakleys certain specified "reasonable costs and expenses." The Hoffmans filed a timely appeal to this Court.

I

The Hoffmans first argue that the trial court erred in admitting the testimony of the Oakleys' expert on accident reconstruction. They contend that the witness gave impermissible opinion testimony regarding the speed Catherine Hoffman was traveling. We disagree.

Typically, an expert witness may testify in the form of an opinion if that expert's "scientific, technical or other specialized knowledge will assist the trier of fact to understand the evidence or to determine a fact in issue...." N.C.R. Evid. 702(a). "[E]xpert testimony in the field of accident reconstruction has been widely accepted as reliable by the courts of this State." State v. Holland, 150 N.C.App. 457, 463, 566 S.E.2d 90, 94 (2002), cert. denied, 356 N.C. 685, 578 S.E.2d 316 (2003).

Nevertheless, our appellate courts held, prior to the amendment to add Rule 702(I), "that 'with respect to the speed of a vehicle, the opinion of a[n] ... expert witness will not be admitted where he did not observe the accident, but bases his opinion on the physical evidence at the scene.'" Marshall v. Williams, 153 N.C.App. 128, 135, 574 S.E.2d 1, 5 (quoting Hicks v. Reavis, 78 N.C.App. 315, 323, 337 S.E.2d 121, 126 (1985), cert. denied, 316 N.C. 553, 344 S.E.2d 7 (1986)), appeal dismissed and disc. review denied, 356 N.C. 614, 574 S.E.2d 683 (2002). Accordingly, unless an accident reconstruction expert actually observed the accident, the expert may not testify as to the speed a vehicle was traveling. See Van Reypen Assocs., Inc. v. Teeter, 175 N.C.App. 535, 542, 624 S.E.2d 401, 405 (noting that, under this rule, "our trial courts are forced to exclude accident reconstruction testimony regarding speed"), disc. review improvidently allowed, 361 N.C. 107, 637 S.E.2d 536 (2006).

Here, the Oakleys' expert, Sean Dennis, testified that he had performed several "skid test[s]" at the accident scene using a 1997 two-door Honda Civic that Mr. Dennis considered to be a "sister or clone" of the 1996 four-door Honda Civic that Catherine Hoffman was driving at the time of the accident. Because the speed limit at the scene of the accident was 35 miles per hour, Mr. Dennis' skid tests included "full, panic-stop application of the brake pedal" at 33, 34.2, 40, 46, and 50 miles per hour. According to Mr. Dennis, his test results indicated that if a vehicle like the one driven by Catherine Hoffman was traveling at 35 miles per hour, it would be able to stop "in just under 54 feet." The Hoffmans argue that this testimony, when viewed in conjunction with that of a responding police officer who found skid marks at the scene measuring 80 feet in length, was merely "evidence of speed through the 'back door.'"

Our Supreme Court has, however, specifically held that such testimony about stopping distances is admissible. See State v. Gray, 180 N.C. 697, 702, 104 S.E. 647, 650 (1920) ("Admitting, then, that each of the particular witnesses was an expert in regard to the matter about which he was examined, testimony as to the distance within which such a truck, as [the] truck [at issue,] could be stopped when going at a rate of speed 20 to 25 miles an hour was plainly admissible."). See also Draper v. Atl. Coast Line R.R. Co., 161 N.C. 307, 312, 77 S.E. 231, 232–33 (1913) (holding that testimony was competent when witness testified that train traveling at particular speed could have stopped within 200 yards). Under Gray and Draper, Mr. Dennis' testimony about stopping distances at various speeds was admissible.

These decisions are consistent with subsequent Supreme Court decisions holding that expert testimony about speed is inadmissible. In Shaw v. Sylvester, 253 N.C. 176, 180, 116 S.E.2d 351, 355 (1960), the Court held that "[a] witness who investigates but does not see a wreck may describe to the jury the signs, marks, and conditions he found at the scene, including damage to the vehicle involved. From these, however, he cannot give an opinion as to its speed. The jury is just as well qualified as the witness to determine what inferences the facts will permit or require." The Court stressed, however, that "[t]he qualified expert, the nonobserver, may give an opinion in answer to a proper hypothetical question in matters involving science, art, skill and the like.... An automobile, like any other moving object, follows the laws of physics...." Id.

This Court has held that the restriction on expert testimony set out in Shaw "is limited to opinions regarding speed; it does not apply to opinions concerning other elements of an accident." State v. Purdie, 93 N.C.App. 269, 276, 377 S.E.2d 789, 793 (1989). Thus, an expert's testimony is properly admitted when he gives no opinion as to the actual speed of a vehicle. Id. See also McKay v. Parham, 63 N.C.App. 349, 353, 304 S.E.2d 784, 786–87 (1983) (holding admissible expert testimony that applied the law of physics to post-collision movement of two cars), disc. review denied, 310 N.C. 477, 312 S.E.2d 885 (1984).

Here, Mr. Dennis never gave an opinion as to the speed that Catherine Hoffman was traveling. He used his scientific expertise to perform an experiment that demonstrated stopping distances at various speeds. See, e.g., Addison v. Moss, 122 N.C.App. 569, 571–73, 471 S.E.2d 89, 90–92 (holding result of experiment involving vehicle admissible on question of contributory negligence), disc. review denied, 345 N.C. 179, 479 S.E.2d 203 (1996). It was left up to the jury to determine Catherine Hoffman's stopping distance—which was a subject of dispute at trial—and make the ultimate determination of the speed of her car, precisely as required by Shaw. The trial court, therefore, did not err in admitting Mr. Dennis' testimony. (The Hoffmans also argue that the trial court erred by refusing to instruct the Oakleys' attorney not to argue the issue of speed in his closing argument. Because Mr. Dennis' testimony was properly admitted, and the jury could infer from that testimony and evidence of the skid marks that Catherine Hoffman was exceeding the speed limit, the trial court properly denied the Hoffmans' request. The Hoffmans further contend that counsel's actual argument—that the amount of damage to the car suggested Catherine was speeding—was improper. Shaw, however, indicates that a jury may draw inferences regarding speed from "the signs, marks, and conditions" at the scene "including damage to the vehicle involved." 253 N.C. at 180, 116 S.E.2d at 355. See also King v. Bonardi, 267 N.C. 221, 227, 148 S.E.2d 32, 37 (1966) (holding that extent of damage, along with other evidence, was sufficient to support inference that car was being operated at dangerous and unlawful rate of speed). Under the circumstances of this case, counsel's argument was not improper.)

II

The Hoffmans next argue that the trial court erred by denying their motions for a directed verdict on the issue of contributory negligence and for judgment notwithstanding the verdict ("JNOV"). When considering a motion for a directed verdict, a trial court must view the evidence in the light most favorable to the non-moving party, giving that party the benefit of every reasonable inference arising from the evidence. Clark v. Moore, 65 N.C.App. 609, 610, 309 S.E.2d 579, 580 (1983). Any conflicts and inconsistencies in the evidence must be resolved in favor of the non-moving party.

If there is more than a scintilla of evidence supporting each element of the non-moving party's claim, the motion for a directed verdict should be denied. Clark, 65 N.C.App. at 610, 309 S.E.2d at 580–81. The same standard applies to motions for JNOV. Smith v. Price, 315 N.C. 523, 527, 340 S.E.2d 408, 411 (1986).

Because contributory negligence is an affirmative defense, the burden was on the Oakleys to prove that there was more than a scintilla of evidence supporting each element of contributory negligence. Snead v. Holloman, 101 N.C.App. 462, 464, 400 S.E.2d 91, 92 (1991). Evidence that a party was exceeding the posted speed limit is sufficient to send the issue of contributory negligence to the jury. See, e.g., Whisnant v. Herrera, 166 N.C.App. 719, 723, 603 S.E.2d 847, 850 (2004) (evidence that plaintiff was "exceed[ing] the speed limit" justified submission of issue of plaintiff's contributory negligence to the jury).

In the present case, the parties stipulated at trial that any negligence by Catherine Hoffman was to be imputed to Paula Hoffman. The speed limit on Brooks Avenue was 35 miles per hour. Shawn Oakley testified that Catherine Hoffman told her that she was traveling "about" 40 miles per hour. In addition, as discussed in the prior section, the Oakleys' evidence included expert testimony that a car like the one driven by Catherine Hoffman would be able to stop "in just under 54 feet" if it was traveling at 35 miles per hour, the road's speed limit. Some of the measurements taken at the scene of the accident indicated that the skid marks from the Hoffman car measured 80 feet in length. If the jury accepted the accuracy of those measurements, then the jury could draw the inference, based on the accident reconstruction expert's testimony, that Catherine Hoffman was exceeding the speed limit. The issue of her contributory negligence was, therefore, properly submitted to the jury.

The Hoffmans nevertheless argue that Catherine Hoffman's speed was not a proximate cause of the collision. "In order for a contributory negligence issue to be presented to the jury, the defendant must show that plaintiff's injuries were proximately caused by his own negligence." McGill v. French, 333 N.C. 209, 217, 424 S.E.2d 108, 113 (1993). In other words, "'[t]here must be not only negligence on the part of the plaintiff, but contributory negligence, a real causal connection between the plaintiff's negligent act and the injury, or it is no defense to the action.'" Whisnant, 166 N.C.App. at 722, 603 S.E.2d at 850 (quoting West Constr. Co. v. Atl. Coast Line R.R. Co., 184 N.C. 179, 180, 113 S.E. 672, 673 (1922) (emphasis original)).

According to the Hoffmans, because the jury found that Shawn Oakley had been negligent, and Shawn herself testified that she did not see the Hoffmans' car before the collision, she must have backed into the roadway without looking. Therefore, the Hoffmans argue, regardless of Catherine Hoffman's speed, Shawn Oakley's negligence must have been the sole proximate cause of the collision. In support of their argument, the Hoffmans point to Ellis v. Whitaker, 156 N.C.App. 192, 576 S.E.2d 138 (2003), in which this Court noted that a plaintiff is not required to anticipate a defendant's negligence and "'has a right to assume that any motorist approaching from his left on the intersecting street will stop in obedience to the red light [or a stop sign] facing him unless and until something occurs that is reasonably calculated to put him on notice that such motorist will unlawfully enter the intersection.'" Id. at 196, 576 S.E.2d at 141 (alteration in original) (quoting Cicogna v. Holder, 345 N.C. 488, 490, 480 S.E.2d 636, 637 (1997)).

The Ellis Court concluded that, although the evidence suggested that the plaintiff may have been speeding, the defendant had failed to show a "real causal connection" between the plaintiff's speed and the accident, and, therefore, the plaintiff's speed was not a proximate cause of the collision. Id.

Unlike Ellis, however, in which no evidence was presented indicating that the collision could have been avoided had the plaintiff been traveling the posted speed limit, the evidence in the present case was sufficient to allow a jury to find that had Catherine Hoffman not been speeding, she would have been able to stop in less than 54 feet, which would have brought her vehicle to a halt prior to any impact. This is sufficient to demonstrate a causal connection between Catherine Hoffman's excessive speed and the resulting accident. See Whisnant, 166 N.C.App. at 723–24, 603 S.E.2d at 851 (distinguishing Ellis and concluding that defendant demonstrated real causal connection between collision and plaintiff's speed when evidence showed plaintiff was speeding while approaching defendant's vehicle and, by the time plaintiff saw defendant, plaintiff was unable to stop).

The trial court, therefore, properly denied the Hoffmans' motion for a directed verdict and motion for JNOV. (For these reasons, we also reject plaintiff's contention that the trial court erred in instructing the jury on contributory negligence and on speed as a basis for finding contributory negligence.)

III

(Discussions pertaining to costs and fees are omitted)

No error.

Judges TYSON and ELMORE concur.

Notes and Problems

1. At the new trial of *Addison v. Moss*, what arguments would you make (and what equations would you use) on behalf of the plaintiff? On behalf of the defendant?

2. Newton's First Law was restated in *Prince v. Bennett*, 322 S.W.2d 886 (Mo., 1959), an action alleging the wrongful death of the driver of a skidding dump truck, which was struck by defendant's approaching tractor-trailer truck:

> It is a well known law of motion or mechanics that an object which is moving over a surface in a straight line will continue to so move unless compelled by impressed force to change the direction.

Perhaps, Newton's First Law could be stated even more simplistically: If something very big (such as a tractor-trailer truck) is out of control and coming directly at you, it will probably keep coming.

3. In *Thurmond v. Pepper*, 119 S.W. 2d 900 (Tex. Civ. App., 1938), a wrongful death action, the driver of a truck carrying heavy pipe was well aware of Newton's First Law when he consciously elected to sideswipe another vehicle instead of hitting the breaks. The driver claimed a defense of "discovered peril." Explain the "peril."

4. Basic relativity, as opposed to Einstein's concepts involving the speed of light, applies to basic Newtonian Physics. At the airport, for example, there is a moving walkway (i.e., a flat escalator) that carries a standing pedestrian at the rate of approximately 4 miles per hour. If you were to walk on the moving walkway at 2 miles per hour as you passed a lazy pedestrian (standing still on the walkway), how fast would you be going? [Hint: A question such as this makes no sense. All motion is relative.] A more appropriate set of questions would ask: "How fast are you traveling with respect to the lazy pedestrian?" or "How fast are you traveling with respect to the ticket counter that you just passed by?"

At the retrial of *Addison v. Moss*, might the defendant argue that the bundle of tobacco, as it zipped along the ground should have been treated as just another moving vehicle by the plaintiff? Would this be a valid argument in a frictionless world? If so, what would be the "relative" speed of plaintiff's vehicle with respect to the bail of tobacco.

5. Stupid Highway Tricks. Friction keeps our cars on the road, and feet on the sidewalks. We need it. Detroit, commonly referred to as the model for automotive science, is far from a model when it comes to highway design.

Man can not drive by friction alone. This is particularly true in the Midwest where fresh fallen snow can reduce the coefficient of friction dramatically close to zero, especially around a curve. Without a bank, cars would slide into the ditch. Blame Newton and his first law of motion. The first law implies that an automobile rounding a curve has a mind to keep going straight ahead even though the wheels may point elsewhere. Friction keeps the tires on the road, causing the car to curve.

In a northern suburb of Detroit, some quasi-engineers designed a curved entrance to I-75. It has a very wide curvature but no bank at all. So when ice or snow reduced the coefficient of friction, cars going around the curve would shoot off into the ditch (At times, the ditch would be mistaken for a used car lot). Careful drivers would go very slow when rounding the curve; they would also sail into the ditch but more slowly. You see, without friction, velocity doesn't matter—it's friction or it's the ditch. To increase friction, the bank around the curve should be slanted or angled. Too bad Detroit can't recall bad freeway entrances as they do cars. The Michigan Department of Transportation is often the target of successful lawsuits based on negligent highway design.

6. *Holder v. Travelers Ins. Co.*, 159 So.2d 292 (La.App. 1964), an automobile negligence wrongful death action, employed Newton's First Law of Motion, banking and the coefficient of friction to a determine whether or not one of the vehicles involved had exceeded the posted speed limit for a curved portion of highway where a collision occurred. The opinion, affirming a judgment for the defendants, stated in part:

> Admittedly, the normal skidding direction of both vehicles, occasioned by the application of centrifugal force outward from the curve, would be toward the outside of this curve and the skidding, if any took place, would normally be toward the outside of the curve. Therefore, if the Wheeling vehicle [Defendant's vehicle] or any part thereof skidded in making this ninety degree curve, its direction of skid would be toward the east-bound lane and into or in the direction of the passing Holder vehicle. Likewise, there could be no doubt but that the normal direction that the Holder vehicle [containing Plaintiffs' deceased] would skid would be toward the outside or shoulder of the road in obedience to Newton's first law of motion.
>
> There are at least two factors which keep vehicles which are negotiating the circular motion around a curve on the curve and which keep it from following the outward force toward the outside of the curve. The first is the banking or superelevation of the highway itself, and the second is the friction between the tires of the vehicle and the surface of the road. Thus, if the speed of the Wheeling vehicle exceeded the speed at which the counter forces of these two factors were overcome by the centrifugal force (which seeks to have the vehicle follow Newton's first law of motion), then the vehicle would skid toward the outer or eastbound lane of said highway.

7. A potentially tricky issue involves only the difference between velocity and speed. While the two are closely related, speed is nothing more than distance divided by time ($s = d / t$). Velocity is similar, but it is a vector quantity as opposed to a scalar. A scalar is nothing more than a simple "number." A vector, however, combines speed with direction. For example, a retarded dog might chase its tail around in a circle at a very fast speed, but if it ends up where it started, its velocity would be zero. So, don't fall for the trick where someone bets you that he can predict the exact velocity of the winner of the next Indy 500. It will always be zero.

§ 7.20 Thermodynamics

Whereas Newton's Laws deal primarily with mechanics and the physical movement of objects, the Laws of Thermodynamics (which are every bit as fundamental and important) concern less visible but equally critical matters involving temperature and energy.

There are three basic laws. The zeroeth law, obviously an afterthought, mirrors the basic transitive law of mathematics: If A = B and B = C, then it is clear that A = C.

> **0. Zeroeth Law (Transitive Relation):** Two thermodynamic systems are each in thermal equilibrium with a third, then they are in thermal equilibrium with each other.
>
> **1. First Law (Conservation of Energy):** Energy can be neither created nor destroyed. It can only change forms. In any process in an isolated system, the total energy remains the same. $\Delta U = Q - W$ (The change in the internal energy of the system = Added heat - Work expended by the system).
>
> **2. Second Law (Entropy):** Entropy > 0. You might think of entropy as a process whereby matter spreads out, slows down and cools off.
>
> **3. Third Law (-273.15 Celsius):** As the temperature approaches absolute zero, the entropy of a system approaches a constant minimum.

§ *7.21 The First Law: The Conservation of Energy*

The First Law, also known as the *Law of The Conservation of Energy* is one of the most significant and far reaching principles of physics. Simply stated: "Energy can neither be created nor destroyed, but rather can only be transformed from one state to another." The same principle holds true with respect to matter which, like energy, can neither be created nor destroyed. The conservation principle has many important interdisciplinary consequences. Several of Einstein's theories and discoveries were a direct result, not the least of which was his General Theory of Relativity (See *infra*, § 8.14).

On the more practical side of the spectrum is a determination of the pre-crash velocity of vehicles involved in a crash. The proper assignment of fault often hinges upon such information. In § 7.14 we considered equations based upon skid marks to determine vehicular velocity. But nice linear skid marks are not always reliable or available. Vehicles may be pushed sideways, or become temporarily airborne thereby disrupting an otherwise smooth linear skid mark. Where curves are involved, the physics may become slightly more complex.

In the event that experts are unable to determine the pre-crash velocity from skid marks alone, the law of the conservation of energy provides another method. It tells us in particular that the total energy of the system prior to the collision must equal the energy of that system immediately after the collision, plus or minus any work (or energy) added or subtracted as the case may be. In § 7.13 we encountered the following important law of physics which tells us the pre-crash energy:

$$E = \frac{1}{2} m v^2$$

The E in the above equation refers to Kinetic Energy, m = the mass of the subject vehicle, and lastly v = the velocity of the vehicle.

The mass of the vehicle as we discussed earlier, will equal its weight multiplied times the acceleration attributable to gravity (32 feet/second^2). The reconstruction expert should include the weight of the driver and passengers and any pieces or parts that may have become separated as a result of the impact.

Other factors can be obtained from a variety of sources.[5] The energy used or dissipated prior to the collision and after the impact (the latter is sometimes referred to as "separation movement") are acquired by the expert from a variety of sources including skid marks or the lack thereof. Much of the energy will be expended in the process of damaging, crushing, compressing, bending and breaking different parts of the subject vehicle. From a detailed examination of parts, the reconstruction expert will establish a crush profile based upon their deformation following impact.

ADMISSIBILITY OF ACCIDENT RECONSTRUCTION TESTIMONY

Always remember that the admissibility of testimony by traffic reconstruction experts is not without its controversy.[6] The use of such an expert is never a foregone conclusion, particularly when the factors are abstruse, abstract or difficult to measure. Measuring the energy of a system from the factors considered above is particularly difficult and, therefore, always subject to challenge.

Although currently, accident reconstruction experts are permitted, subject to *Daubert* hearings, the courts were slow to accept what was once derogatorily referred to as "accidentologists."[7] With the promulgation of Rule 702 of the Federal Rules of Evidence, the judicial gatekeepers, both state and federal have opened the doors to admissibility (See *supra*, § 1.22).

§ 7.22 The Second Law: Entropy

The Second Law of Thermodynamics states that heat cannot naturally be transferred from a colder to a hotter body. As a result, natural processes that involve energy transfers have only one direction of movement—from hot to cold. This law also predicts that the entropy of an isolated system always increases with time. Entropy is the measure of disorder or randomness of energy and matter in a system.

In conjunction with the Second Law, both energy and matter in the Universe are becoming less useful as time goes on. Perpetual motion is impossible; so basically our Universe is constantly winding down and losing its energy. A common expression of the Second Law is embodied in this equation, which suggests that entropy (disorder) is increasing as opposed to decreasing: $\Delta s > 0$.

Perfect order in the Universe allegedly occurred the instant after the Big Bang when energy, matter and all of the forces of the Universe were unified. Thereafter, as a result of entropy, there has been a decrease in knowledge that can be obtained about any given system. Consider, for example, an exploding soda bottle (such as the one in the case that follows). We have less knowledge about the separate fragments of a broken bottle than about an intact one, because when the fragments are separated, one does not know exactly whether they will fit together again, or whether perhaps there is a missing piece.[8]

---

5. See Baker, *Open Forum: Scientific Reconstruction of an Automobile Accident*, 25 Ins. Counsel J. 438 (1958); Zurmuhlen, *Automobile Accident Reconstruction*, 18 Trial Law. Q 49 (Winter 1987).

6. The proponent of such evidence is likely to encounter the hurdles of a Daubert hearing. *Daubert v. Dow*, 509 U.S. 579 (1993). See also *Kumho Tire Company, Ltd. v. Carmichael*, 526 U.S. 137 (1999).

7. See Zachery v. Wheeler, 511 Fed.Supp. 591 (E.D., Tenn., 1981).

8. "Whatever it is that hits the fan will not be evenly distributed."—Roger Angell.

A.B. Chapman v. Redwine

Supreme Court of Colorado, 149 Colo. 515, 370 P.2d 147 (1962)

HALL, Justice.

The parties appear here in reverse order to their appearance in the trial court. We refer to them as they there appeared or by name. The plaintiff in her complaint, filed December 24, 1957, alleges that on July 2, 1955, the defendant, Rocky Mountain Beverage, Inc., herein referred to as Beverage Co., was engaged as a bottler and distributor of a carbonated beverage known as Vess Cola; that at that time Chapman owned and operated a grocery store known as Evans Super Market, located at 527 West Evans Avenue, Denver, Colorado; that on said date plaintiff entered said store for the purpose of buying merchandise. While plaintiff was standing near a display rack in said store on which bottles of a carbonated beverage known as Vess Cola were displayed, one of the bottles of said carbonated beverage known as Vess Cola, which bottle was on said display rack, exploded and caused pieces of glass to be projected through the air. One or more of these pieces of glass struck plaintiff causing the injuries hereinafter described.

At no time did plaintiff by her actions or omissions do anything, or omit to do anything, which proximately caused said explosion of said bottle of Vess Cola. (Emphasis supplied). Plaintiff further alleges that the Beverage Co. bottled the bottle which exploded; transported the bottle which exploded to Chapman's market, placed said bottle which exploded on the display rack in Chapman's market; prior to and at the time of the explosion of the bottle of Vess Cola the display rack was owned, maintained, and stocked by the Beverage Co.

She further alleges that 'this bottle of Vess Cola' which exploded, which was on the display rack with other bottles of carbonated beverages, was being offered for sale by Chapman.

Next plaintiff alleges that the explosion of the bottle of Vess Cola was caused by the negligence of the Beverage Co. in: "[I]mproperly bottling, inspecting, handling, storing, transporting, distributing or placing upon the display rack the said bottle of Vess Cola; or the explosion and injuries were directly and proximately caused by the negligence of defendant Rocky Mountain Beverage, Inc., or defendant A. B. Chapman, or both defendants, in that the bottle of Vess Cola which exploded was improperly handled or offered for sale by either or both of the defendants after the bottle was placed on the display rack."

Next she alleges that defendants at all times had exclusive possession and control of the bottle which exploded, and that all of the facts and circumstances of the bottling, etc., are exclusively within the knowledge of defendants and beyond the knowledge of plaintiff. Then follow allegations of plaintiff's injuries, for which she seeks damages in the amount of $25,000. The defendants filed separate answers in which they admit many of the allegations of plaintiff's complaint, but each denies negligence and each denies that any bottle exploded. The case was tried to a jury which returned a verdict in favor of the plaintiff and against both defendants, and on which judgment was entered.

Defendants are here by writ of error seeking reversal.

There is no substantial conflict in the testimony. The plaintiff testified that she and her husband went to the Chapman store sometime after 7:30 P.M., on July 2, 1955; she was carrying two cartons, one in each hand, each containing six empty pop bottles. She proceeded through an aisle for the purpose of purchasing some 'Dr. Peppers'; on one side of the aisle was a fruit rack, on the other side beverage racks; the Dr. Pepper bottles were at the end of the fruit rack, a fact which she had not yet discovered. She was followed down the aisle by her husband, who was ten or twelve feet behind her. Not seeing the Dr. Peppers, she looked toward her husband, turning her head, and remarked to him that she did not see any Dr. Peppers. She then heard a noise behind her which sounded like an explosion.

She then realized her left leg was cut just above the ankle. She observed glass and liquid behind her on the aisle floor. There was no liquid on her dress; she did not see any bottle explode; she did not observe glass or liquid on the display rack or other place except on the aisle floor.

Harlan Redwine, plaintiff's husband, testified that he was following his wife in this aisle; that he was about twelve to fourteen feet behind her; she was looking down at the 'six-packs'; it was a hot day; plaintiff turned to him and spoke, and he glanced down and as he glanced down he heard an explosion and his wife scream and he went to help her, and in doing so stepped on glass and liquid in the aisle. He, together with Chapman, examined the display rack, and there was no broken glass or liquid on any of the shelves of the rack. On the concrete floor in the aisle, between where he and his wife were standing at the time of the 'explosion,' they found the bottom of a Vess bottle, the neck with cap attached, broken glass and liquid confined to an area about three feet in diameter near the point where plaintiff was standing.

The foregoing is a fair representation of the testimony offered to prove the allegation of the complaint that a bottle 'on said display rack exploded.' Conditions existing in the aisle near where plaintiff was standing when injured, coupled with conditions on the display rack where the Vess bottles were displayed immediately following the so-called explosion, render testimony of the plaintiff that 'I heard this loud explosion'—the noise 'sounded like a loud explosion'—'there was a loud explosion—or something—an awful loud noise,' and testimony of her husband that 'I glanced down, and as I glanced down I heard an explosion'—incredible and insufficient to support any finding of the spontaneous explosion of a bottle on the rack or elsewhere.

It is common knowledge and within the very nature of things that when and if a bottle filled with liquid resting on a shelf explodes, glass fragments, liquid, other bottles, broken or knocked over, remain on the shelf. On the happening of such an event one does not find all of the glass fragments and liquid in an area about three feet in diameter on a concrete floor some feet away from the shelf where the bottle was resting and where the explosion is alleged to have occurred. The location of the glass and liquid tells the true story much more convincingly than the cry of 'explosion' on which plaintiff predicates her whole case.

To accept plaintiff's story that a bottle exploded on the shelf, one would have to ignore simple facts of life, shut one's eyes to reality, substitute fiction and fantasy for fact, do a complete overhaul job on the law of gravity, and live in a land of make believe. Not all noises are explosions. A filled bottle explodes when the inside pressure is so great as to exceed the containing strength of the bottle.

Calling a noise an explosion does not make it such, though it may be some evidence thereof. More tangible evidence is found in the flying glass and liquid and disturbance at the point where the explosion occurs.

Plaintiff's counsel urge that: "The only reasonable inference that could be drawn from all the evidence is that the bottle of Vess did explode spontaneously in the display rack causing Mrs. Redwine's injury." This statement is contrary to all the physical facts. No such inference is reasonable or warranted by the evidence, and in the light of the undisputed facts is incredible.

All of the physical surroundings point indubitably to the fact that a bottle fell or was knocked from a shelf to the concrete floor and broke, scattering the glass and contents over a limited area of the floor.

In evaluating evidence such as that presented by plaintiff, we find the general rule well stated in 20 Am.Jur. 1033, Evidence, §§ 1183, 1184: "The mere fact that testimony given by a witness in support of an issue is not plausible does not destroy all probative value. Where, however, the testimony of a witness is incredible, inherently or physically impossible and unbelievable, inherently improbable and irreconcilable with, or contrary to, physical facts and common observation, and experience, where it is so opposed to all reasonable probabilities as to be manifestly false or is contrary to the laws of nature or to well-known scientific principles, or where it cannot be said to amount to substantial evidence of the facts testified to or accepted as a basis for liability, it is to be disregarded as being without evidentiary value, even though uncontradicted."

Testimony of a witness, although not directly controverted, which is opposed to the unquestioned laws of nature that lie within the court's judicial knowledge or which is clearly in conflict with scientific principles as established by the laws of physics or mechanics is of no probative value. In McLennon v. Whitney-Steen Co., 63 Colo. 568, 167 P. 771, this court said: "He knew the elevator was in constant operation carrying supplies, and was apt to move at any moment. The floors were only nine feet apart, and the cage platform in plain view but a few feet from his head; also the absence of the rope in the shaft advised him that the elevator was above. Plaintiff's testimony that he looked up and did not see the elevator cannot be accepted as against the physical facts that it was there, in plain view, only a few feet from his head. The legal effect and value of such evidence may be determined by the court as a matter of law." In Jacobsen v. McGinness, 135 Colo. 357, 311 P.2d 696, this court, in disregarding McGinness' sworn testimony, that she was on her side of the road and driving slowly, said: "The physical facts speak more loudly than words in a situation such as that presented by the record before us. The impact of the McGinness and Jacobsen cars raised the Jacobsen car into the air and pushed it into the path of the Holthusen automobile.

Despite Miss McGinness' testimony that she removed her foot from the accelerator and applied her brakes, the undisputed facts are that her car was traveling at a high rate of speed; ran head on into the Jacobsen automobile on the wrong side of the highway; pushed the Jacobsen vehicle back a distance of eleven feet and into the path of the Holthusen car." See, also, Snedden v. Summer, 138 Colo. 144, 330 P.2d 530. The following language of this court, taken from the opinion of Mr. Justice Moore, in Winterberg v. Thomas, 126 Colo. 60, 246 P.2d 1058, is applicable to the facts disclosed by the record:

"A court cannot accept as true that which the indisputable evidence demonstrates is false." Lessig v. Reading Transit & Light Co., 270 Pa. 299, 113 A. 381, 382. The applicable rule was quoted, with approval in that case from [volume 1,] Elliott on Evidence, [page 38,] section 39, as follows:

> "Even though it [an appellate court] may not be authorized to weigh evidence and pass upon the facts, it may, and should, so use its juricial [sic] knowledge as to bring about justice. Thus, there are often undisputed physical facts clearly shown in evidence, and by applying to them a well-known law of nature, of mathematics, or the like, it is demonstrated beyond controversy that the verdict

> or finding is based upon what is untrue and cannot be true. In such case, it is very generally held that the appellate court should take judicial notice of the law of nature or mathematics or quality of matter, or whatever it may be that rules the case and apply it as a trial court should have done."

In Brice v. Miller, 121 Colo. 552, 218 P.2d 746, we find the following pertinent language: "The bare statement that the truck was 'stopped' upon the highway, when considered in the light of all the surrounding circumstances, does not rise to the dignity of evidence sufficient in itself to be the sole basis for liability of the defendants." In Wheeler v. Des Moines City R. Co., 205 Iowa 439, 215 N.W. 950, 55 A.L.R. 473, it was said: "The only evidence on which to base this charge [high speed] is that of the plaintiffs above set forth that the street car started with a jerk, and got up a speed of 15 miles an hour in 5 feet. Her opinion that the car got up a speed of 15 miles an hour in 5 feet is so wanting in foundation, so inherently improbable, and so irreconcilable with the undisputed physical facts, that the court would be without justification in accepting it as a basis of liability."

Evidence offered by plaintiff to prove that a bottle exploded on a shelf of the display rack was insufficient to warrant the trial court in submitting the matter to the jury on the theory of an exploding bottle and it was error to instruct the jury on the doctrine of res ipsa loquitur as applied to exploding bottles. The record shows that the Beverage Co. had neither possession nor control of the premises in which plaintiff received her injuries, or the contents thereof. The record clearly shows that it had not been in the store during the five days immediately preceding the occurrence which is the basis of this action, hence, in the absence of proof of an exploding bottle, there is no evidence of any negligence on the part of the Beverage Co. There being no evidence to establish the spontaneous explosion of a bottle, it follows that the plaintiff's claim against the Beverage Co. must be dismissed.

From the record before us we find a very different situation with reference to the plaintiff's claim against Chapman. The evidence is undisputed that plaintiff was an invitee in Chapman's store, that Chapman was in full, complete and exclusive possession of the building and its contents. As pointed out above, all of the testimony and physical surroundings at the time and place of the plaintiff's injuries indubitably establish the fact that a bottle, under the exclusive possession and control of Chapman, located in a store in his exclusive possession and control, fell, or was knocked from a shelf, and broke, causing injuries to the plaintiff.

Such evidence warrants the conclusion that plaintiff has met the requirements of proof necessary to raise a presumption of negligence as against Chapman under the doctrine of res ipsa loquitur, and the court should have instructed the jury on the doctrine as applied to the foregoing facts rather than on the unproven allegation of an exploding bottle. Gylling v. Hinds, 122 Colo. 345, 222 P.2d 413; Scott v. Greeley Joslin Co., 125 Colo. 367, 243 P.2d 394; Weiss v. Axler, 137 Colo. 544, 328 P.2d 88.

The judgment is reversed and cause remanded with directions to dismiss the complaint as against the Beverage Co.; to grant a new trial as to the defendant Chapman, permitting the parties to amend their pleadings as they may be advised, and to make final disposition of the action consistent with the views herein expressed.

SUTTON, Justice (dissenting).

I respectfully dissent for the reasons hereinafter expressed. Clarice Redwine claimed that her personal injuries, which occurred in defendant Chapman's grocery store, were caused by flying glass from what she believed and alleged to be an exploding bottle of pop which had been bottled by defendant Rocky Mountain Beverage, Inc. and placed by it on Chapman's shelf.

The theory of her case as plead and tried was that of res ipsa loquitur. I have examined the record before us and cannot agree with the statements made in the majority opinion to the effect that Mrs. Redwine's evidence was incredible and insufficient to support any finding of the spontaneous explosion of a bottle on the rack or elsewhere. If the entire evidence were as set forth in that opinion it would be correct; however, when based upon the non-conflicting evidence before the court, most of which supports Mrs. Redwine's thesis, and as determined by a jury, I believe the majority opinion to be incorrect.

Let me briefly review the evidence. Both Mrs. Redwine and her husband testified to the events set forth in the opinion; however, in addition there were the following uncontradicted facts: (1) There is evidence that one of Chapman's employees, a Mrs. Meade, at various times prior to the accident had moved, changed and dusted the pop bottles on the shelves next to where Mrs. Redwine was standing at the time of her injury. (2) The same type of control was exercised over the bottles by defendant Rocky Mountain Beverage, Inc.'s driver when he made his usual weekly deliveries. (3) The bottle was one furnished by Rocky Mountain. (4) Almost immediately after the accident, before Chapman and the others who testified as to the location of the debris on the floor could check the shelves and area in question, a boy who worked for Chapman had swept up the mess. (5) A week or so after the accident a Rocky Mountain Beverage, Inc. delivery driver found the shelf from which it appeared the bottle had come to be 'sticky and dirty' and washed it off. (6) The cashier as well as Mrs. Redwine and her husband testified as to the noise of a bottle breaking. The only difference was as to whether it sounded like an 'explosion' or a 'crash'. Weighing the evidence before the court was the duty of the jury under proper instructions. It did so, and there being support for its verdict it is not our function or power to set its solemn declaration aside because we may disagree therewith.

Based on all the evidence I cannot say that it was physically impossible or technically impossible for the bottle to have exploded. Evidently the jury was of the same view. Clearly Mrs. Redwine was in no position to establish the cause of the explosion or of the fall, whichever it was, that resulted in her injury. It seems to me that this is the type of case described in Fries v. Ritter, 381 Pa. 470, 473, 112 A.2d 189, 190 (1955) (a truck accident case) where it was said: "Where circumstantial evidence is relied upon to prove negligence the evidence must be such as to enable the jury to conclude—not by conjecture or guess but—as a reasonable and legitimate inference that the accident was caused by the negligence of the defendant.

In order to do so the evidence must clearly and sufficiently describe or picture the happening of the accident in such a manner that the only reasonable inference and conclusion from the facts and circumstances which were proved is that defendant was negligent."

The Pennsylvania court applied this same rule later in Klimczak v. 7 Up Bottling Co. of Philadelphia, 385 Pa. 287, 122 A.2d 707 (1956). In Loch v. Confair, 372 Pa. 326, 93 A.2d 451 (1953) the court stated: "Viewing realistically the situation presented by the circumstances of the present case, it is obvious that, so far as the evidence discloses, the wife plaintiff was not injured through any fault of her own, and therefore, if the occurrence of the accident was due to negligence on the part of either of the defendants, plaintiffs should be entitled to redress. It is clear too that the breaking of the bottle could have resulted only from a defect in the bottle itself, or from an excessive internal pressure due to over-carbonation, or from the subjection of the bottle to some extreme atmospheric or temperature changes or some mishandling, as, for example, the striking of it by a hard object. Manifestly it would be entirely beyond the ability of the plaintiffs to ascertain and establish which of these possibilities was in fact the cause of the accident, whereas the defendant Beverage Company could readily explain the equipment and the methods employed by it in bottling the ale and the defendant A. & P. Company was equally in a position to explain the manner in which it handled, displayed and protected the bottles placed on its shelves for sale. It would seem, therefore, notwithstanding the limitations on the applicability of the doctrine of res ipsa loquitur and exclusive control previously referred to, that reason and justice alike should entitle plaintiffs to the benefits of those methods of establishing a prima facie case."

The majority opinion indicates that perhaps the bottle fell or was knocked from the shelf and exploded as it hit. That might have happened but again I point out that this court is not the trier of the facts. In this connection upon retrial it is said that the bottler should be dismissed from the suit and the case, if re-tried, should proceed on the res ipsa loquitur theory as against the store owner only. Naturally, taking the position that the judgment should be affirmed I cannot agree with any such limitation.

In this connection I point out that even though the normal type of joint control of the premises by the two defendants is not shown, nevertheless, the facts call for application of the doctrine of res ipsa loquitur as to both of them in this type of case. My reasons are: 1. The plaintiff was in no position to establish the cause of the explosion or fall of the bottle which resulted in her injury. 2. The bottler could show that the bottle did not explode, if such were the case, and, the store owner could show, if such were the case, that the bottle did not fall from the shelf because of safe construction, ample aisle width and so forth. 3. A suit against joint tort feasors under the doctrine even though finally only one was held liable, has been upheld by our court. Beadles v. Metayka, 135 Colo. 366, 311 P.2d 711 (1957), (doctor held liable when he, hospital and anesthetist were joined as defendants when a patient fell from an operating table). Compare Ybarra v. Spangard, 25 Cal.2d 486, 154 P.2d 687, 162 A.L.R. 1258 (a doctor, nurse, hospital, patient case). My position is that under the facts and pleadings of this case the jury was entitled to find that the bottle exploded and, in the absence of satisfactory proof exonerating them of negligence, could find both defendants liable.

It is well to point out here that the question of liability in cases of this type has been the subject of numerous lawsuits and at least seven theories of decision appear to range the spectrum according to Dean Roscoe Pound in his article on 'The Problem of the Exploding Bottle', Vol. XL, Boston University Law Review, No. 2, Spring 1960. Suffice it to say that some jurisdictions still deny the application of res ipsa loquitur holding that negligence must be proved; others that res ipsa loquitur cannot apply to the retailer; and, the doctrine of implied warranty has been advanced as has the theory of a dangerous instrumentality, together with variations in these theories.

Of course, if my view prevailed we would also have to rule upon other matters raised by the writ of error and by the assigned cross-error. As to these, and without further explanation, I can say I have examined the grounds urged and can see no reason to reverse the judgment; however, an affirmance would require a discussion of several issues raised.

To summarize, I am persuaded that Mrs. Redwine was, on this record, entitled to the presumptions afforded her under the doctrine of res ipsa loquitur as applied by the trial court under the holding in Weiss v. Axler, 137 Colo. 544, 328 P.2d 88 (1958). The evidence showed her presence at the scene of the accident, on premises under the ownership and control of Chapman but serviced as well by Rocky Mountain who furnished the bottle in question; and, there is no evidence Mrs. Redwine knocked the bottle off or caused it to fall. This entitled her to invoke the benefit of the doctrine. Her belief that the bottle exploded is immaterial. Even if she had not shown prima facie evidence sufficient to warrant the conclusion that the bottle did explode, her evidence and pleadings did place the duty of explaining what happened on the defendants below. No such evidence being in the record I would affirm the judgment.

In re Speas

273 Fed.Appx. 945, C.A.Fed., 2008.

PER CURIAM.

Danny E. Speas appeals the decision of the United States Patent and Trademark Office, Board of Patent Appeals and Interferences, which affirmed rejections of his patent application under 35 U.S.C. §§ 101 and 112. In re Speas, No.2007-0044 (Bd. Pat.App. & Interferences, July 30, 2007). We affirm.

The second law of thermodynamics states that in a system without an outside supply of energy, a process can only occur if it increases the level of entropy in the system. For example, a scale with equal weights on each side may balance without outside intervention, but outside intervention is required to move a balanced scale such that one side is above the other. Speas claims as his invention "any and all devices and systems which operate in such a manner as to violate the second law of thermodynamics as it is currently understood and accepted as inviolable by a majority of the worldwide scientific community," and "any and all devices and systems which are adapted for converting thermal energy into other energy forms by contacting a heat source without the necessity of also contacting a thermal medium of lower temperature."

According to the specification, the invention raises a ferrofluid out of a reservoir by a magnetic column into a mass. The ferrofluid then escapes a "gradually decreasing magnetic field which holds it up against gravitational force" and is drawn away via tubular element by a capillary force aided by Brownian motion. At the end of the tubular element, drops of this ferrofluid accumulate and drop back into the reservoir below, spinning a wheel along their downward paths. Thus, the movement of the ferrofluid imparts mechanical energy upon the wheel. Speas claims that because this ferrofluid is moved and adds energy to the paddle wheel "without input into the system other than ambient thermal energy," it is proof that the second law of thermodynamics is not inviolate-an object of the invention.

With such broad claim language, the examiner rejected these two claims and their four dependent claims because they were not supported by an asserted or well established utility, and because the specification did not enable any person having ordinary skill in the art to utilize the invention commensurate in scope with the claims. Speas appealed the latter claim to the board, which affirmed the rejection.

This court will uphold the factual findings of the board where they are supported by substantial evidence. In re Gartside, 203 F.3d 1305, 1316 (Fed.Cir.2000). Substantial evidence "is something less than the weight of the evidence but more than a mere scintilla of evidence." In re Kotzab, 217 F.3d 1365, 1369 (Fed.Cir.2000). Substantial evidence "means such relevant evidence as a reasonable mind might accept as adequate to support a conclusion." Consol. Edison Co. v. NLRB, 305 U.S. 197, 229, 59 S.Ct. 206, 83 L.Ed. 126 (1938).

The board's decision was supported by substantial evidence. As it found, the specification does not enable a person having ordinary skill in the art to make "any and all devices and systems which are adapted for converting thermal energy into other energy forms by contacting a heat source without the necessity of also contacting a thermal medium of lower temperature." This particularly broad claim is nearly limitless. However, to be enabled the claim must be described by a commensurately broad description encompassing how to utilize the invention to create any and all such devices and systems. See In re Hyatt, 708 F.2d 712, 714 (Fed.Cir.1983) (citing O'Reilly v. Morse, 56 U.S.(115 How.) 62, 112, 14 L.Ed. 601 (1853)). Speas does not include such a description.

§ *7.23 The Third Law: Absolute Zero*

Just as the speed of light in a vacuum represents the "speed limit" of the Universe, so too is absolute zero the "cold limit" of the Universe. It represents the absence of all heat.

The Third Law of Thermodynamics states that if all the thermal motion of molecules (kinetic energy) could be removed, a state called absolute zero would occur. Absolute zero results in a temperature of 0 Kelvins or -273.15° Celsius. There is nothing colder. The Universe will eventually attain absolute zero when all of its energy and matter are randomly distributed across space. The current temperature of empty space in the Universe is about 2.7 Kelvins.

Entropy, the focus of the Second Law of Thermodynamics, is closely related to temperature. At absolute zero, the point at which all molecular motion ceases, the change in entropy becomes zero. Expressed mathematically:

$$\lim_{\Delta T \to 0} \Delta S = 0$$

Although scientists have managed to achieve laboratory temperatures extremely close to absolute zero, it has been asserted by many that it is essentially impossible by any known procedure to reduce the temperature of a system to absolute zero in a finite number of operations. This, of course, is tantamount to stating that we will never reach the temperature.

It is evident that heat plays a very important role in thermodynamics, which in turn plays a very important role in the movement of objects and heat. Heat energy, which is measured in terms of calories, has many definitions, ranging from the scientific to those employed by phony weight loss centers.

Scientifically, a **calorie**, also known as a "gram calorie," is defined as the amount of energy required to heat one gram of water by one degree Celsius. The energy is approximately 4.2 joules. A kilogram calorie is equal to one thousand gram calories, or 4.2 kilojoules.

---

§ 7.30 Electromagnetic Energy

One of the most dramatic events involving pure numbers and their ability to uncover the mysteries and secrets of the universe involved a set of calculations by James Clerk Maxwell. For many years scientists were trying to discover the mysterious nature of light. Does it consist of particles or waves? Does it possess mass or does it consist of pure energy? Even more basic—What is it? As a scientist interested solely in electromagnetic energy, Maxwell could not care less about the nature of light. He did, however, conduct extensive experiments and calculations involving the force of static electricity and the speed of electromagnetic waves, which led him to the following equation: $C = 1 / (\varepsilon_0 \mu_0)^{1/2} = 3 \times 10^8$ m/s. The left side of the equation deals solely with the properties of electricity; the right side is velocity. Does it look familiar? Maxwell, who pondered about the number at length and thought "I've seen this before," eventually recognized it as the speed of light. One thing led to another until he discovered that this mysterious thing called *light* was nothing more than a member of the electromagnetic spectrum. His measurements and discoveries led to the construction of a set of extremely important equations that pretty much sums up the nature of electromagnetic waves.[9]

§ 7.31 *Maxwell's Laws*

Newton's laws provided a set of principles that govern the mechanical and gravitational forces throughout the Universe. The Laws of thermodynamics provided a set of rules to govern temperature and pressure throughout the Universe; now it is time to round out the scheme by presenting a set of laws, assembled by Maxwell, to govern the rules involving electromagnetic energy. The topic provides a logical cross-over into the realm of modern physics.

Let E be the electrical field intensity; B is the magnetic field density; ρ is the total charge density; ε_0 is the universal electrical constant; and J is the total current density. "$\nabla \cdot$" is the divergence operator and "$\nabla \times$" is the curl operator. A familiarity with calculus would help.

1. **Gauss's Law:** $\nabla \cdot E = \rho / \varepsilon_0$
2. **Gauss's Law for Magnetism:** $\nabla \cdot B = 0$
3. **Faraday's Law of Induction:** $\nabla \times E = - \partial B / \partial t$
4. **Ampere's Law:** $\nabla \times B = \mu_0 J + (\mu_0 \varepsilon_0) \partial E / \partial t$

The first equation, named after physicist-mathematician Carl Friederich Gauss, describes the strength and shape of an electric field generated by a charged particle or object. Similar to Newton's equation for gravitational strength, Gauss's equation presents an *inverse square law*, meaning that as the charged particles move farther apart, their attraction to one another diminishes very sharply (i.e., the distance factor is squared). For example, the electromagnetic strength of the cell phone that you hold 1 millimeter from your ear would be 9 times the strength of one held 4 millimeters away.

The second and third equations pertain to electromagnetic induction, a concept that is indispensable to the generation of electrical power. The second equation describes how currents produce magnetic fields, while the third equation maps out the process for generating electrical current from a magnetic field. It also describes how light and electricity manage to travel from point a to point b at the speed of light. In a nutshell, this equation describes how a collapse of an electromagnetic field propels a current forward, which in turn alters the surrounding electromagnetic field.

The fourth equation (Ampere's Law) describes the shape and strength of a magnetic field. It establishes and explains a number of magnetic characteristics. Magnetic field lines, for example, are always closed. All magnets must have both a north and a south pole; if you divide a magnet in half, each of the remaining pieces must likewise have a north and south pole. Even though quantum physicist Paul Dirac predicted the existence of monopoles, none have been found to exist.

9. Einstein, who incorporated Maxwell's equations into his theories of relativity, considered the scientist's achievements to be on the same level as those of Newton, stating: "*Any intelligent fool can make things bigger and more complex, it takes a touch of genius and a lot of courage to move in the opposite direction.*"

Reynard v. NEC Corp.

887 F.Supp. 1500 (M.D. Fla., 1995)

Nimmons, District Judge.

This cause comes before the Court on several pending motions: (1) Motion of Defendant GTE Mobilnet of Tampa, Inc. to Dismiss Plaintiffs' Complaint Pursuant to Rule 12(b)(6) and Plaintiffs' responses in opposition thereto; (2) Defendant, GTE Mobilnet of Tampa, Inc.'s Motion to Dismiss Plaintiffs' Complaint Against Defendant, GTE, Inc.; (3) Defendant, NEC America, Inc.'s Dispositive Motion for Summary Judgment and Plaintiffs' response in opposition thereto; (4) GTE Mobilnet's Motion for Summary Judgment and Plaintiffs' response in opposition thereto; (5) Plaintiffs' Motion that Discovery Be Extended, That Case Be Designated Track Three Case and Defendant NEC America's response in opposition thereto; (6) Plaintiffs' Motion for Relief Under Fed.Civ.R.P. 56 and Defendant, NEC America, Inc.'s response in opposition thereto; (7) Allied National, Inc.'s Motion to Intervene and Defendant NEC America, Inc.'s response thereto.

A. Summary Judgment Standard

The Court will enter summary judgment only if the evidence shows "that there is no genuine issue as to any material fact and that the moving party is entitled to a judgment as a matter of law." Fed.R.Civ.P. 56(c). On the issue of materiality, "the substantive law will identify which facts are material." Anderson v. Liberty Lobby, Inc., 477 U.S. 242, 248, 106 S.Ct. 2505, 2510, 91 L.Ed.2d 202 (1986). Material facts are those over which disputes "might affect the outcome of the suit under the governing law." Id. at 248, 106 S.Ct. at 2510.

B. Discussion

The Plaintiffs' Complaint alleges that Susan Elen Reynard had used a cellular telephone sold by Defendant NEC America, Inc. and that Defendant GTE Mobilnet of Tampa, Inc. "engaged in the operation of the microwave system required to be used with the cellular telephone." They allege that the cellular telephone was defective and negligently designed, and that electromagnetic radiation (EMR) from the phone initiated or accelerated the growth of a tumor in her brain, which eventually led to her death. The Plaintiffs sue under the wrongful death statute, §§768.16—768.27, Florida Statutes, alleging that Susan Reynard's portable cellular telephone initiated the brain tumor. Alternatively, the Plaintiffs sue on the theory that the cellular phone accelerated and aggravated the course of an already existing tumor.

Defendant NEC America, Inc. and Defendant GTE Mobilnet of Tampa, Inc. move for summary judgment on the issue of causation. The Defendant NEC America Inc. and Defendant GTE Mobilnet of Tampa, Inc. submit the affidavits of Dr. F. Kristian Storm III, M.D. and Dr. Carl Hudson Sutton, M.D., respectively, to demonstrate the absence of medical causation in this case, and to shift the burden of presenting admissible expert testimony on medical causation to the Plaintiffs.

Dr. Storm's affidavit states that he is past chairman of the American National Standards Institute subcommittee that drafted the U.S. safety standard for exposure to EMR. The affidavit also states that Dr. Storm was familiar with or familiarized himself with the relevant medical and scientific literature, Susan Reynard's medical records, the MRI films of the brain taken in 1989 and 1990, and relevant deposition transcripts. Dr. Storm concludes in his affidavit that:

> No scientific or medical studies have shown that exposure to emissions deposited at the brain from a source such as a portable cellular telephone operating at the

> power level and frequency of the portable cellular phone alleged to have been used [by] Susan Reynard is associated with any adverse biological effects, including initiation of brain cancer or promotion of brain cancer growth....
>
> [E]pidemiology reports establish that the overall incidence of brain cancer has remained fairly constant over time and their location in the brain [has remained] random, evidencing the lack of any association [with the use of portable cellular telephones] ...
>
> Based upon my review of [various relevant] materials, as well as my extensive education, training and experience in electromagnetics and as a physician who specializes in cancer treatment, it is my opinion, within reasonable medical probability, that Susan Reynard's use of a portable cellular telephone did not cause or contribute to the initiation or promotion of growth of Susan Reynard's brain tumor ...
>
> Nothing in her medical or clinical course distinguishes her in any fashion from the thousands of patients that I have either treated with cancer or that have reported with this form of brain tumor who have not used portable cellular telephones.

The affidavit submitted by Dr. Sutton sets forth his impressive credentials indicating that he has extensive experience in researching and treating brain cancer, including, significantly, research and publications on the effects of microwave energy on brain tumors. Further, Dr. Sutton has been a member and served on radio frequency and microwave standards-setting committees from 1978 to the present time. The affidavit also states that Dr. Sutton has reviewed the medical records of Susan Reynard, which he states are the type of records customarily and reasonably relied upon by experts such as himself to form medical opinions. Dr. Sutton's affidavit further states:

My opinion within a reasonable medical probability that Susan Reynard's brain tumor was not initiated or promoted by the use of the cellular telephone is based in part on the following:

a. Susan Reynard started using the cellular telephone on about August 1988. Her first MRI was taken in March 1989 and it showed in my opinion what appears to have been a brain tumor. A subsequent MRI in May 1990 showed growth in her brain tumor. From examining the MRIs of March 1989 and May 1990, it is my opinion within a reasonable medical probability that she had the brain tumor before she ever used the cellular telephone. Thus, the cellular telephone could not have initiated or caused the brain tumor.

b. Susan Reynard's brain tumor was diagnosed as a Grade III Astrocytoma. It was located in the left parieto-occipital region of her brain. Neither the type nor location of her brain tumor is unusual. Further, Susan Reynard lived for approximately two years following her brain surgery. She received cancer treatment for her cancer after her surgery. Her survival time is within the normal range for the type of cancer she had and for the type of treatment she received. Therefore, there is nothing in the type, location, or survival time for her cancer which in my opinion, within a reasonable medical probability, would show that her cancer was promoted by the use of the cellular telephone.

Further, both Defendants assert that the deposition testimony of the Plaintiffs' experts in the state court action does not establish medical causation, and that, based on that testimony, the Plaintiffs cannot meet their burden under Daubert v. Merrell Dow Pharmaceuticals, Inc., 509 U.S. 579, 113 S.Ct. 2786, 125 L.Ed.2d 469 (1993), and on remand, 43 F.3d 1311 (9th Cir.1995). First, they point out that Dr. Jan Leestma, a neuropathologist who examined Susan Reynard's brain after her death, testified that he could not determine from the materials, information, and knowledge he had and could not assign a probability as to whether exposure to EMR from a cellular phone caused or contributed to the growth of Susan Reynard's tumor.

Second, the Defendants assert that Plaintiffs' other expert, Dr. John Holt, has expressed no opinion supporting the Plaintiffs within reasonable medical probability. The Defendants assert that Dr. Holt's testimony was unorthodox and state that Dr. Holt himself admits that his theories are not accepted by the medical and scientific community and that no clinical study has ever been done to test his theories. Further, Defendants assert that even if Dr. Holt is to be believed, the application of his theories in the instant case is worse than speculation, because such application is not even supported by his own observations. They explain that Dr. Holt has observed and believes that human cancer cells, exposed to 434 MHZ EMR at high power, died, and those exposed to MHZ EMR at low power exhibited a phenomenon he describes as fluorescence. They assert that Dr. Holt alone believes that this phenomenon is related to the increased growth rate of cancer exposed to radiation at the low power frequency of 434 MHZ. With regard to the fact that Susan Reynard's phone operated not at 434 MHZ, but between 825 and 845 MHZ, Defendants point out that Dr. Holt stated that "I don't know the answer" as to the effects of the cellular phone's frequency and that he would not "hazard a guess" because "everything is so individual in the cancer field." Further, Dr. Holt stated that he could not quantify how much of the growth rate of her tumor might be related to EMR, "[b]ecause one doesn't know what the cancer was like and her clinical progression rate was like before she was exposed to her hand-held phone," and he therefore agreed that there was no way to determine whether her outcome would have been different had she not been exposed to stray electromagnetic radiation.

In response, the Plaintiffs submit an expert affidavit of Dr. David Perlmutter, M.D. disputing the contentions of Dr. Storm and Dr. Sutton. The Plaintiffs also assert that the Defendants have attempted to ridicule Dr. Holt "by selective out-of-context excerpts from his deposition." Further, the Plaintiffs have submitted a medical journal article concerning Dr. Holt that was written by a professor, Dr. William L. Caldwell, M.D., from the University of Wisconsin. The Plaintiffs point out that Dr. Caldwell's article suggests that Dr. Holt's therapy has been effective and that phase II trials utilizing 434 MHZ radiations in conjunction with ionizing radiations have been initiated at the Indiana University School of Medicine and the Wisconsin Center for Health Sciences. The Plaintiffs have also attached as "Exhibit A" a copy of an article from a publication entitled "Microwave News" that they assert suggests that cellular phone emissions can "act as a cancer-causing agent."

Finally, the Plaintiffs have also submitted two affidavits, those of: (1) Plaintiff H. David Reynard, Jr., and (2) the Plaintiffs' attorney, John A. Lloyd, Jr., as well as one other document, the United States General Accounting Office (GAO) Report to the Chairman, Subcommittee on Telecommunications and Finance Committee on Energy and Commerce, House of Representatives. (3) The Plaintiffs have not responded to the Defendants' arguments regarding Dr. Leestma or Dr. Holt.

In evaluating the Plaintiffs' response to the motions for summary judgment, the Court first notes that in light of the evidence set forth by the Defendants, the burden is on the Plaintiffs to come forward with admissible evidence on the issue of medical causation in order to demonstrate that there is a material issue of fact that should preclude summary judgment. Clark, 929 F.2d at 608. Further, under Florida law, the plaintiff in general, of course, bears the burden of proving causation. See e.g., Reaves v. Armstrong World Industries, Inc., 569 So.2d 1307, 1309 (4th DCA 1990). The plaintiff "must introduce evidence which affords a reasonable basis for the conclusion that it is more likely than not that the conduct of the defendant was a substantial factor in bringing about the result. A mere possibility of such causation is not enough." Reaves, 569 So.2d at 1309, quoting Gooding v. University Hospital, 445 So.2d 1015 (Fla.1984). See also Hessen v. Jaguar Cars, Inc., 915 F.2d 641, 647 (11th Cir.1990) ("Florida has adopted a preponderance standard for causation in both negligence and strict liability actions; a mere possibility of causation is not enough.").

With regard to the medical journal article concerning Dr. Holt by a professor, Dr. William L. Caldwell, M.D., and the article from a publication entitled "Microwave News," the Court finds that they do not satisfy the Plaintiffs' burden of demonstrating a material issue of fact regarding causation in the instant case. First, the Plaintiffs have not discussed the manner in which the statements or conclusions of these articles support their case. Rather, the articles themselves suggest that the Plaintiffs' position is not generally accepted. Dr. Caldwell's article states, "A number of ill-conceived speculations and poorly-based claims included in [Dr. Holt's thesis, 'Cure of Cancer—A Preliminary Hypothesis'] were harmful to his image and subsequently have interfered with his being accepted now after he has accrued an exceptional clinical experience in more than 1,500 patients and has moderated his claims." Although Dr. Caldwell's article indicates that phase II trials utilizing 434 MHZ radiations in conjunction with ionizing radiations have been initiated, the Plaintiffs have not shown, and the article does not suggest, how such theory applies to cellular telephone emissions. Additionally, the article in "Microwave News" contains numerous statements indicating that the hypothesis that microwaves could act as a cancer-causing agent is uncertain. For example, Dr. Lai, one of the researchers, states in the article, "We have a long way to go before we can reach any definitive conclusions."

Further, the Court finds that the only medical evidence on causation submitted by the Plaintiffs is the affidavit of Dr. Perlmutter, which the Court finds does not establish a material issue of fact. Dr. Perlmutter's affidavit points out that the November 1994 report of the General Accounting Office is a call for research because the research findings on exposure to low level radio-frequency radiation are inconclusive. Dr. Perlmutter further indicates that he is in complete agreement with Dr. Storm's finding that Susan Reynard's tumor began many years before she began using a cellular telephone in 1988. Dr. Perlmutter also states in his affidavit that he agrees with the affidavit of Dr. Storm in its finding that "No scientific or medical studies have shown that exposure to emissions deposited at the brain from a source such as a portable cellular telephone operating at a power level and frequency of the portable cellular phone alleged to have been used (by) Susan Reynard is associated with any adverse biological effects, including initiation of brain cancer or promotion of brain cancer growth." Dr. Perlmutter states that:

> [I]t is certainly quite evident that both animal studies and human brain tumor cell studies ... provide strong inferential data that the use of this device represents a clear health hazard and would likely accelerate the growth of brain tumors in humans.
>
> It is my opinion, within a reasonable degree of medical certainty, that the use of the hand-held cellular telephone did in fact accelerate the growth of Susan Reynard's brain tumor.

Additionally, Dr. Perlmutter disagrees with several statements in Dr. Storm's affidavit. Dr. Perlmutter states that some patients with Grade III astrocytomas like Susan Reynard live substantially longer than two years and that some have even been cured of their disease, and he therefore believes that the suggestion that her two year survival after surgery was expected is inappropriate. Further, he states that brain tumors of the type at issue in this case "are clearly on the increase as described in a recent article appearing in the New England Journal of Medicine."

The Court first finds that there is no genuine issue of material fact with regard to the Plaintiffs' wrongful death claim, and that the Defendants are therefore entitled to judgment as a matter of law on that claim. In an action for wrongful death, Florida law applies traditional causation principles, which require proof under the but-for test. Tappan v. Florida Medical Center, Inc., 488 So.2d 630, 631 (Fla. 4th DCA 1986). The testimony of Dr. Perlmutter indicates that the Plaintiffs' wrongful death action fails the but-for test due to his statement that he is in complete agreement with Dr. Storm that Susan Reynard's tumor began many years before she began using a cellular telephone. Further, the Plaintiffs have not presented any evidence suggesting that but for her use of the cellular telephone, it was more likely than not that she would have survived.

Secondly, the Court finds that Dr. Perlmutter's affidavit does not create a genuine issue of material fact with regard to the claim that the cellular telephone accelerated Susan Reynard's brain tumor. It is well established that "a party may not avoid summary judgment on the basis of an expert's opinion that fails to provide specific facts from the record to support its conclusory allegations." Evers v. General Motors Corp., 770 F.2d 984, 985 (11th Cir.1985). Further, in a wrongful death action based upon allegations of negligence and strict liability:

While the expert may base his opinion on facts or data reasonably relied upon by experts in the field, and this data need not be admissible in evidence, Fed.R.Evid. 703, an expert's affidavit must nevertheless set forth specific facts in order to create an issue of fact for trial. 'Theoretical speculations, unsupported assumptions, and conclusory allegations advanced by an expert ... are [not] entitled to any weight when raised in opposition to a motion for summary judgment.'... [W]hen causation is the issue, 'courts are particularly wary of unfounded expert opinion.' ...'The uncertainty of the evidence in [toxic tort] cases, dependent as it is upon speculative scientific hypotheses and epidemiological studies, creates a special need for robust screening of experts and gatekeeping under Rules 403 and 703 by the court.' Bell v. Swift Adhesives, Inc., 804 F.Supp. 1577, 1579 (S.D.Ga.1992).

Further, Florida courts hold that the party opposing summary judgment must present nonspeculative evidence as to what extent, if any, the defendant's actions decreased life expectancy. See e. g., Noor v. Continental Casualty Co., 508 So.2d 363, 365 (Fla. 1st DCA 1987). In Noor, for example, the plaintiff was unable to present any nonspeculative evidence regarding the extent to which the doctor's failure to immediately diagnose her disease added to plaintiff's decreased life expectancy. Id. at 365. The court therefore held that the defendants were properly awarded summary judgment as a matter of law. Id.

Additionally, under Florida law, the proffered evidence on causation must meet the more likely than not standard. This standard may be met, for example, by evidence establishing that the patient "would have survived" with prompt diagnosis and surgery. Gooding v. University Hosp. Bldg. Inc., 445 So.2d 1015, 1019 (Fla.1984), citing Hicks v. United States, 368 F.2d 626, 632 (4th Cir.1966). This standard may also be satisfied with expert testimony that the patient had, for example, a seventy five percent chance of recovery with prompt treatment. Gooding, 445 So.2d at 1019, citing Hamil v. Bashline, 392 A.2d 1280, 1288 (1978). The Florida Supreme Court itself held that a plaintiff fails to meet the more likely than not standard by not presenting evidence of a greater than even chance of survival for the patient in the absence of negligence. Gooding, 445 So.2d at 1020.

In the instant case, Dr. Perlmutter's affidavit does not demonstrate that Susan Reynard's use of the cellular telephone more likely than not accelerated the growth of her brain tumor and shortened her life. Dr. Perlmutter, however, states at the conclusion of his affidavit that "the use of the NEC hand-held cellular telephone by Susan Reynard accelerated the growth of her tumor and, therefore, was in part responsible for her early demise." Dr. Perlmutter makes no attempt to quantify the extent to which the use of the cellular telephone was responsible for her early demise. Further, he does not indicate the specific facts of Susan Reynard's case upon which he bases his conclusion. Although he states that he reviewed the affidavit from Dr. Storm, he does not indicate that he had examined Susan Reynard or reviewed her records. Dr. Perlmutter quotes the recent GEO report and thereby admits that, "No research has been completed on long term human exposure to low levels of radiation specifically from portable cellular telephones. Research findings on exposure to other sources of low level radio-frequency are inconclusive." Rather, Dr. Perlmutter states that animal studies and human brain tumor cell studies provide "strong inferential data" that the use of a cellular telephone is a health hazard and would likely accelerate the growth of brain tumors in humans. Although he states that such studies are "peer reviewed," he neither provides a copy of the studies nor comments on their acceptance in the scientific and medical community. Additionally, although Dr. Perlmutter states that it was inappropriate to suggest that Susan Reynard was expected to survive two years after her surgery, he does not provide specific facts upon which he could have concluded that Susan Reynard should have survived longer. Thus, neither Dr. Perlmutter's conclusions nor the studies from which he gathers "inferential data" are supported by specific facts in the record, and therefore, they are not entitled to any weight.

Additionally, the Court finds that Dr. Perlmutter's affidavit does not satisfy the admissibility criteria under Daubert v. Merrell Dow Pharmaceuticals, Inc., 509 U.S. 579, 113 S.Ct. 2786 (1993). In Daubert, the Supreme Court held that " 'general acceptance' is not a necessary precondition to the admissibility of scientific evidence under the Federal Rules of Evidence, but the Rules of Evidence—especially Rule 702—do assign to the trial judge the task of ensuring that an expert's testimony both rests on a reliable foundation and is relevant to the task at hand." Id. at –, 113 S.Ct. at 2799.

Scientific knowledge that will assist the trier of fact in understanding or determining a fact in issue will satisfy those demands. Id. The Supreme Court further stated that although it was not setting out a definitive checklist or test, the following questions were appropriate: (1) whether the theory or technique has been or can be tested; (2) whether the theory or technique has been subjected to peer review and publication; (3) whether there is a known or potential rate of error in the technique; and (4) whether the theory or technique is generally accepted in the scientific community. Id. at ——, 113 S.Ct. at 2797.

The Ninth Circuit Court of Appeals recently rendered its decision on remand in Daubert v. Merrell Dow Pharmaceuticals, Inc., 43 F.3d 1311 (9th Cir.1995), interpreting the Supreme Court's ruling in Daubert. The Court first noted that in performing their "gatekeeping role," judges: "must satisfy themselves that scientific evidence meets a certain standard of reliability before it is admitted. This means that the expert's bald assurance of validity is not enough. Rather the party presenting the expert must show that the expert's findings are based on sound science, and this will require some objective, independent validation of the expert's methodology." Id. at 1316.

Further, the Court considered three significant factors for determining whether the expert's testimony is admissible. The first consideration "is whether the experts are proposing to testify about matters growing naturally and directly out of research they have conducted independent of the litigation, or whether they have developed their opinions expressly for purposes of testifying." Id. Second, if the expert is not testifying based upon research independent of the litigation, "the party proffering it must come forward with other objective, verifiable evidence that the testimony is based on 'scientifically valid principles.' " Id. at 1318. One means of showing this, as suggested by the Court, "is by proof that the research and analysis supporting the proffered conclusions have been subjected to normal scientific scrutiny through peer review and publication." Id. A third consideration, where there is no evidence that the expert's proffered testimony grows out of research conducted independent of litigation, or that the expert's research has been subjected to peer review, is testimony of other experts. Id. The Court explained that for such other expert testimony to be sufficient:"[T]he experts must explain precisely how they went about reaching their conclusions and point to some objective source—a learned treatise, the policy statement of a professional association, a published article in a reputable scientific journal or the like—to show that they have followed the scientific method, as it is practiced by (at least) a recognized minority of scientists in their field." Id.

Applying the Supreme Court's analysis of Daubert and the Ninth Circuit's interpretation to the instant case, the Court finds that Dr. Perlmutter's affidavit is not admissible. The affidavit contains no reference to any scientific or medical research by Dr. Perlmutter independent of this litigation or any indication that he examined Susan Reynard or reviewed her medical records. Dr. Perlmutter also makes no reference to any independent research regarding the use of the type of cellular telephone that Susan Reynard used or the use of any cellular telephone, or research regarding the type of radiation to which she was allegedly exposed. Further, as mentioned above, there is no proffered evidence that the studies, analysis, and conclusions of Dr. Perlmutter's affidavit have been subjected to the normal scientific scrutiny through peer review and publication. Additionally, the conclusions of the affidavit are not supported by any objective source, such as a treatise or a published article in a reputable scientific journal.

Upon consideration of the foregoing, it is hereby ORDERED and ADJUDGED that:

Defendant, NEC America, Inc.'s Dispositive Motion for Summary Judgment is GRANTED;

GTE Mobilnet's Motion for Summary Judgment is GRANTED; Defendant, GTE Mobilnet of Tampa, Inc.'s Motion to Dismiss Plaintiffs' Complaint Against Defendant, GTE, Inc. is GRANTED.

The Clerk is DIRECTED to enter judgment for the Defendants.

DONE AND ORDERED.

§ *7.32 Ohm's Law*

In 1826 George Ohm published one of the most important equations for describing and measuring the key elements of an electrical circuit. Known as Ohm's law, the equation ties together three indispensable characteristics: Current, voltage drop and resistance. The current (I) is commonly measured in terms of amperes (most commonly referred to simply as "Amps"). Voltage (E) is expressed in terms of volts and resistance (R) is measured in Ohm's. The Law basically states that the voltage drop is equal to the product of the current and the resistance, or: $E = I R$. Another common representation of the Law spotlights the fact that current and resistance are inversely related:

$$I = \frac{E}{R}$$

Notice that a minimal amount of voltage could produce an unlimited amount of current if the resistance were small enough. Likewise, an AM broadcast with only a single watt of power could be transmitted and received throughout the entire world if there were no interference.

Problem: Billy Bob is rumored to have stole Willy Fred's horse; and the state of Texas means to execute Billy by means of electrocution. Texas wants to pump 20 amps of current through him. The electric chair has old circuitry carrying 30 Ohms of resistance. How much voltage will be required for Texas to get rid of the rumored horse thief?

Solution: E (Volts) = I(Amps) × R(Ohms).
E = 20 Amps × 30 Ohms = 600 Volts

§ *7.33 Shocking Information*

From a perspective of a personal injury attorney, an electrical current can accomplish many of the same injurious functions as a heavy, moving object. As is the case with momentum, the magnitude of the current is often proportional to the size of the injury. For electricity, it is most often the current that kills. Voltage is nothing more than the potential difference of electromotive force between two points. Without a wire or medium for transporting the voltage in the form of current, there is little fear of electrocution. The bad news is that it only takes a few amps to kill you.

Coughlin v. Great Western Power Co.

Supreme Court of California, 183 Cal. 548, 191 P. 920 (1920)

WILBUR, J.

Plaintiff recovered judgment in the lower court for damages sustained by him November 3, 1913, by the short-circuiting of a 22,000-volt current through his body while he was engaged in the work of cleaning certain bushings and wires in an electric substation operated by the defendant, his employer. Defendant appeals.

At the time of the injury the Roseberry Act (Stats. 1911, p. 796) was in force, and both plaintiff and defendant had elected to be bound by the provisions thereof. Plaintiff, however, ignored the compensatory provisions of that statute as therein authorized, in cases of gross personal negligence, and brought this action, in which the jury found as a fact by necessary implication that the defendant was guilty of such gross personal negligence. Unless such negligence has been proved, the plaintiff cannot recover. The claim of negligence is based upon the failure of the employer to furnish a safe place to work. Briefly stated, the admitted facts are as follows:

In the substation were three stationary transformers, by which electricity entering at 100,000 volts was transformed by induction into a current of 22,000 volts. To prevent the grounding of the high-voltage current, insulators, called bushings, extended 4 feet 6 inches above the upper surface of the transformers, which surface was 11 feet 5¼ inches above the floor of the substation. In order to prevent the current from leaking over the surface of these bushings, it is necessary to clean the surface thereof of dust and other accumulations about once a month, and for this purpose the current was shut off from the 100,000-volt circuit. Plaintiff was engaged in cleaning this surface and the uninsulated copper wire leading therefrom at the time of his injury, and for that purpose the current had been shut off from the high-voltage circuit. The injury resulted from the short-circuiting through his body of the 22,000-volt current coming from a station bus distant from the wire being cleaned, at its closest point, between 28¼ and 31 inches. It was unnecessary to clean the surface of the copper wire, as the purpose of the cleaning was to avoid the danger of short-circuiting the current over the surface of the bushing. The plaintiff had been working in the substation a month, and knew the voltages carried by the respective wires, and believed that the arcing distance from the 22,000-volt wire was only three-eighths of an inch. He had arranged the switches at the substation to shut off the current on the 100,000-volt wire and retain the current on the 22,000-volt wire, and at the time he was engaged in cleaning the former the 22,000-volt wires were 'hot.' The arcing distance of a 22,000-volt current to ground under the most favorable conditions is 1.2 inches.

Assuming that it was properly a part of the duty of the employee, not only to clean the bushing, but the copper wire arising therefrom, the question at issue resolves itself into this: Was it negligence to so arrange the substation that a man was required to clean the wire distant 28¼ inches from an uninsulated wire carrying a current of 22,000 volts? It must be obvious that such a wire could be cleaned by wiping it with a rag, without bringing any part of the body sufficiently near the station bus to produce an arc. If this was a usual place in which a workman was required to perform duties which would distract his attention from the perils involved in working in proximity to an uninsulated high voltage wire, and there was a possibility of contact by reason of unconscious or involuntary action, it might readily be concluded that a finding of gross negligence by the jury was sustained by the evidence.

But this was not a place where workmen were usually required to work. At most but once a month were they called upon to perform this service, and then under conditions which necessitated their careful attention to the condition of the various wires in the substation. To have come in contact with the 100,000-volt wire, or within arcing distance thereof (10¼ inches), meant instant death. It is evident that a man, engaged in the occasional task which occupied the plaintiff, would naturally be alert to the peril of his position.

At this point it is well to give the plaintiff's statement as to the conditions attending his injury. He testified that the current jumped across a space of 21 inches to the back of his hand. If this were a contestable fact, we would have to assume the correctness of this testimony in determining the question of defendant's negligence. But it is a scientific impossibility for a current of that voltage to jump through the air for that distance, and it is shown that a lightning arrester with a perfect ground reached within one inch of the wire. It is a scientific fact, of which we take judicial notice, that if a 22,000-volt wire had been surcharged with a sufficient voltage to jump for a distance of one inch, such additional voltage would have discharged into this ground wire at 1 inch, rather than to ground through the hand of the plaintiff at 21 inches. Moreover, the iron beams of the substation were within 11 inches of the 22,000-volt wire, and connected with the corrugated iron surface of the substation so as to form a perfect ground, and this would receive the discharge before plaintiff could have been injured as he states. Plaintiff testified that the rag in his hand was moist and dirty; that he had been flipping the rag about in an endeavor to clean to 100,000-volt wire at a point higher than he could reach, and for the purpose, also, of shaking the dust out of it.

It is defendant's contention that his injury resulted from the rag coming in contact with the 22,000-volt wire, and thus forming an arc which, once formed, may extend a distance of 2 or 3 feet. If the injury thus resulted, it is evident that it was occasioned by the carelessness of the plaintiff rather than the gross negligence of the defendant. In determining the credibility of the plaintiff's testimony, we may freely concede that he testified with the utmost good faith. It is doubtful if any one who has survived a stroke of lightning or a short-circuit from a high-voltage wire could either observe, or, if he did, could remember, the circumstances. But if we assume that his testimony is absolutely true, and, that the current did leap over this intervening space of 21 inches, without short-circuiting into the lightning arrester or any of the nearer metal conductors in the building, it would at once follow that this extraordinary violation of all known rules of the conduct of electricity of that voltage was such as no human being could anticipate, and the failure to anticipate it could not be considered negligence.

The defendant, in arranging its substation, was only bound to make provision for those laws of nature of which one of ordinary prudence engaged in a like enterprise would be required to take notice. Whether, therefore, we regard the circumstances detailed by the plaintiff as an inherent impossibility, or as an extraordinary and unprecedented electrical phenomena, the defendant cannot be held guilty of gross negligence in the arrangement of its electrical devices in the substation. A good deal is said in the evidence and in the briefs concerning the distance which should be maintained between wires carrying a 22,000-volt current and those carrying 100,000-volt current. This testimony has no significance in this action, for the reason that such testimony is directed to a condition where both wires are charged with the current they are designed to carry, and has no relevancy to a situation where the 100,000-volt wire is 'dead.' We have therefore directed our attention to that situation.

It is proper to say that the testimony of an expert witness to the effect that it was customary and desirable that a partition should be placed between wires carrying these voltages, and that such partitions would have made the place a safe one in which to work, has no significance, other than to point out that, if there had been a partition between the plaintiff and the 22,000-volt wire, he would not have been injured.

This, of course, is a self-evident fact. Mr. Van Norden, an electrical engineer, was asked his opinion as to what precautions could be made which would make the situation perfectly safe to a man working where plaintiff worked. He replied:

> The only safe method of construction in a point of that kind, and I think it is general practice, would be to have a barrier or cell wall outside of the 100,000-volt line to protect the 100,000-volt line from any possible circuit that might be near it. In other words, there should have been a barrier between the 100,000-volt line and the 22,000-volt line.

He further testified that the ordinary form of barrier is a thin reinforced concrete wall, and that the practice was to place each 100,000-volt transformer in a separate cell, or the three in a single cell; that is, a room closed at both ends, at the back and top, and open in front. It developed, however, that the reason this witness considered the place where plaintiff was working dangerous was because:

> It is always dangerous to work within reaching distance of a 22,000-volt line, because a man might unconsciously put his hand out, or his coat might fly out, or in this case the rag that was in the man's hand, the point of the rag might fly out, as he moved around unconsciously, and either come very close to the wire, or actually strike it, in which case there is no question but he would get a short-circuit falling down into his hand, and hence over his body—as it probably was in this case, because, if it had gone through his body, it probably would have killed him—down through his shoes, and into the transformer, and to the ground.

Q. Then, as I understand it, it is because the man might come close to the 22,000-volt wire?

A. He might; if he was constantly keeping his presence of mind, he might work up there every day for years, and not actually come close enough to get a shock off of that wire. Then, again, in an unconscious moment, he might do it. It does not seem to me that there is any particular question of negligence; it might be unconscious.

In response to a question of a juror, this witness also testified that the accident would have been impossible if such a barrier had been erected between the two wires. The arcing distance of a 22,000-volt current having been established by expert testimony, the question as to whether or not plaintiff was working in a safe place was not a question for an expert witness, and although the evidence came in, in the first instance, without objection, motion was made to strike it out. It is sufficient, however, to say that the witness do no more than to call the attention of the court and jury to the obvious fact that a man working within 31 inches of a live wire might carelessly or thoughtlessly come in contact with it. Whether or not this constituted either negligence or gross negligence was a question of law and fact.

Where, as here, all the facts are either shown without dispute, or by scientific principles of which the court takes judicial notice, the question of whether or not the evidence was sufficient to support a finding of gross negligence is a matter of law for the court.

In the view we take of the case it becomes unnecessary to determine whether the plaintiff was guilty of negligence, or that the defendant was guilty of lack of ordinary care; for we hold, as a matter of law, that the plaintiff failed to establish that gross negligence upon which his cause of action was predicated. For the same reason it is unnecessary to determine whether such negligence is personal, within the meaning of the Roseberry Act, supra, and it is unnecessary to pass upon the question of whether the defenses of contributory negligence or assumption of risk are available under the Roseberry Act, in cases based upon the gross personal negligence of the employer.

The judgment is reversed.

We concur: ANGELLOTTI, C. J.; OLNEY, J.; SHAW, J.

LENNON, J.

I dissent. In this action for damage for personal injuries sustained by the plaintiff while in the employment of the defendant, a corporation engaged in the business of generating, transmitting, and distributing electricity for lighting, heating, and power purposes, the prevailing opinion reversed the judgment in favor of the plaintiff upon the theory that the evidence adduced upon the whole case did not support the finding of the jury, implied from their verdict, that the defendant was guilty of gross negligence in the maintenance and operation of that part of its power plant where the plaintiff was injured during the course of his employment. The prevailing opinion holds that defendant corporation was not guilty of gross negligence. I cannot concur in this conclusion. The facts of the ease, as I find them from a perusal of the record, are these:

Along defendant's main transmission line, which carries electricity at 100,000 volts from its generating plant or power house in Butte county to San Francisco Bay, are several substations. These substations are installed for the purpose of reducing the voltage carried by the main transmission line for distribution from the substations for commercial use in the surrounding districts. At the time plaintiff was injured he was employed as an operator at one of the substations of defendant located at Isleton, Cal. It is the duty of an operator in such a substation to watch the apparatus and keep it in proper working condition. In this substation, where plaintiff was employed, were three transformers, standing in a row, north and south. They were 11 feet in height, 6 or 7 feet in diameter, and consisted of coils of copper wire in a metal casing, into which the electricity was carried at 100,000 volts by means of wires connected with the main transmission line.

After being reduced in voltage, the electricity was carried from the transformers at 22,000 volts by wires issuing from the covers of the transformers. The covers of the transformers, which were of boiler iron, were rounded, sloping from the center toward the edges, the only flat portion of the surface being a manhole cover about 18 inches wide and 2 feet long. Upon the top of each transformer, projecting at an angle therefrom, and close to the edge of the transformer shell, were two bushings, 4½ feet in height and about 10 inches in diameter at the bottom, tapering toward the top. The purpose of these bushings, which were made of fiber rings alternating with fiber discs, was to insulate the 100,000-volt wires so as to prevent the electricity running to ground over the metal shell of the transformer.

There were also two smaller bushings, for the purpose of insulating the 22,000-volt wires issuing from the tops of the transformers; but they are of no importance for the purposes of this case. In addition to the connection with the main transmission line, the several substations were also electrically connected between themselves, so that, in case of a shut-down at any one substation, electricity could be fed to the district thus affected from other substations.

It was the custom to shut down the substations once a month in order to clean the apparatus of the substations, and this was ordinarily done after midnight, when the demands for electricity were at a minimum. There were three operators at defendant's substation at Isleton-Dorman, the chief operator, Post, and the plaintiff. It was arranged to have a general cleaning of this substation after midnight, November 3, 1913, and plaintiff, in compliance with a direction from Dorman, opened the switches for the purpose of shutting off the current from the wires leading into and out of the substation. All of the wires and apparatus in the substation were thus entirely freed from electricity, with the exception of the station bus, which consisted of three parallel wires near the roof and about 20 feet from the ground, running east and west (at right angles to the row of transformers) across the south end of the building, and feeders dropping therefrom into oil switches set on the shelf against the south end of the building.

This station bus was fed from other substations and carried electricity at 22,000 volts. No cleaning was to be done on the station bus, and it was left charged in order to furnish electricity to the neighboring towns and light for the substation while the high-tension bank was shut down during the cleaning process. After the substation was thus shut down, the three operators went upon the roof of the building to clean some of the apparatus. They then began to work inside the building, and Dorman told plaintiff to 'take No. 3,' meaning thereby to clean the bushings and wires on the top of the transformer furthest toward the south, known as 'No. 3.'

Plaintiff, standing on the top of the transformer, cleaned its north bushing, and then proceeded to work on the south bushing. After cleaning the south bushing, he began to dust the wire above the bushing. He had a cotton rag, about 2 feet square, and with this rag reached up as high as he could to clean the upper part of the wire, and then, holding the greater part of the rag in his hand, wiped down the wire. Just as he was dropping his hand to his side, the back of the hand was struck by a flash from a live 22,000-volt wire of the station bus, which was just to the south of the transformer upon which plaintiff was standing. The electricity severely burned the back of his right hand and right forearm, and passed through his body, burning the sole of each foot. The shock threw him off the transformer to the floor beneath.

On the night in question there was 'a very heavy fog, everything was just wringing wet,' and this condition prevailed inside as well as outside the substation, owing to the doors of the substation being left open for ventilating purposes. At the time he was injured plaintiff was 21 years of age, and had been working at Isleton about a month. Although he had worked in electricity to some extent before coming to Isleton, he had no experience with high voltage, other than one month's work as an operator at defendant's substation at Cowell. The jury rendered a general verdict in favor of plaintiff for damages in the sum of $6,735.83. On defendant's motion for a new trial, the court made an order granting the new trial unless plaintiff stipulated to deduct from the damages awarded the sum of $1,735.83. This was consented to by plaintiff, and the motion for a new trial was thereupon denied, whereupon the defendant appealed from the judgment.

It is contended by appellant, and the prevailing opinion holds, that gross negligence has not been proven. The distance between the dead 100,000-volt wires, which plaintiff was cleaning, and the nearest point of the live 22,000-volt wires of the station bus, was 28¼ inches, according to appellant's measurements; 31 inches, according to the estimates of witnesses for respondent. Plaintiff's hand, owing to the thickness of the rag he held and the movement in dropping his hand to his side, was about 21 inches from the nearest live wire at the time it was struck. It is pointed out by appellant that it was testified to at the trial by the expert witnesses, including those who testified for plaintiff, as well as those called by the defendant, that the arcing distance of 22,000 volts of electricity is, under no conditions, more than 1.2 inches; that is to say, the air space across which a current of electricity will jump in its effort to reach the ground is ordinarily less than 1 inch, and, under any circumstances, never greater than 1.2 inches. Opposed to this expert testimony is the testimony of the plaintiff that the current 'jumped right over through the air, and it caught me on the back of the hand.' It is further pointed out that, since electricity seeks the path of least resistance, and there were other conductors nearer the wire than plaintiff's hand, had the electricity jumped from the wire, it would have done so, not a distance of 21 inches to plaintiff's hand, but to the conductor, the shortest space from the wire, such as the lightning arrester outside the substation.

The theory is advanced by appellant, in support of the contention that the evidence fails to show any negligence on the part of the company, that, since plaintiff's hand was admittedly 21 inches away from the live wire at the time it was struck, plaintiff must have carelessly brought his rag, by shaking or otherwise, or some other conductor, within an inch of the live wire, and thus drawn an arc. It is further argued that, if the 22,000 volts did arc 21 inches directly to the back of plaintiff's hand, the occurrence was so unheard of and opposed to every known principle of electricity that it was purely an accident, and not a danger which appellant was required to guard against in providing for the safety of its employees. However that may be, and assuming that 22,000 volts could not jump 21 inches, there is no reason thus presented for disturbing the verdict in this case. There were no special findings, simply a general verdict in favor of plaintiff. 'It is, of course, the duty of this court to treat all of the facts brought, out by the evidence which are necessary to support the verdict as having been found to be true by the jury' (Clark v. Tulare Dredging Co., 14 Cal. App. 414, 429, 112 Pac. 564, 570), and, conversely, the verdict of the jury cannot be assumed to be based on inherently improbable testimony, if there is other testimony sufficient to support the verdict.

Plaintiff's action was based upon the alleged personal gross negligence of defendant in maintaining a live 22,000-volt wire in too close proximity to the place where plaintiff was required to work, because of which plaintiff was injured by a current of electricity which 'jumped and passed' from the said live wire 'and came against and in contact with plaintiff's right hand and arm.' Consistently with these allegations of the complaint, plaintiff might have drawn an arc by a close approach to the live wire with his rag, and the arc having once been formed, it could, according to the testimony of the expert witnesses, be extended for 2 or 3 feet. The jury was entitled to infer, from the evidence adduced upon the whole case, that this was the manner in which the current 'jumped and passed' to plaintiff's hand, despite plaintiff's general statement that the spark jumped directly to the back of his hand. Follmer v. Rohrer, 158 Cal. 755, 112 Pac. 544; Bancroft-Whitney Co. v. McHugh, 166 Cal. 140, 134 Pac. 1157; Lynch v. Lynch, 22 Cal. App. 653, 135 Pac. 1101; Turner v. Bush (App.) 185 Pac. 190. It is not, therefore, essential to the general verdict in favor of plaintiff to assume that the jury found that the electricity jumped directly through the air and, without the aid of any intervening substance, the entire 21 inches to the back of plaintiff's hand.

Moreover, there was, in my opinion, evidence tending to show the existence of gross negligence, irrespective of the question of the arcing distance of the 22,000 volts of electricity. While the employer is not an insurer of the safety of his employees, he must furnish a place to work in which is 'reasonably safe, having regard to the character of the work itself.' Spivok v. Independent Sash & Door Co., 173 Cal. 438, 440, 160 Pac. 565, 566. The jury, when arriving at a general verdict, were warranted in taking into account the dangerous nature of electricity, the high voltage which the wires carried, the youth of plaintiff, his inexperience in high-voltage work, and the fact that he had never before performed the particular duties upon which he was engaged at the time of his injuries. It further appears that at least one expert witness, a consulting engineer, testified that, although 22,000 volts could not arc over 1.2 inches, the place in which plaintiff was working was not a safe working distance from the live wires, and that, under the circumstances, 28¼ inches, or even 31 inches, was in dangerous proximity to the 22,000-volt wire, because of the likelihood that a person working in that position would come sufficiently close to the live wire to produce a short circuit.

It was stated by this witness, and others, that the only safe method of construction in a situation of that kind was the erection of a concrete barrier in the form of a concrete wall about the transformers; that this was the general practice; that the wires were thus property protected in other substations belonging to the defendant; and that this 'barrier should be there for a twofold purpose—one for operating purposes, and the other for the safety of anybody working in the cell.'

There was testimony, and it is conceded, that if the barrier had been erected in this case the occurrence now under consideration could not have happened. The evidence also shows that the standard distance between 100,000 and 22,000 volt wires is between 36 and 42 inches; but in one or two cases, where there were building difficulties, as at Isleton, this distance was cut down. There is also evidence that, by installing a switch at a cost of less than $100, it would have been possible to render the substation entirely free from current by shutting off the current from the station bus and obtaining light for the substation from a small transformer house outside the substation.

While most of the evidence above set forth was by no means uncontradicted, the weight of the testimony was, of course, to be determined by the jury. Viewing the record as a whole, and irrespective of any conclusion this court might reach from the same evidence, it cannot be said that the jury was unwarranted in concluding that, even assuming that the electricity could not directly arc a distance greater than 1.2 inches, defendant failed to exercise the required degree of care for the safety of its employee in requiring him, under the above-described conditions and in an inconvenient location such as that presented by the transformer cover, to work within 28¼, or 31, inches of a live 22,000-volt wire and unprotected therefrom.

This, I take it, must be so in keeping with the rule generally accepted and uniformly invoked in the measurement of the degree of care required of electric companies for the safety of their employees, and which is to the effect that, while such companies are not required to operate their plants to the point of perfection in so far as concerns the material construction of the plant or the character of the apparatus and appliances, for the reason that they are not insurers of their employees against accident, nevertheless they are obligated to use reasonable care; that is to say, such care as, under all the circumstances, would obviously be required in a given situation in the construction, maintenance, and operation of their plants. Falling short of this, they will generally be held responsible for resultant injuries to employees.

It is a reasonable corollary of this rule that the degree of care to be employed by electrical companies in the maintenance and operation of their plants varies with the degree of danger dependent upon a particular situation, for obviously a prudent person, individual or corporate, must augment the character and quantity of the care according to the contingencies of differing and varying situations. Thus, if the danger of injury to an employee is but slight, as, for instance, in a situation involving the likelihood of contact with transmission wires carrying merely a harmless current of electricity, as do ordinary telephone and telegraph wires, ordinary care only will be demanded of the company in the maintenance and operation of such wires. On the other hand, where the transmission wires are heavily loaded with high voltage, a very high degree of care—indeed, the highest degree of care commensurate with human prudence and caution and consistent with the practical conduct of business—is required to guard against accident or death which inevitably follow contact with such high-voltage wires. Crosswell on Electricity, § 234; Curtis on Electricity, § 405. The rule is, perhaps, better stated in the case of Uggla v. West End Street Ry. Co., 160 Mass. 351, 35 N. E. 1126, 39 Am. St. Rep. 481, where it is said:

> The vigilance and attention required must conform to the nature of the emergency and the danger to which others may be exposed, and is always to be judged of according to the subject-matter, the danger and force of the material under the defendant's charge.

This being so, then the question of whether or not the required care has been used, is, in the majority of cases, necessarily one for the jury, except when the court can say positively that no reasonable man would have acted in the manner complained of, or that a reasonable man must have acted in the manner complained of. Lee v. Electric Light Co., 140 Pa. 618, 21 Atl. 405; Southwestern Tel. & Tel. Co. v. Woughter, 56 Ark. 206, 19 S. W. 575.

With this rule in mind, and having in mind, also, the facts above referred to, namely, the high voltage, the insecure footing afforded by the cover of the transformer, the youth and inexperience of plaintiff, among other circumstances, I cannot concur in the conclusion of the prevailing opinion that the corporation defendant was not guilty of negligence in the maintenance of its plant, or that part of its plant in which the plaintiff was injured while at work for the defendant. And surely, if in keeping with that rule it can be fairly said, as I think it can be so said, that the corporation defendant was, under all of the circumstances of the shown situation, negligent in not making a known highly dangerous place reasonably safe to work in with the means at hand to do so, then in view of the further fact that the expenditure of the paltry sum of $100 by the corporation defendant was all that stood between the plaintiff and reasonable safety from serious injury, it is not difficult to see and rightly say that the corporation defendant was indeed negligent out of all measure and beyond all allowance; in a word, that it was guilty of gross negligence.

I would find no difficulty in demonstrating that such gross negligence was the personal negligence of the corporation itself. But, in view of the fact that the prevailing opinion rests a reversal of the judgment solely upon the ground that the evidence shows no gross negligence on the part of the defendant, it would be an idle act to discuss the question of personal negligence or any of the several other questions involved upon the appeal.

LAWLOR, J. I dissent.

I am in accord with the opinion of Mr. Justice LENNON, and shall confine myself to some additional observations. The main opinion rests upon a negative answer to the question:

> Was it negligence so to arrange the substation that a man was required to clean the wire distant 28¼ inches from an uninsulated wire carrying a current of 22,000 volts?

This conclusion is arrived at by a line of reasoning that we will take judicial notice that it is a 'scientific impossibility' for a current of 22,000 volts to 'jump' a distance of 21 inches; that this was not the place where the plaintiff was usually required to work, and that the known proximity of the wires made it imperative for him to 'be alert to the perils of his position'; that, inasmuch as he admitted 'flipping' the cleaning rag about in order to reach higher on the wire which he was wiping, and also to shake the rag free from dust, 'it is evident' that the injury was occasioned by his carelessness rather than by any negligence of the defendant; that, moreover, in view of the 'scientific impossibility' already referred to, it cannot be considered negligence for the defendant to fail to provide against such an extraordinary violation of all known rules of the conduct of electricity as the 'jumping' of the current over a space of 21 inches; and, finally, again calling attention to the 'scientific principle' that electricity will not 'jump' a distance of 21 inches, it is declared that the evidence does not sustain the implied finding that defendant was guilty of negligence in failing to provide a safe place for the plaintiff to work.

May we take judicial notice that a 22,000-volt current will not 'jump' 21 inches? The general rule as to when a court will take judicial notice is stated in Dunphy v. St. Joseph Stockyards Co., 118 Mo. App. 506, 523, 95 S. W. 301, 306: "Courts should observe the utmost caution to avoid assuming knowledge of natural facts and laws that are beyond the scope of common positive knowledge."

This rule is thus expressed in 15 R. C. L. 1127:

> Judicial notice will be taken of scientific facts which are universally known ... but they must be of such universal notoriety and so generally understood that they may be regarded as forming part of the common knowledge of every person ... Cognizance may not be taken of scientific matters of uncertainty or dispute or of insufficient notoriety.

And in Curtis, The Law of Electricity, p. 3, it is said: "So dangerous a force as electricity must legally be considered in the same class as high explosives and other treacherous and destructive agents ... The power and manifestations of which are fully comprehended only by experts. See, also, Brown v. Consol. Light, etc., Co., 137 Mo. App. 718, 109 S. W. 1032; Campbell v. United Ry., 243 Mo. 141, 147 S. W. 788. Several witnesses did testify that they had never known a current of 22,000 volts to 'arc' more than

1.2 inches, but Reuben D. Bennett, an electrician, stated: "Electricity is an unknown quantity. They presume that it will 'jump' so far. It never acts the same twice; there is no knowing how far it will go."

I do not think that the 'arcing' distance of 22,000 volts has been proved to be so invariable as to attain the certainty of a scientific fact, or that it is a matter of such common knowledge, or so generally understood, as to entitle it to recognition under the theory of judicial notice.

The majority opinion, in discussing the plaintiff's duties in the cleaning of the transformer, states that—

> It was unnecessary to clean the surface of the copper wire, as the purpose of the cleaning was to avoid the danger of short-circuiting the current over the surface of the bushing.

The evidence was in conflict on this point. There was some testimony that it was unnecessary to clean the wires, but the chief operator at the substation testified that it was the practice to wipe the wire, as well as the bushing, and the plaintiff stated that such was the practice, and that on the occasion in question he had been ordered by the chief operator to clean the wire. It seems to me, from this state of the evidence, that the jury may have found that it was customary to clean the wire, or, at any rate, that the plaintiff, in cleaning it, acted under the instructions of one in authority.

I am particularly impressed by the evidence from which the jury may have decided that the injury was the result of a false policy of economy in failing to erect a concrete wall between the station bus wires and the transformer. The complaint alleged that defendant's attention had, prior to the time when plaintiff was injured, been repeatedly directed to the dangerous proximity of the 22,000-volt wire. The chief operator testified that, in a conversation with defendant's electrical engineer, who controlled and supervised changes in substation construction, said, referring to the spot, where plaintiff was injured:

> We have one awfully dangerous place to work here, and I showed him this bus, how close it was to the 100,000, and where we had to get up in there to work … I told him it was absolutely unsafe … He says: "I know it is pretty close. This place is not arranged just as we would like to have it, but we have not got money to spare now to make any changes."

This witness testified that he had had similar conversations with the superintendent of substations and the superintendent of operations, both of whom acknowledged to him that the place was unsafe, but said that the company could not afford to install safety appliances. As Mr. Justice LENNON points out, the expenditure of $100 would have rendered the place safe. The defendant was under the duty of providing a reasonably safe place for the plaintiff to work, and it must be implied from the verdict, and the order refusing to grant a new trial on the claim of insufficiency of evidence, that the defendant failed to perform this duty. There is nothing in the Roseberry Act to indicate that the term 'personal gross negligence' implies that actual knowledge of the dangerous condition of the premises must be brought to the attention of the officers of a corporation before the failure to remedy such condition becomes the negligence of the corporation.

I therefore conclude that it cannot be held, without invading the province of the jury, that, upon the application of the theory of judicial notice, or otherwise, 'the plaintiff failed to establish that gross negligence upon which his cause of action was predicated,' nor that the plaintiff was careless in wiping the wires, nor that the defendant was not guilty of personal gross negligence in failing to provide a barrier between the station bus wires and the transformer.

Notes and Problems

1. In *Coughlin,* do you agree with the majority or the dissenting opinion? What are your reasons?

2. When mathematics and physics leave their respective realms and wander into the medical arena, attempts to employ judicial notice are certain to draw fire. We may, for example, be able to judicially notice certain aspects about voltage, current and electricity in general, but their effects upon the human body may be subject to dispute. See *Prestige Homes, Inc. v. Legouffe,* Supreme Court of Colorado, No. 81sc263 (1983) holding that where the Court of Appeals, in reviewing a workmen's compensation claim, judicially noticed medical treatises not offered or admitted into evidence (and not cited by either party's medical expert) and where it relied on those treatises for its finding that electric shock caused by contact with a 220-volt power line can cause serious injury without leaving a visible burn mark, the Court thereby assumed the role of an expert medical witness based on independent research and interpretation of medical texts which properly should be interpreted only by experts in the appropriate field. Even if the proposition were widely recognized within the relevant community, the fact that two medical experts had a reasonable dispute over the conclusion to be reached from that proposition, namely, whether the claimant's heart attack was caused by such shock, made the Court of Appeals' disregard of the referee's findings clearly erroneous.

§ 7.34 Current Ownership

From a legal perspective, the distinction between mechanical and electronic devices is not terribly significant. As demonstrated in prior cases, electricity can maim and injure just as effectively as automobiles and other mechanical devices. With respect to patentability, however, 35 U.S.C.A. § 101 requires the invention seeking patent protection fall into one of four categories: process, machine, manufacture, or composition of matter. See also *State Street Bank & Trust Co. v. Signature Financial Group, Inc.*, 149 F.3d 1368, 1375 (Fed.Cir.1998). From this you conclude that an electronic invention lacks the patentable stature of a mechanical device. See *In re Nuijten*, 500 F.3d 1346 (C.A. Fed., 2007), involving the Patent Office's rejection of an improved digital signal with an embedded digital watermark encoded according to a given encoding process. The Federal Circuit held that:

> The Supreme Court has defined the term "machine" as "a concrete thing, consisting of parts, or of certain devices and combination of devices." Burr v. Duryee, 68 U.S. (1 Wall.) 531, 570, 17 L.Ed. 650 (1863). This "includes every mechanical device or combination of mechanical powers and devices to perform some function and produce a certain effect or result." Corning v. Burden, 56 U.S. 252, 267, 15 How. 252 (1853). A transitory signal made of electrical or electromagnetic variances is not made of "parts" or "devices" in any mechanical sense. While such a signal is physical and real, it does not possess concrete structure in the sense implied by these definitions. No part of the signal—the crests or troughs of the electromagnetic wave, or perhaps the particles that make it up (modern physics teaches that both features are present simultaneously) is a mechanical "device" or "part." A propagating electromagnetic signal is not a "machine" as that term is used in § 101.

Electrical current is not patentable. In *The Telephone Cases*, 126 U.S. 1, 534, 8 S.Ct. 778, 782, 31 L.Ed. 863 (1888), the Court explained a prior case (involving Morse) as follows: "The use of magnetism as a motive power, without regard to the particular process with which it was connected in the patent, could not be claimed, but that its use in that connection could.' Bell's invention was the use of electric current to transmit vocal or other sounds. The claim was not 'for the use of a current of electricity in its natural state as it comes from the battery, but for putting a continuous current, in a closed circuit, into a certain specified condition, suited to the transmission of vocal and other sounds, and using it in that condition for that purpose.' Ibid. The claim, in other words, was not 'one for the use of electricity distinct from the particular process with which it is connected in his patent.' Id., at 535, 8 S.Ct., at 782. The patent was for that use of electricity 'both for the magneto and variable resistance methods.' Id., at 538, 8 S.Ct., at 784. Bell's claim, in other words, was not one for all telephonic use of electricity. In Corning v. Burden, 15 How. (56 U.S.) 252, 267–268, 14 L.Ed. 683, the Court said, 'One may discover a new and useful improvement in the process of tanning, dyeing, etc., irrespective of any particular form of machinery or mechanical device.' The examples, given were the 'arts of tanning, dyeing, making waterproof cloth, vulcanizing India rubber, smelting ores.' Id., at 267, 14 L.Ed. 683. Those are instances, however, where the use of chemical substances or physical acts, such as temperature control, changes articles or materials. The chemical process or the physical acts which transform the raw material are, however, sufficiently definite to confine the patent monopoly within rather definite bounds.

§ 7.35 *Radar and Lidar*

For most of us, the first thing that comes to mind when law and physics are mentioned in the same breath is our last speeding ticket. Radar and laser guns are used by law enforcement to measure the speed of automobiles.

Radar (acronym for "Radio direction and ranging") guns are simply radio transmitters combined with receivers. They emit a radio signal, then receive that same signal back as it bounces off of the object whose speed is being measured. However, the radar frequency of the emitted signal is different that of the signal received back. Based upon that difference, the radar gun is able to calculate the speed of the moving object.

A radar beam is similar to a beam of light in that it spreads out as the distance from the signal origin increases. The signal then bounces off of objects in the path of the beam and are reflected back to the gun. The gun uses the Doppler effect to calculate the speed of the object in the beam's path. Using a comparison of frequency shift between received images instead of the frequency shift between sent and received frequencies creates what is known as moving radar. Here is one of several equations used for translating the relevant information into a speeding ticket:

$$F_{\text{Doppler}} = \frac{2\,v\,f}{c}$$

where F_{Doppler} = the changed frequency of the reflected wave, v = the speed of the moving object being measured, f = the frequency of the wave originally transmitted, and c = the speed of light.

All bands of radar operate on different frequencies, work differently, and are often very complicated. X band guns are becoming less common due to the fact the beam is strong and too easily detectable. Also, most automatic doors utilize radio waves on the X band and can possibly affect the readings of police radar. As a result the K band and Ka band are most commonly used by law enforcement agencies.

Traffic radar comes in many models. Some are hand held; others are stationary and moving radar instruments. Hand held units are mostly battery powered, and for the most part are used as stationary speed enforcement devices. Stationary radar is mounted in police vehicles, and may have one or two antennae. They are deployed when the vehicle is parked. Moving radar is employed, as the name implies, when the police vehicle is in motion; such devices can be very sophisticated and able to track vehicles approaching and receding, both in front of and behind, the patrol vehicle. They can also track the fastest vehicle in the selected radar beam, front or rear. When defending a speeder, you might consider demanding:

a. Copies of manufacturers names, including makes, models and serial numbers of all radar/laser guns in use by the Town/City/State Police Department.

b. Copies of manufacturers recommended maintenance for all of the above stated radar guns.

c. Copies of any manufacturer literature as it relates to the correct use, including but not limited to mounting, aiming, weather and traffic limitations, for all radar guns in use by the Town/City/State Police Department.

d. A copy of the past six months' maintenance records for all of the above stated radar guns, including, but not limited to, calibration specifications.

e. A copy of the Authorized Certification of Training, issued to the Officer who gave you the ticket, in the proper use of all radar guns in use by that Town/City/State Police Department he/she works for.

f. A copy of the patrol car assignments for the date you got the ticket.

Modern variations of "radar" include "lidar" (light detection and ranging) a/k/a "laser radar," though technically lidar does not employ microwaves or radio waves; lidar uses only light, normally a laser. A laser speed gun measures the round-trip time for light to reach a car and reflect back. It shoots a very short burst of infrared laser light and then patiently waits for it to reflect off the vehicle. The gun counts the number of nanoseconds that it takes for the round trip; then, by dividing by 2 it can calculate the distance to the car. If the gun takes 1,000 samples per second, it can compare the change in distance between samples and calculate the speed of the car. By taking several hundred samples over the course of a third of a second or so, the accuracy can be very high. The size of the cone of light that the gun emits is very small, even at a range like 1,000 feet (approximately 300 meters). The cone at this distance might be 3 feet in diameter. This allows the gun to target a specific vehicle. A laser speed gun is also very accurate. The one thing here is the laser gun must be aimed at a car—normal police radar with a broad radar beam can detect Doppler shift without aiming.

State v. Wojtkowiak

170 N.J.Super. 44, 405 A.2d 477 (1979)

WELLS, J. S. C.

This is an appeal from defendant's municipal court conviction for speeding 68 m.p.h. in a 55 m.p.h. zone. At the request of defendant, and with the State's consent, the court opened the record below for the sole purpose of determining the scientific reliability of the K55 Radar Speed Detection Device. Defendant's conviction rests on a speed reading displayed by the K55 in State Trooper Albert J. Dempster's troop car as he and defendant approached one another, Trooper Dempster in a northbound lane and defendant in a southbound lane, on Route 295.

Speed-measuring radar in various forms has been accepted since State v. Dantonio, 18 N.J. 570, 115 A.2d 35 (1955). Four experts, two called by each side, testified. In addition, the State called Trooper Dempster. The State's witnesses were principals of MPH Industries, manufacturer and distributor of the K55. The first witness, Robert E. Patterson, who built a crystal radio at age 10, is "entirely responsible for the total technical design and construction and manufacture of the K55 radar." He is a high school graduate with two years of college, graduated from the Army's Signal School at Fort Monmouth and ended his military career as head of maintenance of school equipment at the Missile Guidance School at Redstone Arsenal. This military career gave Patterson extensive theoretical and practical knowledge of radar of all types. Patterson, after his military discharge, held successive jobs as chief electrical engineer at several companies in the electronics and radar industry in each of which degreed engineers reported to him. He holds five patents in various types of electronic circuitry design and built the first solid state cardiac monitors; he has also designed or contributed to the design of music amplification equipment, police radar (before the K55) and cardiac telemetry equipment. These devices depend on a common thread of theory and practical technology applicable to radar-informed speed measuring equipment and, in several instances, Patterson was in the forefront of the developing technology which finally emerged in various speed measuring radar devices.

The second State's witness was Edward Walker Sergeant. His qualifications are detailed in State v. Musgrave, supra. In this case Sergeant testified solely about the training programs offered by MPH, and, in particular, about the one he gave to a class of New Jersey State Troopers among whom was Trooper Dempster.

The defense's experts were Andrew L. Soccio and Dr. Leo Nichols. Soccio's qualifications are outlined later in this opinion. Dr. Nichols possesses a B.S. in Electrical Engineering from Virginia Military Institute, a Master of Science in Electrical Engineering from Ohio State and a Ph.D. in the same subject from Virginia Polytechnic Institute. Although Dr. Nichols holds a license as a professional engineer in Virginia, his primary profession is that of teacher, having risen from an instructor to his present position as head of the Department of Electrical Engineering at Virginia Military Institute, a job he has held for 11 years. He has taught basic electric circuits, thermodynamics and microwave theory and techniques. He has testified as an expert many times in several states on the theory and operation of traffic radar.

It is from the testimony of these witnesses, and Trooper Dempster, that the court makes its findings on all aspects of the theory and operational characteristics of the K55 and, finally evaluates its reliability.

Principles of Doppler Radar:

It is necessary to review certain basic ideas in order to understand the K55 and to describe its limitations. Engineers generally depict radiant energy as moving in an undulating wave form pattern called "cycles." The number of cycles passing a given point in a given period of time is called "frequency." In music, the frequency at which sound reaches the ear determines the pitch people hear. Middle C on the piano, for example, when played, produces a sound wave at a frequency of X cycles a second, and the C one octave above middle C will generate a sound wave at 2X cycles a second. The higher the frequency, the higher the pitch will be. Note at the outset that this relationship between frequency and pitch is direct and not affected by other factors. For instance, it makes no difference how loudly or how softly you play middle C, or if you play it loudly and the octave C softly, the two sound waves produced will still reach the ear at the same frequency, X and 2X cycles a second. Nor does distance from the source of the sound make a difference. If a pianist in a concert hall plays middle C, that note is heard as middle C whether a listener is in the front row or in the last balcony row.

So far, we have considered a stationary source of sound waves. Let us now take a moving source. Assume a passenger at a station awaiting a train. Down track the train approaches at 50 m.p.h. The engineer blows his whistle, which sounds a single high-pitched note. The awaiting passenger will hear that single note slide up the scale or, as we have just learned, its frequency increases. This natural phenomenon, imparting the vaguely romantic wailing sound of approaching or receding train whistles, is the Doppler effect (or "shift") at an audible frequency. Note once again the strength of the source (I. e., whether the whistle blows loudly or softly) or its distance from the listener has no bearing on the effect the pitch will slide up the scale, regardless of these factors as the train approaches. Furthermore, by definition, the Doppler effect cannot be heard from a stationary source. The frequency changes only with movement of the source.

All radar, including the K55, transmit or broadcast high frequency microwave energy which emanates from the transmitter in the cycle pattern and "echoes" back to the source. Sound, of course, does this too, but the reflector must be quite large whereas, with microwaves, small objects can and do reflect them.

If either the source of the microwave transmission is moving or a reflecting object is moving, the Doppler shift occurs. Thus if a person sitting in the moving source could "hear" microwaves, he would note a change in frequency of the returning cycles: increased frequency if the reflector is moving toward him (as in the case of the road or an approaching car) and decreased frequency if the reflector is moving away. Physicists have measured the frequency change and expressed the measurements in one of those simple, elegant formulae to which all nature appears ultimately reducible:

Where f dop is the frequency of the returning microwaves, v is the velocity (speed) of the reflector, f is the transmitted frequency from the microwave source and c is the speed of light. Since c is always constant and the transmitter sends out microwaves of a known f, the only variable in this formula is v. Note again, that distance between source and reflector is absent from the formula as is any factor for the strength of either the transmitted or echoed energy.

The K55 Implementation of Doppler Principles:

The K55 unit is a small rectangular instrument which resembles a digital clock radio with switches, buttons and two windows on its face, one for patrol car speed and one for target vehicle speed. These readings appear as lighted, red digital numbers on a black background much as a digital clock or modern calculator. When no readings are being displayed, only the black background can be seen. The set is secured to the dashboard of the patrol car directly behind the steering wheel and can be readily seen either over or through the wheel. It is so small that it cannot obstruct the driver's vision through the windshield. The set can be plugged into the vehicle's power source in several ways, one of which is through the cigarette lighter, and it comes with an antenna and transmitter-receiver which is secured to the center of the dashboard.

The transmitter-receiver and antenna are constructed of standard components based on well known and accepted technology for the transmission and reception of high-frequency energy. The K55 does not use any experimental, new or patentable component or process in the antenna or transmitter-receiver. Every transmitter is factory-tested with instruments which derive their accuracy from the National Bureau of Standards in Washington, D.C. The transmitter—receiver has also been tested by an independent concern to verify that it transmits at the frequency assigned to enforcement radar by the F.C.C.

When the set is turned on and calibrated it may be operated in two modes: stationary and moving. Calibration is accomplished with the use of two tuning forks, the accuracy of which must be the subject of documentary proof. Use of the K55 does not eliminate the need for such proof. Cf. State v. Overton, 135 N.J.Super. 443 (Cty.Ct.1975). Each mode may be controlled in either manual or automatic position. With the antenna pointed straight out the windshield over the imaginary centerline of the patrol car and the power on, the transmitter will emanate a large lobe of high frequency microwave energy down the road. The set may be adjusted to give audible warning of a motorist approaching in excess of a predetermined speed; but the warning signal is not a prerequisite to reliable operation. The machine displays readings continuously for speeds between 20 m.p.h. and 99 m.p.h.

As the patrol car moves over the road, the microwaves reflected from the road will arrive back at the receiver at a frequency, predicted by the Doppler formula, varying with the speed of the patrol car. Let us call that frequency, frequency "X". Now assume a target vehicle approaches the patrol car. Microwaves reflected from the target vehicle will arrive back at the receiver at a frequency predicted by the Doppler formula varying with the sum of the speeds of both cars. Let us call that frequency, frequency "Y". Virtually by definition, frequency X will always be less than frequency Y. In the stationary mode only the high frequency "Y" is received and computed into a readable speed.

Thus, in the case of vehicles approaching one another ("closing") there are two returning frequencies, X and Y, both higher than the frequency originally transmitted, but Y always being greater than X. Because the transmitted frequency is constant, it can be dropped from further consideration and attention focused on the difference between X and Y, and they can now be called the low frequency (X) and high frequency (Y) returns.

We are now finished with the Doppler effect. All the rest of the K55 is devoted to processing the low and high frequency return signals into a readable speed in miles per hour of the patrol car and target vehicle. But review what has been accomplished: Because of a single variable, speed, in one case relative to the road and, in another case, relative to a closing vehicle, two completely distinct frequencies have been generated, which can be computed into the speeds that generated them.

Description of the K55 Signal Processing & Digital Units:

The high and low frequency returns from the receiver flow into the signal processing and digital unit of the K55. A precise technical description of the components and circuitry of the signal processor would take hundreds of pages, but the court finds that all of its individual components are standard and used in the K55 within their respective design limitations. The components are manufactured by reputable concerns such as RCA and Texas Instruments and are available generally on the wholesale or retail market for a wide variety of uses in electronic circuits, from radios and pacemakers to satellite telemetry. They all have known and accepted parameters of performance and durability in the industry.

Although the exact combination of components comprising the circuitry of the signal processing unit may be unique to the K55 (in one case a patent is pending with respect thereto), the overall circuitry design is also within industry-known, and accepted, electronic theory and practice. Neither the function nor the efficiency of any component is frustrated by its particular use or placement within the design of the signal processing unit.

The signal processing unit has ten functions, all of which it performs more or less simultaneously and in milliseconds: (1) it amplifies the high and low frequency returns from the receiver; (2) it filters out frequencies resulting in patrol car speed of less than 20 m.p.h.; (3) it separates the signal into low and high frequency channels of parallel design, and as to each channel, (4) it filters random frequencies (commonly referred to as "noise") from the Doppler frequency being returned; (5) it tracks and locks on the Doppler frequencies and verifies them as the frequencies to be computed into vehicle speeds and, in the case of the high frequency channel; (6) it passes the now clear and verified signal into a subtractor, and in the case of the low frequency channel; (7) it passes one part of the signal directly to a digital computer unit which computes and displays patrol car speed in miles per hour; meanwhile (8) it passes the other part of the low frequency signal into the subtractor where, (9) it subtracts the high and low frequency signals and (10) it computes and displays the target speed in miles per hour.

The processing unit sweeps through these functions so quickly that a newly computed reading arrives at the viewing window about 15 times a second. In the manual mode the reverification and computing process is continuous throughout the time and distance the vehicles are closing and thus permits the officer to lock in a reading manually when, in his judgment, the target has been properly identified and the reading stabilizes. In contrast, in the automatic mode, the K55 itself, automatically, locks in the first reading computed from a discerned signal. MPH does not recommend the use of this position and present State Police policy forbids it.

One of the most important overall operations of the machine that emerges from the ten-step process described above causes it to focus on or to tune to, precisely and automatically, the strongest frequencies being passed to it by the receiver. To use a mundane simile, also from the law enforcement field, the K55 acts like a well-trained bloodhound, who once given an example of the target scent, wants to, and can, filter out or discriminate between all the other distracting scents its nose senses and track the target scent. This simile, however, should not be misunderstood. The "strongest" signal to a radar may not mean the fastest car. Thus, as we shall see, signal strength depends on proximity to the transmitter and thus a slower target may well dominate the radar to the exclusion of a faster, but more distant, target.

The Technical Challenge:

After extensive examination, Andrew L. Soccio testified for the defense as a highly experienced technician in the field of electronic circuits. Soccio is not a physicist, scientist or engineer; he has not designed a radar-informed speed measuring device from scratch. But he is entirely familiar with electronic circuitry, the various components that comprise modern circuitry and the various methods common in testing such circuitry. Soccio can build a complex electronic circuit from a schematic diagram; he can analyze its function; test performance of that function and evaluate its performance. The court believes that in certain cases Soccio has the experience to design or redesign electronic circuitry to improve its intended function or increase its efficiency. After a wide variety of experience in various jobs in the electronics industry, Soccio opened his own business which is devoted to troubleshooting and repair of traffic radars in South Jersey. He has not, under contract or otherwise, serviced the K55.

Soccio's qualifications did not include extensive or intimate knowledge of the K55. He disassembled one in Florida this winter and examined certain parts of its signal processing unit for about an hour; he observed K55s in operation on police cars for about six hours, and he obtained and copied a partial schematic diagram of the signal processing unit issued by MPH to police departments for making repairs. Using this schematic, Soccio built, in his home laboratory, out of standard components, two elements of the circuitry in the signal processing unit and tested them. These elements were the "phase lock loops" and related components used in the high and low frequency channels. Based on these tests it was Soccio's opinion that the phase lock loops in both the high and low frequency channels possessed too wide a latitude in which to monitor and lock on an incoming frequency. Depending on minute transient errors in voltage, the phase lock loops operating in such a latitude could lock on a wrong frequency and pass it through to be computed. Patterson, called in rebuttal, challenged both the design of Soccio's home-made circuits because the schematic was incomplete, and his tests because they were inadequate. It was his view that the overall design of components aligned after the phase lock loops in the circuit functioned together to nullify both the underlying assumptions of the Soccio tests and the test methods and equipment used.

Thus was framed the most important and difficult single issue in this case: to believe the architect of the K55 or the technically astute opposing witness. After due consideration the court is satisfied that the Soccio tests do not raise a reasonable doubt as to the technical capability of the signal processing unit of the K55 to turn the high and low Doppler frequencies into accurate readings of patrol car and target speeds. The court is not satisfied that there was sufficient similarity in all necessary respect between the Soccio prototypes and the actual, complete circuitry of the K55. Furthermore, the court does not believe the test methods used by Soccio were sufficiently accurate to ground the broad opinion he gave on the K55's potential for displaying undetected, erroneous readings to a reasonably skilled operator.

The Operational Challenge:

The operational challenge to the K55 was grounded in the testimony of Dr. Nichols. He highway-tested the K55 in both Ohio and Florida for about 20 hours altogether. On these occasions, accompanied by others, he operated the K55 himself and observed others do so. All the participants were civilians with engineering training and experience. No K55-trained policemen assisted, nor had anyone performing the tests received training from MPH. In at least one instance the test was financed by Fuzzbuster, a manufacturer of traffic radar countermeasures.

In Florida six makes of radar were operated in a single vehicle, calibrated and tested for interference with internal noises such as generators, fans and the like. Then the vehicle, with the radars switched on in pairs, was driven on randomly selected highways in Miami and its environs. No written record of the tests were made, nor were any targets of known speed run through the radars. The radars were operated in both stationary and moving modes and in manual and automatic positions. Based on this experience and his own academic background Nichols' thoughts fell into four general problem areas: (1) problems of target identification where there were several possible targets approaching;

(2) problems of cosine error resulting in computation of lower-than-actual patrol car speed; (3) problems of internal and external mechanical, radio or microwave frequency interference, and (4) problems of a subjective nature. In all of these areas the K55 seemed to be most prone to potential error where it was switched to the automatic position. Dr. Nichols identified two sources of difficulty relating to target identification. Both problems arise from the necessity to decide which of several possible targets is actually generating the speed readout.

One of the design characteristics of the K55 is to track, hold and verify the strongest frequency being echoed to it. In practice, the strongest return frequency is usually that which is closest to the receiver. Thus, MPH in its Operator's Manual directs: Care should be taken by the operator that he recognize the violator is traveling at a higher rate of speed than the norm, that the vehicle is out in front, by itself, nearest the radar. (Emphasis supplied) But Nichols testified that a large target, rather than one closer to the patrol car, may actually be reflecting the strongest signal. Thus, if a truck is following a Volkswagen the K55 may compute the truck's speed rather than the Volkswagen's, contrary to the above-quoted identification rule. He also pointed out that some materials and shapes are better reflectors than others and could result in a more distant target, producing the reading, if made of such materials or in such shapes.

The court is satisfied that this problem could and probably does occur in a very close-following situation, and policemen should be alerted to the possibility of an incorrect identification. But Dr. Nichols gave no guidance whatsoever as to what was "close following." He apparently could not and did not say whether at 50 feet, 25 feet or at what distance between the approaching truck or Volkswagen, the rearward truck would begin to produce the stronger signal. Thus, the court finds the above-quoted rule a reasonably reliable guide on which to base identifications in the situation described.

Dr. Nichols also sought to describe target-identification problems where a possible violator was closer to the centerline of the transmitted microwaves than a target nearer in distance to the transmitter. He described the imaginary extended centerline of the patrol car (with the antenna so aligned) as the line of maximum transmitted power, and suggested that a return signal along that line would also be strongest. He pointed out that signal strength falls off far less as one moves away from that centerline than it does as one moves away from the transmitter. It has been found that one-half of the energy transmitted lies within an angle of 8o on either side of the centerline and, Dr. Nichols estimated, if the sides of that angle are carried out as much as 1500 feet from the patrol car, 85% of all the transmitted energy would lie within it. According, to Dr. Nichols, a target physically closer to the centerline, but more distant from the patrol car than the target nearest the patrol car, might produce a reading. An unwary policeman might attribute that reading to the closest vehicle, thus resulting in an unjust arrest.

But Dr. Nichols' views in this respect were unsupported by any precise measurements or detailed tests. None of his highway tests was performed under controlled conditions. No experimental results or theoretical models demonstrated Dr. Nichols' theories, to any degree of scientific probability, that a nearer target would produce a signal less strong than a more distant one, notwithstanding such distant target's proximity to the centerline of the micro-move-energy lobe. In fact, in most situations presented him, Dr. Nichols appeared to believe the nearest vehicle to the patrol car would generate the reading. Because it is true that signal strength (not frequency and hence not accuracy of a reading) falls off by the square of the distance from the transmitter, the court finds it more probable that that distance is, in fact, the key one, and therefor, once again, finds the quoted rule of identification reasonably reliable.

Dr. Nichols described cosine error as that error in target speed, which results from the fact that the Doppler frequency reflects along the line between the target to the patrol car, whereas the target is actually moving, not straight at the patrol car, but at an angle to it. Because of the trigonometric function of the cosine, which is always less than 1, the error always favors the target, i.e., displayed target speed is less than the actual target speed. But Dr. Nichols also theorized cosine error could reduce the display of patrol car speed. This is, potentially, a serious matter since reduced patrol car speed would result in false, higher target speed readings. Thus, if a huge sign or a long building with a long reflective surface on the side of the road reflected a signal stronger than that of the road and thereby generated the low frequency return for computing the patrol car's speed, the angle of such an object relative to the motion of the car might theoretically reduce the patrol car's speed reading by the cosine error. However, the court determines such situation would be necessarily rare, and the reading generated, transitory. Moreover, an alert officer, checking his patrol car readout against him speedometer, should note dropping speed and reject any target speeds obtained in that situation. Accordingly, the court determines that while cosine error as it affects patrol car readout may be a theoretical problem, its significance in routine traffic situations is not such as to raise a reasonable doubt as to the reliability of the K55 when operated by a trained and experienced officer.

Dr. Nichols also testified to internal and external sources of interference with the K55 which can produce spurious speed readings. The K55 possesses circuitry designed to filter out most internal and some external interference. Like all radar devices, the K55 can, and does, receive signals other than that which it originally transmitted. A wide variety of devices in relatively common use transmit electromagnetic energy at frequencies at or close to those assigned by the F.C.C. to police radar. Depending on the strength of these signals, which again depends on how close the source is to the K55, a speed reading can appear.

The signal processing unit may itself reject the signal and cause the reading to go blank, or the reading will be erratic, or so high or low as to tip the operator off at once that electronic interference may be present. Nonetheless, troopers are trained not to use their own radios when patrolling for speeders and to be mindful of the use of CB radios in the vicinity of the patrol area. Airports and even low strung high tension wires may also be common trouble spots. State and local police may well be advised to begin to catalog, within their respective jurisdictions, the existing sources, strength, frequency, range and direction of radiated energy which might intersect with or flow over the roads and highways. Armed with such information, tests could be run to determine the effect, if any, on the K55.

By so suggesting, however, the court does not wish to leave the impression that external radio or microwave interference seriously detracts from the reliability of the K55. The court is satisfied that the chance of an undetected interference increasing a speed reading, to the detriment of a motorist, is so remote as not to raise a reasonable doubt under the "high degree" of reliability standard here at issue. Absolute perfection, of course, is not required.

As a result of his Florida tests, Dr. Nichols expressed certain subjective reservations about the K55. He expressed concern over the practical difficulties of target identification in the multiple-lane, heavy traffic situation, and surprise at the variety of response times in the various radars and the differences between them in the speed with which they repeated their readings. He stated the K55 did not give him "the personal satisfaction that it was as stable and as reliable in the quickness with which it reached a reading that seemed to satisfy the physical situation as the others." He also stated, "it seemed to be more erratic to me. It went up too far or down too far. The swings were more violent. Ultimately it would stabilize and come back and give an adequate and satisfactory reading."

Considering the qualifications of Dr. Nichols, these observations are entitled to respect. In fact, to the extent he notes that the K55 did come to an "acceptable" reading as compared with other radars, he points to the sensitivity of the K55 in the manual position in rejecting spurious readings or other transitory phenomenon. Since Dr. Nichols was as critical of the automatic position, with its ability to lock in a transitory reading, as Patterson, the court finds it difficult to understand his subjective feeling about the K55. In any event, such feelings do not raise a reasonable doubt in the court's mind as to the reliability of the K55 when operated in the manual position.

Ghosts and Shadows:

Every law enforcement tool, whether it be a radar set or a bloodhound, must be understood and used within its inherent limitations. The K55, as all radar, has such limitations. Transmitted signals echo randomly from anything the microwaves reach and sometimes that signal, echoed from a tree, fence or billboard, will a second time be reflected from a moving object out of the trooper's visual range and be received and processed as a speed reading. These are "ghost" readings spurious speed readings of unseen vehicles or stationary objects. He who seriously reports tracking speeding trees with the K55 is either a fool or a knave, since such a report presents an inherent conflict with the underlying Doppler principle.

Such ghosts will be either so transitory or display such erratic readings that any experienced operator will at once recognize them. Moreover, ghosts will always be banished by a stronger signal because the K55 functions to find the lock on the strongest signal returning to it. Ghost readings cannot add to or detract from the speed of a real target. None of the experts challenged its reliability in that respect.

"Shadowing" occurs when the patrol car closely follows or is overtaken by a truck or large car with the K55 on in the moving mode. The signal usually returning from the road surface may now be temporarily supplanted by the stronger signal returning from the vehicle in front. Since that vehicle is traveling slower relative to the patrol car than the road was, the viewer will notice an apparent decrease in the speed reading of the patrol car. Any reading taken on a target vehicle under such conditions would be inaccurate. See, also, the discussion, at 58, 405 A.2d at 484, regarding reduced patrol car speed as the result of cosine error. But the evidence is clear that State Troopers are trained to check continually, while in the moving mode, the K55 display of patrol car speed with the speedometer. Any difference between the K55's reading and the speedometer is a trouble sign that will suggest caution on the part of the trooper in relying on a target speed reading.

Training

The court finds that the operational reliability of the K55 is largely dependent upon the training and experience of the policemen who use it. In State v. Dantonio, 18 N.J. 570, 573-574, 115 A.2d 35 (1955), the court quotes a law review article stating that "the average person engaged in traffic control work can learn to use the radar speedometer after about one and one-half hours of instruction." Judged by that standard, the State Police get ample instruction in theory and practice on the K55. Trooper Dempster, for instance, received a full day of classroom and "hands on" practical instruction from MPH, Inc., representatives, and then "practiced" with the machine in his own patrol car 80 hours before he arrested violators.

Such a level of instruction and experience acquaints the officers with the technical capability of radar as well as practical use in everyday traffic situations. For instance, like our bloodhound who cannot, having found the source of the scent, identify it as a vicious criminal or a lost child, so also cannot the K55 identify the speeding vehicle: the officer using it must do that. Troopers are taught a three-step procedure when the K55 displays a target speed: (1) identify the probable target producing the reading; (2) lock in the speed, and (3) compare visually the speed displayed with the officer's own estimate of the target's speed. This must be a complex procedure requiring well-coordinated eye and hand movement as well as the exercise of quick judgment. The officer must, also, be monitoring his own speedometer with that of the K55's readout on patrol car speed and driving his car with safety. With some traffic patterns, such as heavy approaching traffic in multiple lanes where no one car is clearly in front, it will always be difficult if not impossible to identify a target. But experience should quickly expose such situations.

In view of the above, however, and the high degree of skill and judgment required to operate the K55 reliably, it is the court's view that periodic follow-up training be instituted in order to verify continuing qualification as a K55 operator. Moreover, the skill and judgment of troopers who themselves instruct in its use should be most carefully evaluated.

Finally, it is clear from the testimony of all the witnesses, that the K55 should not be operated in the "automatic" position in either the moving or stationary mode. That is present State Police policy and should remain so. By "automatic" in the sense used here, is meant that position on the K55 which "automatically" locks on the first echo it receives and processes that echo to a readout and will not then process further echoes. Thus may be instantaneously captured an interfering signal or a ghost which would not be reflected from the visible target.

Conclusion:

For all the reasons stated above the court determines that a properly calibrated K55 Speed Detection Device installed in a car with a calibrated speedometer has a high degree of scientific and operational reliability when used in either stationary or moving mode, in the manual position by a person having at least three hours of classroom training and two to three hours of practical instruction together with some minimum experience prior to use in actual law enforcement. The court, on the record before it, cannot specify what the minimum experience should be but holds that 80 hours is, beyond doubt, sufficient.

When operated in the automatic position the operational reliability of the K55 is subject to greater question, and acceptance of readings while in that position must hinge to a far greater extent on detailed examination of the surrounding circumstances as well as the experience and training of the operator.

In view of the holding above, the court will shortly schedule argument on any and all other issues defendant may choose to raise on the record below relevant to his conviction.

With respect to admissibility, radar and its more modern offspring, lidar, have witnessed fits and starts over the years, but have nevertheless been accepted, often achieving a status that is subject to judicial notice. See *People v. Abdallah*, 82 Ill.App.2d 312, 226 N.E.2d 408 (1967). A separate issue may arise as to whether the readout by the device itself, without other evidence is sufficient to establish a conviction. To the misfortune of most of us, that question has been settled in the affirmative. *State v. Thompson*, 168 Ohio Misc.2d 3, 966 N.E.2d 320 (2012); see also *Watson v. U.S.*, 43 A.3d 276 (D.C., 2012) and *People v. Dusing*, 5 N.Y.2d 126, 181 N.Y.S.2d 493, 155 N.E.2d 393 (1959).

§ 7.36 *Chaos*

Chaos, many would say, has its place in the lawschool classroom, not in the world of mathematics or physics; they would be wrong. In truth "chaos" is an area of study in applied mathematics, not to mention physics, economics and biology. Chaos is present in many natural systems; the weather is a good example. Chaos theory always brings to mind the famous quote from poet Robert Burns' *Ode to a Mouse*: "Mousie thou art no thy lane in proving foresight may be vain; the best laid schemes of mice and men often go astray."

Chaos theory is predicated upon the notion that small differences in initial conditions or events often result in large variations down the road. Even so-called deterministic systems are, according to chaos theory not very predictable or deterministic. "*Sensitivity to initial conditions*" is a key principle. Firing a pistol at distal target is a good example. A tiny variation or twitch by the person holding the gun could result in a large range of potential consequences with respect to the target.

For another example, consider a relatively tiny rounding error in a missile guidance system's computer. Due to memory limitations, large decimal numbers must be rounded. During the course of the 1991 Gulf War, our Patriot missile system apparently failed to function properly due to a small round-off error. It appears that a computer performing the tracking calculations had an internal clock whose integral values were converted to decimals by multiplying the binary approximation to 0.1 as follows:

$$0.1_{10} = 0.00011001100110011001100_2 = 209715 / 2097152.$$

The error resulting from the calculation is particularly tiny (only 0.00001%). In fact, the quantity of the error could be calculated as follows:

$$1/10 - 209715 / 2097152 = 1 / 10485760.$$

Because the timing errors incremented at intervals of 0.1 seconds, the accumulated errors were proportional to the time on the clock; so after H number of hours, the accumulated rounding error was:

$$1 / 10485760 (3600 \times 10 \times H) = 225/65536 \times H$$

After 8 hours, the system's accumulated error amounted 0.0275 seconds. And after the system had been operating for 100 hours (accumulating an error of 0.3433 seconds), an enemy Scud Missile launched from Iraq found its way through the defense system and into allied barracks, killing 28 people. The accumulated error caused the defense missile to miss its target by potentially more than half of a kilometer.[10]

Fans of Mathematical Chaos may be interested in Oliver Stone's 1991 film, *JFK*, which purported to reenact the trial of *People v. Shaw* wherein Clay Shaw and a number of prominent New Orleans citizens were accused of taking part in an alleged conspiracy to assassinate former President John F. Kennedy. Physics apparently "took a hit" from Kevin Costner, who played New Orleans Prosecuting Attorney Jim Garrison (Closing Argument):

> To prove there was a conspiracy involving Clay Shaw we must prove there was more than one man involved in the assassination. To do that we must look at the Zapruder film which my office subpoenaed. The American public has not

10. Chartier, *Devastating Roundoff Error*, Math Horizons: Published by the Mathematical Association of America, April 2006 at 11.

seen the film. It's been locked in a vault the last five years in the Time-Life Building in New York. There's a reason for that. Watch. The picture speaks 1,000 words. The Warren Commission thought they had an open-and-shut case. Three bullets, one assassin.

But two unpredictable things happened that day to make it virtually impossible. One, the film shot by Abraham Zapruder while standing near the grassy knoll. Two, the third wounded man, James Teague, nicked by a fragment while standing near the triple underpass.

The time frame, 5.6 seconds, established by the Zapruder film left no possibility of a fourth shot. So the shot of fragment that left a superficial wound on Teague's cheek came from one of the three bullets fired from the Depository's sixth floor. That leaves just two bullets. We know one was the fatal head shot that killed Kennedy. So a single bullet remains. A single bullet must account for the seven wounds in Kennedy and Connally.

Rather than admit to a conspiracy or investigate further the Warren Commission endorsed the theory put forth by an ambitious attorney, Arlen Spector. One of the grossest lies ever forced on the American people. It's known as the "magic bullet" theory ... Some bullet. Any combat vet can tell you, never in the history of gunfire has there been a bullet this ridiculous.

The government says it can prove it with some fancy physics in a nuclear lab. Of course they can. Theoretical physics can prove an elephant can hang from a cliff with his tail tied to a daisy. But use your eyes, your common sense. The Army wound ballistics experts fired some comparison bullets. Not one of them looked anything like this. Take a look at CE-856. An identical bullet fired through the wrist of a human cadaver just one of the bones smashed by the magic bullet. Seven wounds, gentlemen. Tough skin dense bones. This single-bullet explanation is the foundation of the Warren Commission's claim of one assassin. And once you conclude the magic bullet couldn't create all seven wounds you must conclude there was a fourth shot and a second rifleman. And if there was a second rifleman then by definition there had to be a conspiracy.

Much of the preceding information was predicated upon the famous or infamous Warren Report (Report of the President's Commission on the Assassination President Kennedy), even though in the actual trial, Garrison made no such comments during closing argument.

Clay Shaw was found innocent and Garrison's discredit rose to Saturday Night Live levels. Nevertheless, in 1979, the House Select Committee on Assassinations stated that available records: "lent substantial credence to the possibility that Oswald and David Ferrie had been involved in the same [Civil Air Patrol] C.A.P. unit during the same period of time." Ferrie was alleged to have been involved in the conspiracy.

As far as chaos theory is concerned, if truth be told, there was no way to predict the trajectory of a bullet as it passed though two individuals. It is doubtful that any studies in that regard were ever conducted.

In any event, it is now time to leave chaos behind and turn attention away from the otherwise deterministic predictable precepts of classical physics and enter the world of modern physics, wherein chaos is an understatement.

Chapter 8

Modern Physics

"We live in a Newtonian world of Einsteinian Physics ruled by Frankenstein logic."

—David Russell

There exist two fundamental explanations as to the manner in which the universe is governed. The two theories are inconsistent and yet, surprisingly, both are accepted as true. On the one hand, there is the predictable, deterministic universe of Newton, as amended by Einstein, measurable through a calculus which requires waves of continuity and infinite divisibility. For the other side of the coin, there is a universe where nothing is completely predictable, but rather statistically probable or improbable, where *uncertainty* rules the day and particles as opposed to waves are free to dart about uncontrollably. It is the universe of quantum mechanics, whose ironic and unintended architect was also Albert Einstein.

Prior to the turn of the nineteenth century, scientists were erroneously confident that they possessed answers to all of the major questions of life, evolution and physics. Basic chemistry, atomic theory, Newton and Darwin, they thought, pretty much explained how everything worked. There were, however, a few serious unanswered questions.

The 1887 Michelson-Morley experiment (performed at the site of Case Western Reserve University) yielded results that conflicted with some very basic principles of classical physics. The logical theory of a luminiferous ether designed to explain the movement of light waves throughout the universe was being tested. Conventional wisdom figured that light waves going with the flow of the ether should go faster than light traveling against the flow; this seems to make sense. But the experiment, to everyone's upset, said that they both traveled at the same speed. At the time, no one could explain why something going against the current traveled at the same speed as something going with it.

Meanwhile, a different experiment was causing its own grief. It was designed to pull together in a single demonstration the principles of thermodynamics, heat emission and electromagnetism. The experiment, known as the "ultraviolet catastrophe," also known as the "Rayleigh-Jeans Catastrophe" (named after its creators and conductors) represented a simple prediction of classical physics that an ideal black body at thermal equilibrium would emit radiation with infinite power. This later became based upon the notion that the energy emitted by the black box would continue to increase to infinity as the frequency of the applied light increased. The classical theory, were it true, would result in the incineration of all sunbathers from the high frequency ultraviolet light of the sun.

Unlike the predictions of classical physics, the energy levels for the Rayleigh-Jeans experiment did not increase to infinity, but instead fell to near zero. Thus, the only catastrophe was to be found in the teachings of classical physics.

Enter Einstein. At the turn of the century, he determined that there were four unsolved mysteries of physics (which included the above mentioned dilemmas). A solution to any one would have guaranteed a Nobel prize along with a place in history as one of the great scientists of all time. Einstein would solve, not one, but all four, the first three in a single year, 1905.

§ 8.10 Einstein

No single proper noun name, not even that of Newton or Galileo, carries more intellectual mass than the name *Einstein*. There are other great scientists to be sure. Newton, the most prominent, gave us basic mechanics, physics and the calculus to handle them. Einstein, however, modified and corrected Newton's Laws, then carried scientific knowledge way beyond anything that even Newton could have imagined.

§ *8.11 The First Paper: The Birth of Quantum Mechanics*

In 1894 and for the next several years, the great theoretical physicist Max Planck turned his efforts toward the black body problem. Since the experimental results could not be explained by classical physics, Max turned instead to a "statistical interpretation" of the second law of thermodynamics in an attempt to explain the inconsistency. In the process, Max developed the radical notion that electromagnetic energy could be emitted only in "quantized" form. In other words, energy (E) could only exist in a multiples of an elementary unit. In particular: E = nhv.

The h is Planck's constant (representing the smallest divisible particle) and v is the frequency of the radiation. The n, sometimes referred to as the "quantum number," is particularly significant (n = 1, 2, 3 ...) because it implies that E will take on discrete as opposed to continuous values. To appreciate the significance of this "n" factor, and its anti-intuitive nature, pretend for the moment that your bathroom scale became "quantized" in packets of 5 pounds each. In other words, the scale could only read, and you could only weigh-in for groups of "5 pounds." You could weigh, for example, 150 pounds, 155 pounds, 255, etc., but you could never weigh 156 or 157. That is precisely how quantized energy works. Plank did not fully understand the significance of his discovery. He, nevertheless, received the nobel prize in 1918 for his work in quantum mechanics.

In the first of his three papers published in 1905, Albert Einstein carefully examined the phenomenon uncovered by Max Planck, indicating that electromagnetic energy seemed to be emitted from radiating objects in quantities that were ultimately discrete as opposed to continuous. The energy of these emitted quantities (the so-called light-quanta) was directly proportional to the frequency of the radiation not the intensity (as it was once believed). This was confusing because classical electromagnetic theory, based on Maxwell's equations and the laws of thermodynamics, had assumed that electromagnetic energy was composed of waves propagating in the hypothetical, all-pervasive medium called the luminiferous ether (of Michelson-Morley), and that the waves could contain any amount of energy no matter how small.

Einstein, unlike Plank, thoroughly understood the quantum hypothesis and used it to accurately describe visible electromagnetic radiation and light. According to Einstein's viewpoint, light could be imagined to consist of discrete bundles of radiation. Einstein used this interpretation to explain the "photoelectric effect," by which certain metals emit electrons when illuminated by light at a particular frequency. Einstein's theory, and his subsequent elaboration of it, formed the basis for much of quantum mechanics. His paper on the "Photoelectric Effect" earned a Nobel Prize.

§8.12 *The Second Paper: The Theory of Special Relativity*

The second paper that Einstein published in 1905 attracted the most media attention, due in part to its shattering effect on accepted classical physics and in part due to its prediction about the massive amounts of energy to be generated by atomic fission.[1]

At the time of publication Einstein knew that, according to Lorentz's theory of electrons, the mass of an electron increased as the velocity of that electron approached the velocity of light. Einstein was also aware that the electron theory, based on Maxwell's equations, carried along with it the assumption of a luminiferous ether, but that attempts to detect the physical properties of the ether had failed miserably. Einstein further realized that the equations describing the motion of an electron in fact could describe the non-accelerated motion of any particle or any defined rigid body.

Consistent with the unexpected results of the Michelson-Morley experiment, Einstein postulated that the laws of the laws of physics operated the same in any frame of reference, without regard to any ether. He further explained that the speed of light would remain constant throughout all frames of reference. Einstein disclaimed the existence of the so-called either, despite the fact that waves require a medium for transmission. As the amount of energy varied, if the speed of light was to remain constant, then mass and time itself would have to vary with velocity. Because the speed of light was to remain constant in all frames of reference (as required by classical Maxwellian theory), Einstein revitalized Lorentz's theory of electrons and the consequence thereof involving time dilatation.

$$\Delta t' = \gamma \Delta t = \frac{\Delta t}{\sqrt{1 - v^2 / c^2}}$$

Time, analogous to length and mass, is a function of the velocity of a frame of reference. From his theory, several now famous truths emerged. Matter and energy were indeed equivalent. And, of course:

$$E = mc^2$$

Where E is energy, m is mass and c, a constant in the extreme sense of the term, is equal to the speed of light in a vacuum. The speed of light, according to Einstein, is the true speed limit of the universe.

The mathematical proof of Einstein's theory of special relativity flows from the equation that measures energy in terms of mass and velocity for moving objects (See §7.13): $E = ½ mv^2$ combined with Maxwell's theory pertaining to momentum and light energy.

1. Einstein: "The release of atom power has changed everything except our way of thinking ... the solution to this problem lies in the heart of mankind. If only I had known, I should have become a watchmaker."

EXPLAINING AND PROVING SPECIAL RELATIVITY:

Explaining relativity is never as easy as physicists, mathematicians and writers think that it should be. Most educated individuals accept but never quite understand its basics holdings: The speed of light is constant for all observers; energy equals mass times the speed of light squared; clocks slow down and things become shorter as they go faster; etc. A complete understanding as to how these things happen eludes most of us, especially when math enters the picture. Cute cartoons and stories about traveling twins at beauty pageants sometimes help; but they can also confuse matters. For that reason, it may be best to use the simplest unembellished fact pattern. So let's put aside the twins and deal with a single person Albert (Al) and a simple ball (Ball).

We know from basic Newtonian physics (Chapter 7) that the speed or velocity (v) of the ball will equal the distance traveled (d), divided by the time (t) that it takes for the ball to cover that distance. In general:

$$\text{Velocity} = \text{Distance} / \text{Time}$$

This simple rule of physics together with other laws of Newton will remain unchanged regardless of the point of reference (or inertial frame) under consideration. Assume that Al, who is standing still, tosses the ball to a height of "d" feet. It takes "t_0" seconds for the ball to reach its highest point before it begins its descent. Note that "t_0" instead of "t" is being used to represent this "initial" constant time. (By contrast, the variable "t" without the subscript is going to represent a change or dilation of time.)

d ↑  ↓ d

The total distance traveled by the ball (i.e., going up and coming down) will equal 2d. Therefore, the speed (v) of the ball would be represented by: $v = 2d / t_0$. This equation and the path of the ball with respect to Al and Al's viewpoint, would not change regardless of where Al was or how fast Al was traveling. He would always perceive the path of the ball as it is represented by the arrows above.

The fact that Al would always observe the ball's motion to be the same (regardless of his inertial frame of reference) is significant though not surprising. For Al, it would make no difference whether he tossed the ball while on the ground or while in the aisle of a jet plane traveling at supersonic speeds, or while standing on the floor of a train's box car rushing down the tracks. The path of the ball would always appear *to him* to be straight up and down.

Now assume that Al happens to be on a train traveling past you as he bounces his ball. If *you* stand still, a short distance from the tracks, watching the train whiz by (much like the unlucky Mrs. Paslgraf or the unluckier Mr. Tompkins), you would see a different path; it would resemble an upside down "v" or pyramid (∧), instead of the straight vertical up-and-down path (↑↓) that Al would see.

We know that regardless of the velocity of the train, the distance traveled will appear to be greater from your stationary perspective as compared to Al's perspective. The legs of the ∧ will always be longer than ↑↓, in much the same fashion that each of the equal legs (d') of an isosceles triangle will always be greater than the height (d) of the triangle.

For Al, we already agreed that: $v_0 = 2d / t_0$. Therefore, we could represent the distance the ball traveled by solving for "d" as follows:

$$v_0 = 2d / t_0 \quad \Rightarrow \quad t_0 v_0 = 2d \quad \Rightarrow \quad d = t_0 v_0 / 2 \quad \Rightarrow \quad d = \tfrac{1}{2} t_0 v_0$$

Concerning your perspective or that of anyone observing the train go by (we presume that the boxcar is made of glass to help visualization), the height (d) of the ball would be the same, but the distance traveled would *not* be 2d, but rather the distance 2d', where:[2]

d' ∧ d' is a longer path than d↑↓d (because d' > d).

Next, for the moment, forget about vertical speed of the ball and shift attention to the velocity of the train (v). As the train moves, from your perspective, the ball will travel distance d' twice, whereas, for Al, it will appear to travel only distance d twice (a shorter distance than d' twice). In addition, from Al's perspective the lateral (horizontal) distance traveled by the ball would be 0; but the lateral distance of the ball from *your* perspective would be: v t (because Distance = v t_0 which follows directly from the simple fact that Velocity = Distance / Time).

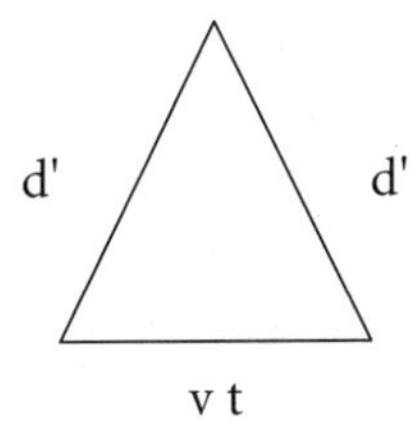

We can create a right triangle by slicing the above isosceles triangle in half. Notice that the height of both triangles is a constant equal to d.

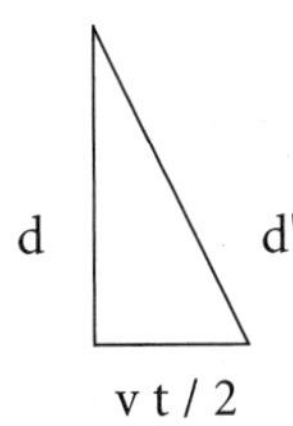

The next step involves calculating the length of "d'" using the Pythagorean Theorem (See § 2.60) which, as we know, states that the square of the hypotenuse of a right triangle equals the sum of the squares of the other two sides. Therefore: $d'^2 = d^2 + (\tfrac{1}{2} v t)^2$.

2. Strictly speaking, the ball's path would more closely resemble an arc as opposed to a pyramid.

All of our discussions up to this point involve nothing more than "classical" pre-Einsteinian relativity. Now comes the "big" substitution. Instead of a normal ball, assume that our ball bounces at the speed of light. Einstein proved that the speed of light is a constant (186,000 miles per second). Herein lies the lynchpin of "Einstein's Theory of Relativity," which today is merely referred to as "relativity." If the speed of light is constant for **both** you and Al, and if "time is time" and does vary just because it happens to be measured by two different people, then we come to the following paradox:

Premise 1: With respect to Al: $v = 2d / t_0$

Premise 2: With respect to you: $v = 2d' / t_0$

Conclusion 1: $2d' = 2d$

Conclusion 2: $d' = d$ (which is impossible)

The distance that the ball travels from *your* perspective (2d') will be longer than the distance traveled from Al's perspective (2d). But if the distance 2d' is going to be traversed at the same velocity as a shorter distance (2d), then the distance 2d' *must* take longer to traverse. Time must stretch to make the accommodation. Think about it. If you and a friend set out for a journey and travel at identical speeds, the person traveling the longer distance must take more time to complete the journey. Therefore, despite common sense and intuition, Conclusion 2 is false; time must in fact stretch from your perspective, and correspondingly shrink for Al.

For Al, as compared to you, the faster he travels, the greater the shrink (contraction) in time. It will, therefore, take longer for him to age as compared to you. That is why it is said that a person traveling in space at near light speed would return to Earth young while everyone else on the planet would have aged or keeled over and dropped dead.

The amount by which time must stretch proportionately depends upon how fast the train is going. Therefore, the time dilation factor would equal d / d'. Also, now that we know that v is a constant, we can use the traditional notation: c = the speed of light. A lengthy and somewhat annoying algebraic rearrangement of our Pythagorean equation, solving for the variable t will eventually lead us to:

$$t = \frac{t_0}{\sqrt{1 - \left(\frac{v}{c}\right)^2}}$$

Notice that unless the velocity (v) is very large, the dilation of time will be "relatively" insignificant (no pun intended). In particular even if our train were moving at 1,000,000 miles per hour, v/c would still be very small. There are 3600 seconds in one hour. Therefore:

1,000,000 miles per hour = 1,000,000 / 3600 = 278 miles per second. Therefore:

$v/c = 278/186{,}000 = .0015 \Rightarrow$

$(v/c)^2 = .0000022 \Rightarrow$

Because $(v/c)^2$ is extremely small, $1 - (v/c)^2 = .9999978$, which is extremely close to $1 \Rightarrow$

The denominator will be so close to 1 that there would be no noticeable effect on time dilation.

The above equation led Einstein in many directions particularly since the results had an Earth shattering impact on every major field of physics, including Electromagnetics (§ 7.30), Thermodynamics (§ 7.20) and, of course, all of Newtonian Physics (§ 7.10). The equation also applies to other physical properties in addition to time. As objects increase in speed, they become shorter, and more massive.

Thanks to Maxwell's landmark discovery we know that light is part of the electromagnetic spectrum (§ 7.30). Both electromagnetic radiation and light travel at the same speed (c). Einstein determined that because light had mass, the *Law of the Conservation of Momentum* (§ 7.10) would apply. Recall that momentum (ρ) = mass (m) · velocity (v).

Velocity (v) = d / t

Energy = m · d^2 / t^2

Prior to Einstein, physicists knew that a beam of light could exercise radiation pressure against matter. This meant that light had both mass and momentum. A beam of light would exhibit energy (E) with momentum E/c. Einstein used this fact to show that radiation energy has an equivalent mass. For a particle of light, Einstein reasoned, the Conservation of Momentum would require that ρ = E/c. You can verify this equivalence algebraically:

$$\rho = E / v$$

$$\frac{m \cdot d}{t} = \frac{m \cdot d^2}{t^2} \cdot \frac{t}{d}$$

From the fact that light has momentum, Einstein showed that light energy has the characteristics of mass also. In other words, energy has inertia. It turns out that all energy has this feature. That's because one form of energy can be transformed into another. So, if one kind of energy has this characteristic, all forms of energy do. The following is an oversimplification of the thought process that Einstein used to arrive at his most famous equation:[3]

$\rho = E/c \Rightarrow$

$\rho\, c = E \Rightarrow m\, v\, c = E \Rightarrow$

Since v will equal the speed of light ⇒

$E = m\, c\, c \Rightarrow$

$E = mc^2$

§ 8.13 The Third Paper: Brownian Movement

In the early 1800's, biologist Robert Brown noticed that pollen grains in water jumped and jiggled. Termed "Brownian movement," Robert spent a great deal of time looking through a microscope in a futile attempt to figure out what was causing it; apparently, Robert had no life.

---

3. Hear his famous lecture at: http://www.drphysics.com/syllabus/energy/energy.html.

In his third paper, Einstein explained the nature of atoms and molecules, an idea that had been around for a long time. In fact, the famous chemist John Dalton had intimated that all chemicals were made of tiny invisible molecules, which in turn were made of even tinier atoms. The problem, however, was that no one had proved their existence until Einstein looked into the problem of Brownian motion.

Einstein recognized that the jiggling motion of the pollen grains was attributable to molecules of water hitting the tiny pollen grains. The pollen grains were visible but the water molecules were not. It looked like the grains were bouncing about on their own, which of course makes no sense. Einstein also demonstrated that it was possible to calculate how many molecules were hitting the grains and how fast the water molecules were traveling.

Einstein's paper also made significant predictions about the properties of atoms that could finally be tested. The French physicist Jean Perrin used Einstein's predictions to work out the size of atoms and remove any remaining doubts about their existence. Einstein's paper also provided convincing evidence for the physical existence of atom-sized molecules, which had already received much theoretical discussion. His results were independently discovered by the Polish physicist Marian von Smoluchowski and later elaborated by the French physicist Jean Perrin.

§ 8.14 *General Relativity*

Published in 1915, and considered as the single most advanced concept in the history of the human species, Einstein's theory of general relativity, which flowed from his special theory, accurately describes the manner in which space and time bend according to the mass of the objects in the universe. In developing and explaining general relativity, Einstein modified another cherished concept of Newton—Gravity.

For the moment, forget what Newton taught you about gravity. It's not 100% accurate. There are no mysterious invisible arms that reach up from the ground to pull down apples on your head. The apple does not fall because of an invisible force; rather it merely falls (or moves) into the dent that the massive Earth causes to appear in the time-space continuum. In particular, time and space are curved by the objects that populate them.

General relativity is a geometric theory of gravitation that unifies special relativity and Newton's law of universal gravitation. In addition, it describes gravity as a geometric property of space and time, or spacetime. In particular, the curvature of spacetime is directly related to the mass and energy at a particular location. The relation is specified by Einstein's field equations, a system of partial differential equations. Among the field equations, which are beyond the scope of this text is the common notation: $G_{im} = - k\ (T_{im} - \frac{1}{2}\ g_{im}\ T)$. The left side of the equation encompasses a set of factors that characterize the geometry of space; the right side describes the distribution of energy and momentum.

Every piece of matter in the universe bends time and space, to a lesser or greater degree depending upon its mass. In turn, time and space determine the movement of the planets, the stars and every piece of matter in the universe. The universe, therefore, does not consist of isolated objects traveling through space in strict adherence to the laws of Newtonian physics.

The new views of physics have not been lost on all legal scholars. Some see a connection between general relativity and jurisprudence.[4]

Notes and Problems

1. Classical relativity took center stage in the famous "Monkey Trial," involving the prosecution of high school biology teacher John Scopes for teaching Darwin's Theory of Evolution in the state of Tennessee.[5] The court prevented famous defense attorney Clarence Darrow from using the testimony of experts in biology, anthropology and geology. As it turned out, all that Darrow had left was relativity.

The Scopes case was prosecuted by the *three-time* unsuccessful presidential candidate and Christian fundamentalist, William Jennings Bryant.[6] Bryant charged that teaching evolution was deemed contrary to the Bible, and therefore, contrary to Tennessee law. The trial, as portrayed in Stanley Kramer's classic film *Inherit the Wind* pitted the notorious Clarence Darrow [Played by Spencer Tracy] against Bryant.[7] The arrogant Bryant, who was at times assisted by co-counsel, took the stand as an expert on the Bible, permitting himself to be cross-examined by the legendary Darrow.[8] Darrow's goal was to show that *Evolution*, or *Science* in general, did not *necessarily* conflict with biblical doctrine. During the classic cross examination that follows, Darrow ["Drummond" in the film] managed to shake the otherwise unshakable Bryant [Brady] and his intractable dogmatic claim that the unholy "*Evil-Lution*" as he termed it, could never be reconciled with biblical teachings:

Drummond: [Holding up a divided piece of stone] How old do you think this rock is?

Brady: [Sarcastically] I am more interested in the *Rock of Ages* than I am in the age of rocks.

[Portions Omitted]

---

4. Laurence H. Tribe, *The Curvature of Constitutional Space: What Lawyers Can Learn from Modern Physics*, 103 Harv. L. Rev. 1 (November, 1989).

5. *Scopes v. State*, 289 S.W. 363 (Tenn. 1927).

6. His famous *Cross of Gold Speech* decrying the country's shift from the gold standard became a "standard" for *over-the-top* bombastic rhetoric.

7. In the film, Darrow (Drummond) was retained by the American Civil Liberties Union to defend Scopes in a Tennessee courtroom, in front of a town and jury harboring a lynch mob mentality. The case was actually tried in Dayton Ohio.

8. This wasn't *just* good Hollywood, it actually happened. Subject to the court's discretion and the stipulation of the counsel, an attorney for a party may take the stand to be examined and cross-examined.

Drummond: Dr. Page of Oberland College tells me this rock is at least 10,000,000 years old[9] ...

Brady: That rock is not more than 6,000 years old.

Drummond: How do you know?

Brady: A fine Biblical scholar, Bishop Usher has determined for us the exact date and hour of the creation; it occurred in the year 4004 B.C.

Drummond: Ah ... well ... That's ... Bishop Usher's opinion.

Brady: It's not an opinion. It's a literal fact, which the good Bishop arrived at through careful computation of the ages of the prophets as set down in the *Old Testament*. In fact, he determined that the Lord began creation the 23rd of October, 4004 B.C.... at 9:00A.M.[10]

Drummond: Was that Eastern Standard Time? Or Rocky Mountain Time? ... It wasn't daylight savings time, was it? Because the Lord didn't make the Sun until the 4th Day?

Brady: That is correct.

Drummond: That first day ... What do you think it was? 24 hours long?

Brady: The Bible says it was a *day*.

Drummond: Well ... but there was no sun out. How do you know how long it was?

Brady: The Bible says it was a day!

Drummond: But was it a normal day? A literal day? A 24 hour day?

Brady: I don't know.

Drummond: What do you think?

Brady: I do not think about things that I do not think about.

Drummond: Do you ever think about things that you *do* think about?[11] Isn't it possible that it could have been 25 hours? There was no way to measure it, no way to tell ... Could it have been 25 hours?

Brady: It's possible.

Drummond: Then you interpret that the first day as recorded in the book of *Genesis* could have been a day of indeterminate length?

Brady: I mean to state that it is not necessarily a 24 hour day.

---

9. Under normal circumstances, this should, of course, draw a hearsay objection (or at least an objection for assuming unestablished facts). See Rule 801 of the *Federal Rules of Evidence.* Here, however, it might be urged that Page's out of court statement was merely being used as a foundation for Brady's testimony about the Biblical 6,000 year limit.

10. It is worth noting that despite Hollywood's spin, the fundamentalist Bryant was not anti-science. In fact, he was a Biologist (no less successful than he was as a politician).

11. This is as a good a place as any to plug a reminder that you are not entitled to argue with a witness, even during cross-examination. See generally, *Rule* 611(a) of the *Federal Rules of Evidence.* However, in criminal matters, greater latitude is permitted when cross-examining a witness for the prosecution. See, for example, *Commonwealth v. Gagnon*, 557 N.E.2d 728, 408 Mass. 185 (1990) and *In re W.D.*, 141 Ill.Dec. 364, 551 N.E.2d 357, 194 Ill.App.3d 686 (1990).

Drummond: It could have been 30 hours![12] Could have been a week! Could have been a month! Could have been a year! Could have been a hundred years ... Could have been 10,000,000 years![13]

---

2. If you represented the defendant in a moving rear end collision case, how would you use relativity to your advantage?

3. How is the Doppler effect used by police officers sitting at speed traps?

4. Modern physics has provided some powerful evidentiary tools for law enforcement, including Atomic Absorption Spectrometry (AAS), Infrared Spectrometry (IR), and Ultraviolet (UV) Analysis.

§ 8.20 Quantum Mechanics

The subject of Einstein and quantum mechanics are by no means separate topics. Many credit Einstein with being the father of quantum mechanics. But as the theory advanced, it became apparent that uncertainty, probability and statistics were taking aim at the pure absolutistic nature of physics and mathematics. Had Einstein helped to create a monster?[14]

Quantum mechanics, often referred to as "quantum physics," happens to be based upon extensive scientific study and experimentation. Its principles and precepts often lead to some strange, anti-intuitive conclusions, but that does not render it Voodoo or junk science. Nevertheless, whenever crazy or adventive allegations present themselves in court, judges are quick to call upon the world of quantum physics. In fact, the treatment of quantum physics by our court system is as strange as the subject matter itself. It's as if there is this *other* quantum universe, separate and apart from reality, that forms a suitcase full of every half-baked, unsupported allegation that civil and criminal litigants dream up. For example, see: *Sommers v. Household Int'l, Inc.*, 1999 WL 1285858 (N.D.Ill., 1999) ["When event A precedes event B, it is sometimes questionable whether A caused B, but only in the realm of quantum physics can B cause A."]; *Driver v. Sentence Review Div. in Supreme Court of State of Montana*, 227 P.3d 1018 (Mont., 2010) ["Obviously, equity cannot simultaneously be both part of the calculus and not part of the calculus—except, perhaps, in the world of quantum physics."]; *In re Meyer*, 120 F.3d 66 (7th Cir., Ill., 1997) ["Only in the realm of quantum physics can an entity exist in two places at the same time, and Meyer is no electron."]. In *Edwards v. Groner*, 124 F.R.D. 605 (D. Virgin Islands, 1989), an attorney who claimed during a closing argument that human consciousness functioned according to quantum physics principles, was characterized as bazaar and somewhat imbalanced. (He may have been correct, but the fact that the eccentric attor-

---

12. Time, of course, is relative, and has no meaning without some moving object as a reference point. Therefore, from a relativistic stance, at least, Darrow was on rock-solid ground.

13. Where you face a barrage of unfavorable testimony from a stubborn adverse witness, try to aim for a single target or point. The more time that it takes to unseat the witness, the greater the significance that the jury may attach to it. In other words, give your winning points, small though they be, as much air-time as possible.

14. When confronted with the notion that probability ruled the universe, Einstein replied: "I am convinced that He (God) does not play dice." When confronted with Einstein's statement, Woody Allen replied: "God doesn't play dice; he plays hide and go seek."

ney routinely sent copies of his pleading to President Ronald Reagan, Prime Minister Yitzhak Shamir and Mikhail Gorbachev probably suggested imbalance).

In *Wilson v. Shea*, 104 Cal.Rptr.2d 880 (Cal.App., 2001), as part of a tirade regarding some disfavored guidelines pertaining to custodial-parents (which the bench referred to as "a sad state of affairs"), the court's opinion cited *In re Marriage of Schulze,* 70 Cal.Rptr.2d 488 (19987) which characterized the guidelines as: "a kind of hybrid of quantum physics and Zen philosophy." In *Sanchez v. State,* 926 S.W.2d 391 (Tex.App.–El Paso, 1996), an appellate court went so far as to declare the science to be un-Texan, by describing a particularly unacceptable circumstance as being: "a condition perfectly possible in quantum physics, but not in Texas criminal law." *Sanchez* dealt with judicial bias and its relation to court funding, while noting: "Like Schrodinger's cat, we view the motions as being two things at once: a challenge to the Impact Court's method of funding, and a challenge to the neutrality of a trial judge who relies upon that method of funding for his supper. Unlike that paradoxical cat, however, we do not view those two challenges as mutually exclusive."

The hypothetical cat referred to in *Sanchez* was owned (and simultaneously not owned) by quantum physicist Erwin Schrodinger. The cat is a favorite pet used to demonstrate an important principle of quantum physics, even though ironically, it was intended by its creator to demonstrate the absurdity of that very principle. The famous thought experiment places one ordinary cat in one large box with a particle detector, a radiation source and a bottle of cyanide gas. The detector is set up so that if it detects a particle from the radiation source, it will open the cyanide gas and kill the cat. It is configured in such a way that there will be a 50% probability of a particle being detected from the radiation source within a five minute period. Add the cat to the box. Quantum theory says that the cat will enter a quantum state where it is 50% completely alive and 50% totally dead until the experimenter looks inside the box; at that moment, the cat will either be alive or dead. Until then, the cat is both dead and alive at the same time. The purpose is to demonstrate that reality only exists if there is something or someone to observe or measure it.

The experiment has been both ridiculed and praised (and condemned by the ASPCA). Comedians teach us that the severely pissed off cat will escape from the box before the 5 minute period ends and scratch the heck out of the physicist.

In truth, the scientific and philosophical implications of quantum physics are very far reaching, even to the extent that they occasionally shine some light on our antiquated judicial system. In their law review article "*The Legality of Early Retirement Incentive Plans: Can Quantum Physics Help Resolve the Current Uncertainty?,*" 84 MARQLR 587 (Spring 2001), Michael Van Sistine and Bruce Meredith used quantum physics to illustrate the need to revise our classic, outdated discrimination laws and rules, to better conform to current statistical trends with respect to early retirement. Unlike prior judicial statements which praise precedent while eschewing quantum physics, these authors urged legal thinkers to follow in its footsteps by rethinking underlying premises that no longer work. They thoughtfully point out the manner in which the classic laws of Newton and Einstein are very accurate when dealing with matters at a tangible or even a celestial level. But those laws, without modification, do not work for subatomic particles. A new way of thinking was needed or as the writers said, a "paradigm shift" was required. They said: "It was not until physicists developed a new paradigm—quantum physics, which recognized that subatomic events required a separate and distinct analysis—that they achieved greater accuracy in understanding and predicting subatomic phenomena. A similar state of confusion and anomaly now exists in the area of age discrimination in general, and early retirement incentive plans in particular. Some courts are behaving much like the physicists who unsuccessfully attempted to impose their 'classical' Newtonian notions of physical reality on the subatomic world. These courts continue to analyze ERIP's under strict 'classical discrimination principles,' even while acknowledging the seemingly arbitrary and unfair results stemming from such an analysis."

A more simplistic approach to subatomic particles (electrons, in particular) was provided in *GFI Wisconsin, Inc. v. Reedsburg Utility Commission*, 440 B.R. 791 (W.D.Wis., 2010), by the District Court in a bankruptcy battle over the nature of electric currents; the court held that electricity received by the debtor qualified as a "good," for which utility companies were entitled to assert an expense claim, notwithstanding the debtor's contention that electricity did not satisfy the "movability" requirement for a "good" under the Uniform Commercial Code (UCC). The creative debtor claimed that the movement of electrons was so fast as to be simultaneous with its consumption. The court, nevertheless, concluded that "electricity is a 'good' within the meaning of § 503(b)(9) of the Code. "Determining the physical nature of electricity is complex." According to the opinion, it requires an understanding of the nature of electrons and a grasp of quantum physics and special relativity. "For the purpose of determining administrative priority under the Bankruptcy Code, the meaning of 'goods' under the UCC should not depend on quantum physics, how fast electrons are moving at a particular time or even where a debtor's meter is located on an electrical circuit. Rather, determining whether a particular thing qualifies as a good should be a straightforward assessment, taking into consideration the nature and common understanding of the thing, but also considering its similarities to goods that fall undisputedly under the UCC."

Dr. Prentice, a physics professor at the University of Michigan, stated that no one should be permitted to graduate high school without knowing the six basic postulates of quantum mechanics. Nevertheless, at a time when a recent survey among high school students demonstrated that most could not even name a single living scientist (Most common response: "Chemical Ali"), we needn't be too concerned about any of them knowing the postulates. It would, however, seem that during a discovery deposition, any self-styled "scientific *expert*" might be fair game for the request to recite, or at least paraphrase them:

> **Postulate 1:** The state of a physical system at time t is represented by an abstract unit vector $|\Psi(t)\rangle$ called a state vector, belonging to a vector space called a "state space."
>
> **Postulate 2:** Every measurable (observable) physical quantity is represented by a linear hermitian operator acting on the state space.
>
> **Postulate 3:** The only possible result of the measurement of a physical quantity is one of the eigenvalues of the corresponding operator.
>
> **Postulate 4:** If a physical system is in the state $|\Psi\rangle$ and a measurement of an observable, A, is performed then the probability of observing the eigenvalue λ_i is: $P(\lambda_i) = |\langle e_i \mid \Psi\rangle|^2$ where $|e_i\rangle$ is the eigenvector corresponding to λ_i.
>
> **Postulate 5:** If a measurement yields the result λ_i, then the state of the system immediately after the measurement is: $|e_i\rangle$.
>
> **Postulate 6:** The time evolution (dynamics) of the state vector $|\Psi(t)\rangle$ is determined by the equation (Schrodinger's Equation):
>
> $i\hbar\, d/dt\, |\Psi(t)\rangle = H\, |\Psi(t)\rangle$, where $\hbar$ = Plank's Constant, H is the Hamiltonian Operator corresponding to the total energy of the system, and i is the complex number, the square root of -1.

Notes and Problems

1. The creative argument advanced by the debtor in the *GFI Wisconsin* case (*supra*) was based upon UCC § 2-105(1) which defines goods as all things which are "movable" at the time of identification to the contract for sale. How would Einstein have reacted to the debtor's claim that electricity was so fast as to be "simultaneous" with its consumption?

2. In the game show *Deal or No Deal* (See *Notes and Problems*, § 4.32), contestants select one from 25 brief cases. That selected case remains closed until all of the others are opened, thereby indicating the contents of the closed case by process of elimination. Assume that 40% of the briefcases contain $10,000 and 60% of the briefcases contain $20,000. According to the laws of quantum physics, as soon as the contestant selects his unopened briefcase, what are its contents? How does the contents of the selected closed case "change" as the other cases are opened?

3. As might be expected quantum mechanics and nanotechnology are closely related. Size-wise, the relationship was quite predictable. When describing nanoparticles, scientists and manufacturers are discussing measurements in the range of one billionth of a meter. Nanoparticles and nanotechnology are working their way into our products and our environment at a dramatically increasing rate. Computers, batteries and electric automobiles represent few in a growing number. This is apparently significant from the perspective of the products liability bar; like the radiation of the 1950's, the dangers of nanotechnology are still unknown, mysterious and unregulated. See Lewis L. Laska, *What Every Personal Injury Lawyer Should Know About Nanotechnology*, Trial Magazine (September 2012). Perhaps the subject matter will attract attention one day when a nanotech handler gives birth to a 20 pound foot.

4. Rene Descartes is sitting in a bar doing what he does best—philosophizing. He's had a few too many and it's now last call. The bartender asks him if he wants another drink. Descartes says, "I think not," and promptly vanishes. That same night, Heisenberg left the bar driving drunk and fast. He was stopped by a traffic cop. The cop asked, "Do you know how fast you were going?" Heisenberg proclaimed: "No, but I know where I am." (With respect to mathematicians and senses of humor, this is about as good as it gets.)

§ 8.30 Going Nuclear

The following case, *Allen v. U.S.*, deserves a special word of introduction. Even though it has been extensively edited (cut by approximately 60%), it may still hold the record as the longest ever casebook case. *Allen*, however, is more than a "case." Judge Jenkins' decision is a self-contained course in nuclear physics, radioactive fallout and the biological effects of both. It is a treatise of modern scientific analysis and decision-making.

The opinion presents and discusses some of the most fascinating subjects, previously mentioned throughout this text, ranging from a discussion of Heisenberg's Uncertainty Principle, through Einstein's theories of relativity.

Not all of the discussion involves the more esoteric subjects of modern physics, however. The basic rules pertaining to scientific notation and calculation along with a consideration of the chemical elements that comprise the entire periodic table are included.

As far back as July 16, 1945 the government of the United States and the literate population of the world were aware of the fact that nuclear bombs were capable of destroying buildings and people in the immediate blast area. At the time and shortly thereafter, the rule of thumb for witnessing an atomic bomb explosion directed that you position yourself at least three miles away from ground zero, turn your back to the blast site, cover your eyes with your elbow, then close your eyes. You then wait patiently for the explosion, at which time, the intensity of the heat and light will allow you to see your bones through your closed eyelids.

Unlike the rapid dramatic visual display presented by the bomb, radiation, though silent and invisible, became a more insidious killer. Its effects could continue depending upon the radioactive half-lives of the elements subjected to nuclear fission. Due to the top secret nature of the government's nuclear program (which is subject to the exclusive jurisdiction of Nuclear Regulatory Commission), it became clear that the government knew more about the dangers of radioactivity than it was telling the public.

Allen v. United States

588 F.Supp. 247 (D.C.Utah, 1984)

JENKINS, District Judge.

I. INTRODUCTION AND STATEMENT OF THE CASE

In a sense this case began in the mind of a thoughtful resident of Greece named Democritus some twenty-five hundred years ago. In response to a question put two centuries earlier by a compatriot, Thales, concerning the fundamental nature of matter, Democritus suggested the idea of atoms. This case is concerned with atoms, with government, with people, with legal relationships, and with social values.

This case is concerned with what reasonable men in positions of decision-making in the United States government between 1951 and 1963 knew or should have known about the fundamental nature of matter. It is concerned with the duty, if any, that the United States government had to tell its people, particularly those in proximity to the experiment site, what it knew or should have known about the dangers to them from the government's experiments with nuclear fission conducted above ground in the brushlands of Nevada during those critical years.

The complaint in this action alleges that each plaintiff, or his predecessor, has suffered injury or death as a proximate result of exposure to radioactive fallout that drifted away from the Nevada Test Site and settled upon communities and isolated populations in southern Utah, northern Arizona and southeastern Nevada. Each of the plaintiffs or their decedents resided in that area. Each claims serious loss due to radiation-caused cancer or leukemia.

Each asserts that the injury suffered resulted from the negligence of the United States in conducting open-air nuclear testing, in monitoring testing results, in failing to inform persons at hazard of attendant dangers from such testing and in failing to inform such persons how to avoid or minimize or mitigate such dangers.

A. JURISDICTION

This Court has jurisdiction of this action pursuant to 28 U.S.C. § 1346(b) (1976) and the Federal Tort Claims Act, 28 U.S.C. §§ 2671–2680 (1976). Venue of this action is proper pursuant to 28 U.S.C. § 1402(b) (1976). The Federal Tort Claims Act (FTCA) is the exclusive legal remedy for claims against the United States for money damages for personal injury or death caused by the negligent or wrongful act or omission of any employee of the Government while acting within the scope of his office or employment.

This action was tried to the Court, without a jury, pursuant to the requirement of 28 U.S.C. §2402 (1976) that "[a]ny action against the United States under section 1346 shall be tried by the court without a jury, ..." See O'Connor v. United States, 269 F.2d 579 (2d Cir.1959). While the court may in its discretion empanel an advisory jury, responsibility for making findings of fact and conclusions of law remains with the court. See e.g., Coffland v. United States, 57 F.R.D. 209 (N.D.W.Va.1972).

B. NATURE OF THE ACTION

This action is a consolidation of the individual claims of the 1,192 named plaintiffs in this lawsuit. This is not a class action. Cf. Annot., 48 A.L.R.Fed. 860 (1980). Trial was held in this action beginning September 14, 1982 and concluding with final arguments on December 17, 1982.

The trial encompassed 24 of the claims in their entirety. Pursuant to the suggestion of the court these cases were selected by plaintiffs' and defendant's counsel as "bellwether" cases. The effort was to provide a selection of "typical" cases which when decided and reviewed may provide a legal and factual pattern against which the remaining issues in the pending cases may be subsequently matched.

The trial was conducted as well so as to make a full and complete record concerning legal, historic, and scientific matters common to all of the 1,192 plaintiffs with the idea in mind of avoiding future duplication of effort. See Park Lane Hosiery Company, Inc. v. Shore, 439 U.S. 322, 99 S.Ct. 645, 58 L.Ed.2d 552 (1979). Other than as noted in this opinion, this court has not decided the remaining issues in the claims of the more than 1,100 plaintiffs that are still pending in the consolidated case.

This opinion decides claims of individuals, each with his own history and relationship to the open air nuclear tests. It fully decides 24 separate cases, tied together by common legal, historic and scientific threads of unique importance.

C. THE TRIAL

This action has been pending since August 30, 1979. It has been the subject of extensive pre-trial motions, dealt with on a preliminary basis by this Court's earlier opinion. See Allen v. United States, 527 F.Supp. 476 (D.Utah 1981). That opinion in a general way defined the framework for the trial that followed.

During the course of trial, this court received into evidence the testimony of 98 witnesses as well as more than 1,692 documentary exhibits. The evidence provides testimony of witnesses ranging from those who participated in the testing program and related operations to highly trained and gifted "experts" offering conflicting opinions, to claimants who seek solace for their test-blamed sorrow.

The trial transcript [hereinafter cited as "Tr."] extends to more than 7,000 pages. The exhibits now in evidence before this court amount to over 54,000 pages of written material contained in 19 cardboard boxes. Depositions filed with the court in this action fill an additional 5 boxes. In addition, the court has deemed additional material submitted by the parties to have been offered and received, e.g., S. Glasstone & P. Dolan, The Effects of Nuclear Weapons (3d ed. 1977), DX-1242, see Tr. at 3276. The court has also taken judicial notice of elementary principles of physics, chemistry, and other sciences, e.g., Handbook of Chemistry and Physics (64th ed. Weast 1983), see Part II, infra, of specific learned treatises, and of government compilations of scientific and historical data, e.g., Committee for the Compilation of Materials on Damage Caused by the Atomic Bombs in Hiroshima and Nagasaki: The Physical, Medical, and

Social Effects of the Atomic Bombings (Engl. trans. 1981), an official publication of the two cities. The record contains historic documents, internal agency memoranda newly declassified, agency directives and correspondence, epidemiological studies, scientific texts and articles, as well as extracts from news media of the day and public information pamphlets.

D. THE PROBLEM OF UNCERTAINTY

We all seek to simplify and to order. The mind eschews the uncertain. We strain for certainty and perfect knowledge in an infinitely complex and dynamic universe. In doing so, we often fail to distinguish between those things which we directly experience, see, feel, hear, taste, smell—facts we experience—from those facts we infer. Each is a form of knowledge. Each we say we know. But we must be constantly aware of the nature of that which we say we "know."

For example, we "know" of the existence of the atom. We "know" of the existence of gamma rays. We have never seen an atom or a gamma ray. We infer that atoms exist. The atom is a mind-created abstract model which provides a convenient, coherent and consistent explanation of an immense collection of perceived effects.[15] It is a model fashioned by many minds after a meticulous sifting of the observed and the reported.[16] But, we remain uncertain still of our scientific certainties.

This court has attempted to formulate an ordered theory of decision. While the effort lacks the mathematical purity of physical theory, it is the judicial resolution of the questions raised by this case with which the court is concerned. The theory of decision melds the method of science with principles of law and public policy. In doing so, the court endeavors to follow the suggestions offered in a recent address by Chief Judge Howard T. Markey of the U.S. Court of Appeals for the Federal Circuit: "The differences between the judicial and the scientific-technological processes are profound and pervasive. Failure to recognize that difference has led to judicial expressions of frustration and an unfortunate tendency to rest judicial decisions on current, and often transient, truths and facts of science and technology. The purpose and function of science is to learn physical facts ..."

In the law, as in science, one always faces uncertainty. This court, faced with the duty of judgment in this case, does not have the luxury of the zealous absolutists who "know beyond doubt" that each and every cancer in the Great Basin is the result of open air atomic testing, or of their absolutist counterparts who "know beyond doubt" that none resulted. The court is disciplined by the record and the application of rules of law.

Consider, for example, Heisenberg's Uncertainty Principle, which is expressed in the following equation (Where p represents uncertainty of position, x represents uncertainty of momentum and h represents Planck's constant):

$$p\, x \approx h$$

[This equation] informs us that it is not possible to determine both the position and the momentum of a subatomic particle simultaneously and with unlimited accuracy. The

15. See e.g., O. Frisch, Atomic Physics Today (1961); I. Asimov, Understanding Physics: The Electron, Proton, and Neutron (1966); S. Hecht, Explaining the Atom 3-94 (rev. ed. Rabinowitch 1954).

16. See e.g., V. Weisskopf, "Atomic Structure and Quantum Theory," in The Mystery of Matter 95–120 (L. Young, ed. 1965); T. Ashford, "Rutherford's Theory of the Nuclear Atom," in id., 84–94; R. Moore, Niels Bohr: The Man, His Science, and the World They Changed (1966).

closer we come to one value, the greater our uncertainty of the other. I. Asimov, Understanding Physics: The Electron, Proton, and Neutron 106–107 (1966).]

The court's findings of fact have a certainty that is relative to the evidence presented to the court by others. They are not fixed in absolute terms. Judicial determination of facts in this case is indistinguishable from fact-finding in other cases no matter how "complex" the facts here might be.[17]

Thus, this opinion speaks in terms of "natural and probable" consequences, "substantial" factors and things "more likely than not." In the pragmatic world of "fact" the court passes judgment on the probable. Dispute resolution demands rational decision, not perfect knowledge.

II. BACKGROUND: BASIC PRINCIPLES OF RADIATION AND NUCLEAR PHYSICS

Evaluation of the risks and consequences of exposure to atomic radiation in this case demands some familiarity with the concepts of radiation physics and its basic language. Such familiarity is a prelude to the knowledgeable application of rules of law. It does not come easily. It did not come easily for the court.

The synopsis found in the next section of this opinion is part of the judicial effort to understand this case. It does supply some of the reasons why this court has found as it has found and has decided as it has decided. It does supply some of the reasons why the rules of law discussed and applied in subsequent sections have been applied as they have.

A. SCIENTIFIC NOTATION AND MATHEMATICAL PREFIXES

Nuclear physics explores the universe using numbers and quantities which range from the extremely large to the infinitesimally small. To simplify the task of using numbers larger than 10 or less than 1, a shorthand system of expression has been devised that relies upon exponential powers of ten. Scientific notation, as this system is called, works in this fashion: Consider, for example, the number 100. One hundred is one way of expressing the quantity 1×100, or $1 \times (10 \times 10)$. Using exponents, this expression is shortened to 1×10^2, which still means 100. Similarly, 1,000 ($1 \times 10 \times 10 \times 10$) may be expressed as 1×10^3 (or simply 10^3) one million (1,000,000) as 1×10^6 (or 10^6) and so forth. More complex numbers may be expressed in this fashion:

$$6{,}205{,}000{,}000 = 6.205 \times 10^9$$

$$1{,}899{,}205 = 1.899205 \times 10^6$$

$$89{,}450{,}000{,}000{,}000 = 8.945 \times 10^{13}$$

Decimal fractions may be expressed in scientific notation as well. For example, 0.01 equals 1/100 or, as we have seen $1/10^2$. The common form of stating this fraction in scientific notation is 1×10^{-2} or 10^{-2}. Thus a negative exponent indicates a fractional quantity while a positive exponent indicates a value greater than 1. In scientific notation, $10^{\circ} = 1$.

Consider the following examples:

---

17. See generally, W. Bishin & C. Stone, Law, Language and Ethics (1972); B. Cardozo, The Nature of the Judicial Process (1921); R. Wasserstrom, The Judicial Decision (1961); L. Green, The Litigation Process in Tort Law (2d ed. 1977).

(1) $0.0032 = 32/10{,}000 = 32/10^4 = 32 \times 10^{-4}$ or, more commonly, 3.2×10^{-3}

(2) $0.65 = 65/100 = 65/10^2 = 65 \text{ x } 10^{-2}$ or 6.5×10^{-1}

(3) $0.000042 = 42/1{,}000{,}000 = 42/10^{-6}$ or 4.2×10^{-5}

Scientific notation proves extremely useful in performing mathematical operations involving very large or very small numbers. Multiplication is accomplished through simple multiplication of the initial terms and through addition of the exponential terms. For example,

(1) $6{,}500{,}000 \times 42{,}120{,}000{,}000 = ?$ Switching to scientific notation gives us:

$(6.5 \text{ x } 10^6) \times (4.212 \text{ x } 10^{10}) = ?$ which is computed in this fashion:

$(6.5 \text{ x } 4.212) \times 10^{(6+10)} = 27.378 \times 10^{16}$ or, expressed in simplest form, 2.7378×10^{17}.

This expression seems far simpler to write than 273,780,000,000,000,000, which is the more conventional form. Fractions are multiplied in the same fashion:

(1) $1.2 \times 10^{-4} \times 6.88 \times 10^{-12} = ?$

$(1.2 \times 6.88) \times 10^{(-4 + -12)} = 8.256 \times 10^{-16}$ (expressed conventionally as .0000000000000008256)

Division of numbers expressed in scientific notation is accomplished in a parallel way:

(1) $6.2 \times 10^6 / 3.8 \times 10^2 = ?$

$(6.2 / 3.8) \times 10^{(6-2)} = 1.6315789 \times 10^4$

(2) $3.2 \times 10^{18} / 4.5 \times 10^{11} = ?$

$(3.2 / 4.5) \times 10^{(18-11)} = .71111 \times 10^7$ or 7.1111×10^6—a short way of saying 7,111,100.

(3) $1.62 \times 10^{-2} / 6.04 \times 10^{-6} = ?$

$(1.62 / 6.04) \times 10^{(-2 + -6)} = .2682 \times 10^{-8}$ or 2.682×10^{-9}—a short way of writing 0.000000002682.

Numbers such as these are commonplace in nuclear physics. For example, Planck's Constant, a fixed number defining the proportional relationship between the frequency of a light wave and its energy, is expressed as $6.6261965 \times 10^{-34}$.

Thus:[18] $h = 6.6261965 \times 10^{-34} = 0.00000000000000000000000000000000006626196$

The illustration is Planck's Constant expressed in joule seconds. Expressed in erg-seconds, the constant is 6.261965×10^{-27}. B. Taylor, W. Parker & D. Langenberg, 43 Rev.Mod.Phys. 375 (1969). Scientific notation makes it possible for a small electronic calculator with an eight-digit or ten-digit display to calculate numbers ranging from 10^{99} to 10^{-99}. Mathematical operations beyond either limit of that range have no practical value; nothing in our experience is either that large or that small.[19]

The mathematics used in nuclear physics is simplified for practical purposes through use of scientific notation. Another system of simplification involves units of measurements routinely used by humans in describing matter, energy and their interactions. At

18. Equation inserted by author.

19. Mathematicians sometimes venture beyond these limits for purely theoretical purposes, dealing in numbers as large as 10^{100} (popularly known as a "googol"), or 10 raised to the power of 10, raised again to the power of 100 (a.k.a. "googolplex") or simply, infinity. See generally, E. Kasner & J. Newman, Mathematics and the Imagination (1953). But even if measured in atomic mass units—a very, very small unit—the entire mass of all matter in the known universe amounts to less than 10^{80} amu—a mere one-hundred-quintillionth of a googol. See C. Sagan, Cosmos 219 (1980).

PREFIX	SYMBOL	POWER
Tera	T	10^{12} 1,000,000,000,000
Giga	G	10^{9} 1,000,000,000
Mega	M	10^{6} 1,000,000
Kilo	K	10^{3} 1,000
Lecto	L	10^{2} 100
Deka	da	10^{1} 10
Deci	d	10^{-1} 1/10
Centi	c	10^{-2} 1/100
Milli	m	10^{-3} 1/1,000
Micro	μ	10^{-6} 1/1,000,000
Nano	n	10^{-9} 1/1,000,000,000
Pico	p	10^{-12} 1/1,000,000,000,000
Femto	f	10^{-15}
Atto	a	10^{-18}

this point in history, science seems to maintain a preference for units in the metric system. Mass is measured in grams, length in units called meters, volume in liters, energy in units such as ergs or joules.

Often, calculations are made in terms of very large or very small quantities—thousands of units or infinitesimal fractions of units. One may be working with a thousand grams or a million grams or with a billionth of a gram. A system of word prefixes has been devised under the International System of Units to adjust units to more closely relate to the actual quantities being utilized. The Prefix Table lists these prefixes and their definitions.

Prefixes for the Units in the International System (McGraw-Hill Dictionary of Physics and Mathematics at A7 (Lapedes ed. 1978).)

Thus, when one is working with 6×10^{3}, or 6,000 grams, one is also working with 6 kilo grams. Likewise, if the quantity is 0.000000082 grams, or 8.2×10^{-8} g, a more workable expression may be 8.2×10^{-2} or .0820 micro grams (g). As will be seen, it is not uncommon to speak of micrograms, or pico curies (a tiny unit of radioactivity) of fallout material deposited in human bodies, or of kilograms of material, or mega curies of fallout radioactivity generated by detonation of a nuclear weapon.[20]

The difference between a picocurie of radioactivity and a megacurie of radioactivity is a factor of 10^{18} or 1,000,000,000,000,000,000.[21] Yet both units are meaningful to the evidence in the record before this court.

20. The common measurement units of the explosive power of atomic or hydrogen bombs are the kiloton or megaton.

21. In rough terms a curie represents the radioactivity of one gram of pure Radium-226, an important radioactive element. A mega curie of material would have the activity of 106 grams or 1,000 kilograms of pure Radium; a pico curie would have the activity of a millionth of a micro curie of pure Radium—quite a difference. The common measurement units of the explosive power of atomic or hydrogen bombs are the kiloton or megaton.

A kiloton is approximately the amount of energy that would be released by the explosion of 1,000 or 10^3 tons of trinitrotoluene, a high explosive. Similarly, a megaton represents the explosive power of 10^6 or 1,000,000 tons (103 or 1,000 kilotons) of TNT. In energy units, a kiloton is equivalent to approximately 10^{12} calories, or 4.2 x 10^{19} ergs. A megaton approximates 10^{15} calories, or 4.2 x 10^{22} ergs. See S. Glasstone, The Effects of Nuclear Weapons 635, 636 (3d ed. 1977).

In practical terms, all of the bombs dropped on all of the cities bombed in World War II represent an explosive power of nearly two megatons. See C. Sagan, Cosmos 320 (1980). In more precise terms, a curie is an amount of radioactive material that is decaying at a rate of 3.700 x 10^{10} disintegrations per second. See S. Glasstone, Sourcebook on Atomic Energy ¶ 17.37 at 521n (2d ed. 1958).

Simply by referring to the Prefix Table, one can determine what multiple or fraction of a standard unit is being discussed in the text. Quick reference to the Prefix Table prefixes will aid the reader in identifying the units used and in making a meaningful comparison of quantities.

B. THE ATOM: PROTONS, NEUTRONS AND ELECTRONS

After centuries of careful observation, our best answer to Thales' 2,500-year-old question "What is the world made of?" seems to be that all matter is composed of atoms. See G. Amaldi, The Nature of Matter (1966). Atoms are very tiny packages of mass which have specific physical qualities. In the natural world around us, science has identified 92 species of atoms commonly referred to as elements. Each element has unique physical and chemical properties. Each element's atoms differ slightly in structure and composition from the atoms of any other element. Some elements are familiar: copper, iron, oxygen, gold, carbon, calcium and two dozen others are well known as part of our own chemical makeup, or as part of the everyday world around us. Others, such as praseodymium, rubidium, and polonium, are far more obscure. Each, however, represents a different type of atom.

To say that atoms are very small is almost an understatement; the diameters of atoms (considered as spheres) generally range from 0.1 to 0.3 nm (i.e., 1 x 10^{-10} meters to 3 x 10^{-10} meters), or one to three ten-billionths of a meter. In a grain of sand, for example, one can easily fit 10^{18} atoms; a tablespoon of water holds 10^{22} atoms without spilling. See C. Yoder, et al., Chemistry 34 (2d ed. 1980).

In addition to the 92 "natural" elements, scientists have produced a dozen more "synthetic" elements, such as plutonium, which are heavier, and often more unstable and short-lived than their "natural" brethren. A complete list of the known chemical elements is found in Periodic Table.

Mendelevium-256,[22] one of the heaviest synthetic elements, has a half-life of 1.5 hours, (meaning a sample of mendelevium-256 created first thing in the morning would be half gone in an hour-and-a-half, and would largely vanish by the end of the day, all but 1/64 of the sample having decayed into lighter elements.)

22. Periodic Table of Elements provided by About Chemistry, http://chemistry.about.com.

Periodic Table of the Elements

1A	2A	3B	4B	5B	6B	7B		8B		1B	2B	3A	4A	5A	6A	7A	8A
1 H 1.00794 Hydrogen																	2 He 4.002602 Helium
3 Li 6.941 Lithium	4 Be 9.012182 Beryllium											5 B 10.811 Boron	6 C 12.0107 Carbon	7 N 14.0067 Nitrogen	8 O 15.9994 Oxygen	9 F 18.9984032 Fluorine	10 Ne 20.1797 Neon
11 Na 22.989769 Sodium	12 Mg 24.3050 Magnesium											13 Al 26.9815386 Aluminum	14 Si 28.0855 Silicon	15 P 30.973762 Phosphorus	16 S 32.065 Sulfur	17 Cl 35.453 Chlorine	18 Ar 39.948 Argon
19 K 39.0983 Potassium	20 Ca 40.078 Calcium	21 Sc 44.955912 Scandium	22 Ti 47.867 Titanium	23 V 50.9415 Vanadium	24 Cr 51.9961 Chromium	25 Mn 54.938045 Manganese	26 Fe 55.845 Iron	27 Co 58.933195 Cobalt	28 Ni 58.6934 Nickel	29 Cu 63.546 Copper	30 Zn 65.38 Zinc	31 Ga 69.723 Gallium	32 Ge 72.64 Germanium	33 As 74.92160 Arsenic	34 Se 78.96 Selenium	35 Br 79.904 Bromine	36 Kr 83.798 Krypton
37 Rb 85.4678 Rubidium	38 Sr 87.62 Strontium	39 Y 88.90585 Yttrium	40 Zr 91.224 Zirconium	41 Nb 92.90638 Niobium	42 Mo 95.96 Molybdenum	43 Tc [98] Technetium	44 Ru 101.07 Ruthenium	45 Rh 102.90550 Rhodium	46 Pd 106.42 Palladium	47 Ag 107.8682 Silver	48 Cd 112.411 Cadmium	49 In 114.818 Indium	50 Sn 118.710 Tin	51 Sb 121.760 Antimony	52 Te 127.60 Tellurium	53 I 126.90447 Iodine	54 Xe 131.293 Xenon
55 Cs 132.9054519 Cesium	56 Ba 137.327 Barium	57-71 Lanthanides	72 Hf 178.49 Hafnium	73 Ta 180.94788 Tantalum	74 W 183.84 Tungsten	75 Re 186.207 Rhenium	76 Os 190.23 Osmium	77 Ir 192.217 Iridium	78 Pt 195.084 Platinum	79 Au 196.966569 Gold	80 Hg 200.59 Mercury	81 Tl 204.3833 Thallium	82 Pb 207.2 Lead	83 Bi 208.98040 Bismuth	84 Po [209] Polonium	85 At [210] Astatine	86 Rn [222] Radon
87 Fr [223] Francium	88 Ra [226] Radium	89-103 Actinides	104 Rf [267] Rutherfordium	105 Db [268] Dubnium	106 Sg [271] Seaborgium	107 Bh [272] Bohrium	108 Hs [270] Hassium	109 Mt [276] Meitnerium	110 Ds [281] Darmstadtium	111 Rg [280] Roentgenium	112 Cn [285] Copernicium	113 Uut [284] Ununtrium	114 Uuq [289] Ununquadium	115 Uup [288] Ununpentium	116 Uuh [293] Ununhexium	117 Uus [294] Ununseptium	118 Uuo [294] Ununoctium

57 La 138.90547 Lanthanum	58 Ce 140.116 Cerium	59 Pr 140.90765 Praseodymium	60 Nd 144.242 Neodymium	61 Pm [145] Promethium	62 Sm 150.36 Samarium	63 Eu 151.964 Europium	64 Gd 157.25 Gadolinium	65 Tb 158.92535 Terbium	66 Dy 162.500 Dysprosium	67 Ho 164.93032 Holmium	68 Er 167.259 Erbium	69 Tm 168.93421 Thulium	70 Yb 173.054 Ytterbium	71 Lu 174.9668 Lutetium
89 Ac [227] Actinium	90 Th 232.03806 Thorium	91 Pa 231.03588 Protactinium	92 U 238.02891 Uranium	93 Np [237] Neptunium	94 Pu [244] Plutonium	95 Am [243] Americium	96 Cm [247] Curium	97 Bk [247] Berkelium	98 Cf [251] Californium	99 Es [252] Einsteinium	100 Fm [257] Fermium	101 Md [258] Mendelevium	102 No [259] Nobelium	103 Lr [262] Lawrencium

Other synthetic elements persist for much longer periods: of an existing sample of plutonium-239, half would still exist 24,400 years from now. Plutonium-244 decays with a half-life of 8×10^7 years; curium-247 persists with a half-life of 1.6×10^7 years.

The differences between atoms of different elements are accounted for by variations in their composition and structure. All atoms are thought to be composed of three types of smaller particles: protons, neutrons and electrons. A proton is a tiny particle with a mass of approximately 1.672×10^{-24} gm, or 1.007 atomic mass units.[23] Each proton carries a positive electric charge.[24] An electron is a particle with a negative electric charge equivalent to that of a proton but with 1/1837 the mass of a proton, or 5.5×10^{-4} amu. An electron has a diameter of approximately 1×10^{-12} cm. A neutron is a subatomic particle having no electric charge (hence the name) yet having a mass of 1.0087 amu—slightly heavier than a proton. An atom of a particular element represents a specific combination of protons, neutrons and electrons. Far from being randomly distributed within an atom, these particles are ordered according to specific principles of structure:

(1) Each atom contains a nucleus at its center;

(2) Protons and neutrons are located within the nucleus;

(3) The nucleus is tiny in relation to the atom itself, comprising about 1/100,000th of the volume of the atom, yet containing almost all of its mass;

23. The mass of atoms and their component particles is often expressed in AMUs rather than grams. An approximate conversion factor is 1 AMU = 1.66×10^{-2} gm. Handbook of Chemistry and Physics at F-303 (64th ed. Weast 1983). The mass given in the text corresponds to a proton at rest, i.e., not one with a significant velocity.

24. Each proton has a positive electrostatic charge of 4.80×10^{-10} absolute electrostatic units, equal to that of the electron, but of opposite sign.

(4) Electrons orbit the nucleus in constant motion, and in patterns better described using the physics of standing waves;[25]

(5) An electrostatically neutral atom is one with an equal number of electrons and protons; an atom with an imbalance of protons and electrons will itself act as a charged particle, called an ion. Ions—electrically charged atoms are often far more reactive with other atoms than are neutral, unionized atoms;

(6) Electrons are distributed in the space around the nucleus in stable orbitals, which correspond to a discrete amount of energy, often called an energy state;

(7) Only certain energy states, or orbitals, are allowed in atoms of a given element;

(8) An electron may move from one energy level to a higher energy level, by absorbing energy from an outside source in an amount equal to the difference between the two energy levels; an electron may fall to a lower energy level by emitting energy in the amount of the difference between the two levels. The energy is emitted in the form of a photon.[26]

An electron may absorb energy, (i.e., become "excited") to a degree sufficient to allow it to leave the atom altogether. This phenomenon is called ionization, and is the key to innumerable chemical reactions.

The simplest element is hydrogen, whose atoms consist of a single proton in the nucleus and a single electron in the surrounding orbitals. The number of protons in the nucleus determines the number of electrons in the atom's orbital shells, which has significant effect on the atom's chemical characteristics. The number of protons is often referred to as the atomic number and is identified in the literature by the symbol Z. An element's atomic number determines its place in an important scheme of classification known as the Periodic Table.

The elements in each column grouping, or period, share similar electron structures, physical and chemical properties. See I. Asimov, Understanding Physics: The Electron, Proton, and Neutron 13–18 (1966). In addition to the atomic number, atoms are described according to atomic weight, symbolized by the letter A, which quantifies the total mass of an atom as expressed in atomic mass units (amu). The atomic weight of hydrogen is approximately 1.00797 amu, the single proton being the source of almost all of its mass. Heavier elements have nuclei containing neutrons as well as more protons.

Helium, for example, has two protons and (usually) two neutrons in its nucleus, giving it an atomic number (Z) of 2 and atomic weight (A) of approximately 4.0026. While varying the number of protons will change the atom from one element to another, varying the number of neutrons changes the atomic weight of the atom with-

25. See H. Semat & J. Albright, Introduction to Atomic and Nuclear Physics 228 (5th ed. 1972). Underlying the text discussion is the deeply complex field of physics known as quantum mechanics. See e.g., id., at 153–308, 325–360; D. Livesay, Atomic and Nuclear Physics 172–271 (1966); L. Pauling & E. Wilson, Introduction to Quantum Mechanics (1935); J. von Neumann, The Mathematical Foundations of Quantum Mechanics (Beyer trans. 1955); M. Born, Atomic Physics (1959). For a simplified discussion of the central concepts, see H. Pagels, The Cosmic Code: Quantum Physics as the Language of Nature (1982); G. Zukav, The Dancing Wu Li Masters (1979).

26. For example, an electron in an orbit having an energy of -1.36×10^{-12} ergs, may fall to an orbit having an energy of -2.41×10^{-12} ergs. The differences in energy, 1.05×10^{-12} ergs, is released in the form of a photon, a parcel of electromagnetic radiation, having the energy 1.05×10^{-12} ergs. To return to the higher orbit, the electron must absorb 1.05×10^{-12} ergs of energy from some external source. C. Yoder, Chemistry 107 (2d ed. 1980).

out radically affecting its physical or chemical properties. Atoms of the same element which have different numbers of neutrons in the nucleus are referred to as isotopes of the element.

Hydrogen, for example, has three isotopes: the most common form, protium, has no neutrons; deuterium, or heavy hydrogen, has one proton and one neutron; tritium, the heaviest, has two neutrons for each proton. Isotopes are often referred to by an abbreviation of their atomic weights called a mass number. Radium (Z = 88, A = 226.054) becomes simply radium 226 or 226Ra. Radium 224 (A = 224.0202) is another isotope of the same element having two less neutrons.

In lighter elements the number of neutrons and protons in the nucleus tends to be equal, or nearly so. In the heavier elements toward the bottom of the Periodic Table, neutrons outnumber protons. Lead, for example, has 82 protons (Z = 82) and an average of 115 neutrons (A = 207.19) in its nuclei, compared to Calcium (Z = 20, A = 40.08) or Neon (Z = 10, A = 20.183).

C. RADIATION AND RADIOACTIVITY

Perhaps the simplest definition of radiation is the one most easily understood: radiation is a transfer of energy through space (or some other accommodating medium). Usually energy is radiated in the form of light or heat, though as we shall see, energy can be radiated through emission of particles having momentum.

In conventional terms, momentum represents a quantity of motion determined by the product of the mass and velocity of an object or particle. For a relativistic particle, i.e., one moving at a velocity near the speed of light, momentum is determined by the formula:

$$\text{Momentum} = (mv)\ (1 - v^2 / c^2)^{1/2}$$

Where momentum = $mv\ (1 - v^2 / c^2)^{1/2}$ where m is the rest mass of the particle, v the velocity and c the speed of light (2.98×10^{10} cm/sec).

Despite its outward appearance, light—or, more broadly, electromagnetic radiation—is not transmitted as a continuous flow of energy. When light is radiated, its energy is packaged in discrete units,[27] tiny bundles of energy called photons. Careful observation has disclosed that photons have some properties which are best explained if they are considered as waves; other properties are best explained if photons of light are thought of as particles. The energy of a particular photon, or light wave-particle, is related to its frequency as follows:

$$E = h\,v$$

Where E is energy (in ergs), v is the frequency (in Hertz) and h is a number known as Planck's Constant. The relationship may also be described in terms of wavelength (λ) where c represents the speed of light:

$$\lambda = hc / E$$

In short, the higher the frequency (or the shorter the wavelength) of light, the greater its energy. Visible light falls roughly midway on the spectrum of light energy with wavelengths ranging from 7×10^{-5} cm for red light through 4×10^{-5} cm for violet. Wavelengths longer than those for visible light fall toward the infrared end of the spectrum: radar

27. These units are often referred to as quanta (singular, quantum) of energy. This quantum theory, as it is called, originated with the work of the German physicist Max Planck, ca. 1900. W. Michels et al., Foundations of Physics 363 (1968). Albert Einstein received the 1922 Nobel Prize in Physics for his theoretical expression which describes the interaction of light with matter in terms of discrete quanta of energy.

waves, radio waves, and microwaves have photons of less energy than visible light. Light wave-particles of greater energy range from ultraviolet light having sufficient energy to cause sunburn to x-rays, gamma rays and high energy cosmic rays (> 10^{-12} cm) which are of particular interest to this lawsuit.

Differences in photon energy are crucial. We are constantly awash in an invisible ocean of radio waves which pass by—and through—our bodies with no perceivable harmful effect. However, exposure to gamma rays, which easily may be one-hundred trillion (10^{14}) times more powerful than broadcast radio waves, raises serious human health concerns.

At the end of the last century, scientists in Europe led by Wilhelm Roentgen discovered basic techniques for producing high energy photons—"x-rays," Roentgen called them—in the laboratory. By bombarding a metal plate within a sealed glass vacuum tube with high-voltage electrons, invisible radiation was produced which would cause certain chemicals to fluoresce brilliantly, and which could fog or darken photographic plates wrapped in paper, or even concealed within a box. See S. Glasstone, Sourcebook on Atomic Energy 48–51 (2d ed. 1958); I. Asimov, Understanding Physics: The Electron, Proton and Neutron 33–35 (1966). In 1896, the French physicist Henri Becquerel discovered that the same kind of penetrating radiation emanated from uranium salts.[28] In 1898, Marie Curie gave this phenomenon of constant emission of penetrating, ionizing radiation the name radioactivity. S. Glasstone, Sourcebook on Atomic Energy, at 53; see Mme. Sklodowska-Curie, 126 Comptes Rendus 1101 (1898). One of the first properties observed in both x-rays and radioactive substances was the induction of an electrical charge in the air surrounding an x-ray tube or immediately in contact with a sample of radioactive material. This ionizing effect enabled early researchers to detect such radiation using very simple devices. "[T]o detect nuclear particles we detect ionization." E. Pollard & W. Davidson, Applied Nuclear Physics 43 (1942) (emphasis in original). Radiation detection and measurement techniques are still very heavily dependent upon this particular quality. See e.g., N. Tsoulfanidis, Measurement and Detection of Radiation (1983).[29] As we shall see, the adverse health effects of exposure to radiation are also a product of ionization. See Part IV infra.

Similarly, the pocket dosimeters in use at the time of the testing, as well as currently, are not different in principle from the original gold-leaf electroscopes used by Becquerel and Mme. Curie in the late 1890s. See S. Glasstone, Sourcebook on Atomic Energy 53, 143, 600 (2d ed. 1958).

D. IONIZING RADIATION: ALPHA, BETA, AND GAMMA RAYS

Early research work by Ernest Rutherford and others using heavy radioactive elements such as radium, polonium, thorium and uranium, disclosed that ionizing radiation emanating from radioactive materials could be resolved into three different

28. H. Becquerel, 122 Comptes Rendus 501,689 (1896). As early as 1867, it was known that uranium salts could expose wrapped photographic plates, but the effect had always been ascribed to prior exposure of the salts to light. S. Glasstone, Sourcebook on Atomic Energy 53 n. (2d ed. 1958).

29. While modern instrumentation is far more sophisticated and far more accurate, it represents in many cases a refinement of original technique. Film badges—photographic film wrapped in opaque paper—are currently used by health physicists to estimate with fair accuracy the radiation exposure of those who wear them. N. Tsoulfanidis, supra at 505; S. Glasstone & P. Dolan, The Effects of Nuclear Weapons ¶ 8.26 at 331 (3d ed. 1977), DX-1242; D. Halliday, Introductory Nuclear Physics 189–190 (1950); H. Yagoda, Radioactive Measurements with Nuclear Emulsions (1949); H. Yagoda, 2 Nucleonics 2 (May 1948). Film badges were available and were used at the Nevada Test Site in conjunction with nuclear testing. See Part VIII, infra.

types: alpha rays (α), beta rays (β) and gamma rays (γ). Application of a strong magnetic field to a stream of ionizing radiation emitted by a sample of radium salts readily deflected beta rays in a fashion indicating that beta radiation carries a negative charge.

Less easily deflected in the opposite direction were alpha rays, which were deduced to have a positive electrostatic charge and greater momentum than beta rays. Gamma rays passed undeflected by the strongest magnetic fields; their highly penetrating qualities led researchers to conclude that gamma rays and x-rays were very similar (which they are). See e.g., J. Cork, Radioactivity and Nuclear Physics 11 (1947). Soon it was determined that beta rays exhibited all the properties of high-energy electrons, while alpha rays were identified as consisting of the nuclei of helium atoms stripped of both outer electrons. Both gamma rays and x-rays were found to consist of very short wavelength high energy photons. While at one time, gamma rays were thought to be distinguishable from x-rays by their higher energy, the more recent production of x-rays of equivalent energy has mitigated that difference. "The difference in nomenclature has, however, been preserved to indicate their origin: we call x-rays those originating from outside the nucleus, while by -rays [gamma rays] we mean rays coming from the nucleus." Jones, Rothblat, & Whitrow, Atoms and the Universe 31 (3d ed. 1973).

Of crucial importance was the discovery that radioactivity is the product of internal processes within the nucleus of the atom rather than the result of excitation by some undetected outside source of energy. Radioactivity properties are unaffected by external forces, such as heat, light or pressure, or by any chemical reaction.[30] Instead, they operate according to specific principles of nuclear physics.

The basic principle finds simple expression in the work of Pollard and Davidson: "This process, the passage from one nearly stable nucleus to one which is stable, is the underlying process of radioactivity." Applied Nuclear Physics 102 (1942). Recalling that the nuclei of atoms are made up of various combinations of protons and neutrons, and that within atoms of a given element, the ratio of neutrons to protons may vary from isotope to isotope, careful observation of radioactive properties reveals that some proton/neutron ratios tend to be far more stable than others. Radioactivity represents the mechanism of internal adjustment by which less stable nuclei transform their composition into a proton/neutron ratio having greater stability. Emission of an alpha particle, for example, lightens the nucleus in evenhanded fashion; the numbers of neutrons and protons are each reduced by 2.

Radioactive transformation by particle emission is observed most often in the decay of heavy elements ($Z > 84$), such as uranium ($Z = 92$, $A = 238$), which ultimately "decay" to the very stable nuclear structure found in lead 206 ($Z = 82$, $A = 206$). At this point, we know of no element with an atomic weight greater than 209 which can be described as "stable," i.e., exhibiting no radioactivity properties.

A nucleus which has a greater ratio of neutrons to protons than is found in more stable forms may emit a beta particle (-)—a high energy negative electron—as one of the neutrons is transformed into a proton. Strontium-90, with 38 protons and 52 neutrons, is radioactive. It decays by β-emission into Yttrium-90, with 39 protons and 51 neutrons.

30. There is some evidence that in a fairly rare sort of spontaneous nuclear reaction known as an orbital electron capture, or K-capture (in which the nucleus captures on of its own orbiting electrons, reducing Z by 1), the chemical state of the atom has some impact. See S. Glasstone, Sourcebook on Atomic Energy 301 (2d ed. 1958). The far more common processes of, and emission, however, are not perceptibly affected. Id. at 124.

Yttrium-90 in turn decays into Zirconium-90 by β-emission. Zirconium-90 with 40 protons to 50 neutrons, is the stable, naturally predominate isotope of that element. Radioactivity ceases.

Atoms with an overabundance of protons may decay by emitting a positively charged electron, or positron (+), or by capturing an electron from a nearby orbital. Either process reduces the number of protons and increases the number of neutrons by 1. When the radioactivity properties of the nuclides are plotted on a graph of numbers of neutrons (N) and protons (Z), a pattern emerges. The term "**nuclide**" refers to a species of atom characterized by the composition of its nucleus. A "**radionuclide**" is a nuclide having observable radioactivity properties.

Radioactive decay enhances nuclear stability not only by adjusting the total number of particles, but also by carrying away discrete bundles, or quanta, of energy from the nucleus. Alpha particles don't quietly drift away from the nucleus of Plutonium-239; they are propelled away at high speed with an energy of more than 5 MeV—five million electron volts.

While 5 mev is a trifling amount of energy in the sphere of human endeavors, at the atomic level it is enormous. In 1919, Rutherford bombarded nitrogen atoms with particles of lesser energy. The particles slammed into the nitrogen nuclei, yielding oxygen nuclei—a different element—and a free proton.

A nucleus may also expel energy through emission of a gamma ray of a particular wavelength. Likelihood of gamma emission depends on the specific nuclides. Strontium-90 does not emit gamma rays during decay. Iodine-131 does. Cobalt-60 emits two gamma rays in the MeV range, and has been used as an important source of radiation in the treatment of cancer.

The detailed properties and mechanics of radioactivity and nuclear transformations are as awe-inspiring as they are complex. Why it is, for example, that unstable heavy nuclei emit particles poses a challenging theoretical question, the answer to which is beyond the scope of this inquiry.[31] Yet a specific understanding of the statistical behavior of radionuclides (e.g., concepts such as radionuclide "half-life") and of the ways in which ionizing radiation interacts with other matter is crucial to adequate analysis of the causation and negligence issues raised in this action.

E. STATISTICAL NATURE OF RADIOACTIVITY

Current theory holds that events at the atomic or sub-atomic level occur according to statistical probabilities rather than any strict determinism.[32] Radioactive decay in any given quantity of a radioactive element occurs at a constant rate, yet no one can tell at what time any particular atom will decay. We know only the probability that it will decay within a chosen time period, or viewed another way, that over a chosen period of time, number of nuclei in our sample will decay into "daughter" nuclides. The most important expression of this statistical approach is found in the concept of half-life. The half-life of a radionuclide is the specific length of time during which half of the nuclei in any amount

31. See e.g., 1 E. Hyde, I. Perlman & G. Seaborg, The Nuclear Properties of the Heavy Elements: Systematics of Nuclear Structure and Radioactivity 211-302A (1971 ed.), and references cited therein; H. Semat & J. Albright, Introduction to Atomic and Nuclear Physics 440–442 (5th ed. 1972). Again the answer seems to be found in analyzing subatomic particles in terms of wave functions and probabilities.

32. The probabilistic nature of nuclear physics is a product of analysis of matter in terms of quantum mechanics, a revolution in theoretical physics which has exploded in this century. See e.g., H. Pagels, The Cosmic Code: Quantum Physics as the Language of Nature 3-165 (pap. ed. 1982); A. Pais, "Subtle is the Lord ...": The Science and the Life of Albert Einstein 55–110, 357–469 (1982).

of the radionuclide will have decayed. For example, if you start with a one gram sample of radium-226 (226Ra), after 1,600 years only half of the sample (0.5 gm) remains 226Ra. The other half has undergone radioactive decay.

After another 1,600 years, only half of that remainder is still 226Ra (0.25 gm). Passage of 1,600 more years witnesses decay of one-half of that fraction, leaving 1/8 gram (0.125 gm) of 226Ra. The process of decay continues at this rate until no more 226Ra remains. To say, therefore, that a radionuclide such as strontium-90 has a half-life of 28.1 years means that half of the quantity existing at the beginning of the 28.1 years remains. To say, therefore, that a radionuclide such as strontium-90 has a half-life of 28.1 years means that half of the quantity existing at the beginning of the 28.1 years remains Sr at the end of that time.

Half-life varies dramatically from radionuclide to radionuclide. Some isotopes last for half-lives of only a few seconds; others last far longer. Uranium (235/92U), although radioactive, has a half-life of 4.51×10^9 years—nearly the age of the earth itself. See R. Heath, 'Table of the Isotopes,' in Handbook of Chemistry and Physic B-232 to B-316 (64th ed. Weast 1983).[33] Half-life is indicative not only of the persistence of a radionuclide in the laboratory or the environment, but also of the intensity of its radioactive decay. Radiation yielded by one gram of iodine-136, an isotope found in the fireball of a nuclear explosion, is intense: emitting particles and rays in the MeV range, 136I decays rapidly, reflecting a half-life of 83 seconds. Consequently its radioactivity fades rapidly. After 15 minutes, less than 1/2000 of the original amount is identifiable as 136I. After an hour more, the fraction remaining is nearly 2.28×10^{-15} of the original amount. In contrast, Plutonium-239, a radionuclide present in every nuclear fallout cloud to date,[34] persists with a half-life of 24,400 years. Consequently, all but a fraction of the estimated 3 tons of 239 Pu deposited on the earth by fallout can still be found somewhere.

Half-life is not an average lifetime for a radionuclide; it is an expression of rate, of the statistical probability of decay over a specific time period. The average lifetime of a given radionuclide may be computed from the half-life.

The half-life of a radionuclide gives some immediate insight into its behavior and the nature of the hazard that might be associated with it. A small quantity of radionuclide with a half-life in the order of minutes will not persist long enough to present a significant hazard a few days later and it is not likely to become dispersed very far by natural forces.

In contrast, a radionuclide with a half-life on the order of several years may represent a long-term hazard and become widely dispersed if not held in containment. 1 F. Whicker & V. Schultz, Radioecology: Nuclear Energy and the Environment 44 (1982).

Quantity of radioactive material may be expressed in either of two ways: (1) the mass of the material; or (2) its radioactivity. To say, however, that a sample of Strontium-90 has a mass of grams does not by itself offer any perspective on the amount of ionizing radiation being emitted by the sample at any given moment. The activity of a known mass of radioactive material may be computed as follows:

---

33. Unless otherwise specified, the data concerning half-life and the nature and energy of decay of the radioactive isotopes which are relied upon by this court are derived from this Table. This court takes judicial notice of this Table and of R. Heath, "Gamma Energies and Intensities of Radionuclides," a Table in the same Handbook, same edition, pages B-137 through B-355. See Rule 201, Federal Rules of Evidence.

34. While 239Pu is not a fission product, see Part III, infra it is still present, whether as unused fuel from a plutonium-based bomb, or as a product of the reaction between neutrons and uranium found somewhere in the bomb's structure, such as a reflector.

$$A^* = 0.693 \text{ m } N_A / T_{1/2} \text{ A } F_t$$

Where A* is the activity of the sample; m is the mass of the sample; A is the atomic weight of the radionuclide; $T_{1/2}$ is the half-life in years or whatever units; F_t is the conversion factor from years (or whatever units) to a desired time period, e.g. minutes; seconds, etc. N_A is a constant, known as Avogadro's number, expressing the number of atoms in a mole (a mass in grams equal to the nuclide's atomic weight) of material. See 1 F. Whicker & V. Schultz, Radioecology: Nuclear Energy and the Environment.

For example, a 1.2 gram sample of sodium-24, an important radioactive by-product of nuclear weapons testing having a half-life of 15.0 hours, would have an initial activity of

$$A^* = \frac{(0.693)\ (1.2 \text{ grams})\ (6.022 \text{ x } 10^{23} \text{ atoms/mole})}{(15.0 \text{ hours})\ (24 \text{ grams/mole})\ (3600 \text{ seconds/hr.})} = 3.864 \times 10^{17} \text{ disintegrations per second}$$

This activity represents a rate of emission paralleling that of a sample of pure radium-226 having a mass of 1.0557×10^{-7} grams, or 10.5 metric tons (sic). Of course, 24 Na activity would fade quickly in comparison to the 226 Ra; the mean life of 24Na atoms would be 21.64 hours, while atoms in our 10.5 metric tons would persist for a mean life of 2,308.8 years, making it a far more serious hazard overall. Expression of radioactivity in terms of rate of decay (e.g., disintegrations per second) is a common measurement. The most frequently used unit of activity is the curie (Ci), which represents a mass undergoing 3.7×10^{10} disintegrations per second. Practical quantities of radioactive materials are more easily expressed in fractional units, such as millicuries (mCi), microcuries (Ci), or picocuries (pCi), representing 10^{-3}, 10^{-6} and 10^{-12} curies, respectively.

A tiny new unit, the Becquerel (Bq), represents an activity of one disintegration per second, or the equivalent of 27 pCi. Yet the radioactive fallout yield of even a "nominal" nuclear device is more easily expressed in megacuries-millions of curies of activity.

The comparison to radium is not purely coincidental. The unit of activity known as the Curie (Ci) was originally based upon the activity of one gram of radium, approximately $3.7 \text{ x } 10^{10}$ disintegrations per second. S. Glasstone, Sourcebook on Atomic Energy 521 & n. (2d ed. 1958). Direct comparison with radium itself is tricky business.

Consumer products such as luminous watch dials or home smoke detectors may contain 100 Ci or so of promethium-147 or americium-241, or 20 Ci of plutonium-238. See UNSCEAR Report (1977), at 95, PX-706/DX-605. Measurement in curies does not, however, define the type or energy of the radiation being emitted. That information is specific to each radionuclide and is important to the evaluation of the risk created by exposure to curie, millicurie, or microcurie amounts of material.

F. INTERACTIONS OF RADIATION WITH MATTER

Each type of radiation emitted by radioactive materials interacts with other matter in important yet distinct ways. Alpha particles carry an electrostatic charge of +2 units. When a particle passes near another atom, the electrons in its orbital shells are attracted to the particle by virtue of their own negative (-) charge. Some electrons are merely excited by the event, i.e., they move from a lower to a higher energy state while remaining in orbit around the nucleus. For many others, however, the coulombic attractions of the particles are too strong; these electrons are stripped away from their original atoms and travel freely for a time, leaving the mother atom in an ionized state. Balance of electric charges is soon restored, but in reaching neutrality, some atoms form new combinations with other atoms—new molecules are born. In many ways, a particle can be visualized as a tiny, speedy electromagnet, sweeping free electrons into its path.

The electrostatic attraction is a mutual one; the electrons in nearby atoms are attracted by—and attract—the speeding particles. The coulombic forces energize the electrons, pulling them away from atoms by giving them sufficient energy to escape. At the same time the electrons act as a drag on the particle. Its energy is reduced by the amount that the electrons' energy is increased. According to fundamental physical law,[35] the energy is conserved, neither created or destroyed. The ionization that occurs in the path of a particle represents a transfer of energy. The amount of energy transferred per unit of distance traveled by the particle is known as a linear energy transfer (LET). See e.g., Committee on the Biological Effects of Ionizing Radiations, The Effects on Populations of Exposure to Low Levels of Ionizing Radiation: 1980, at 13 (1980), DX-1025 (hereinafter cited as the "BEIR-III Report"). Alpha radiation is "high-LET" radiation; so many electrons are excited or ionized by particles that a great deal of energy is transferred in a short distance. The range of a particle traveling through matter thus tends to be fairly short.

As soon as the particle has transferred the bulk of its energy, it slows down sufficiently to capture two electrons of its own, becoming a neutral, inert helium atom.[36]

The range of a particle depends largely upon the energy of the particles and the density of the medium through which it passes. Radioactive - emitters shower their surroundings with particles having an energy between 0.1 MeV and 10 MeV. Such energies give the alpha radiation a range in air of less than 10 cm.

The range in more dense material, for example paper, aluminum foil, or human cell tissue is far less, on the order of a few m. The range of particles, or conversely, the stopping power of material bombarded by them, is computed with good accuracy through a system of mathematical formulae not directly relevant here. For the purposes of this case, the general range of particle radiation, i.e., a few centimeters in air, a few microns (m) in living tissue is important information to use in assessing the risks presented by potential and actual exposure to - emitting radionuclides.

While the risks posed by - emitters is popularly minimized by pointing out the stopping power of simple materials such as paper, or aluminum foil, particles in the MeV range may penetrate from 30–60 m in cell tissue. This distance encompasses the diameters of 2–6 cells. A particle may carve a path through sensitive internal tissues that is strewn with hundreds of thousands of ionizations per m.

Beta particles (-) interact with matter somewhat differently than radiation. As a consequence of -particles having far less mass than a particle (M = 1/7360 M) and a negative rather than positive electric charge,[37] relatively low linear energy transfer (LET)

35. The law of the conservation of energy, which is often called the First Law of Thermodynamics, may be stated in this way: energy can neither be created nor destroyed and therefore the total amount of energy in the universe remains constant. See e.g., Handbook of Chemistry and Physics F-76 (64th ed. Weast 1983); see also R. Feynman, The Character of Physical Law 68–77 (pap. ed. 1965). According to this principle, the energy released in nuclear reactions must go somewhere. It does not quietly dissipate into nothingness. It may change form—even become matter under certain circumstances—yet it is not lost.

36. One of the early discoveries (ca. 1903) about the element radium, an emitter, was that helium gas is continuously liberated from samples of radium salts. A sophisticated experiment by E. Rutherford and others in 1909 demonstrated that particles were the source of the helium. S. Glasstone, Sourcebook on Atomic Energy 55–56 (2d ed. 1958).

takes place per unit of distance traveled. Beta particles are far more easily deflected by collisions with other electrons or nuclei, and cause ionization of atoms by colliding with other electrons or by passing near enough to repel electrons (having the same negative charge). Having a low LET factor, however, means that -particles in the 100 Kv-2 MeV energy spectrum have significantly greater range and penetrating power than do particles carrying even twice as much energy. They may traverse several meters in air at speeds approaching that of light. J. Cork, Radioactivity and Nuclear Physics 118 (1947). In cell tissue, -particles may leave a trail of ionization several millimeters in length.

Beta particles, like other ionized electrons, also radiate energy in the form of gamma rays as they are deflected by nuclei and slow down, eventually returning to orbitals of other atoms. This "braking" radiation, called Bremsstrahlung, may in turn strike other electrons, exciting them into further ionizations. Gamma radiation interacts with matter in several different ways:

(1) Photoelectric absorption;

(2) Compton scattering;

(3) Pair production;

(4) Rayleigh scattering.

Or, it may not interact at all, passing through the exposed material completely unimpeded. Since gamma rays have no electric charge, ionization by gamma radiation occurs only when a gamma ray strikes an orbital electron. If a gamma ray striking an electron is totally absorbed, the electron is stripped away from the atom, and possessed of the full energy of the gamma ray, careens through surrounding matter in a fashion very much akin to a particle. Thousands of ionizations result. This "photoelectric" effect is the dominant form of interaction for gamma rays whose energy is less than 1.0 MeV.

For gamma rays of greater energy (0.3–3 MeV), the phenomenon known as Compton scattering is predominant: a gamma ray striking an electron may only impart a portion of its energy to the particle. The ionized electron is driven in one direction, the gamma ray photon, now of lesser energy, is deflected on a new path. In Compton scattering, a gamma ray may thus ionize several electrons before its energy is wholly transferred.

For gamma rays of an energy of 1.02 MeV or greater another interaction may occur if the photon strikes a nucleus: the gamma ray can disappear, leaving a positron (+) and an electron (-) in its place. This phenomenon, known as pair production, ultimately generates further gamma radiation: when the positron comes in contact with another electron both particles are annihilated, yielding two gamma photons of an energy approximating 0.51 MeV each. These go on to ionize other electrons, or pass out of the irradiated material entirely. At energies above 10 MeV, pair production is the predominant interaction. Finally, gamma rays of low energy (< 100 Kev) may experience Rayleigh scattering, the elastic deflection of the gamma photon by an electron with no energy being transferred to the elec-

37. Some - emitters are overbalanced in favor of protons and therefore seek greater stability through emission of + particles, or positrons. A positron is an electron that carries one unit of positive charge (rather than the usual negative charge). A positron (sometimes termed an anti-electron) is a particle of antimatter; if a + collides with a negative electron, both particles are annihilated with the release of rays.

tron. Gamma radiation by itself does no damage to exposed matter—if it passes through without collision.[38]

When the gamma rays collide with electrons, as they may at any point along their path, ionizations occur just as if the matter had been bombarded with high energy beta particles. As we shall see, ionization of matter in living tissue may cause serious harm to the affected cells, and ultimately, the whole organism.

A fourth type of radiation emitted by nuclear explosion, free neutrons, also causes ionization in a somewhat indirect fashion. A neutron may be absorbed by a nucleus with which it collides, yielding a free proton, a particle, another neutron, or a gamma ray. These charged particles and gamma rays interact with matter as previously described. Neutrons may also collide with nuclei and scatter them without being absorbed. Neutrons remaining in the free state decay into free protons and -particles at a rate giving them a half-life of roughly 10.8 minutes:

Rather than attempt a catalogue of all types of gamma ray and neutron interactions with matter, this court has chosen merely to summarize the major type of reactions which are significant to human health and safety, or to the measurement of radiation exposure.

While neutron interactions are a significant health concern in cases of direct exposure to a nuclear explosion or to a nuclear reactor, they are important to this case in only one respect: free neutrons from a nuclear explosion will interact with the surrounding air or with soil drawn up into the mushroom cloud, forming radioactive isotopes through absorption reactions. This additional activity adds to the radioactive burden of the fallout cloud, increasing the total risk of radiation exposure from the event.

G. THE MASS-ENERGY RELATIONSHIP

Mention has already been made of the Laws of Conservation of Mass and Energy—the concepts that in any chemical reaction the total mass of the system remains constant, and that in any physical interaction the total energy of the system remains constant. In an absolute sense, neither matter nor energy is created or destroyed in any reaction. Yet an atomic bomb, a device containing only a few kilograms of metallic material yields energy—heat, light, and explosive force—in incredible amounts.

The key is found in a simple, fundamental relationship between matter and energy, a relationship which Albert Einstein originally defined and which most people have at some time seen expressed as

$$E = mc^2$$

E represents energy, expressed in ergs; m represents mass, expressed in grams; c is the speed of light, expressed in cm/second ($c = 2.9979250 \times 10^{10}$ cm sec^{-1}). Put together in the simplest terms, matter and energy are equivalent. Matter may be converted to energy by a factor of approximately 9×10^{20} ergs per gram. 9 ×x 10^{20} ergs, if released all at once, generate the explosive power of approximately 21.5 kilotons of TNT—more than either of the bombs dropped on Japan. Cf. S. Glasstone & P. Dolan, The Effects of Nuclear

---

38. Unlike or - particles, which can be totally attenuated (i.e., reduced) by small thicknesses of many common materials, total attenuation of radiation is a more difficult task. While the intensity of a beam of photons decreases exponentially with the thickness of the bombarded material, some fraction of the rays will pass all the way through materials of significant thickness without interaction. See N. Tsoulfanidis, Measurement and Detection of Radiation 149–152 (1982) for treatment of attenuation calculations.

Weapons 13 & Table 1.45 (3d ed. 1977) DX-1242. A "nominal" yield nuclear device (approx. 20 kt TNT) derives its energy from the annihilation of a little less than one gram of matter.

The Einstein equation may also be expressed in terms of joules of energy, kilograms of mass and $c = 3 \times 10^8$ meters per second. See e.g., W. Michels, et al., Foundations of Physics 233 (1968).

Far more mundane energy-producing processes also annihilate matter, though in much smaller amounts. Eating a 200-Calorie candy bar (metabolism yields 200 Calories of energy [1 Calorie = 1000 calories as used in physics]), for example, converts a small amount of matter into energy:

$$m = \frac{(2.0 \times 10^5 \text{ calories}) \times (4.184 \times 10^7 \text{ ergs/calorie})}{(9 \times 10^{20} \text{ cm 2/sec 2})}$$

$$= 9.2977 \times 10^{-9} \text{ grams, or } 0.000000009297 \text{ grams.}$$

If $E = m\,c^2$, then $m = E/c^2$. Using this equation and a conversion factor (1 kt TNT equals 4.18×10^{19} ergs) then the quantity of matter consumed can be calculated:

$$m = \frac{(4.18 \times 10^{19} \text{ ergs}) \ (20)}{(9 \times 10^{20} \text{ cm 2/ sec}^2)}$$

$$= 9.288 \times 10^{-1} \text{ grams, or } 0.9288 \text{ gm.}$$

Similarly, the energy released in radioactive decay can be precisely accounted for by finding the difference in mass between the mass of the original nuclide and the mass of the daughter nuclide and emitted particles. The decay products will be a tiny bit lighter. In 1907, Albert Einstein ventured an interesting speculation based upon the relationship $E = mc^2$:

> It is possible that radioactive processes may become known in which a considerably larger percentage of the mass of the initial atom is converted into radiations of various kinds than is the case for radium [decay]. A. Einstein, 4 Jahub. Rad. Elektr. 411, 443 (1907) quoted in A. Pais, "Subtle is the Lord ..." The Science and the Life of Albert Einstein 149 (1982).

H. THE NUCLEAR FISSION PROCESS

The discovery of radioactivity at the turn of the last century led quickly to further discoveries. The nature and properties of alpha, beta and gamma radiation were closely scrutinized. In 1920, physicists predicted the existence of a new particle—the neutron—having roughly the same mass as a proton, but no electric charge. In 1932, experimental bombardment of beryllium metal with particles yielded a new highly penetrating form of radiation that British physicist James Chadwick correctly identified as neutrons.

By 1937, reports were published indicating that elements heavier than uranium could be synthesized through bombardment of uranium with neutrons.[39] By early 1939, Otto Hahn and Fritz Strassman published evidence that bombarding uranium with neutrons produces an unprecedented reaction: fission of the uranium nuclei into two lighter fragments.[40] It was soon determined that fission of uranium yielded two fragments of unequal mass, at least two additional free neutrons, and a shower of gamma rays and neutrinos. The atomic numbers (Z) of the two fragments add up to 92, the atomic number of uranium. The weights of the fission products and the three neutrons released by fission add up to an amount slightly less than the mass of the uranium atom plus one neutron: a difference of approximately 0.18 amu.

Fission of 235U will release two or three neutrons immediately ("prompt" neutrons); some fission products will release additional neutrons—about 1% of the number of prompt neutrons (165/10,000 fissions)—over a period of a few seconds following fission. While the emission of delayed neutrons is too slow to affect the rate of reaction in an uncontrolled chain-reaction "bomb", delayed neutrons are important to calculations involving sustained reactions in nuclear reactors.

Fission of 239Pu results in fewer delayed neutrons (61-63/0,000 fissions) being produced. Conversion of that mass difference to energy according to $E = mc^2$ gives an energy (Q) of nearly 175 MeV. The two fission products are both highly radioactive nuclides which soon decay to stable elements (141Pr and 92Zr), releasing 22 MeV of additional energy in the process. The sum of these energies is nearly 200 MeV per fission of 235U—close to the energy value determined by theoretical calculation. See S. Glasstone, Sourcebook on Atomic Energy 390–393 (2d ed. 1958); H. Semat & J. Albright, Introduction to Atomic and Nuclear Physics 509–510 (5th ed. 1935); 3 E. Hyde, et al., The Nuclear Properties of the Heavy Elements: Fission Phenomena 5 (1971).

If gauged according to everyday human activity, 200 MeV is a small amount of energy, hardly noticeable. Yet at the atomic level, 200 MeV is enormous; even the radioactive decay of powerful - emitters such as 226Ra or 239Pu barely releases 1/40 of the energy yielded per event by fission. Normal chemical processes, such as those involved in the combustion of fossil fuels, involve energies per atom on the order of a few eV, not MeV. See H. Semat & J. Albright, supra at 570.

The phenomenon of uranium fission generated great excitement from the moment of its discovery. Not only was fission a fascinating theoretical novelty once thought to be totally impossible, and not only did each fission of an uranium nucleus release dramatic quantities of energy, but the process offered a key to something much greater: couldn't the free neutrons released by one fission event be used to trigger two or three others?

Scientists imagined a fission chain reaction in which each fission event leads to one or two more, ultimately yielding energy in practical, even explosive quantities. Assume, for

39. See e.g., O. Hahn, L. Meitner & F. Strassman, 106 Zeits. F. Phys. 249 (1937). But see O. Frisch, Atomic Physics Today 20–21 (1961).

40. See O. Hahn & F. Strassman, 27 Naturwiss. 11, 89 (1939). The evidence was first analyzed in terms of splitting uranium into nearly equal fragments by Lise Meitner and Otto Frisch, who accurately calculated the energy released to be in the range of 200 MeV—and who first described the process as nuclear "fission"—in a paper dated January 16, 1939, which was sent to the British publication Nature. The paper was published in Nature in the February 11, 1939 issue. R. Moore, Neils Bohr 224–228 (1966); S. Glasstone, Sourcebook on Atomic Energy 385–389 (2d ed. 1958). See L. Meitner & O. Frisch, 143 Nature 239, 471 (1939).

example, that a chain reaction could be produced in which each fission of 235U would lead to two more (2:1).

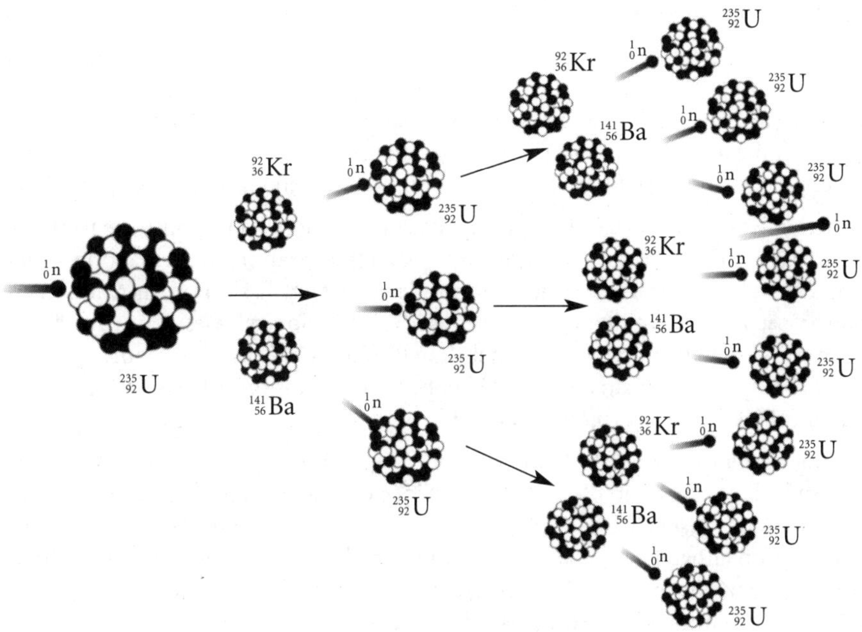

[Table added by author]

While a single fission instantly yields nearly 180 MeV, or 3.2 × 10^{-11} joules of energy, a chain reaction growing at the 2:1 rate would produce after 40 generations nearly 1.1 × 10^{12} fission events, yielding nearly 2 × 10^{14} MeV, or 32 joules—enough power to set a small light bulb aglow for a second or two. After 37 more generations, the total yield approaches 4.3 × 10^{12} J, or roughly the equivalent of 1 kiloton of TNT. Five additional generations would produce a total energy equivalent to 33 kt of TNT. Cf. S. Glasstone & P. Dolan, The Effects of Nuclear Weapons 13 (3d ed. 1977) DX-1242. Assuming a perfect 2:1 chain reaction, an explosive yield of 33 kilotons would require fission of 4.8 × 10^{24} atoms, or approximately 1,887 grams of 235U—less than five pounds.

In a 2:1 chain reaction, assuming ideal conditions and no loss of neutrons to non-fission activity, such "generation" would be twice as large as the last. The number of fission events is easily calculated;

$$Nf = 2^{X}$$

Where Nf is the number of nuclei experiencing fission and is the number of generations, assuming a reaction started by a simple neutron at = 0. This calculation is offered as a highly simplified model of the chain reaction process for illustration purposes only.

This amount is close to the actual mass of fissionable 235U that would be consumed in a device of 33-kt yield. Under normal conditions, however, a 1.8 kg lump of 235U will not be "critical," i.e., undergo a chain reaction that will result in an explosion. The actual "critical mass" of fissionable metals, whether 235U, 239 Pu, 233U or whatever, which will spontaneously produce such a reaction has been estimated to be in the range of 3–30 kg[41] depending on a number of factors. In smaller amounts, too many neutrons escape through the surface of the metal to sustain the reaction. A nuclear fission bomb, therefore, will contain a larger subcritical mass of fissionable material than is actually consumed in the subsequent fission reaction.

In a weapon, a "critical" or "supercritical" mass of 235U may be achieved in one of two ways: (1) through violently forcing two subcritical quantities of metal together; or (2) by compressing a subcritical mass upon itself to the point that it becomes supercritical. This second method, the implosion technique, is the far more efficient—and far more technically difficult-design. See also J. McPhee, The Curve of Binding Energy 75–95, 108–10 (1974).

The reaction would take place very quickly. Each generation would complete its fission in 0.01 micro seconds (millionths of a second). An 82-generation chain reaction would take approximately 0.82 microseconds to complete. In fact, fission chain reactions in 235U can proceed more rapidly; on the average, 235U fission releases 2.5 neutrons per event, permitting a chain reaction of greater than the 2.0 rate. Glasstone reports that one kiloton of TNT equivalence can be reached by the 51st generation. 58 generations would yield roughly 100 kiloton TNT equivalence.

According to data in The Effects of Nuclear Weapons, Fission of 1.45×10^{23} nuclei yields the equivalent of 1 kt of TNT. Id., at 13. Since one neutron is required per fission, 1.45×10^{23} neutrons must be made available to the fission process. Some neutrons inevitably escape the reaction altogether, while others are absorbed in non-fission nuclear reactions. The number of neutrons produced by one generation of fission (f) is reduced to that extent (l). For a given number of neutrons (N) present at a given time the increase in the number of neutrons by the next generation of fission is:

$$X = N\,(f - l) - N$$

(The N number of neutrons having been absorbed
to cause the fission in that generation)

If g is the generation time when the rate of neutron increase is $N = N_{0e}{}^{xt/g}$ when N_0 is the initial number of neutrons present, n is the number at a time (t) later, e is the base of natural logarithms (2.7182818), and is the quantity (f -1). To simplify, let t/g (the number of generations that have elapsed during time t) be represented by:

$$N_{0e}{}^{xn}$$

For 239Pu, f averages 2.9 or so. J.C. Giddings, Chemistry, Man, and Environmental Change 424 (1973). It is seen, therefore, that 99.9 percent of the energy of a 100-kiloton fission explosion is released during the last 7 generations, that is, in a period of roughly 0.07 microseconds.

Clearly most of the fission energy is released in an extremely short time period. The same conclusion is reached for any value of the fission explosion energy. S. Glasstone & P. Dolan, The Effects of Nuclear Weapons ¶ 1.57 at 17 (3d ed. 1977), DX-1242. Thus, for a 100-kiloton 235U bomb, the chain reaction starts and ends in 0.58 microseconds, the last 0.08 microseconds of that process determining the gross energy yield of the fission device.

The time-scale of a nuclear chain-reaction explosion is not merely curious atomic trivia. The energy yield in the last 0.1 microsecond of fission is a critical factor determining weapons design, as are considerations such as rate of neutron escape per "gener-

41. See W. Laurence, The Hell Bomb 13 (1950) citing an estimate in 1946 by Dr. M.L.E. Oliphant, a noted British physicist. Actual critical masses and theoretical formulations for uncontrolled fission remain secret.

ation." As a fission reaction proceeds, the tremendous release of nuclear energy creates immense heat and pressure; the temperature at the core of the mass of uranium or plutonium reaches millions of degrees Celsius by the 51st fission "generation." Intense heat causes intense pressure, forcing the mass apart violently.

To achieve maximum explosive yield, the device—by now a superheated mass—must somehow hold together long enough for the next seven or eight generations of fission to take place. Consequently, nuclear weapons designers have worked on ways of encasing the fissionable portion of a device in a "tamper," a shell of heavy material which by mere inertia will hold the mass together for the crucial 0.07–0.08 microseconds. Using tamper material that will also reflect neutrons serves two purposes: (1) containment of the fissionable material long enough to achieve high yield; and (2) increasing the number of free neutrons available for fission by reducing the rate of neutron escape from the reacting mass. Uranium-238, for example, proves useful for this purpose.

Inertia is the term applied to the "tendency of undisturbed bodies to stay at rest" or to keep moving in a given direction. W. Michels, et al., Foundations of Physics 83 (1968). The greater mass, the greater the inertia. Because of inertia, a densely packed mass of heavy atoms will resist a force that would tend to move it—even the force of a nuclear explosion. Even 0.25 microseconds of resistance would permit many more fission reactions, producing a much higher yield.

The testing of various materials as tampers for neutron reflectors was among the many purposes of the series of open-air detonations at the Nevada Test Site. The value, for example, of beryllium as a neutron reflector might be one of many technical questions answered by a specific test. This type of testing has varying impact on the radioactive fallout burden in surrounding areas; use of a heavy tamper of 238U leaves far higher quantities of plutonium in the fallout cloud than does the fission process itself. The 238U in the shell, bombarded with neutrons from the chain reaction and vaporized by its heat, is converted in significant part to 239Pu.

Some unreacted 238U would remain as well. If the neutrons emitted by the bomb's initial processes are of sufficient energy, 238U will itself undergo fission, adding to the energy of the explosion and to the quantity of highly radioactive fission products in its fallout.

The greater time-scale permitted by use of efficient tamper/reflector material as well as improved design technique is indispensable to another process: the nuclear fusion reaction of a hydrogen bomb.

[Parts I and II omitted]

III. NUCLEAR FALLOUT

A. NATURE OF FALLOUT RADIOACTIVITY

The processes of nuclear fission and nuclear fusion release tremendous amounts of energy as predicted by Einstein's postulate $E = mc^2$. Both processes directly produce reaction products, gamma radiation and free neutrons of significant energy. Many of the gamma rays are absorbed by air surrounding the bomb which along with the reactants themselves, generate tremendous amounts of heat.

A luminous "fireball" forms around the exploding device within the first second following detonation, glowing at many times the apparent brightness of the sun. The fireball quickly expands outward as a toroidal (i.e., doughnut-shaped) cloud of superheated material, drawing great amounts of cooler air up into the familiar "mushroom" cloud associated with atomic explosions.

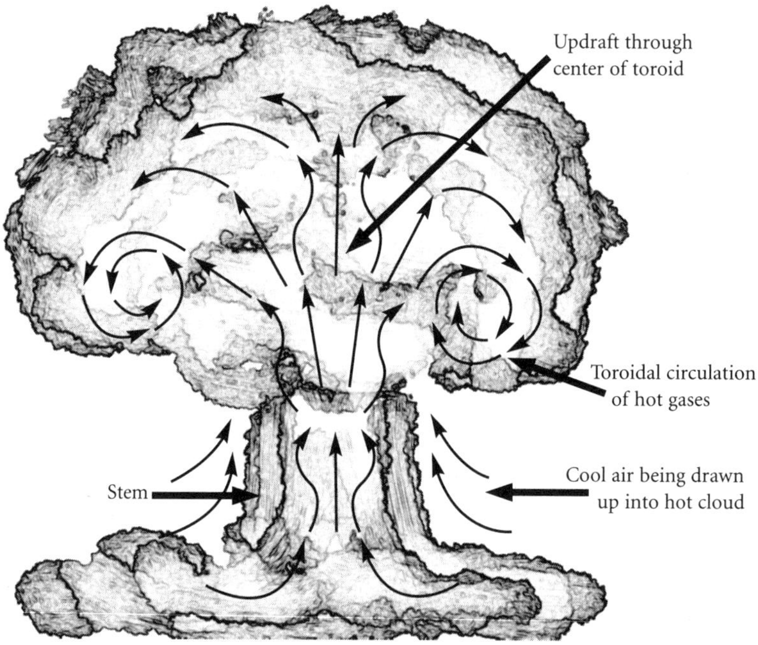

As much as 70–80% of the energy of a fission device may be released in this way as thermal radiation. Such temperatures—in the millions of degrees Celsius—instantly vaporize all matter within the radius of the fireball. See S. Glasstone & P. Dolan, The Effects of Nuclear Weapons ¶ 1.78, 2.03-2.04 (3d ed. 1977).

For those who are exposed directly to the explosion of a fission or thermonuclear device, the heat, the blast, the gamma and neutron radiation pose serious and often lethal risks of injury. See Comm. for the Compilation of Materials on Damage Caused by the Atomic Bombs..., Hiroshima and Nagasaki: The Physical, Medical, and Social Effects of the Atomic Bombings 105–334 (pap. ed. 1981) [hereinafter cited as "Hiroshima and Nagasaki"]. The plaintiffs in this action, however, do not allege direct exposure to atomic blast effects. Instead they are concerned with exposure to the residual radiation present in the smoky ashes of the fireball.

The rapid expansion of the "fireball" generates a tremendous shock wave which can crush buildings and structures and hurl objects about with considerable force. Shock wave effects are not at issue in this lawsuit. Nor are other effects such as the electromagnetic pulse generated by a nuclear explosion.

The residual radioactivity of an atomic explosion—the material which becomes "fallout"—is derived from three sources: (1) fission products, i.e., the nuclear fragments resulting from the splitting of uranium or plutonium nuclei; (2) unused nuclear fuel (esp. plutonium), and (3) induced activity, i.e., radionuclides generated by bombarding stable elements in the bomb materials, air, tower metal, soil, seawater, etc. with neutrons. Any detonation of a nuclear device produces residual radioactivity in these three ways. The amount of material contributed by each process, however, varies with the specific qualities of each explosion.

[The Court then considered data pertaining to the three sources of fallout.]

The direct products of the nuclear fission process are the most important source of fallout radioactivity. We may infer that the explosion of a fission device produces an intensely radioactive cloud of fallout which releases tremendous beta and gamma radiation in the first few hours and which declines at a rapid rate until only the longer-lived radionuclides and stable isotopes remain. These inferences are borne out by the experimental evidence derived from atomic testing.

All nuclear devices tested in the atmosphere have produced some quantity of highly radioactive fission products that return to the surface as "fallout." Fortunately, the rate of fission product activity declines very rapidly.

Plutonium, uranium and other heavy element residues represent a significant long-term component of nuclear fallout. In fact, "in the period from 20 hours to 2 weeks after the burst, depending to some extent upon the weapon materials, those isotopes can contribute up to 40% of the total activity of the weapon debris." S. Glasstone & P. Dolan, The Effects of Nuclear Weapons ¶ 9.32 at 405 (3d ed. 1977).

B. ISOTOPES AND ENVIRONMENTAL IMPACT

Because of their rapid rate of decay, some of the fission and activation products of a nuclear explosion are of minimal importance to assessment of the risks of injury posed by fallout. They die too quickly. Isotopes such as iodine-136, with its 83 second half-life, have all but vanished within minutes after detonation, long before the hot gases and dust of the mushroom cloud could reach offsite communities. Other isotopes, such as 8-day iodine-131, or 30-year cesium-137, have come to be recognized as posing serious risks of injury to the downwind environment.

Furthermore, the chemical properties of fallout radionuclides affect their hazard potential. Isotopes such as krypton-85 (T½ = 10.5 yrs.) or xenon-135 (T½ = 9.2 hours) are inert gases by nature; they don't react with other chemicals, and they literally do not "fall out." Other products such as strontium-90 (T½ = 28.5 years) or plutonium-239 (T½ = 24,000 years) settle as dust on foodstuffs and upon ingestion, are absorbed into bone tissue, where they bombard sensitive marrow tissue with ionizing particles. The hazard presented by each component isotope of the fallout cloud is thus a product of both nuclear and chemical factors. The nuclides that are of interest to our inquiry are identified according to a number of factors such as:

(1) total yield;

(2) half-life;

(3) radiation energies;

(4) refractory or volatile qualities;

(5) chemical characteristics;

(6) uptake and concentration in the food chain;

(7) degree to which the nuclide is retained in the body; and

(8) concentration of the nuclide in sensitive tissue.

The refractory (i.e., heat resistant) qualities of a given isotope, for example, are an important factor determining whether it will be found mainly in the early fallout at the site of the explosion, or whether it will escape in the cloud of hot gases and debris that becomes "delayed" fallout affecting offsite communities. The fallout yield of any nuclear device experiences fractionation, the uneven distribution of radioactive products in early and delayed fallout debris.

The phenomenon of fractionation is due to both chemical and physical separation of the radionuclides. Chemical separation occurs in the first few minutes. At zero time, everything in the vicinity of the device is vaporized. As the fireball cools to about 3000° C the Fe0 [iron oxide] and soil components form liquid droplets in which the refractory elements dissolve. The fireball cools to the melting point of soil and iron oxides to 1500° C in about 20 seconds. The solidified droplets contain the refractory elements, while the volatile elements and their radioactive daughters remain in the gas phase. In 6–8 min., when the cloud has cooled to ambient temperature, -50° C, the volatile elements (except for Kr and Xe) and their daughters condense. H. Hicks, "Calculation of the Concentration of Any Radionuclide Deposited on the Ground by Offsite Fallout from a Nuclear Detonation," 42 Health Physics 585, 586 (1982), DX-1162. (citations omitted). Hicks lists elements important to fallout in two categories: [1] Volatile at 1500° C, in a gaseous state: Germanium, Krypton, Silver, Tin, Cesium, Iodine, Gold, lead, et al., and [2] [1] Refractory at 1500° C, condensed as a solid or liquid: Sodium, Iron, Cobalt, Uranium, Plutonium, Curium, Strontium, et al.

Elements in the "refractory" column will predominate in early fallout debris; those in the "volatile" column will more likely be found in delayed fallout residues. A number of variables affect how efficiently the fractionation process separates nuclides of the two categories as the fallout cloud develops. For example, strontium-90, a refractory isotope produced in significant quantities by fission explosions, does not condense in the initial refractory particles because it does not yet exist in quantity. Strontium-90 is produced primarily as a "daughter" isotope of the volatile elements krypton and rubidium.

Where external exposure to fallout is of concern, the pathway to man is a simple one: direct contact with fallout materials in one's immediate surroundings. Fallout particles may come to rest upon skin or clothing, on houses or automobile or other nearby surfaces, or fallout—particularly the tiniest particles—may be inhaled, a direct pathway resulting in internal exposure.

These examples illustrate that the risks of internal exposure to fallout radioactivity are a product both of the specific properties of each radionuclide and the characteristics of the organisms—including humans—that ingest them. Dr. John Gofman has detailed some of the most important variables:

> For each radionuclide there are several crucial factors that will determine how much energy is actually deposited in the tissue, including:
>
> a. the route of entry of the radionuclide into the body.
>
> b. the fraction of the administered dose that actually reaches the tissue of interest.
>
> c. the biological rate of removal of the radionuclide from that tissue.
>
> d. the amount of radiation the tissue of interest receives from the portion of the radionuclide deposited in tissues other than the one of interest (so-called crossfire radiation).
>
> e. the number of microcuries of the radionuclide taken in (crucially dependent on the correct measurement of the strength of the radionuclide source).
>
> f. careful calculation of the average energy of the beta [or alpha] particles emitted, of any ancillary gamma rays emitted and of any loss of radiation out of the specified tissue.
>
> g. metabolic or other factors that might alter the distribution of the radionuclide in the various tissues of the people studied.

One of the other factors affecting the distribution of radionuclides in various tissues is one wholly independent of either nuclide or organism: in food chains that contain an abundance of stable nutrient elements (e.g., calcium, iodine-127, potassium), absorption of the chemically analogous radionuclides (e.g., strontium-90, the iodine isotopes, cesium-137) is significantly reduced in comparison to ecological systems that are nutrient-poor.

A person whose diet provides an abundant supply of potassium, for example, will absorb and retain less cesium-137 from contaminated food than will someone whose diet lacks that nutrient. Similar effects are observed in livestock, poultry and fish. Fresh water fish, for example, will retain cesium-137 and strontium-90 in their tissues at a rate 200 times greater than will salt water fish.

Radioactive isotopes of nutrient elements, or of elements that are chemically similar to nutrients, are far more likely to be absorbed and retained than many of the relatively insoluble metal oxides found in fallout debris. A frequently used method of comparing relative rates of retention is computation of the biological half-life. The biological half-life is simply the time period in which one half of a given quantity of absorbed material leaves the body. The rate at which a particular radionuclide leaves tissue varies from organ to organ:

The biological excretion rate of an element from the human body is not necessarily the same for the whole body and for a particular organ. In fact, in most cases, the biological elimination rates are different for different organs and for the body as a whole. For example, the biological half-life of iodine is 138 days for rejection from the thyroid, 7 days for the kidneys, 14 days for the bones, and 138 days for the whole body.

Before the internal exposure attributed to a particular contaminant may be evaluated, the biological half-life of the nuclide must be adjusted to account for radioactive decay. While iodine-131 has a biological half-life (in the thyroid) of 138 days, its 8-day radioactive half-life determines that an absorbed quantity of iodine-131 will have decayed almost entirely before a meaningful fraction of it is eliminated by the body.

The energy released by radioactive decay will largely be diffused along thousands of ionization tracks in the absorbing tissue. Adjustment of the two half-lives, radioactive and biological, yields an effective half-life of 7.6 days. The radionuclides are listed in the chart herein titled: *Radioactive, Biological and Effective Half-Lives of Certain Common Isotopes.*

If the radioactive and biological half-lives are known, the effective half-life may be calculated according to the formula:

$$Te = (Tr)(Tb) / (Tb + Tr)$$

Where Te is the effective half-life, Tr the radioactive half-life and Tb the biological half-life. J. Gofman, Radiation and Human Health 422–24 (1981), PX-1046. As a result of the combined radioactive and biological elimination of a radioisotope from the whole body or from an organ, the dose rate to the body or the organ is not constant over time.

The complexity of the problem yields a great temptation for everyone to generalize. Abstract computations derived from hypothetical models are far easier than painstaking calculations based upon exacting observation and analysis. It is far easier to project a hypothetical dose of radiation received worldwide from an evenly distributed and known quantity of one radionuclide than it is to calculate in 1984 the total dose, that Sally Jones or Timmy Smith received by 1962 from eating unknown quantities of food or milk from unknown sources contaminated with unknown yet real quantities of fallout radioactivity of largely unknown

Radiological, Biological, and Effective Half-Lives of Certain Common Isotopes

Isotope	Organ of Reference	Radiological Half-Life	Biological Half-Life	Effective Half-Life
3 H	Total Body	2.3 y	12 d	12 d
14 C	Total Body	5,700 y	10 d	10 d
	Fat		12 d	
	Bone		40 d	
32 P	Total Body	14.3 y	257 d	13.5 d
	Liver		18 d	8 d
	Bone		1,115 d	14.1 d
	Brain		257 d	13.5 d
40 K	Total Body	1.28×10^9 y	58 d	58 d
88 Fe	Total Body	1,1000 d	800 d	463 d
	Spleen		600 d	388 d
	Lungs		3,200 d	819 d
	Liver		554 d	368 d
	Bone		1,680 d	665 d
57 Fe		45.1 d		
99 MTC	Total Body	0.25 d	1 d	0.2 d
	Kidneys		20 d	0.25 d
	Lungs		5 d	0.24 d
	Skin		10 d	0.24 d
	Liver		30 d	0.25 d
	Bone		25 d	0.25 d
129 I	Total Body	1.726×10^7 y	138 d	138 d
	Thyroid		138 d	138 d
	Kidneys		7 d	7 d
	Liver		7 d	7 d
	Spleen		7 d	7 d
	Testes		7 d	7 d
	Bone		14 d	14 d
131 I	Thyroid	8 d	138 d	7.6 d
238 U	Total Body	7.12×10^8 y	100 d	100 d
	Kidneys		15 d	15 d
	Bone		300 d	300 d
238 U		4.66×10^9 y		
229 Pu	Total Body	24,000 y	175 y	175 y
	Liver		82 y	82 y
	Kidneys		87.7 y	87.7 y

composition. Consequently, hypothetical (especially worldwide) projections are common in the literature; analyses of specific persons or particular small communities are rare.

The record in this case is replete with examples of analysis based upon worldwide or nationwide projections. More localized studies are far fewer in number. Of the group concerning the Nevada/Utah/Arizona area, several were undertaken only after the commencement of this lawsuit. A common characteristic of each study, however, is the presence of at least one hypothetically derived generalization about phenomena which are incurably variable—and unpredictable at least to some extent. How much fallout radioactivity follows the complex network of environmental and biological pathways to reach man is determined in the first instance not by the dynamics of the pathways but by the initial amount of the fallout debris that is deposited at a particular place in time. This initial distribution pattern is itself a product of important and often unpredictable variables. In the standard handbook on the subject—published in 1950, a year before Nevada testing began—the authors observed:

> With regard to the internal radiation hazard, it is not possible to make any sound estimate of the amount of material which is likely to be ingested in various circumstances. A person working under normal indoor conditions, for example, would absorb much less than one engaged in an occupation in which there was much dust. Children, because of their habits and closeness to the ground, would be expected to ingest more than adults. These factors would greatly complicate a rehabilitation program, and make it almost impossible to attempt to assess universal permissible contamination levels.

C. FACTORS AFFECTING FALLOUT DISTRIBUTION

The fallout yield of a nuclear explosion is predictable with some degree of accuracy, so long as certain information about the device is known. How airborne fallout is ultimately dispersed in the environment, however, is far less amenable to abstract calculation. Fallout distribution is influenced by a series of variables which seemingly multiply without end. Yet fallout dispersal has been the subject of repeated generalization at the theoretical level. Monitoring techniques have assumed and relied upon the soundness of various dispersal models. The validity of many radiation exposure and dosage estimates in turn has depended in large part on the accuracy of those monitoring techniques.

> [The Court considered several factors contributing to the scattering of nuclear fallout, including: Weather generally, wind conditions, rain, and even human activities, such as farming and ranching. The Court noted that the uncertainty factor could be considerable.]

From the moment of detonation, radioactive material is dispersed unevenly; while the head of the mushroom cloud contains 90% of the fallout activity, distribution within the cloud is anything but uniform. As the cloud expands, large fallout particles begin falling to the surface, while winds propel smaller particles of radioactive dust away from the site of the explosion. Hot gases diffuse quickly into the surrounding air, blending with the environment in the eddies of numberless air currents.

D. FACTORS AFFECTING TOTAL EXPOSURE TO FALLOUT RADIATION

The most significant single factor determining the extent of radiation exposure from nuclear fallout is time. As noted above, the fission and induced activity products of a nuclear explosion are often very unstable and decay very rapidly. Theoretical estimates by Glasstone and others predicted an overall decay rate expressed as t -1.2, where t represents the time elapsed since detonation. Actual measurements following some of the Nevada tests indicated that the decay rate in the short period immediately following the explosion was even higher (e.g., t-

1.3 or t-1.4). A large portion of the radiation emitted by fallout materials will thus be spent in the first hours and days following the detonation. While some fallout components such as plutonium, cesium-137, strontium-90, and iodine-129 will persist in the environment for many years following a test, most of the 200 or more radionuclides decay and disappear far more quickly. While the high rate of decay of fallout materials materially reduces the risk of exposure over a long period, it also indicates that contact with fallout in the first minutes or hours following detonation can easily result in serious radiation exposures.

Besides the mere passage of time, other factors operate to reduce, or attenuate the degree of radiation exposure experienced from fallout. From an external exposure standpoint, distance from radioactive materials reduces the effective dose, particularly from alpha and beta radiation. The highly ionizing quality of alpha radiation allows it only a short range in air before the particle energies are spent, perhaps a distance of a few centimeters, or an inch or two. Beta particles, as low-LET radiation, can travel greater distances by virtue of their weaker ionizing qualities; "[m]any of the beta particles emitted by the fission products traverse a total distance of 10 feet (or more) in the air before they are absorbed. However, because the particles are continually deflected by electrons and nuclei of the medium, they follow a tortuous path, and so their effective (or net) range is somewhat less." More dense materials, e.g., water, wood, glass, paper, and brick, experience far greater ionization along the particle tracks, which consequently are much shorter, perhaps one-thousandth of the range observed in air.

Where, however, the material exposed is living tissue, either through deposit of particles on the skin or inhalation or ingestion of fallout materials into the body, alpha and beta radiations can pose a serious risk of acute and long-term injury. Clearly, radiation exposure from fallout can be materially reduced simply through physical processes, i.e., removal of fallout particles from skin, clothing, hair and immediate surroundings by thorough washing, use of face masks around contaminated dust to avoid inhalation, and careful handling of food and water to avoid or minimize contamination by radioactive materials. Gamma radiation ionizes far fewer air molecules as it passes through, permitting it a far greater range in air than the charged -particle radiations. As described previously, gamma radiation is a form of electromagnetic radiation similar in its physical qualities to visible light, radio waves, and ultraviolet rays.

The far greater energy of gamma radiation permits it to penetrate and pass through dense materials with a range considerably greater than either alpha or beta particles. Clothing stops alpha particles and most beta particles, but is poor shielding against gamma rays, many of which will pass through both fabric and wearer relatively unimpeded. More substantial materials, such as building materials or soil, can significantly reduce or attenuate gamma radiation exposures when placed between the source and the person at risk.

In terms of alpha, beta and gamma radiation, the potential exposure that may be experienced during a radioactive fallout event can be materially reduced simply by staying indoors (assuming the contamination is kept outdoors) until the exposure hazard has abated. Washing of fallout materials into the soil, either by rainfall or deliberate effort, will further attenuate the rate of gamma exposure, as does turning under the soil itself when the surface is contaminated with fallout. The shielding effect of buildings and dwellings is augmented simply by washing down the roof and large wall surfaces, reducing the effect of being geometrically surrounded by radiation sources.

Internal exposure arising from ingestion, absorption or inhalation of radioactive fallout materials presents some different problems governed by additional variables. The complex pathways by which radioactive materials make their way into biological and environmental systems have already been mentioned in this Part. Where nuclear contamination is present, great care must be taken to avoid inhaling or consuming fallout materials in food and water. Once ingested or inhaled, the degree of exposure actually experienced depends upon the highly variable physical and chemical qualities of each individual radionuclide. Many fallout contaminants inhaled or ingested as dust are in the form of relatively insoluble metal oxides. These may pass quickly through the body, radiating tissues as they pass by, but not persisting in any of them for extended lengths of time. Others, whose solubility is greater and which may have chemical properties similar or identical to important nutrients, can quickly find their way into sensitive human tissue. As noted above, the radioactive isotopes of iodine quickly migrate to thyroid gland tissues; strontium and barium isotopes find their way to important bone tissue. Once absorbed, the fallout radionuclides diminish in accordance with a combination of radioactive and biological half-lives. Some contaminants, such as plutonium with its effective half-life in the bones of 400 years, become practically permanent sources of ionizing radiation within the body itself.

The fission products can produce injury either as an external source of radiation or, if they gain entry into the body, by acting as an internal radioactive poison, quite analogous to radium poisoning. This latter consideration is a major concern, since the amounts required within the body to produce injurious effects are minute compared to the quantities necessary to induce damage by external beta and gamma irradiation.

Prediction of internal exposure from such varied toxins becomes an exceedingly complex task, for it is apparent that "no radionuclide that is produced should be ignored until a critical analysis demonstrates that it is insignificant." Y. Ng, et al., Prediction of the Maximum Dosage to Man From the Fallout Nuclear Devices: -IV. Handbook for Estimating the Maximum Internal Dose from Radionuclides Released to the Biosphere, Lawrence Radiation Laboratory, UCRL-50163 Part IV (1968), PX-720, at v. [hereinafter cited as "LLRL Handbook for Estimating Maximum Internal Dose"]. While many fission products decay and nearly vanish before making their way into the food chain, others persist with tremendously variable effect upon the degree of internal radiation exposure received. Even where metabolic or ecological variables can be ignored, as in estimation of external alpha, beta, and gamma dose from exposure to fallout debris, the calculations are not simple.

Clearly, where exposure to radioactive fallout is concerned, the complexity and the variability of the multitude of factors makes careful and comprehensive monitoring and measurement a genuine necessity, if external and internal doses are to be accurately evaluated. Assessment of the risks posed by the infiltration of nuclear debris into the daily life of off-site residents in communities in Utah, Nevada and Arizona is a demanding task at best. Where the available data generated from actual measurements of fallout activity and contamination is grossly inadequate and incomplete, the difficulty of the task is compounded a thousand-fold.

IV. BACKGROUND: BASIC PRINCIPLES OF HEALTH PHYSICS

A. UNITS USED IN RADIATION DOSIMETRY

The testimony and exhibits in the record as well as the scientific literature concerning radiation and health make use of a variety of units to express or describe the amounts of radiation people are exposed to or have absorbed into their tissues. In evaluating information and analysis of radiation hazards, it is useful to identify the specific units being used and the concept and quantity that each unit expresses.

Radiation-induced damage to biological tissue results from the absorption of energy in or around the tissue. The amount of energy absorbed in a given volume of tissue is related to the type, energy, and number of radiations traversing the tissue volume, and the interactions which occur between the radiations and the atoms and molecules of the tissue.

(1.) Rad

The most commonly used unit to describe radiation dose, that is, the actual amount of ionizing radiation absorbed by living tissue, is the rad. Measurement of radiation dose in rads is accomplished by dividing the total radiation energy delivered by the mass (in grams) of the tissue which absorbs it. A rad expresses an energy/mass ratio:

> Historically, the erg has been used to describe energy delivered by ionizing radiation. If 100 ergs of energy are deposited in 1 gram of tissue, the tissue is said to have received 1 rad of radiation.

The definition of the rad says nothing about the time it takes to deliver 100 ergs per gram of tissue. In some cases good reason exists to be concerned about the rate of delivery of radiation, and in these cases we would express radiation doses in terms of rads per minute, rads per hour, or rads per year. For many purposes we shall be concerned with total rads delivered, and not with their rate of delivery. Like other units discussed so far, the rad may be more conveniently expressed in fractional form:

1 milli rad (mr) = 0.001 rads (or 10^{-3} rads)

1 micro rad (r) = 0.000001 rads (or 10^{-6} rads)

Larger multiple units, such as kilo rad (1,000 rads) have little application to measurement of doses to specific individuals; absorbed in a short time, such a dose would likely be fatal. Recently, however, more use has been made of a unit from the international system of units, the Gray. A Gray (Gy) represents the absorption of 1 joule (10^7 ergs) of energy in 1 kilogram (1,000 grams) of tissue. Simple calculation disclosed that 1 Gray represents 10^4 ergs per gram of tissue, or a dose equaling 100 rads. J. Gofman, Radiation & Human Health, supra, at 46. A dose of one rad would equal 0.01 Gray. 50 millirads would equate to 0.0005 Gray. On paper, a dose of radiation expressed in Grays looks smaller than the same dose expressed in rads or millirads. Units must, however, be kept in perspective. A lethal acute dose of radiation would be a "mere" 6 Gy.

Except for a few recent scientific publications, the testimony and exhibits in this case speak in terms of rads rather than Grays, as will the text of this opinion. Radiation dose expressed in rads can be related to radioactivity expressed in curies through simple mathematical calculation. Assuming that all of the energy released by the beta decay of 1 picocurie (PCI) of radioactive material evenly dispersed in 1 kilogram of tissue is actually absorbed in that tissue, the dose rate to that tissue, expressed in rads per hour, would equal:

Dose rate (rads/hr) = $(2.134 \times 10^{-9}) \times$ (average decay energy [MeV])

The total dose received to the kilogram of tissue (e.g., the lungs of an adult person), may be calculated using an equation previously discussed:

Total Dose (rads) = (dose rate/0.693) $\times$ (T e) $\times$ (k)

Where T e is the effective half-life of a specific radionuclide and k is a mathematical term which converts T e into hours.[42]

---

42. The equation gives a fair estimate where the total period (T) considered is much greater than the effective half-life (Te). For nuclides with long effective half-lives, a different formula is used. See Tsoulfanidis, supra, at 498.

For example, 1 pCi of iodine-131, with an average decay energy approximating 0.6 MeV, would deliver ionizing radiation in the form of beta particles at the rate of 1.28×10^{-9} rads per hour, for a total dose of nearly 3.37×10^{-7} rads, or 0.33 micro rads, based upon an effective half-life of 7.6 days. This is a very small dose of beta radiation. The picocuries can add up, however, as illustrated by an example:

> By drinking milk contaminated with radioactive fallout debris, a child absorbs 20,000 PCi of iodine-131 into the tissue of her thyroid gland. The thyroid gland of a child is far smaller than a kilogram, weighing in the neighborhood of 2–4 grams. Rather than being distributed throughout a full 1,000 grams of living tissue, the energy is concentrated in far fewer radiosensitive cells. Correcting our calculation for the differences in contamination and tissue mass, we find:
>
> $(2.134 \times 10^{-9})(0.6)(20{,}000)(1000/4) \times (7.6 \text{ days})(24 \text{ hr/day})\ 0.693$
>
> = 1.68 rads to the child's thyroid gland from 131I.

From a quart or so of contaminated milk, our example child could receive a radiation dose to a sensitive organ significantly higher than "background" levels due to the presence of a single fission product. Besides iodine-131, nuclear fission produces various quantities of 11 other radioactive iodine isotopes, with half-lives ranging from a few seconds (e.g., 5.9 sec. 138I) to 16 million years (129I).

The example calculation is far from totally abstract. One study of milk sampled from Utah dairies shortly after the 100-kiloton SEDAN test at the Nevada Test Site in 1962 estimated the concentration of 131I to vary between 0 and 800,000 pCi per litre (1.05 quarts). Further, it was estimated as a result of an accident involving the Windscale nuclear reactor in the United Kingdom in October 1957, 20,000 curies of 131I was released into the atmosphere:

Sufficient pasture contamination occurred to require the confiscation of milk for several days in a 200-square-mile area downwind of the reactor. The milk produced in a much smaller area remained contaminated above the confiscation level chosen by the British for more than a month. As a result of this confiscation and other precautionary procedures, the mean absorbed dose to the thyroid glands of children downwind of the Windscale reactor was estimated to be 16 rad, and the mean adult absorbed dose was estimated to be 4.0 rad.

While radioactive fallout debris from Nevada was dispersed over a much wider area than was contamination at Windscale, a "nominal" 20-kiloton fission device yields 250,000 curies of 131I, representing between 2.9 and 3.1 percent of the total fission products. The calculation performed above demonstrates that rads of radiation dose to internal organs may arise from picocuries of contamination. 250,000 curies (Ci) of 131I is equivalent to 2.5×10^{17}, or 250,000,000,000,000,000 picocuries (pCi) of 131I—still barely 3% of the total fission product yield. At the very least, one can say that where fallout is concerned, a lot can go a very long way. The example figure is based on a 20-kt explosive yield. Larger devices produce proportionately larger total yields of radioactive fission product. Estimation of radiation dose from external sources, particularly multiple sources of gamma radiation, requires far more complex calculations.

Radiations of external origin that may be absorbed in living tissues include natural or man-generated rays from the atmosphere, earth, water, and other tissues or organisms. The type of external exposure that is probably the simplest to evaluate is that resulting from a "point source" of photons. Photon emitters have a gamma ray exposure constant, termed I, which has the units of roentgen per hour per curie at an air distance of 1 m. The value of I, may be calculated for any gamma- emitting radionuclide and it is dependent upon the number of photons emitted per disintegration and the energies of the photons.

Nonpoint sources and causes in which several gamma emitters are more difficult to evaluate. The case in which the radionuclide is distributed within a mass of substance is even more difficult to evaluate because of absorption and scattering of photons prior to their exit from the substance. If such complexities are great enough, direct measurement of exposure rate may be necessary. The tremendous variability of potential gamma and beta radiation exposures from nuclear fallout renders accurate and thorough measurement of radiation intensity an unavoidable necessity.

(2.) Roentgen

The complexity of dose calculation involving gamma exposure often makes actual measurement of radiation in rad units impractical.

Frequently, it is not convenient to measure dose in rads. In the particular case of irradiation of organisms or tissues from an external X- or gamma ray source, it is more convenient to measure the radiation exposure at the surface of the tissue of interest using an appropriate instrument or device. The dose to the tissues can be estimated from the measured exposure from theoretical considerations. The fundamental unit of radiation exposure which is applicable only to X- or gamma radiation, is the roentgen (R), which is defined as:

$$1 \text{ R} = 2.58 \times 10^{-4} \text{ coul/kg air}$$

This definition of the roentgen is equivalent to the production by X- or gamma rays of 1 electrostatic unit of charge of either sign per cubic centimeter of dry air at 0° C and 760 mm mercury. The roentgen quantity is defined on the basis of production of ionization in air because it is normally measured with an air-filled ionization chamber.

Simply expressed, the roentgen unit describes a certain degree of ionization caused in air by gamma rays. In the process of ionization, of course, some gamma energy is absorbed by the air, approximately 88 ergs per gram, or 0.88 rad. N. Tsoulfanidis, Measurement and Detection of Radiation 485 (1983). The greater density of living tissue results in greater absorption of the same amount of gamma radiation; a roentgen of gamma radiation in air equates with approximately 94 ergs absorbed per gram of tissue, i.e., 0.94 rads at the surface. Gamma rays with energies ranging from 0.3 MeV to 3.0 MeV permit a rough equivalence between 1 roentgen (R) of exposure and 1 rad (R) of absorbed dose in soft tissues.

(3.) rem

In strict terms, a rem of ionizing radiation, or roentgen equivalent man, describes "[t]hat amount of ionizing radiation of any type which produces the same damage to man as 1 roentgen of about 200- = kv X radiation." The rem describes a quality of ionizing radiation not encompassed by either of the first two units: neither rad nor roentgen say anything about the penetrating qualities of particular radiation or differentiate between high-LET and low-LET radiation. We know that the linear energy transfer (LET) is much higher for some radiations than for others. For those biological effects where LET matters, the rad unit must be modified to a unit that shows the greater effectiveness of one kind of radiation over another. A term has been introduced, the relative biological effectiveness, or RBE, to express this difference is effectiveness.

Defined in terms of a dose of gamma radiation producing a certain biological effect, e.g., chromosome lesions, cell-killing, etc., a rem represents the proportionate amount of other ionizing radiation that will produce the same biological effect, i.e., the same number of chromosome lesions, dead cells, etc. The specific proportion of one to the other, the RBE, varies by type and energy of radiation, the type of cell exposed, the biological effect being studied, the total radiation dose, the dose rate, and other factors. N. Tsoulfanidis, supra, at 486. The same kind of radiation may have many RBE ratios.

As a general rule, biological damage increases as the amount of energy deposited per unit of distance traveled, the linear energy transfer (LET), increases. High-LET radiation, such as alpha particles, neutrons and fission fragments are often significantly more damaging than low-LET radiation, such as gamma ray photons, beta particles, and positrons, which have far less mass, and travel far greater distances, leaving less ionization in their wake. Id. J. Gofman, supra, at 26–29. In modern health physics, the concept of quality factor, or Q, is used to describe distinctions between radiations based upon linear energy transfer values. N. Tsoulfanidis, supra, at 486–87. While low-LET radiations may have a "Q" of 1, 5-MeV alpha particles may have a "Q" of 10 or even 20. Id. Whether expressed in terms of Q or RBE, the relative effects of different radiations is critical to accurate calculation of biological risk associated with radiation exposure.

Once the task of evaluating a particular exposure to radiation in terms of its relative biological effectiveness has been accomplished, the expression of dose in rem units is easily transferred or compared with other dose estimates expressed in rems. The risk arising from exposure to various radiations—alpha, beta, gamma or neutrons, among others, may be standardized and directly compared. A dose expressed in rems identifies a certain quantum of biological effect, regardless of source. Recent years have seen increasing use of the equivalent international unit, the Sievert (Sv). A dose of 1 Sievert equates with 100 rems.

In much of the literature, testimony and documentary evidence now before this court, however, doses of low-LET radiation expressed in rads are treated as equivalent to doses expressed in rems. Only as to alpha radiation is a distinction usually made. To be fully effective, analysis of data and measurements concerning exposure of human beings to ionizing radiation must be made with particular reference to the units in which the information is expressed. Whatever unit is used, be it rad, roentgen, rem, or some fractional or multiple variant of these units, it describes the exposure to ionizing radiation or absorption of its tremendous energies by living tissue.

B. RADIATION ENERGY: QUANTITY VS. QUALITY

So far this discussion has dealt with the units of radiation measurement in terms of each other, and other measurements of radioactivity, such as the curie, or the concept of half-life. It is important, however, to make at least a general comparison between the kinds and quantities of energy infused into biological systems by ionizing radiation and those imparted by more common, more familiar physical processes. It is one thing to define a unit called "rad" in terms of 100 other units of energy called "ergs," or an international unit known as the Joule. How much energy does an erg represent, or a rad? One Joule (J) is the energy equivalent of 1.000165×10^7 ergs.

Perhaps the best comparison of amounts—and qualities—of energy can be drawn between ionizing radiation and heat. To do so, though, requires definition of one more unit describing energy:

> The calorie is the familiar unit in chemistry that describes energy transfers involving heat. One calorie is that amount of energy which will raise the temperature of one gram of water by one degree centigrade [or Celsius]. (This definition does change some at different temperatures of water, but for our purposes here we can neglect those small changes.).

The best estimates are that approximately 400 rads of whole-body radiation, if delivered rapidly, are sufficient to cause 50% of the exposed humans to die within a period of days to weeks. This is the so-called acute radiation sickness. Is this a great deal of energy in heat terms? Some simple calculations show that it is not. Biological tissue is quite comparable with water in the amount of heat required to raise its temperature by one degree centigrade [or Celsius]. So we shall say that the required amount is one calorie per gram for biological tissue too. Therefore, our 10^{-3} calories/gram from the absorption of 400 rads of ionizing radiation energy would be enough to raise the temperature of biological tissue by 0.001 centigrade. Not much of a fever! We tolerate fevers of several degrees centigrade (not thousandths of a degree) in a variety of infectious diseases. Yet the amount of ionizing radiation that can kill half of the humans exposed to it, would—if converted first into heat—raise temperatures only by 0.001° centigrade.

This points up the biologically deadly difference between energy in the form of heat versus the same amount of energy in the form of ionizing radiation. The difference resides in the fact that the energy of ionizing radiation is not distributed the way the thermal energy of a fever is, the latter being distributed among all the molecules of a gram of tissue. Instead, the energy of ionizing radiation is transferred from photons to single electrons [or from charged alpha or beta particles], which in turn transfer all their energy to relatively few electrons in relatively few molecules. The transfer occurs in extremely concentrated fashion compared with the even diffusion of heat energy. Therefore, the energy delivered by ionizing radiation is energetic enough to break any chemical bond, even the strongest ones in living tissue.

It is not the quantity of energy penetrating the cells of the human body, but rather the quality of the energy, the nature of its interaction with cellular chemistry, which makes exposure to ionizing radiation an extraordinary hazard. As described in Part II, supra, alpha, beta and gamma radiations streak through matter at high energy and high speed, stripping electrons away from any chemical compound or individual molecule with which they come in close proximity or direct contact.

The ionized molecules may reunite as before, with no detectable damage. They may also react with each other or with other molecules nearby, forming new and not necessarily useful compounds, reflecting a temporary, or perhaps a permanent injury to the chemical structures originally present.

From the standpoint of biological injury, energy absorbed in the ionization of air, or a multitude of inanimate objects, is essentially meaningless. A glass of water which has been irradiated by alpha, beta or gamma rays poses no risk to health greater than any other water—unless, of course, it has become contaminated with radioactive material that would bombard internal organs with additional radiation if ingested. Far more important is the interaction of ionizing radiation with the tightly ordered, extremely complex and relatively fragile molecules of life. Of particular concern is the effect of ionization on the lengthy and critically important strands of genetic material found in each living cell.

C. IONIZING RADIATION AND CHROMOSOMES

At the interior of each living cell is found a nucleus of specialized proteins, enzymes, and other chemicals. The nucleus appears for most purposes to be the control center of the cell. It contains the long, spiralling molecules of DNA (deoxyribonucleic acid), which carry encoded within their delicate patterns the entire genetic "blueprint" controlling the structure, composition and numberless chemical activities of the individual cell, the tissue in which it is found, and—somewhere within its strands—the remainder of the instructions followed by all of the other specialized cells that make up an incredibly detailed living organism. The DNA molecules are joined together in discrete units called genes, which in turn are grouped together in a series of genetic superstructures, called chromosomes. Each of the billions of cells in the human body, with the exception of red blood cells, contains an invariable normal count of 46 chromosomes within its nucleus. Within those chromosomes are carried more than 5 billion individual bits of genetic information, carefully stored in the chemical patterning of the DNA molecules. C. Sagan, Cosmos 276 (1980).

Beginning with the work of Muller in 1927—many years before the structure of the DNA molecule was known or its function could be described, it was quickly discovered that ionizing radiation in all of its forms could effectively disrupt, impair or alter the genetic code.

D. HEALTH EFFECTS OF ACUTE RADIATION EXPOSURE

The use of the atomic bomb upon the populations of Hiroshima and Nagasaki, Japan presented a tragic opportunity for the detailed study of illnesses experienced by an entire population, young and old, men and women, following varying degrees of radiation exposure. This data, combined with information generated over the years by unfortunate experiments involving too much radiation, provides a fairly complete picture of the short-term effects of acute doses greater than 25 rads.

Summary of Clinical Effects of Acute Ionizing Radiation Doses

Range	**0–100 Rems**	**100–200 Rems**	**200–600 Rems**	**600–1,000 Rems**	**1,000–5,000 Rems**	**> 5,000 Rems**
Status	Sub-clinical	Surveillance	Therapy Effective	Therapy Promising	Therapy Palliative	Therapy Palliative
Vomiting	None	100: Infrequent 200: Common	300: 100%	100%	100%	100%
Initial Phase Onset	–	3–6 Hours	1/2–6 H	1/4–1/2 H	5–30 Min.	Immediate
Duration	–	< Day	1–2 Days	< 2 Days	< 1 Day	Immediate
Latent Phase Onset	–	< 1 Day	1–2 Days	< 2 Days	< 1 Day	Immediate
Duration	–	< 2 Weeks	1–4 Weeks	5–10 Days	0 Days	Immediate
Final Phase Onset	–	10–14 Days	1–4 Weeks	5–10 Days	0–10 Days	Immediate
Duration	–	4 Weeks	1–8 Weeks	1–4 Weeks	2–10 Days	Immediate
Leading Organ*	H	H	H	H	G	C
Prognosis	Excel.	Excellent	Guarded	Guarded	Hopeless	Hopeless
Convalescence	None	Several Weeks	1–12 Months	Long	–	–
Death Within	–	–	2–12 Weeks	1–6 Weeks	2–14 Days	< 1–2 Days
Cause of Death	–	–	Hemorrhage or Infection	Hemorrhage or Infection	Circulatory Collapse	Respiratory Failure or Brain Edema

* H refers to Hematopoietic Tissue; G refers to Gastointestinal Tract; C refers to Central Nervous System.

Though several of the witnesses in this action testified concerning symptoms such as reddening of skin and loss of hair (epilation) which are characteristic of acute radiation illness syndrome, no one in the present 24 cases has sought an award of damages in compensation for acute radiation injuries. The testimony was offered for a different purpose. Besides being notable for the dose ranges at which specific effects occur, the symptoms and injuries reflected in acute radiation illness are important herein for historical reasons; on a number of occasions the operational treatment of off-site radiation safety at the Nevada Test Site seemed to have been geared almost entirely to avoidance of only these most visible—and traceable—effects.

Other somatic effects traced to radiation exposures include loss of fertility, development of cataracts, and premature aging as well as permanent scarring and other persistent effects of acute radiation injuries. See e.g., Hiroshima and Nagasaki, supra, at 187–251.

E. RADIATION DOSAGE AND DOSE RATES

The illness syndrome accompanying acute radiation exposures clearly establishes that the rate at which an accumulated dose of radiation is received may have a material impact on the biological consequences of exposure. Except for narrowly perceptible changes in blood cell counts and characteristics which can now be detected after short-term exposures of 1 rad or less, the symptoms of acute illness indeed reflect the presence of a "threshold" level of exposure below which they do not appear.

Where long-term effects are concerned, however, the importance of dose rate seems greatly diminished. In 1980, the Committee on Biological Effects of Ionizing Radiation of the National Research Council (the "BEIR-III Committee") reported that "dose rate may affect the risk of cancer induction, but believes that the information available on man is insufficient to adjust for it." BEIR-III Report, supra, DX-1025, at 3. Other researchers are far more adamant on the subject. See J. Gofman, Radiation and Human Health, supra, PX-1046, at 404–407.

If, in fact, cancer and leukemia result from some genetic injury induced by radiation, or chemical substances—or both—the case for the proposition that low rate of exposure greatly reduces risk is weakened by the observation made in the earliest research that radiation injuries to genes and chromosomes appear to be cumulative, except to the limited extent that they are correctly repaired by processes within the cell. While the extent of radiation injury to cells inflicted at "high" dose rates may perceptibly affect the functioning of the organism more dramatically than a series of "low" doses imparted to cells over a period of time, injury may nevertheless result. At the level of the individual cell, ionization is ionization, and a linear energy transfer is a linear energy transfer. Id. PX-1046, at 42, 404–07. If a single alpha particle does carcinogenic damage to the genetic machinery of a cell, it may be irrelevant whether additional particles crash through the cell chemistry or not. Indeed, the 1980 Report of the BEIR-III Committee, quoted above, observes that "[t]here appear to be mechanisms ... pertaining especially to exposure to high-LET radiation, that increase the observed effect per unit dose when the dose rate is reduced." BEIR-III Report, supra, DX-1025, at 3. The failure of the human epidemiological studies to persuasively identify a "threshold" dose below which the risk of cancer is not at all increased lends additional support to the view that even at low doses, critical biological injuries accumulate.

F. RADIATION EXPOSURE: INTERNAL, EXTERNAL, WHOLE-BODY, PARTIAL-BODY

Prior sections of this opinion discuss and emphasize the fact that the potential for radiation injury from nuclear fallout debris is not limited to the dose received from gamma-emitting (or beta- emitting) radionuclides outside the body. There also exists the potential for serious exposure from radioactive debris inhaled, ingested or absorbed into the human body from contaminated food, water, milk, or dust. At the cellular level, however, the distinction between external and internal dose disappears. The amount of chemical disruption that a cell experiences from whatever source of radiation in any direction or at any distance is the important consideration. The phenomenon of internal exposure simply allows shorter-range alpha and beta particles greater access to more sensitive cells and tissues.

An important distinction must be drawn however, between radiation dose to the whole body and estimated radiation doses to specific organs. The biological impact of 300 rads absorbed by the skin is distinctly different in nature and severity from the impact of a 300-rad whole-body dose. Where whole-body exposure is concerned the absorbed energy is being visited upon nearly every gram of living matter in the body. A partial-body dose absorbed by a particular organ, such as the thyroid, the bone marrow, or the lung may be of considerable importance to the organ involved, but signifies far less relative damage to other organs and tissues.

In reviewing articles, studies, documents and testimony, considerable error in risk estimation may be avoided simply by paying heed to whether exposure and dose are being expressed in terms of the whole body or a particularized organ or system.

G. OTHER SOURCES OF RADIATION EXPOSURE

Fallout from atmospheric nuclear testing in Nevada is only one of a collection of sources of exposure to ionizing radiation, both man-made and naturally occurring. First, every person living on the earth receives some amount of radiation from the natural sources generally blanketed under the term "background" radiation. Natural background radiation derives from a number of sources: naturally occurring heavy elements, such as uranium, thorium, radium, radon, and polonium, and a few lighter radionuclides which are readily incorporated into living tissue in trace amounts, e.g., potassium-40 and carbon-14. Natural radioactivity is all around us, in soil, in building materials, and in trace amounts in meat, vegetables, fruits, grains and drinking water.

Additional background exposure is traced to cosmic rays, high-energy gamma rays and atomic particles which rain down from the sun through the atmosphere. People living at higher altitudes receive slightly more radiation than those who live at sea level, as do people whose occupations take them on frequent airline trips.

The most prominent source of man-made radiation exposure is the use of X-rays and radioisotopes in medical diagnosis and treatment. A significant dose of X-rays may be received from older or poorly maintained equipment. A number of current efforts are addressed to the accurate evaluation of risks arising from present methods and practices using radiation in medicine.

V. DISCRETIONARY FUNCTION

At the end of the trial the United States moved to dismiss this matter on the grounds that the actions of the United States of which plaintiffs complain are insulated from complaint by the discretionary function exception found in the Tort Claims Act. 28 U.S.C.A. 2680(a) (1976) states:

> The provisions of this chapter and section 1346(b) of this title shall not apply to—
> (a) Any claim based upon an act or omission of an employee of the government, exercising due care, in the execution of a statute or regulation, whether or not such statute or regulation be valid, or based upon the exercise or performance or the failure to exercise or perform a discretionary function or duty on the part of a federal agency or an employee of the government, whether or not the discretion involved be abused.

The United States first raised this question during the preliminary stages of this case. At that time, this court denied the motion, awaiting the development of a full and complete record. The history of the discretionary function exception, its rationale, and cases construing the exception were discussed in the court's opinion found at Allen v. United States, 527 F.Supp. 476 (D.Utah 1981).

The matters discussed in that opinion are equally pertinent at this stage of the proceeding. At this stage we do know enough to determine whether the acts complained of were discretionary and thus beyond the jurisdiction of this court. The early opinion provides the analytical framework against which we may now view the actions of the United States as developed in the full record.

> [The "discretionary function" exception to the Federal Tort Claims Act is designed to eliminate lawsuits that challenge the decision-making functions of government as opposed to its ministerial duties. By way of an extreme example, consider the mail carrier who runs over a pedestrian (driving the truck was merely ministerial) as compared to a president who declares a war (discretionary). Both cause injury, but only the former is subject to redress. In this case, the court held that the conduct of open-air atomic testing on behalf of the Atomic Energy Commission at the subject Nevada test site was not within the "discretionary function" exception to the Federal Tort Claims Act.]

VI. STATUTE OF LIMITATIONS

> [Because of the Government's failure to tell persons downwind what it knew about the long-range hazards of radiation fallout and how to minimize them, and further because public announcements downplayed the hazards, plaintiffs were not charged with knowledge of the connection between radiation and cancer so as to accrue claims under Federal Tort Claims Act; and, therefore claims were not barred for failure to file claims within two years after accrual.]

VII. THE DUTY ISSUE

Before the court in a case such as this may reach the question of whether the evidence demonstrates any negligence on the part of the defendant, it must first determine a crucial question of law: whether there exists a legal duty of care on the part of that defendant which extends to the plaintiff. If not, there can be no actionable negligence and no liability. If the court finds that a duty does indeed exist, then the court must determine whether the duty of care owed by the defendant extends to the specific risks that resulted in the plaintiff's injuries. In essence, the court is called upon to define the scope of the law's protection as applied to the facts of the particular case.

In performing this task, the court often has the benefit of guidance from the legislature in the form of statutory language, and from other courts who in the process of deciding prior cases have preserved their reasoning in the form of written opinions. In addition, however, the court is called upon to use its own judgment in evaluating the important ethical, economic, practical and remedial factors—often treated in wholesale fashion as "public policy"—which must be satisfied in every court decision. Finally, the court seeks to do justice between the parties, however "justice" may ultimately be defined.

The decision must be made, within constitutional limits, whether plaintiff will be granted the legal system's protection—that is, will the defendant be required to have met a specified standard of conduct in the case at issue or be subject to liability. This is a policy decision in purest form.

The fact that the defendant is the Government does not substantially alter the analysis of the duty issue; indeed, the Federal Tort Claims Act expressly provides that:

> The United States shall be liable, respecting the provisions of this title relating to tort claims, in the same manner and to the same extent as a private individual under like circumstances, but shall not be liable for interest prior to judgment or for punitive damages.... 28 U.S.C. § 2674 (1982 ed.). With a few exceptions, the duty issue as against the Government is examined in the same fashion as it would be as against the individual or corporate defendant in a negligence action.

The Atomic Energy Act of 1946, Act of Aug. 1, 1946, c. 724, 60 Stat. 755, makes repeated, express reference to the protection of health and safety as a significant goal for activities of the Atomic Energy Commission (AEC) created by that Act. In an effort to promote a program of private and governmental atomic energy research leading to maximum scientific progress in the nuclear field, the Act provided:

> Sec. 3. (a) Research Assistance—The Commission is directed to exercise its powers in such manner as to insure the continued conduct of research and development activities in the fields specified below by private or public institutions or persons and to assist in the acquisition of an ever-expanding fund of theoretical and practical knowledge in such fields. To this end the Commission is authorized and directed to make arrangements (including contracts, agreements, and loans) for the conduct of research and development activities relating to—
>
> (1) nuclear processes;
>
> (2) the theory and production of atomic energy, including processes, materials, and devices related to such production;
>
> (3) utilization of fissionable and radioactive materials for medical, biological, health, or military purposes;
>
> (4) utilization of fissionable and radioactive materials and processes entailed in the production of such materials for all other purposes, including industrial uses; and
>
> (5) the protection of health during research and production activities.

[Other portions of the Atomic Energy Act are discussed and considered with respect to public safety.]

At the operational level of nuclear testing activities, the availability of judicial review is manifest. For example, it has already been held that persons residing downwind from the proposed venting of radioactive gases from an underground nuclear detonation have standing under the Administrative Procedure Act to challenge operational activities of the AEC "which allegedly disregard the congressional directive to protect the public health and safety" under the Atomic Energy Act. Crowther v. Seaborg, 312 F.Supp. 1205, 1217 (D.Colo.1970). See also Nielson v. Seaborg, 348 F.Supp. at 1373. Recent nuclear detonations, particularly the 5-megaton range CANNIKIN test on Amchitka Island in 1971, generated important litigation testing the applicability of the National Environmental Policy Act to those activities. The courts have held that nuclear testing, like other federal agency actions affecting the environment, is subject to the requirements of NEPA; the agency must make an appropriate evaluation of environmental impact before proceeding with tests such as CANNIKIN. See Committee for Nuclear Responsibility, Inc. v. Seaborg, 149 U.S.App.D.C. 385, 463 F.2d 788, 796 (1971) injunction denied, 404 U.S. 917, 92 S.Ct. 242, 30 L.Ed.2d 191 (1971); Aleut League v.

Atomic Energy Commission, 337 F.Supp. 534 (D.Alaska 1971). See also Committee for Nuclear Responsibility, Inc. v. Seaborg, 404 U.S. 917, 92 S.Ct. 242 (1971) (Douglas, J. dissenting).

The question relevant to the duty issue in this case is whether the interests asserted by the plaintiffs are protected by the safety and health provisions of the Atomic Energy Acts, as amended, i.e., whether the acts impose a duty upon the Government in favor of these plaintiffs to avoid the specific risks of injury at issue in this lawsuit. If so, then the next question is whether the acts define a standard of conduct against which the defendant's conduct may be measured.

In Crowther v. Seaborg, 312 F.Supp. 1205 (D.Colo.1970), the court held that the interests of persons residing downwind from AEC activities in avoiding harmful exposure to radioactive fallout released from an underground test site were sufficiently within the scope of the Atomic Energy Act's protection to afford them standing to challenge the proposed release under the Administrative Procedures Act.

A sensible reading of the Atomic Energy Acts readily proves (1) and (2); the type of harm alleged by the plaintiffs, i.e. physical injury resulting from exposure to ionizing radiation, is precisely the type of harm that the requirements and guidelines of those acts were intended to minimize as a protection for downwind residents and the general population. (Whether there was a violation of the statutory duty that proximately caused the plaintiffs' injuries and losses remains to be determined. See Part VIII, infra.) Although the criteria set forth by Hall v. Warren and related cases are generally applied to state statutes and local ordinances, no compelling reason presents itself which would justify exclusion of a federal statute which meets the criteria, particularly in the context of the Federal Tort Claims Act. The FTCA in hybrid fashion applies state tort law to judge conduct of governmental agencies and employees whose activities and duties will most often be defined initially by federal statute or regulation.

In Crowther, the court found that promulgation and enforcement of adequate radiation protection standards by the AEC adequately defined the scope of the AEC's duty under the relevant acts. The duty imposed upon the Government in this action is at least as extensive as that in Crowther. It is the same basic duty of care owed to a similar group of people. In the context of a negligence action, it may well be determined that the law imposes additional duties upon the defendant by virtue of the nature of its conduct toward the plaintiffs. Or it may not. At a bare minimum, however, the duty of care imposed by the Atomic Energy Acts establishes that some legally significant relationship exists between the defendant and the plaintiffs in this action.

While in some jurisdictions, violation of a statute designed to protect those in the plaintiff's position from the type of injury incurred by the plaintiffs as a result of defendant's action establishes negligence as a matter of law, in Utah, such violation is prime facie evidence of negligence only.

A number of factors bear upon the question of the issue of standard of care. The significance of the risks created by the defendant's conduct and the relative utility of measures which minimize or mitigate those risks are important considerations. As Cardozo instructs us, "The risk reasonably to be perceived defines the duty to be obeyed, ..." Palsgraf v. Long Island R. Co., 248 N.Y. 339, 162 N.E. 99, 59 A.L.R. 1253 (1928); "the orbit of the danger as disclosed to the eyes of reasonable vigilance would be the orbit of the duty." This holds true not only in determining to whom the defendant's duty of care reasonably extends, but also in determining the degree of care to be exercised. In Meese v. Brigham Young University, 639 P.2d 720 (Utah 1981), the Utah Supreme Court highlights an important corollary to the reasonable man standard: "[i]n the exercise of ordinary care, the amount of caution required will vary in accordance with the nature of the act and the surrounding circumstances." 639 P.2d at 723.

Under Utah law, the standard of conduct to be attained ordinarily is that expected of the reasonable man, i.e., the standard of reasonable care under the circumstances.

This court has defined negligence as a failure to exercise the degree of care which a reasonable person would have exercised under the same circumstances, whether by acting or by failing to act.... The care to be exercised in any particular case depends upon the circumstances of that case and on the extent of foreseeable danger involved and must be determined as a question of fact.

One important source of practical guidance in setting standards of conduct is the standard of care established and applied by government nuclear laboratories apart from the Nevada Test Site. It is one thing if responsible persons at Oak Ridge, or Hanford, or Los Alamos react with indifference to all but the most acute radiation exposures; it is quite another if in these laboratories, where much smaller quantities of material are much more easily controlled, all leakages and exposures are painstakingly monitored and analyzed in an effort to keep risks as low as is reasonably achievable. If stringent care is exercised, for example, in monitoring and confining curie quantities of radioactive material escaping into the ponds near Oak Ridge, why should not at least as much care be exercised in dealing with the mega curies of radioactivity in the pink-orange clouds of dust, gases and ash drifting eastward from southern Nevada? The safety standards of the nuclear research community provide important background and landmarks for evaluation of conduct as have the "community standards" of practice in other professions in other cases defining appropriate standards of care. While certainly the "laboratory" functions of the Nevada Test Site differ considerably from those of the other AEC research facilities, the risks created by the materials under study differ only in degree, not in kind. The value of nuclear community practice is further enhanced by the common control exercised during the relevant time periods by the AEC. Each was an arm of the same government.

The immediate danger, i.e., exposure of humans to ionizing radiation, involves direct contact with the invisible. Alpha particles, beta particles and gamma rays fall beyond the range of human senses. Specialized instruments or materials must be used to detect their presence. Those without special technical skills are hard pressed to exercise the degree of care needed to protect themselves from such a hazard. A stringent duty of care to minimize such hazards is deliberately imposed on the party who acts with vastly superior knowledge, in favor of those having less information. The information gap between the plaintiffs and the Government in this action approaches the absolute; a duty of care adjusted according to that vast difference is a very stringent duty indeed.

Another reason for imposing a high standard of care is supplied by precedent: in dealing with conduct of potential danger to children, there is a duty to observe extra caution for their safety. See e.g., Rivas v. Pacific Finance Co., 16 Utah 2d 183, 397 P.2d 990 (1964). Of course, children are less able to understand even commonplace risks than are adults. This is one justification for increased care. In this case, however, it has long been recognized that exposure to ionizing radiation inherently creates an increased risk of injury to children because of their greater sensitivity and vulnerability to radiation damage. See Part IX, infra. It was suggested during the open-air testing program, for example, that exposure limits for children be reduced to one-tenth of the adult limits. See e.g., Memorandum, Lt. Col. R. Campbell to Gn. K. Fields, May 8, 1953, PX-90.

The evidence now available simply confirms the early hypotheses of increased risks to children arising from radiation exposure. See e.g., J. Gofman, Radiation and Human Health (1981), PX-1046; Part IX, infra. Under Utah law, as in most American jurisdictions, increased risk of injury imposes a greater duty of care on the party who creates the risk. See e.g., Black v. Nelson, 532 P.2d 212, 213 (Utah 1975); Erickson v. Bennion, 28 Utah 2d 371, 503 P.2d 139 (1972); Brigham v. Moon Lake Electric Ass'n, 24 Utah 2d 292, 470 P.2d 393 (1970).

The weighing of these factors, i.e., the relative degree of risk of serious injury, the tremendous imbalance of knowledge and skill, the increased hazard to children, accompanied by the increasing foreseeability of the potential hazards, see Part VIII, supra, readily leads one to conclude that the appropriate standard of care must be one requiring great caution.

This court now holds that in its conduct of open-air atomic testing, and in its offsite radiation safety programs—including its public information activities—the Government was bound by a legal duty to act with the highest degree of care in light of the best of available scientific knowledge. Government employees need not have been perfect. But the course of action followed must have been that most protective of health and safety if the requisite standard is to be met.

A number of contemporaneous documents disclose that the Atomic Energy Commission was conscious of its duty to protect health and safety from the beginning of the continental testing program. Other documents and testimony in the record also attest to the Atomic Energy Commission's abiding concern for its "obligation" and its "responsibility" for public safety. Even had the record been wholly silent on this subject, the law would nevertheless impose a stringent duty of care upon the defendant, as it certainly would upon a private individual releasing megacuries of radioactivity into the air. The record, however, is not silent; responsibility, obligation—duty of care was acknowledged and accepted from the beginning.

The only remaining questions relating to the duty issue are (1) whether the particular risks which have taken effect in the alleged injuries to the plaintiffs fall within the scope of the defendant's duty of care; and (2) whether the defendant's conduct breached that duty.

The risk that off-site residents near the Nevada Test Site might develop cancer or leukemia following exposure to ionizing radiation from nuclear fallout in amounts significantly in excess of normal background radiation was one well within the scope of the Government's duty of care, perhaps even closer to the core of that duty than the risks of purely acute injury; the claims litigated in this action deal with injury and loss far more serious than reddening of skin or temporary loss of hair due to acute exposure. If it can be established that the Government negligently or wrongfully breached that duty, and that a rational factual connection exists between the Government's conduct and the plaintiffs' injuries, then law and public policy require that liability be imposed upon the party creating this particular risk. Even where there remains the possibility that the injury would have occurred in the absence of defendant's conduct—which is always a possibility in every case of cancer or leukemia—doing justice between the parties requires that the party creating a materially increased risk of that harm bear at least the economic burden of its consequences.

Where by a preponderance of the evidence, the plaintiffs establish that the defendant's conduct negligently or wrongfully breached the duty of care, and that the conduct materially augmented or increased the risks of injury, and contributed to the harm suffered by the plaintiffs, the defendant shall be held liable.

VIII. BREACH OF DUTY

A. State of Scientific Knowledge

In holding that the Government is bound to exercise "great care in light of the best of available scientific knowledge," one necessarily must determine what the best of available scientific knowledge was during the period between 1951 and 1962. Knowledge and information was not constant; it was persistently expanding during those years and has continued to grow ever since. Consequently, the conduct of those who were responsible for public health and safety on the operational level must be measured against a standard of care that grew more stringent as time passed. In evaluating that state of knowledge, one confronts "the most difficult task in studying past science: to forget temporarily what came afterward." A. Pais, 'Subtle is the Lord': The Science and the Life of Albert Einstein 9 (1982).

> [In the course of a lengthy discussion, the Court determined that there was adequate evidence of negligence on the part of the Government, including a failure to exercise the required care in light of the best of available scientific knowledge of the time. The Government failed to monitor exposure resulting from the inhalation or ingestion of fallout material; with respect to the dangers of radioactive fallout, the Government failed to warn and failed to educate residents and persons coming in contact of the dangers.]

Both in monitoring and information activities, the off-site radiation safety program at the Nevada Test Site served largely to check the possibility of an immediate, acute exposure crisis resulting from nuclear fallout and to reassure the off-site residents that one would not occur. Long-term consequences of exposures below the acute symptom "threshold" were measured, analyzed and explained in terms of nationwide or worldwide populations placed at small risk, not in terms of local communities placed at greater risk. In both regards, monitoring and information, the employees of the defendant negligently and wrongfully breached their legal duty of care to plaintiffs as off-site residents placed at risk.

IX. THE QUESTION OF CAUSATION

Before any findings as to duty or negligent breach of duty may be applied to determine the question of liability, each plaintiff must show that he has suffered injury as a result of the defendant's conduct, at least in part. L. Green, et al., Cases on the Law of Torts 3 (2d ed. 1977).

The plaintiff's starting point on the road to a tort recovery is to be able to pick the defendant out of the crowd; that is, to demonstrate factually that there is a reason why this particular person is the defendant. This is usually called the causation or factual causation issue. I find "factual connection" to be a more accurate term. Factual connection in the manner in which the term is used herein, carries no connotation of fault or of liability. It is the means of selecting a particular defendant on whom to focus the process of the legal system. It is the statement of what happened between plaintiff and defendant. Whether the factual connection between plaintiff's injury and defendant will lead to liability depends upon plaintiff successfully establishing the remainder of the issues that are relevant to the determination of liability.

In most cases, the factual connection between defendant's conduct and plaintiff's injury is not genuinely in dispute. Often, the cause-and-effect relationship is obvious: A's vehicle strikes B, injuring him; a bottle of A's product explodes, injuring B; water impounded on A's property flows onto B's land, causing immediate damage.

In this case, the factual connection singling out the defendant as the source of the plaintiffs' injuries and deaths is very much in genuine dispute. Determination of the cause-in-fact, or factual connection, issue is complicated by the nature of the injuries suffered (various forms of cancer and leukemia), the nature of the causation mechanism alleged (ionizing radiation from nuclear fallout, as opposed to ionizing radiation from other sources, or other carcinogenic mechanisms), the extraordinary time factors and other variables involved in tracing any causal relationship between the two.

At this point, there appears to be no question whether or not ionizing radiation causes cancer and leukemia. It does. Once more, however, it seems important to clarify what is meant by "cause" in relation to radiation and cancer: "When we refer to radiation as a cause, we do not mean that it causes every case of cancer or leukemia." Indeed, the evidence we have indicating radiation in the causation of cancer and leukemia shows that not all cases of cancer are caused by radiation. Second, when we refer to radiation as a cause of cancer, we do not mean that every individual exposed to a certain amount of radiation will develop cancer. We simply mean that a population exposed to a certain dose of radiation will show a greater incidence of cancer than that same population would have shown in the absence of the added radiation.

> [The court recognized that even though cancer and leukemia may be caused by radiation from other sources, nevertheless, exposure to nuclear fallout proves to be a likely "substantial factor" in causing cancer where the radiation dosage materially exceeds that which is received from other sources. The court determined that if, from a preponderance of the evidence, the conduct of the United States significantly increased the risk of bodily injury to a plaintiff (in the form of a biologically and statistically consistent somatic injury such as cancer or leukemia), an inference may be drawn that the Government's conduct of open-air atomic testing was a substantial factor contributing to the plaintiff's injury and, unless facts are proven otherwise by sufficient evidence, the inference provides a rational basis for imposing liability. Evidence in three of federal tort claims actions failed to establish a relationship between the particular cancers of the off-site residents and exposure of the ionizing radiation from open-air atomic testing. On the other hand, the evidence in 21 of the claims arising out of open-air atomic testing established that the decedent or the living plaintiff having cancer was probably exposed to fallout radiation significantly in excess of "background" radiation rates; the injury was of the type consistent with those known to be caused by ionizing radiation; and that the person injured had resided in geographical proximity to the Nevada test site for some if not all of the years of atmospheric testing between 1951 and 1962.]

X. DAMAGES

After careful examination of the factors discussed above in all of the preceding sections, it appears that ten of the twenty-four bellwether cases merit compensation. Eight are wrongful death cases, 2 in Arizona and 6 in Utah. Heirs or survivors seek compensation for themselves for the wrongful deaths of their predecessors. (Specific awards ranged between $15,000 and $400,000.)

XI. FINDINGS OF ULTIMATE FACT

The court finds that defendant failed to adequately warn the plaintiffs or their predecessors of known or foreseeable long-range biological consequences to adults and to children from exposure to fallout radiation from open-air atomic testing and that such failure was negligent.

The court finds that defendant failed to measure adequately and concurrently with open-air atomic testing the actual fallout in communities and population centers near the Nevada Test Site on a person-specific basis, or its equivalent, and that such failure was negligent.

The court finds that contemporaneously with atmospheric atomic testing the defendant failed to adequately and continuously inform individuals and communities near the test site of well-known and inexpensive methods to prevent, minimize or mitigate the known or foreseeable long-range biological consequences of exposure to radioactive fallout, and that such failure was negligent.

The court finds that as a direct and proximate result of such negligent failures, individually and in combination, defendant unreasonably placed plaintiffs or their predecessors at risk of injury and as a direct and proximate result of such failures that each prevailing plaintiff designated in Part X, beginning at page 446, suffered injury for which the sums set opposite their respective names should be paid.

CONCLUSIONS OF LAW

This opinion in its entirety shall constitute the Findings of Fact required by Rule 52 F.R.C.P. Based on the findings of specific and ultimate facts set forth above in this opinion and for the reasons set forth herein, the court concludes that each of the designated prevailing plaintiffs (claimants) named in Part X beginning at page 446 above is entitled to judgment against the United States in the amount set forth in that section. As to the remaining plaintiffs of the bellwether group of 24, this court concludes as a matter of law that the evidence is insufficient in each instance to demonstrate with the requisite weight that the defendant's negligence proximately caused the condition of which each complains and further concludes that the United States is entitled to a Judgment of Dismissal as to each of them.

Since the legal questions dealt with in the 24 bellwether cases have significance to those cases individually and to the remaining cases affecting more than 1,100 plaintiffs, the court will certify in the judgments to be entered that each judgment shall be entered as final pursuant to Rule 54(b), terminating this action as to each plaintiff, and that an appeal shall be immediately available as to each of the plaintiffs in the 24 cases, there being no just cause or reason for delay.

Notes and Problems

1. What is the difference between an element's atomic weight and its atomic number? What word describes atoms with the same number of protons but varying numbers of neutrons? See *Durel Corp. v. Osram Sylvania Inc.*, 256 F.3d 1298 (Fed. Cir., 2001), a patent infringement suit, wherein the Circuit Court mentioned (in a footnote) the lower court's failure to appreciate an important atomic distinction: "The court stated that its calculations were based on the atomic masses of the most commonly found isotopes of aluminum, oxygen, and hydrogen. Durel, 52 USPQ2d at 1429. We question whether it did so, because it apparently used the atomic number, representing the number of protons, rather than atomic mass, which is an average value based on an element's isotopic composition and includes both protons and neutrons. Steven S. Zumdahl, *Chemistry* 75 (2d ed.1989). Aluminum's atomic number is 13 and its atomic mass is 26.98; oxygen's atomic number is 8 and its atomic mass is 16.00; hydrogen's atomic number is 1 and its atomic mass is 1.008. Therefore, using atomic masses, AlO (OH) contains 1.7% hydrogen and Al(OH)3 contains 3.9% hydrogen." There is, however, a happy ending; the error in this particular action was deemed to be immaterial.

2. At 4 a.m. on Wednesday, March 28, 1979, the Three Mile Island Nuclear Facility outside of Harrisburg Pennsylvania suffered a glitch. In particular, a pilot operated re-

lief valve (PORV) got stuck, allowing large amounts of radioactive reactor coolant to escape. The mechanical failures were complicated by the initial failure of the plant's operators to recognize the situation as a loss of coolant due to various human factors. The scope and complexity of the accident magnified over the five day period that followed. Employees of Metropolitan Edison (the utility operating the plant), Pennsylvania state officials, and members of the U.S. Nuclear Regulatory Commission (NRC) tried to understand the problem, communicate the situation to the press and the surrounding community, and most importantly decide whether the accident required an emergency evacuation.

The residents in and around the Three Mile Island Facility commenced a personal injury actions against plant's owners and operators, the companies that provided the applicable designs, engineering, or maintenance services for the plant, and vendors of the equipment or systems installed inside the plant. See *In re Three Mile Island Litigation*, Nuclear Reg. Rep. P 19,183, 193 F.3d 613 (3rd Cir. Pa., 1999). After a number of proceedings were consolidated, the defendants moved for summary judgment. The United States District Court for the Middle District of Pennsylvania, Sylvia H. Rambo, Chief Judge, 927 F.Supp. 834, granted the motion and imposed sanctions against counsel for certain residents for discovery violations. The residents and sanctioned counsel appealed. The Court of Appeals, McKee, Circuit Judge, held that:

> (1) The proffered scientific testimony of various experts was properly excluded as being unreliable or as unhelpful to the finder of fact;
>
> (2) even when the improperly excluded expert testimony was considered, the residents failed to establish that they had suffered the requisite degree of radiation exposure as a result of the reactor accident;
>
> (3) exclusion of the expert's evidence as a sanction for failure to comply with discovery deadlines was not an abuse of discretion; and
>
> (4) the summary judgment granted in defendants' favor against the group of "typical" plaintiffs for which a "mini-trial" was to be held under a case management order could not be extended to the remaining plaintiffs.

3. Can you distinguish the *Three Mile Island Decision* from the *Allen v. United States* case?

4. It is interesting if not ironic that the classic 1979 James Bridges film, "*The China Syndrome*," starring Jack Lemon, Michael Douglas and Jane Fonda, was released 12 days before the Three Mile Island nuclear accident. In the film, which was strikingly similar to the actual event, a physicist stated that the China Syndrome would render "an area the size of Pennsylvania" permanently uninhabitable. The term "China Syndrome" is a metaphorical description of the manner in which a nuclear reactor out of control would "melt through the Earth" and pop out in China. This is, of course, not likely for a number of reasons. The publicity of the accident did, nevertheless, help sell movie tickets even though the producers, in all fairness, went out their way to avoid capitalizing on the near-tragedy. Restraint would not, however, affect the producers of Saturday Night Live, nor keep the show from parodying the situation. Following the accident, when actor Michael Douglas appeared on NBC's The Tonight Show with Johnny Carson, Carson commented to Douglas: "Boy, you sure have one hell of a publicity agent."

5. Working with nuclear materials carries risk. No one will deny this. In order to further study and probe the behavior and existence of subatomic particles, so-called atom smashers accelerate particles at near-light speed and then study the resulting collisions. When physicists first began using these accelerators in the 1950s and 1960s, they discov-

ered hundreds of particles that were smaller than the three classic subatomic particles (protons, neutrons and electrons). As larger accelerators were built, more particles were found. Most of the particles exist for only fractions of a second, but some particles combine to form more stable composite particles.

Subatomic Particles

Quarks	Leptons	Force Carriers
Up (u)	Electron Nutrino (Ve)	Photon (γ)
Down (d)	Electron (e)	Gluon (g)
Charm (c)	Muon Nutrino (Vμ)	Z—Bozon (Z)
Strange (s)	Muon (μ)	W—Bozon (W)
Top (t)	Tau Nutrino (Vτ)	
Bottom (b)	Tau (τ)	

6. In order to save the world from the potential dangers fantasied by theoretical physicists, two private individuals sued to enjoin several federal agencies and a European nuclear energy research center from operating the Large Hadron Collider ("LHC"), a subatomic particle accelerator operating on the French-Swiss border near Geneva, Switzerland. The LHC is designed to collide high-energy beams of subatomic particles into one another for purposes of scientific research into subatomic particles. The plaintiffs alleged that the operation of the LHC could potentially trigger various irreversible processes that could lead to the science fiction-like destruction of the Earth and mankind. The Plaintiffs' implicit argument was that the United States District Court had jurisdiction to enjoin the operation of the LHC because the Defendants were obligated to fulfill the requirements stated within the National Environmental Policy Act ("NEPA"), 42 U.S.C. §§4321–4347, prior ro commencing the operation of the LHC. See *Sancho v. U.S. Dept. of Energy*, 578 F.Supp.2d 1258 (D.Hawaii, 2008).

The Complaint alleges that Plaintiff Luis Sancho is a citizen of Spain, with legal residence in the State of Hawaii. Plaintiff Walter L. Wagner alleges he is a citizen and resident of the State of Hawaii. The Defendants include two federal agencies, the United States Department of Energy and the National Science Foundation, a United States government agency responsible for the promotion of scientific research. The plaintiffs have also named as defendants The Center for Nuclear Energy Research ("CERN"), an intergovernmental European agency that conducts nuclear research, and Fermilab.

The court found that the Large Hadron Collider is a particle accelerator that straddles the French-Swiss border near Geneva, Switzerland. The LHC is comprised of a 27 kilometer ring of superconducting magnets, designed to collide high-energy beams of subatomic particles into one another. The collision fractures the atoms into more fundamental particles that can be observed and studied for purposes of scientific research. Plaintiffs allege that the collisions are unsafe and could potentially result in the destruction of the Earth. Plaintiffs posit three separate theories regarding the outcome of the LHC particle experiments:

(1) the creation of a runaway fusion reaction that would eventually convert all of Earth into a single, large 'strangelet';

(2) the creation of a 'micro black hole' into which the Earth would fall; and

(3) the creation of a runaway reaction due to the formation of a 'magnetic monopole.'

Under all of Plaintiffs' theories, the LHC particle experiments could lead to the end of all mankind. Plaintiffs do acknowledge, however, that various competing scientific theories exist regarding the outcome of the subatomic collisions to be performed at the LHC.

The court further determined: "It is the Federal Defendants' position that the LHC particle experiments have been thoroughly reviewed and are completely safe. Federal Defendants' position regarding the safety of the LHC is fully stated in the Declaration of Bruce P. Strauss. Dr. Strauss is the Program Manager in the Office of High Energy Physics, Office of Science, for the United States Department of Energy. In support of Federal Defendants' position, Dr. Strauss attaches to his Declaration the "Review of the Safety of LHC Collisions," a report authored by CERN's LHC Safety Assessment Group. The "Review of the Safety of LHC Collisions" states as its conclusion: "There is no basis for any concerns about the consequences of new particles or forms of matter that could possibly be produced by the LHC."

The court determined that it lacked subject matter jurisdiction over the plaintiffs' claims: "Upon analysis of the relevant two factors, the Court concludes that Federal Defendants' involvement with the Large Hadron Collider does not qualify as a 'major Federal action' within the meaning of the National Environmental Policy Act. 42 U.S.C. § 4332(2)(c). As a result, the Court does not have jurisdiction to adjudicate Plaintiffs' claims."

7. Wouldn't you feel a little better if the court in *Sancho* took the time to decide the matter on the merits, particularly since the outcome involved the death of mankind and the planet, not to mention all of the dolphins and baby seals? Perhaps the court wasn't bothered by the risk of a reversal; after all, if the ruling was wrong, who would be left to appeal? There is, however, a positive note. The Collider did its thing on schedule and we're apparently still here.

Perhaps the plaintiffs were motivated by Jay Leno's monologue on the *Tonight Show*, wherein he mentioned the plans to operate the collider, stating that scientists hoped to repeat the explosion that created by the Big Bang. According to Leno: "Well hey! (Sarcastically) What could possibly go wrong there? Perhaps we should bring in BP (referring to British Petroleum) to manage the operation."

8. As indicated in the "*Summary of Clinical Effects of Acute Ionizing Radiation Doses*" Chart (See Section IV (D) of *Allen*), "rem" has replaced the "rad" as the most common measure of harmful radiation. Less than 100 rems is potentially harmless, whereas 5,000 rems should assure near-instant death. Those exposed to intermediate quantities should plan to do some serious suffering.

9. Heisenberg's Uncertainty Principle, mentioned briefly in *Allen,* is a fundamental notion of quantum mechanics that has been expanded by philosophers and stretched into several unintended disciplines. In the broadest sense, it suggests that many concepts (other than just velocity and position) come in pairs that cannot *both* be simultaneously measured with accuracy. Try, for example, to recall the last time that you were stopped at a traffic light in a busy intersection of a large city, preparing to make a right hand turn. As you begin to make your turn, you must look to the *right* to avoid flattening pedestrians who may have begun to walk across the intersection. At that very same instant, you must also tun your head to the left to avoid oncoming traffic. Regardless of how quickly you shake your head from left to right, you can never be certain that either position is clear. As the Uncertainty Principle suggests, if you keep your head fixed to the right, your certainty of avoiding pedestrians may rise to close to 100%, but the uncertainty with respect to traffic from the left will rise to infinity.

The concept of "uncertainty" by itself, plays a fundamental role in the extensive use of statistical analysis for solving problems in quantum mechanics. Nothing, according to quantum theory, is exact or predictable; rather, the best that science can hope to determine is the statistical likelihood or probability that given event will or will not occur.

10. Another important tenet of quantum mechanics is the wave-particle duality. Having discovered that light is not just a wave (but is also constructed of particles), quantum mechanics teaches us that all matter shares this duality. We not only consist of particles, therefore, but also waves.

11. According to the great physicist, Richard P. Feynman, a substantial contributor to our knowledge of quantum physics and holder of two Einstein awards said this: "I think I can safely say that nobody understands quantum mechanics."

Johnston v. U.S.

597 F.Supp. 374 (D.C.Kan., 1984)

PATRICK F. KELLY, District Judge.

These consolidated lawsuits have been fully tried before the Court pursuant to 28 U.S.C. § 1346(b) (1976), and the Federal Tort Claims Act, 28 U.S.C. §§ 2671–2680 (1976). Venue of the action is proper pursuant to 28 U.S.C. § 1402(b) (1976).

Simply said, the cases concern the claims of four employees of Aircraft Instrument and Development, Inc. (AID), of Wichita, Kansas, who personally, or through their representatives, contend that their respective cancers and resultant damages were caused by exposure to minute quantums of ionizing radiation that originated from luminous dials and instrument parts sent to the AID Plant. The United States, being the primary supplier of these products, is the sole remaining defendant.

Essentially, this case deals with the science of health physics and the ramifications, if any, of exposure to ionizing radiation. The litigants offered the testimony of 53 witnesses, including that of some of the most eminent persons in the field. Massive documents from both sides were offered. The testimony of each witness has been carefully digested with perceptions of demeanor and expertise noted. Each of the exhibits has been carefully reviewed and studied. The Court's detailed trial notes and much of the transcript (5,409 pages) have been reviewed and reread. Counsel have provided the Court with suggested findings of fact and conclusions of law and have fully argued their causes.

This decision is intended to address and detail as fully and succinctly as possible the factual and legal reasons why the plaintiffs cannot recover in this suit.

I. Introduction of Parties; Some History of the Litigation; and a Brief Resume of the Respective

Claims and Defenses

The estate of Earl E. Johnston claims damages for his wrongful death and conscious pain and suffering in that his exposure caused leukemia; the estate of Don C. Vessels similarly claims damages as a consequence of his lung cancer; Barbara J. Womack claims damages for pain and suffering, for a shortened life expectancy, and for related damages as a consequence of her thyroid cancer. She additionally claims damages for mental suffering from an expectancy of future cancers. Loyd B. Womack, husband of Barbara Womack, claims a loss of consortium and mental suffering from an expectancy of cancer; Lila M. Mewhinney and her husband, Richard, similarly claim damages as a consequence of her colon cancer. Plaintiffs Womack and Mewhinney have claimed damages for an expectancy of cancer in their children. It appears that these children had from time to time visited the AID facility and been exposed. These claims were stricken by the Court in its order dated February 24, 1984. The detailed history of each person's involvement, as to time and extent of alleged exposure and the ramifications of their respective claims, are hereinafter discussed.

From the outset, each has urged that radium–226, a radioisotope which produces ionized radiation, is ultrahazardous per se and inherently dangerous, as it is destructively powerful. They urge here, by example, that should radium-226 be inhaled and ingested into the human body, it gives off alpha particles that alter a cell formation and start the cancer process. Detailed ramifications of radium-226, radon gas, daughter products, etc., are also hereinafter defined and discussed. Plaintiffs claim that since there is no known safe threshold from such exposure, the United States was required to adequately warn of the dangerous properties by clearly labeling each radium-painted instrument, which instruments were then distributed into surplus markets.

The plaintiffs further contend that through the recent years, the United States has released massive quantities of unmarked radioactive aircraft instrument parts into the surplus market; that while stored and used at such places as the plaintiffs' employer's plant, AID, these parts accumulated in storage areas and about the employees' work area, during which time the ionized radiation was permitted to build up to dangerous quantums, giving rise to injury.

Shortly prior to this trial, these cases were also pending against numerous manufacturers and suppliers of these products, who, in turn, directly or indirectly, distributed the products, including those from the United States. This is mentioned here inasmuch as a goodly number of different types of instruments have been discussed, and for the following reasons: The plaintiffs have always maintained, in part, that the United States' liability is absolute, following Rylands v. Fletcher, L.R. 3 H.L. 330 (1868). The Court ruled prior to trial that as it would appear that radium compounds and one's exposure to such compounds are highly dangerous, then in such an event, the doctrine probably should be considered. The Court alternatively suggested a remedy under the concepts embodied in the Restatement (Second) of Torts § 519 (1979). A decision on these issues, however, was reserved pending receipt of the evidence. Coinciding with this theory, if applicable, however, was the Court's additional finding that inasmuch as the plaintiffs' claim asserts a fault concept at the hands of others, it should as a consequence be weighed pursuant to the comparative fault principles of Kansas (K.S.A. 60–258a). Thus, in the course of these cases, the respective comparative fault has been jointly considered. In this regard, the Court has heard the evidence as it may apply to the plaintiffs' employer; perhaps the State of Kansas, whose duty it was to regulate, monitor and/or inspect the premises; and the plaintiffs' acts themselves. In other words, each have been weighed consistent with the law of Kansas. For reasons shown, all of this has been an unnecessary exercise.

For the most part, these cases, considered alternatively as negligence actions, test the traditional duties, if any, of the United States. They suggest that the product is highly dangerous, and that the United States has a high standard of care and is required to warn ultimate users such as the plaintiffs of the consequence or dangerous propensity of radiation exposure; that failure here has caused or contributed to the plaintiffs' injuries.

The United States has adamantly denied any liability to these plaintiffs for sundry reasons. Some of their defenses have been previously addressed, including those of the statute of limitations and the propriety of filing of the plaintiffs' claims. It is first contended that the United States is immune by reason of the discretionary clause found in 28 U.S.C.

§ 2680(a), and as a consequence this Court has no jurisdiction. Again, prior to trial, the Court took this issue under advisement awaiting some developments in the evidence.

When the plaintiffs rested their cases, the Court ruled that the government's decision to use and/or ultimately dispose of these instruments was indeed a discretionary function and no liability will flow from that decision. This issue is additionally addressed herein. As to immunities, if any, for the government's alleged failure to label or warn of the propensities of the products, the Court, sensitive to the Tenth Circuit's rationale in Miller v. United States, 710 F.2d 656 (1983), again took that issue under advisement. The Court, in this decision has also fully addressed its views and findings.

Next, and what develops to be the central issue in these cases, is the United States' insistence, that while there may not, as yet, be a scientifically established threshold of exposure from which these cancers could arise from exposure to ionized radiation, that nevertheless all of the scientific studies, tests, and experiments known to the health physics community suggest that there are no epidemiological studies or findings to support an occurrence of cancer from exposure at less than 50 rad. There probably is, at least, a safe or practical threshold from exposure at about five rem per year, as a maximum permissible low dose (LET). A rem is a unit for measuring a dose of radiation received by an individual. (A rad is defined as a basic unit of radiation dose based upon the amount of energy absorbed in a given mass of tissue.) Even in this event, the kinds of cancers claimed by the plaintiffs would not occur. They have claimed here, under the circumstances shown, the plaintiffs' exposure from the instruments is so minute that each would have to literally eat the paint, and even then only incur the unfortunate result of bone sarcoma or cancer of the paranasal sinuses. Admittedly, as these cases commenced, this Court, a layman in the truest sense, impressed with representations of the plaintiffs' counsel from the outset, received the government's claims with some skepticism. Indeed, the government's counsel was chided from time to time, as these claims were then thought to border on the facetious.

A rem is defined as a unit of radiation dose which takes into account the differing biological effectiveness of different types of radiation. For example, 100 rem of any type of radiation should produce the same biologic effect. Rads times effectiveness factors equal rems.

To give some contrasting perspective, which in great measure directs the course of the Court's findings and reasonings, it is recalled in the course of a previous hearing that the shipment of the parts identified for testing was to have been delivered from plaintiffs' counsel to one of the defendant counsel's office. The plaintiffs' counsel understandably exercised his concern as to the safety of his colleague and wanted to assure care in the delivery of these parts to counsel. Defendant's counsel was not the least bit interested, but as a layman this judge most certainly was! Now, in the course of trial, an aircraft instrument coated with radium paint was placed on the Court's exhibit stand. Again, plaintiffs' counsel exclaimed some concern as to our safety. As of then, the Court had digested the teachings from a superb witness, Dr. John Auxier, and reasonably appreciated the quantums of radiation, i.e., 3.5 microcurie, which each instrument would emit. A microcurie represents 1/1,000,000 of a curie. Additionally, the Court had come to understand alpha emissions, which under the circumstance were harmless. In every event, on this occasion the Court was not so concerned, because as this decision attempts to show, such a notion, standing alone, is nonsense! Lastly, the government denies the factual foundation on which plaintiffs' cases are premised.

II. The Plaintiffs' Case

The plaintiffs' cases describe a state of circumstances, much of which is entirely circumstantial. Since exposure to Mrs. Womack is alleged to have commenced as early as 1967, it is necessary to establish the environmental basis as of that time. There is, of course, no hard data to support this, and as a consequence we commence the cases with comparative studies. In 1968 the Bureau of Radiological Health carried out a national program designed to reduce the exposure of man to hazardous ionizing and non-ionizing radiation. Radium luminous dials had been used since 1919 in military aircraft, particularly during World War II and immediately thereafter. Many of those instruments, declared surplus by the United States Department of Defense, were being utilized by private industry for repair parts and original installation in light aircraft. Significant contamination had been observed in these facilities as they disassembled and repaired the instruments.

Inasmuch as over 75% of the light aircraft produced in the United States are fabricated in Wichita, a total of 15 companies participated in a study. The findings of these studies are referred to as the Geiger-Schmidt Studies. These studies may have included AID. They did include Air Capitol Dial (ACD), which is referred to here as Plant C, an apparent comparable shop. Those findings are used as a basis of comparison from which the plaintiffs premise their foundation. Included in these findings is the fact that, on an average, approximately five percent of the instruments are considered as radioactive. The millirem (mrem) level, being 1/1,000 of a rem, averages less than 2 mrem. The two employees in the study shown to have been exposed to radiation reflected a finding in their body burdens, being a showing of radium residue in the bone. Simply said, the plaintiffs claim that if Plant C was bad then, the AID Plant was always in a worse condition. The plaintiffs also have the findings taken from time to time by representatives of the State of Kansas, Bureau of Radiation Control, as they have regularly and/or routinely monitored the AID Plant. A review of the exhibits and the state's entire file, suggests that the conclusions drawn by these representatives did not present a health hazard, but rather the presence of excessive quantums of alpha contaminants. At the outset of this case, this struck the Court as a "distinction without a difference."

Through the years, AID's management has left much to be desired. While a state-licensed entity and operating pursuant to certain radiation protection regulations, the employees have removed or stripped radiated paint from the products without license or authority. The employers have attempted to conceal their operations from the state and perhaps their employees, and they have deserved strict sanctions and forfeitures from time to time.

In February 1981, on representation of plaintiffs' counsel, and with accompanying affidavits, this Court authorized an unannounced inspection of AID by the plaintiffs' group, including Mr. Robert Gallaghar, an applied health physicist, and Dr. Carl Johnson, a medical health physician. This rather dramatic process is referred to as the "Raid on AID." To no one's surprise, the results of this inspection were remarkable—the place "reeks" with radiation. Admittedly, if asked as of then, the Court, per Rule 65, would probably have temporarily enjoined that operation, ordered a cleanup, and recalled all radioactive products. As often develops, following a full hearing such as that which has transpired here, the Court, however embarrassed, would have set the order aside! In every event, and while detailed more fully herein, the Gallaghar findings become of paramount interest to the plaintiffs' experts for the benefit of their ultimate opinions. The Geiger counter readings revealed certain alpha and/or beta gamma content.

Mr. Gallaghar's film badge revealed levels of 80 mrem per 14 hours, or approximately 6 mrem per hour. These findings were significant. An audioradiograph, simply an x-ray process, taken of certain air filters, was positive. Again, this is suggested to be of significance.

Since the raid on AID, additional state inspections and an OSHA inspection revealing findings of external alpha contaminations, the plant has been cleaned up by AID's present management. Again, plaintiffs logically argue that if the circumstances are bad now, they were most assuredly worse then.

The plaintiffs' first witness and one of their principal experts was Dr. Karl Z. Morgan. This person is presented as a most eminent scientist. Indeed, without further resume, one would easily conclude that he has literally founded the science of health physics. The Court has been reminded, more than a few times, that he and his colleague, Dr. Gofman, come here as the "heroes" of the Silkwood litigation and others. In days past, having dealt with his proffered findings in the course of several hearings and dispositive motions by the government and others, and given appropriate credence to the same, all were overruled. Understandably, this witness was received with anticipation and interest.

The witness's testimony is quite lengthy (500-plus pages), and for the purpose of these findings, the Court will not attempt to resume all of it. Rather, the Court will attempt to recite that which is considered of importance with regard to his opinions.

During Dr. Morgan's early periods, he prepared a paper on procedures to calculate permissible exposure and tolerance levels with regard to the risk of exposure. He has developed the mathematics for calculating this exposure. The witness has discussed formulas and procedures to analyze dosage. (Ex. 12,381). It is his thesis and conclusion that an overwhelming amount of data has been accumulated to show that there is no safe level of exposure. Now, it is not in issue in these cases that an excessive or large amount of radiation creates an increase in the naturally occurring rate of cancer. The National Cancer Institute would suggest that 16% of 100 people will die from naturally occurring cancers. Thus, if 100 people have been irradiated with a large amount of radiation, the increase might be that of 16 to 20 naturally occurring cancers, to a total of perhaps 25 cases, or an excess of five. The witness claims that causation may only be adduced by statistical analysis. Dr. Morgan is clear that he does not hold himself out to be a medically trained person, yet their expertise is not in the field of how much dose will cause a cancer. A physician's expertise is in analyzing cancers and their origin. Dr. Morgan's expertise is in what dose will cause a certain form of cancer. He has taken the data on what he calls the "dose coefficient," that is, the number of cancers per person rem in the numerator, divided by the same factor, plus the natural incidence of cancer.

With this understood, the witness reviews his superlinear chart, which suggests that he can extrapolate from these known high doses to extremely low doses to confirm his view. The witness has lectured the Court on the alteration of cell formation and the difference in humans by reason of age and metabolic conditions. He has established that ionization can provide enough damage to divide the cell and initiate a malignancy. He attributes most of the problems to the existence of radon gas (R222), as it builds up from the source of emission. He concludes that 97.6% of the plaintiffs' exposure and injury is a result of this event. It is his thesis that while alpha particles create no problem to man, once inside the tissue they do create a problem. He suggests, by example, that once in the colon alpha rays will bombard the tissue in millicuries per second, and he tenders his publication in the Encyclopedia Brittanica (Ex. 13,299) to support the claim. He identifies beta particles as fast-moving electrons which, in turn, give off daughter products and penetrate human cells, such as the thyroid. These rays throw off approximately 37,000 beta particles per second per microcurie. He tenders numerous articles, many of his own, to support his thesis. Additionally, radon gas gives off gamma rays. These will penetrate the human body and alter cell formation, i.e., external gamma.

While, as indicated, there is more to his testimony than capsuled here, it suffices that the witness expressed his opinion in these cases. He is asked to assume as true all that is resumed above, including the Geiger findings and those reported by Gallaghar. Additionally, he is asked to assume that we deal here with an approximate 49,000 instruments, of which 50% are radioactive. He has measured seven of the 10 instruments selected by Gallaghar and established that each emits approximately 26.7 microcurie. Inasmuch as ventilation and space are of interest, he has received an air study (Walker Report) of the AID facility, and as contrasted with Plant C. These findings suggest an average air exchange at AID of one per day. This finding is necessary for the witness' ultimate conclusions regarding the buildup of radon gas from the instruments, which he has established at 14 mrem per hour, by average, but constant, throughout the plant, over an approximate 15 years. Without objection, the witness concludes that each of the plaintiffs' cancers are derived from this exposure. He further postulates that each would, or will likely, incur additional cancers. The assumptions underlying his conclusions are contained in his voluminous report, which is handed up without comment or further explanation.

Dr. Morgan's conclusions were enthusiastically endorsed by the next witness, Dr. John Gofman. This gentleman is readily perceived as a most intelligent, persuasive, and articulate spokesman for the plaintiffs' cause. Indeed, he is an alarmist, truly, obsessed with the righteousness of his long espoused concerns regarding exposure to radiation in any setting. The witness is quite clear: the United States' positions are "rubbish." While he professes a sense of hesitancy for having taken up these cases, it is clear from the outset that this man enjoys the "limelight" and is no stranger to the courtroom.

The witness enjoys professor emeritus status at the University of California at Berkeley, the Department of Biophysics and Medical Physics. He has worked with Dr. Glenn Seborg, who received the Nobel Prize for work on plutonium. The witness took on a study of a substance known as Element 91 and was able to prove the existence of Element 91, which was later to become uranium-233. He claims credit with having discovered it in collaboration with Dr. Seborg. He has degrees in nuclear physical chemistry; he has taught courses in dose and dose calculations; he has calculated hypothetical processes with regard to the separation of plutonium and drops J. Oppenheimer's name, who organized the Los Alamos laboratory to produce the atomic bomb. The witness uses the word "we" quite liberally. He claims the patent for the discovery of fissionability of uranium-233, and was the first to separate a milligram quantity of plutonium.

After World War II, he returned to medical school. His training is in the field of internal medicine, and he treated cancer and leukemia victims at the outpatient clinic and as a resident. He became an assistant professor of medical physics at Berkeley. He then worked at the Donner Laboratory, which is funded from the Atomic Energy Commission, Division of Biology and Medicine. He then took up under Dr. Ernest Lawrence, who invented the cyclotron, and he represented him in his concern for radiation safety. He became the first director of the Livermore Laboratory and the Department of Health Physics and Health Chemistry, which did the measurements of the radiation safety. He returned to the Donner Laboratory in 1957, and back to Livermore in 1962, principally in light of recent atomic weapon tests in Nevada. He has been honored by the Livermore Laboratory and makes some point of this inasmuch as he is apparently now in controversy with them. Presently he is associated with no one and beholden to no one. His book, Radiation and Human Health, published by Sierra Book Club in 1981, has been introduced.

In 1969, the witness claims that while working for the United States he made discoveries of a world nature in the field of ionized radiation. First, he suggests that the risk of cancer from ionized radiation was 1 to 10 times worse than anyone had thought, and second, there were three generalizations that could be made. These principles are called the laws of cancer induction by radiation. The first is essentially that all human cancers can be adduced by ionizing radiation; second, that a given amount of radiation increases by a certain percentage the number of cancers over what would have occurred anyway, i.e., if a certain cancer has a 100 cases occurring per year in a certain group, a given amount of radiation would increase that spontaneous cancer by 10%. This is called the relative risk method, and he was the one who assisted in the development of it. Third, children are much more sensitive than adults with respect to radiation-produced cancers. He has found, and in days past has stated on behalf of the United States, that there is no safe threshold. Most clearly, they have seen no evidence of a safe threshold.

He suggests that these cases present a new law for him, in that one must consider the particles of radiation; in other words, one for radiation like alpha as opposed to cancers from gamma radiation, i.e., a much shorter span for alpha and later for gamma. This is called high cap LET ("LET" stands for linear energy transfer). Alpha particles rage through tissue, destroying as they go, and transfer much energy for every unit path, whereas gamma rays produce eons at a bigger space, and we call that low LET. The time of appearance of cancer is totally different.

The witness further lectures the Court on the effects of radiation and its cause of cancer by destruction of cells. He discusses the thesis of disruption of chromosomes, deletions, and/or translocations. He makes an interesting analysis regarding production of radon, in that radon is formed by radium regardless of the physical or chemical state of the radium at the rate of 2.1 microcurie of radon per second per gram of radium. To follow through, the so-called gamma rays of radium arise from these two transformations among the decay products that are not emitted by radium itself. This is why radon is so hazardous, because the short-lived products which really produce the heavy gammas are not from the radium itself—they are from radon and its daughters. It would appear that in light of the United States' obstinancy as regards his claim, his outspoken criticism precipitated the formation of the Committee on the Biological Effects of Ionizing Radiation (BEIR Committee), in that as a result of communication from Senator Muskie to Robert Finch, then Secretary of HEW, the National Academy of Science appointed this committee to look into and update the general knowledge of radiation hazards.

The witness traces massive articles regarding the subject, all contained in Gofman I and II. Many of these are related to the United States' involvement in radiation, and some are geared to the United States' duty to warn of the propensities of radiation. Comments and discussions as to the latter are reserved. These exhibits additionally support the witness' thesis.

As with the resume of Dr. Morgan's testimony, the Court has not begun to trace all that has been said by this witness, but hopefully this resume suffices all that is important in reaching an opinion.

When the pertinent questions are addressed with regard to probability of injury from exposure, the United States appropriately objects. Now, in light of his medical background, and on the condition that all requisite factual material with which he will be concerned is to be adduced, the Court permitted the witness to testify.

Dr. Gofman has no opinion as to plaintiff Vessels, but is prepared to state his opinions within the range of medical probabilities as to the other parties. To form his opinions, he needs the medical records. He needs measurements of radon in the air since 1967; he prefers measurements of ongoing particulate radium in the air, but he doesn't have it, and he supplants this with measurements from other facilities and hard data from British plants. He has the Gallaghar film badge measurements. Additionally, he considers the number of instruments, the size of the plant, ventilation factors, and measurements of the amount of radium in the plant. He draws upon his knowledge of the natural rate of radon escape, i.e., radon escaping into the air to provide gamma radiation. The alpha particle data is of interest to him in relation to its effect on the lungs. He has read Dr. Morgan's report and opines that a reconstructed dose of 15 mrem per hour at the AID Plant is "very reasonable." He has studied the Gallaghar report and reviewed the Kansas summaries. He accepts the Walker report on ventilation. The Geiger studies suggest that 5% of instruments are radioactive. He understands that a Mr. McKinney has also so estimated this, yet Gaughn has reported a 50% ratio of radioactive instruments in the AID Plant. Accordingly, he recalculates that 27% of the instruments are radioactive, this figure representing the difference between the two other estimates.

Essentially, the factors necessary for his opinion are that the Air Capitol Dial Plant had no more than 500 instruments when the Geiger Study was taken; 5% of Air Capitol's instruments were radioactive; the Walker comparative studies of Air Capitol and AID's air exchange rates; the total surplus instruments and the percentage of radioactive surplus instruments at AID.

On the strength of Dr. Gofman's understanding of the evidence, of course, he testifies that each plaintiff's cancer is causally related to the radiation in the plant. In scrutinizing the plaintiffs' medical conditions, we exhaust the field of leukemia, i.e., chronic myelocytic (CML). We distinguish whether it is Philadelphia positive or negative—it's negative! We explore concepts of latency, classic signs and symptoms, and hematology. As to Mrs. Mewhinney, we explore the colon passage, the rectum, and whether her cancer is within the colon or rectum. Even the minute crypts of Lieberkuhn are explored as regards the passage of radiation. Indeed, as in the fields of radiation and health physics, here in the varied field of medicine the witness's expertise seems to be without limits.

The witness prognosticates as to future cancers. He calculates that the plaintiffs have received a certain dose, and this dose is calculated from a vast range of human experiences (the known or valid exposures to radiation). He discusses the cancer coefficient, which is a formula for estimating the number of people who will die of cancer as a result of certain amounts of radiation injury, and he then calculates the increased likelihood of a certain person getting a cancer than it would have been had that person not been irradiated. In other words, though not clinically diagnosed, these cancers have already occurred. The witness is convinced that these cancers can be diagnosed at this time on a microscopic level.

Dr. Carl Johnson has also expressed his opinions as to medical probabilities of causation. Lastly, other local physicians, including Dr. Dennis Moore, an oncologist and hematologist, and Dr. William Padgett, a surgeon, are called. Each are given license to assume the plaintiffs' hypothetical factual basis to be true. Dr. Moore has asserted a relationship between the exposure of Johnston, his patient, and Vessels, and their respective cancers. Dr. Padgett similarly testifies as regards Mrs. Mewhinney. Dr. Ernest Rieger, physician for Mrs. Womack, while called, has no opinion on causation.

Dr. Moore is also impressed by the fact that each of these persons has been exposed to some radiation "above background," which additionally supports his opinion.

Now the Court has synopsized the nature and extent of the plaintiffs' cases with respect to causations, if any. In other words, it is here as if to say that all of the plaintiffs' evidence was in. The record is quite clear that this Court has expressed considerable displeasure with both counsel for having permitted witness Morgan to testify as to his conclusions in response to hypothetical questions at a point at trial when there was absolutely no evidence admitted to support the factual assumptions underlying the hypothesized question. In the course of some banter, initiated by the Court, all of the foregoing was conditionally tendered. Surely, if the United States had timely objected to his testimony, these cases would probably have been over in the course of two days. As it stands, the Court has been required to patiently listen to all of the evidence, both sides, in the interest of ascertaining if, in fact, there is any evidence on which any of the foregoing witnesses are permitted to testify, and if so, to test its reliability and weight.

Before the Court addresses these and other issues, it is probably best now to articulate the scientific lessons learned in these cases from all the evidence as they apply here. It goes without saying that the Court is aware that considerable interest from the health physics scientists is focused on this case, and doubtless this decision will be the subject of some discussion within their various circles. This decision is drafted with such an event in mind, ever sensitive to the Court's own inadequacies, not only with respect to a thorough comprehension of the field of health physics, but also with respect to an ability to articulate the application of the science to the facts of these cases. For those persons the following will probably appear as a "primer".

III. Subject Matter Introduction and Scientific Background

The Court has drawn from the findings of a collection of studies known as the BEIR Committee for the purpose of definitions and indeed for much of what is ultimately concluded by the Court as reasonably probable. This is a committee of the National Research Council that studied the biological effects of ionized radiation. The committee personnel read as a "Who's Who" from the pertinent disciplines. The document, BEIR III, i.e., The Effects on Population of Exposure to Low Levels of Ionized Radiation: 1980, is the result of a massive study. The Court finds the findings of the study to be entirely persuasive and relevant in these cases.

Types of Radiation. For the purposes of this lawsuit, there are three basic types of ionizing radiation. An alpha particle is composed of two neutrons and two protons, like the nucleus of a helium atom. A beta ray is a single electron. A gamma ray is a photon, or bundle of energy which contains some of the properties of both matter and light. None of these forms of radiation can be seen.

Compared to the other two types of radiation, an **alpha particle** is relatively large and heavy. It is stopped by a single sheet of paper and can only travel the distance of a few human cells inside the human body. Yet, because they are large, alpha particles can do considerable "damage" along the short track they follow. The "damage" is referred to as an ionization event, which means that an electron has been taken away from an atom inside a human cell, making that atom an ion with a positive charge. These ionized atoms may then undergo chemical reactions inside the human cell which may damage the cell. Also, an alpha particle may damage a cell by disrupting the DNA inside the human gene. Alpha particles are considered to be high LET radiation which means that they have a high linear energy transfer. That simply is a way of saying that "large" alpha particles transmit a "high" amount of energy to surrounding tissue as they travel a short (linear) distance.

Gamma rays are very different. They are low LET radiation, meaning that they transmit low amounts of energy as they travel long distances. For example, an alpha particle will give up or transfer all of its energy in the first few dead skin cells of a human being while a gamma ray may pass through the whole human body without striking anything to which it could transmit any of its energy. More gamma rays are stopped as the rays penetrate material of higher density. For example, x-rays are essentially the same as gamma rays, and the increased density of human bone over human soft tissue makes medical diagnostic x-rays possible. If a gamma ray does strike a human cell as it passes through the human body, it could cause the same type of "damage" as an alpha particle. An atom in a cell could be ionized or the DNA inside a gene could be disrupted. Alpha particles will cause many more ionizations than gamma rays "per unit track," i.e., over the same linear distance. Thus, alpha radiation is high LET and gamma radiation is low LET.

Beta rays are actually particles like alpha rays but they are much smaller particles. They are electrons with a negative charge. Consequently, their physical properties fall between alphas and gammas although they are similar to alphas in that they are particles. They can penetrate several sheets of paper but not a book. Most beta particles will be stopped by normal clothing or by the thickness of human skin.

The relative power of these various types of ionizing radiation is measured in a unit called an electron volt. This is expressed as ev or MeV for a million electron volts. There are actually many different kinds of alpha particles of different energies (MeVs). The same is true for beta rays and gamma rays although these various MeVs still cluster around a certain range of values consistent with the properties of that particular type and source of ionizing radiation. Scientists generally refer to radiation of higher MeVs as more powerful radiation, or radiation of higher energy.

These three forms of ionizing radiation can harm a human cell in three different ways. If these forms of radiation strike a female egg or a male sperm, they can rearrange the DNA which could cause genetic damage to the next generation. Since this case involves no allegation of genetic damage, this possibility will not be discussed further. The second type of harm to a cell could be a chemical or DNA change which kills the cell. This is an acute form of radiation damage, and if enough cells are killed it causes observable effects such as a loss of hair or nausea. At high enough doses, enough cells will be killed so that the human body cannot survive this impact. That is known as a lethal dose. Since this case does not involve the acute effects of radiation, they will not be discussed further.

The third way that ionizing radiation, whether alpha, beta, or gamma, can harm a cell is to cause a chemical or DNA change that will not kill that cell, but rather cause a change that the cell passes on to its daughter cells. At some future point in time, a granddaughter cell expresses this change as a lost ability to multiply in a controlled manner. Consequently, the cell multiplies wildly until there is an accumulation or a lump large enough to observe and to diagnose as a cancer. The time from the radiation-induced change in that original cell until the cancer is diagnosed is known as the latent or latency period. A part of that latency period, from the time the granddaughter cell starts uncontrolled multiplication until the time the cancer is large enough to diagnose, is also known as the period of silent growth. The cancer is there and is growing at the same rate. It just seems silent because it cannot yet be diagnosed. The time span of this tumor growth is measured in units called doubling times. That is the time it takes for a tumor to double in three dimensional mass, not the time it takes to double in two dimensional size (diameter).

Radium–226 and Its Daughters. This case is really founded upon one particular element, radium–226, and its natural decay products. Radium is a naturally occurring radionuclide. It is not man-made like a radioisotope. In fact, it is the element in pitchblend which Madame Curie first identified as a radioactive material. Since it is a natural element which cannot be used to make weapons or build nuclear reactors, it is not regulated by the United States under the Atomic Energy Act.

All radioactive elements spontaneously emit a type of radiation and then change into a different element. This process is termed a radioactive decay. When an element, such as radium, is followed as it emits radiation and changes from element to element until it becomes stable, one is observing what is known as a decay chain. A lump of radium will not decay and change all at once. Each atom decays in a random manner over a period of time. Science can measure the length of time necessary for one lump of radium 226 to decay into half a lump of radium–226. This period of time is known as the half-life of radium–226.

The plaintiffs have placed great emphasis on the fact that the half-life of radium is 1602 years, as if to infer that radiation is bouncing about everywhere as its consequence. This notion in part premises their thesis for a finding of absolute liability.

For purposes of this case, the 1602 year half-life means that whatever amounts of radium were placed on the instruments during World War II, essentially the same amount would still be there 20 years later when that instrument entered AID. Also, whatever amounts were measured on average instruments today would effectively reflect the amounts on the average instruments when the plaintiffs worked at AID. It would take 1602 years for half of the radium in AID to change to radon (a gas), but from what has been learned here, if AID had reasonably normal air exchanges, the radon gas would be removed about as fast as it was generated. In other words, a little ventilation would go a long way toward control of such a slow generation of radioactive gas. The decay chain shows five primary emissions of alpha particles. It is these alpha particles that constitute the primary concern of health physicists over the safe handling of radium-bearing items. Because of alpha's inability to penetrate human skin, the main concern is to keep these alphas outside of the human body where they can cause no harm. The radon in the chain is radon gas. Like radium, it is commonly found in nature. Since it has a short half-life of 3.8 days, any radon gas which is not removed on a daily basis will quickly change into polonium-218 and then into lead-214. These three decays will all give off alpha particles. Lead-214 will change in minutes into bismuth-214, which also changes in minutes into polonium-214. These two decays of radon daughters give off beta and gamma rays. It is these betas and gammas which can cause external radiation to the whole body from this radon in the air even though none of the particles in the air actually get inside the human body. If the radon gas or these daughter products are inhaled into the lungs, their decay in the human lung can cause damage to the cells in that lung. When certain witnesses in this case referred to "harm" from a "radon cloud" at AID, they were actually referring to these betas and gammas from radon daughters, lead-214 and bismuth-214. If a facility like AID has adequate air exchanges, a "radon cloud" will not exist and there will not be a significant dose from radon daughters in the air.

With this decay chain and the basic facts about types of radiation in mind, it can be understood that a plant can protect its employees from radium and its daughter products by keeping the plant adequately ventilated, and by keeping the employees from ingesting radium paint. Alpha rays coming from the floor, the workbench, or an aircraft instrument dial can certainly make a suitable Geiger counter click, but they will not travel far enough to harm living human cells, i.e., it is external contamination. Beta and gamma rays generated by particles in the air can be greatly reduced to safe levels by simply opening doors and windows, or by using some other method of changing the air. The Court has toured the AID Plant and comment is noted hereafter.

There is a third method of exposure to employees. If the radium in the paint decays to radon and that radon gas decays into the particle polonium-218 before the gas leaves the paint, then the subsequent radon daughters will give off their beta and gamma rays while still in the paint rather than while in the air. Consequently, an employee working with a radioactive instrument dial would be exposed to some beta and gamma rays coming from that dial. Approximately 30% of the radon generated by radium in the paint will escape to the room air while approximately 70% will remain in the paint. This 30% is the radon emanation rate. Although the five alphas of the radium decay chain would not get through the paint or the glass on the front of an instrument, the beta and gamma rays from retained (unemanated) radon daughters will pass through the glass and cause a Geiger counter to click over a foot away. In fact, this physical feature is the best way to separate radium-painted dials from nonradium-painted dials.

Now, if we consider workers at AID, such as the four plaintiffs, we can see that they could receive radiation exposure through three different pathways. First, they could accidentally ingest or inhale paint particles of radium. Second, they could work in the midst of a radon cloud which sends gamma rays through their bodies. Third, they could work in close proximity to large numbers of radioactive dials which would also send gamma rays through their bodies. Much of the expert testimony in this case involved whether or not any of these pathways existed, and if so, whether or not they caused a significant dose to these four plaintiffs. The expert witnesses' opinions on this point were as different as night and day. One group of experts is wrong!

To determine the likelihood of each pathway's existence, one needs to understand how that pathway works and whether or not it leaves any evidence of its existence. If radium is inhaled or ingested, there is an international scientific consensus that 20% of what is taken into the body will pass into the blood. Since radium acts like calcium once inside the body, it will be deposited in bone. Since it has a 1602 year half-life and remains in the body for many decades, if a significant amount of radium finds its way into the human body, it will still be there when and after that person dies. There is a scientific device known as a "whole body counter", that can measure such minute amounts of radium in the human body that it can find the trace amounts of radium all humans have from the radium found in our environment. The four plaintiffs in this case have had numerous whole body counts, each reported as negative, and which conclusively prove that they have no radium in their bodies. Therefore, the evidence before this Court conclusively proves that no significant amount of radium made its way into the bodies of these plaintiffs while they worked at AID. Consequently, whatever the contamination at AID, it is scientifically established that these four plaintiffs neither ingested nor inhaled any significant amount of radium while they worked at AID. However, the whole body counts would not show whether or not these plaintiffs received a significant external gamma dose from either radon in the air at AID or from large numbers of instruments in their immediate working area.

If, however, these four plaintiffs had always worn film badges, devices used to measure gamma and beta rays, then clear evidence would exist as to whether or not they received a significant gamma or beta exposure through either of the two remaining pathways. Without this film badge evidence, the Court will have to look to other factors, explained later in this opinion, to determine whether plaintiffs have proven by a preponderance of the evidence either that a large "radon cloud" existed or that there were large numbers of radioactive instruments in the plaintiffs' immediate work area.

The "Hazardousness" of Radiation. As stated, plaintiffs have attempted to convince this Court that radiation, in general, and radium, in particular, are extremely hazardous. For example, Dr. John Gofman described radium as "this ultimate hazardous material." No attempt was made to separate high levels of radiation from drastically lower levels of radiation. In fact, Dr. Gofman constantly referred to certain handbook statements as setting safety standards for the instruments in this lawsuit. However, he failed to inform this Court that the handbook statements referred to amounts of radium 3,000 to 100,000 times larger than the amount of radium on these aircraft instruments. While this numerical difference is astonishing, it is even more revealing to translate it into something in our common experience such as degrees Fahrenheit. Under such a comparison, where one microcurie is equal to 1 ° F, the average aircraft instrument in this case would be approximately 4° "hot", whereas the "standards" referenced by Dr. Gofman spoke about items which were 10,000 ° F "hot" to 300,000 ° F "hot."

This Court, hopefully using a sense of common sense, cannot accept the plaintiffs' argument that all, or any, amounts of radiation are extremely hazardous. It follows that no person with common sense would argue that the standards for safety in handling a piece of steel which is at 10,000 ° F to 300,000 ° F must be equally applicable to handling a piece of steel which is only at 4° F.

It also follows that if this Court was to consider all, or any, amounts of radium or radon hazardous, and following plaintiffs' thesis, this Court would have to label our natural environment as ultra-hazardous. This seems absurd. Radium and radon commonly occur in nature in quantities greater than the amounts found on an average aircraft instrument dial. For example, enough radium to make one aircraft instrument dial will be found in approximately every five pickup loads of Kansas dirt. Moreover, radium is found in brick, concrete, cement, plaster, and other common building materials. In Kansas, one public water supply contains enough radium to make one aircraft instrument from about every 200,000 gallons of water. Moreover, radium is commonly found in most foods. Radium is also found in coal in such abundance that a model coal-fired power plant will release from its smokestack in one year enough radium to make 200,000 aircraft instrument dials.

Radon gas is so common that average indoor household air contains 5% of the amount allowed occupationally. The main sources of this radon gas are soil and water. Plaintiffs argue that radium—painted surplus military instruments are very hazardous just because they contain radium which will release radon gas and its daughter products, thereby creating some radiation exposure to consumers or employees who handle or use these items. For this Court to now adopt such reasoning would require this Court to find that vast numbers of common consumer products are ultra-hazardous just because they are radioactive. Many common items, such as television sets, smoke detectors, electron microscopes, airport x-ray machines, certain ceramic products, tobacco products, and radium wrist watches, are radioactive. In fact, the aircraft instruments in this case seem to fall into the same general class as luminous wrist watches, pocket watches, or clocks. This Court recalls a demonstration in Court which showed that a common "fiesta" or ceramic plate was twice as radioactive as the average of the ten instruments plaintiffs removed from AID, and that a common Coleman lantern mantle was one-half as radioactive as the average instrument plaintiffs removed from AID.

It was about this time that the plaintiffs' cases began to erode, for indeed, Gallaghar's Geiger readings, or at least the emphasis given to them, appeared misleading when compared with other materials that are radioactive. Also, radiation scientists recognize the existence of natural background radiation. This is radiation all humans receive no matter where or how they live on this earth. It may vary somewhat from place to place to the degree that someone living in Denver, Colorado, gets approximately twice the background radiation as someone living in Washington, D.C.

Although mankind has produced many sources of radiation, natural background remains the greatest contributor to the radiation exposure of the U.S. population today.

Background radiation has three components: terrestrial radiation (external), resulting from the presence of naturally occurring radionuclides in the soil and earth; cosmic radiation (external), arising from outer space; and naturally occurring radionuclides (internal), deposited in the human body.

In arguing that radiation is hazardous without specifying the levels of exposure, plaintiffs have surely painted their case with a broad brush! For this Court to hold that something is hazardous just because it causes a Geiger counter to click is over-simplistic and would force this Court to logically conclude that nature itself is ultra-hazardous. This is not to say that the amount of radiation which can kill the average human within a few days, the lethal dose, is not hazardous. The important factor that must be known before a "hazardous" label can be attached with any degree of scientific validity is the amount of radiation with which one is dealing, and this is what the case is really about.

Radiation Dose. Fortunately, radiation scientists have developed instruments capable of measuring very small amounts of radiation and a system of labeling which can express both very small and very large amounts of radiation. A basic understanding of these units of measurement is important to place the radiation involved in this case into a meaningful context.

The number of rems or mrems a person gets during a specific unit of time, such as during employment at AID, is known as that person's dose. The specialty of measuring or estimating a person's dose is the science and art of dosimetry.

In order to make these units of measurement more meaningful, it is of interest to note what doses some common experiences yield. The earth in Florida gives a person living there a dose of approximately 23 mrem per year. If a person lives there for 64 years, he will receive a dose of 64 x 23 mrem = 1472 mrem from Florida dirt in a lifetime. This is equal to 1.472 rem. If another person lives in Colorado for 64 years, he will receive a dose of 64 x 90 mrem = 5760 mrem from Colorado dirt in a lifetime. This is equal to 5.76 rem. In 1970, approximately 129,000,000 Americans were exposed to x-rays for medical or dental purposes. X-rays are a form of ionizing radiation almost exactly the same as gamma rays except for their source. The average American by age 64 will receive about 6.5 rem of radiation from x-rays. Consequently, total doses of approximately 12 rem would be common for a Colorado resident who had normal exposure to dirt and x-rays.

The question remains of the exact dosage these plaintiffs received while employed at AID, and whether that total dose should be considered hazardous. As will be explored later in this opinion, the answer depends upon just what assumptions are made by the expert witnesses who have attempted to "reconstruct" a precise dose for these plaintiffs. Defendant's experts generally testified that the doses were 7 rem or less, while plaintiffs' experts generally testified that the doses were hundreds or thousands of rems. Obviously, this is a huge, unbridgeable gap. If defendant's experts are correct, the doses these plaintiffs received at AID are less than the doses the average person will get in his lifetime from dirt and x-rays. If plaintiffs' experts are correct, then the doses these plaintiffs received at AID are so large that it is scientifically reasonable to consider them likely to cause harm and to label them "hazardous" or even lethal. The important point to understand at this time is that the label "hazardous" cannot be broadly attached to anything that is radioactive without an understanding of the doses involved. To this extent, plaintiffs' experts' attempts to describe all quantities of radium as "this ultimate hazardous material" must be rejected as unscientific if not outright misleading.

Today, there are three standards relevant to these cases which are nationally and internationally accepted as safety standards. The first is that a radiation worker is allowed to receive 5 rem of whole body gamma exposure for every year he works after the age of 18. This is sometimes known as the 5 rem/yr occupational limit or the 5(N–18) rule where N equals the age of the worker. The second standard is that any individual in the general public is allowed to receive 500 mrem gamma exposure each year. These first two standards are external radiation standards. The third standard is an internal radiation standard expressed as one MPBB of a radionuclide. As already noted, the MPBB of radium is 0.1 microcurie. In order not to exceed this level over a lifetime of occupational exposure the MPCa should not be exceeded. This Court readily adopts these exposure standards since it is apprised of no reason to disagree with the national and international consensus of eminent radiation scientists, the views of Drs. Morgan and Gofman notwithstanding.

The general evidence which does exist for a carcinogenic effect from radiation occurs where doses were approximately 50 to 100 rem or above. Large groups of people exposed to these high doses do show some excess cancers above those normally expected. However, the problem comes in interpreting how this risk at high doses should be extended or extrapolated to a risk at low doses, such as 25 rem and below. If one assumes that the risk per rem is the same at low doses as it is at high doses, one is assuming that if a million people exposed to 100 rem will have 10 excess cancers, then a million people exposed to 10 rem will have one excess cancer and a million people exposed to one rem will have 1/10 of one excess cancer. When such an assumption is drawn on a standard dose response graph, the result is a straight line. This assumption is known as the linear hypothesis.

Twenty-one of the eminent scientists who comprised the National Academy of Sciences 1980 BEIR III Committee reviewed the world's scientific literature and decided that the linear-quadratic hypothesis best estimates the real risks at low dose. One member of BEIR III argued that the linear hypothesis was better, while one other member argued that the quadratic hypothesis was better. This range of scientific opinion "best illustrates the degree to which scientists disagree on this subject in absence of sufficient evidence to compel conclusion." It is important to underscore again that a court of law is not a scientific experiment. When a court of law determines responsibility for human suffering and awards damages, it must do so based upon reasonable evidence, not upon speculation or hypothesis. Just because scientists use hypotheses to describe something they really don't know for sure does not justify a court of law in using speculative hypotheses to determine that one person has caused harm to another. This is not to say that the various hypotheses used by radiation scientists do not have a valid place in court. They are very helpful in setting safety standards. However, an informed court must guard against using hypotheses for more than what they really are. Speculation is not valid and trustworthy evidence even though that speculation is conservative, and comes from eminent scientists. Lawsuits must be resolved by reasonable conclusions based on the evidence, not by educated guesses.

> [The court next presented a lengthy historical perspective along with discussions of the biological effects of ionizing radiation and a set of statistical risk calculations. Relevant testimony of the several expert witnesses were also considered. The following critique of the plaintiffs' witnesses is presented due to the rare nature of the language used by the court:]

Darrell Murphy, a drug-addicted, incompetent witness, testified that 90% of the government instruments received by AID were radioactive. Robert Gallaghar estimated that 50% or more of the instruments he found in AID were radioactive even though he did not conduct a systematic examination of them. Dr. Carl Johnson agreed with Mr. Gallaghar. Dr. Gofman assumed that 27% of the aircraft instruments at AID were radioactive. He reaches this by calculating the difference between 5% and 50%. As of that time, the Court pondered how, if only 5%, could it be 27%? More succinctly, how is it that this scientist, however eminent, can so lightly treat such flimsy factual opinions to reach a scientific opinion? As shown later, in the Court's view, it is because these witnesses just aren't that eminent! Alec Fulks, President of AID, who has 32 years of experience in the surplus instrument business, testified that 4% to 6%, but not more than 10%, of the instruments at AID were radioactive.

The Court finds here that the estimates which have ranged from 70% to 5% are mostly mere speculation. None of the witnesses actually measured all instruments at AID to determine what percentage were radioactive. Simply said, the only sure way to separate a radioactive instrument from a nonreadioactive one is to measure it with a Geiger counter, and since no one ever actually made the necessary measurements, the actual fact of what percentage were radioactive is not known.

At the heart of this case is the question of what amounts of radiation are potentially dangerous enough that people should be protected from such exposures. This question was a subject of scientific concern back in 1927 when Dr. Lauriston Taylor helped found the ICRP. Since then, the group of scientists who have specialized in examining this precise question have been known as the radiation protection community. Although this group now numbers about 6,000 to 10,000 scientists, Dr. John Gofman has never been an active member of the group because he has never made any significant contributions to this field and because his writings on the subject have not been found to be scientifically credible by the radiation protection community. Dr. Morgan was once a contributing member of the radiation protection community. In fact, Dr. Lauriston Taylor appointed him to the chairmanship of ICRP Committee II. However, Dr. Morgan was later removed for cause and he has not been an active participant in the radiation protection community for the last decade. The most eminent scientists in this international radiation protection community are the people who have been selected to serve without compensation on committees which were formed to write reports that would summarize the most knowledgeable scientific consensus about the actual risks of radiation.

The Court finds that the next reason why the opinions of Dr. Morgan and Dr. Gofman must be rejected is because neither man represents the views of the vast majority of competent, respected scientists in this field. Rather, Dr. Morgan and Dr. Gofman represent the views of an extreme minority of scientists. This is not a situation where the scientific community is equally divided between two respected schools of thought. It is a case where there is a very small, but yet very vocal, group of scientists, including Drs. Morgan and Gofman, that holds views which are not considered credible by the experts in the field.

Next, this Court rejects the opinion testimony of Dr. Morgan and Dr. Gofman because they both evidence an intellectually dishonest invention of arguments to protect their opinions. One example is the solubility of radium-226 in the human body. There is no question that the world consensus is that 20% of inhaled or injected radium-226 paint will be absorbed into the blood and eventually much will be deposited in bone. Consequently, any significant inhalation or ingestion of radium-226 will be scientifically observable on whole body counts all during and even after the person's life span. In this case, there can be absolutely no disagreement with the fact that the whole body counts of these plaintiffs show no more radium-226 in their bones than in their lawyer's bones. What do Dr. Gofman and Dr. Morgan do in the face of these facts, which absolutely disprove any significant inhaled or ingested dose from radium-226? They invent the argument that this particular radium paint is insoluble because each particle in the air is coated with plastic. Dr. Morgan apparently forgot that his own health physics textbook unequivocally states on page 348 that this paint is soluble in the human body. Likewise, he apparently forgot that his own ICRP No. 2 unequivocally states at pages 221 and 14 that this radium paint is soluble in the human body. Dr. Gofman thought his plastic invention solved the whole body count problem until he realized that if alpha particles can only travel 50 minus, this plastic coating would create a zero dose to Lila Mewhinney's colon. Then he conveniently decided that his plastic coating just happened to be about 10 microns thick. For all Dr. Gofman's speculation about polymerization, the plaintiffs have totally failed to provide this Court with any scientific evidence for their argument that the world consensus is wrong on the facts of this case. Consequently, the Court finds that the radium paint at AID was soluble in the human body and that Dr. Morgan's and Dr. Gofman's arguments otherwise are simply speculation. The next reason why this Court rejects the opinions of Drs. Morgan and Gofman is because neither is actually an expert in the particular radionuclide, radium, which is at the very heart of this case. Neither Dr. Morgan nor Dr. Gofman participated in the radium dial painters study; in fact, neither man has ever published a peer-reviewed paper of original research on the topic of radium in man. Both men's knowledge about radium depends upon the research of other scientists like Dr. Robert Rowland.

For all the reasons explained previously, this Court now holds that the cancers of Don Vessels, Earl Johnston, Lila Mewhinney, and Barbara Womack were not caused by any radiation they received while employed by AID. Consequently, and for these reasons, the plaintiffs' case should also fall.

IT IS THEREFORE ORDERED this 15 day of November, 1984, that the foregoing findings of fact and conclusions of law are entered consistent with F.R.Civ.P. 52. Accordingly, the Court now having found fully in favor of defendant United States and against the plaintiffs, and each of them, these cases are dismissed with prejudice, and costs are assessed against the plaintiffs.

Notes and Problems

1. Assume that you represent the plaintiff in a case involving exposure to radioactive substances (the exposure involved approximately 400 rads of whole body radiation); how would you respond to the correct statement by defense counsel or his expert?

Biological tissue is comparable to water with respect to the number of calories required to raise its temperature; in particular, only 1/1000 calories/gram would result from the absorption of 400 rads of ionizing radiation energy, which would be enough to raise the temperature of biological tissue only by 0.001 degrees centigrade. Such a small increase in fever would not even be readable on a standard thermometer.

2. During the potential meltdown of several of Japan's nuclear reactors following the tragic March 2011 earthquake and tsunami, the media used millisieverts (mSv) and microsieverts (μSv) to describe the levels of radiation when discussing the biological effects of escaped radiation. Whereas other measurements are designed to describe physical radiation absorption in general, the "*sievert*" specifically measures human absorption. 1 rem = .01 Sv (seiverts) = 10 mSv.

1 sievert = 100 rems

Assume that media reports were correct when they predicted that no more than 1 Sv of radiation from Japan would make contact with the citizens of Los Angeles; according to the teachings of *Allen v. United States*, what would be the worst likely biological effects on exposed Californians?

3. Because of lengthy lapses in time between exposures to hazardous materials and the presence of related symptoms ("latency period"), proving a causal connection can be very difficult. Even those who had previously been exposed to the harsh nuclear tests of the 1950s (as in the *Allen* case) often face difficulty. See *Pace v. Shinseki,* 2011 WL 3792336 (2011); the victim's case was dismissed for failure to link lymphocytopenia (a severe blood disorder) to ionizing radiation.

4. Mining uranium, which involves licensing by the Nuclear Regulatory Commission (and detailed environmental impact statements), tends to stir up the local folks. Regulations promulgated under the Atomic Energy Act limit the amount of airborne radiation from an NRC-licensed operation to 0.1 rem in a year. See 10 C.F.R. § 20.1301(a)(1). See also *Morris v. U.S. Nuclear Regulatory Commission*, 598 F.3d 677 (10th Cir., 2010), involving a minor slip up wherein radiation from an abandoned mine site was erroneously characterized as naturally occurring.

§ 8.40 Tomorrow

Near the beginning of this last century there existed two correct and at the same time inconsistent explanations as to the manner in which the universe is governed. One hundred years later, despite dozens of attempted explanations and as many theories, we are still without a proven resolution of this great dilemma.

How do we reconcile the continuity of bendable time and space with the discrete, discontinuity of quantum mechanics? Had Einstein lived long enough to complete his attempts at a unified gravitational theory, we might have had an answer. Nevertheless, there has not been a shortage of theories, to be sure. String Theory, for example, seemed to deliver a great compromise between points and waves by theorizing that matter consists of vibrating one-dimensional strings that require a universe of ten dimensions. The dimensions that we can't see are curled up into tiny spaces.

From String Theory, we arrive at M-Theory. No one seems to know what the "M" stands for. M Theory employs 10 spatial dimensions and 1 dimension of time and includes various types of particles and membranes. The particles are collectively referred to as p-branes (not "pea brains"). Next comes Hawking's theory involving parallel universes, wherein all of the apparent contradictions are resolved by recognizing the existence of parallel realities.

Which, if any theory is correct? Only time (whether bent, compressed, or dilated) will tell.[43]

43. See Steven Hawking and Leonard Mlodinow, "The Grand Design," 2010.

Chapter 9

Accounting and Finance

"The most powerful force in the universe is compound interest."

—Albert Einstein

As an attorney, your relationship with accountants and financial matters is virtually assured. Even if you happen to fantasize that your only contact with numbers will involve drafting a simple retainer agreement or settlement statement, you must, nevertheless, accept that some knowledge of accounting will be required. In truth, almost everything that an attorney does is accounting related. Damages, fees, costs, expenses, annuities, bills, notes, debentures, life expectancies, sanctions, hourly rates, mortgages, rents, profits, fraud and debts are what lawyers are all about. Even the clients' common cry, "*It's the principle of the thing,*" dissipates into thin air as soon as he or she receives the first bill for attorney fees.

Accounting, often termed the "language of business," might be formally defined as a function designed "to provide quantitative information, primarily financial in nature, about economic entities that is intended to be useful in making economic decisions."[1] There are several different types of accounting activities. Here, we are primarily concerned with financial accounting and tax accounting,[2] in addition to GAAP (Generally Accepted Accounting Practices) associated with them. Financial accounting has the purpose of providing information to officers and investors in addition to other groups not directly involved in the business itself.[3] Clearly, however, the basic concepts and principles that flow from financial accounting are relevant and important to everyone concerned with a business, including its managers, regulators and taxing authorities.

There is an often-confused, but nevertheless important distinction between financial accounting and tax accounting. To reflect the difference, multiple sets of records may be employed. The purpose of financial accounting is to provide an accurate picture or economic description a business, whereas tax accounting is meant to satisfy federal, state and local taxation obligations. Consider, for example, a 50 year old building that is profitably responsible for generating rental income. Even though for purposes of financial accounting, such a building might be listed as a multi-million dollar asset, it may simultaneously carry a zero value on the tax ledgers due to depreciation. This is not uncommon or illegal. In order to encourage investment, the tax codes typically permit taxpayers to depreciate, amortize and otherwise write-off assets that have yet to diminish in value. Thus, while our 50 year old building may be depreciating by 10% of its purchase price

---

1. The American Institute of Certified Public Accountants, "*Basic Concepts of Accounting Principles Underlying Financial Statements of Business Enterprises*" (New York, October 1970).

2. Other types of accounting include: Managerial Accounting, Cost Accounting, Regulatory Accounting, and Auditing.

3. Or, as in the case of Enron (and its deceptive use of "mark-to-market" accounting), the purpose might be to cheat investors. See Rapaport and Dharan, "*Corporate Fiascos and Their Implications,*" Foundation Press, 2007.

every year on the tax ledgers, it may in reality be increasing in value. There is nothing cynical or unlawful about maintaining multiple ledgers for a single set of assets.

§ 9.10 The Cost of Money

The "cost" associated with borrowing, lending, paying and receiving money has many names and personalities, but mathematically they all boil down to calculations of "interest." Interest, according to *Black's Law Dictionary* is: "The compensation allowed by law or fixed by the parties for the use or forbearance of borrowed money." A better definition[4] suggests that interest is simply the fee paid to borrow an asset. For the legal profession, interest is an indispensable concept because of the manner in which legal obligations are always delayed, postponed, bargained for and bequeathed.

In the Middle Ages, charging interest was termed "usury," condemned, and considered sinful. Later, only interest rates that were excessive (over 9%) were considered illegal and usurious. Today, however, banks pretty much charge whatever they wish; it's not uncommon to see credit card rates triple those charged by loan sharks in the 1950s.

For those interested in how long it will take for invested money to double at a particular interest rate, there is the so-called "Rule of 72." Mathematically, it is far from precise but it does provide a down and dirty method. If, for example, you want to know how long it will take money to double at 6% interest compounded annually, you merely divide 72 / 6 = 12. Answer: 12 years.

§ *9.11 Simple Interest*

The purpose of **simple interest** is to provide consideration for the use of a fixed amount of money known as the **principal** (P) at a fixed percentage or **rate** (**r**), for a fixed period of **time** (t), normally (but not necessarily) a year. Simple interest (I) is calculated as follows:

$$I = P\,r\,t$$

Simple Example 9-1: Sammy the Shark loaned Freddy the Fallguy 20 big ones ($20,000) for 6 months at the rate of 9% per annum. How much interest must Freddy pay Sammy to make certain that his (Fred's) legs and arms will continue to work properly?

Solution: P = $20,000
r = .09
t = 6 months / 12 months = ½
I = ($20,000) (.09) (.5) = $900

It is common to request information concerning the amount of money required to pay off an entire loan (as opposed to calculating "interest only"), particularly where simple interest is involved. Normally, the borrower does not intend to be trapped into paying perpetual interest for the rest of his life. Even loan sharks don't stoop that low. (Such conduct is reserved for banks.)

4. It's pretty much accepted that compared to *Black's Law Dictionary*, *all* other definitions are better.

Therefore, a more common calculation is designed to include both principal and accrued interest based on the following formula for **future value** (V_F), which is essentially equal to the principal (P) plus accrued interest (I):

$$V_F = I + P$$

A common form of this equation results from the following derivation:

$V_F = I + P$
$I = P\ r\ t$
$V_F = P + Prt = P\ (1 + rt)$ or
$V_F = P\ (1 + rt)$

Example 9-2: In Example 9-1, what is the total amount required to pay off Sammy the Shark?

Solution: $V_F = P\ (1 + rt)$
$P = \$20{,}000$
$r = .09$
$t = 6 \text{ months} / 12 \text{ months} = 1/2$
$V_F = (\$20{,}000) \text{ x } ((1 + (.09)(.5)) = (\$20{,}000) \text{ x } (1 + .045) =$
$V_F = \$20{,}000 + \$900 = \$20{,}900$

§ 9.12 *Compound Interest*

Most every type of investment or debt that exceeds one year will involve compound as opposed to simple interest. Sometimes referred to "interest on interest," the concept of compound interest is nothing more than a recognition that once interest has accrued, it should be treated the same as principal, and thereby draw interest accordingly.

Of particular significance is the number of **periods** (n) for which the interest is to be calculated. Each period accumulates its own discrete interest (i). Recall Chapter 8, which highlighted the difference between continuous measurement (classic physics) and discrete quantities (quantum mechanics); we might think of simple interest as focusing on a continuous period, whereas compound interest deals with discrete periods. For calculating **compound interest**, we use the following to compute the future value (V_F) of a given amount of principal (P) using the periodic interest rate (i) over n periods of time:

$$V_F = P\ (1 + i)^n$$

It is important to remember that "i" does not represent the annual interest rate, but rather the interest rate for a single cycle. Only if interest were to be compounded annually, would the annual interest rate r would equal the periodic interest rate i.

Example 9-3: Assume that you deposited \$10,000 in a time savings account at a still-solvent bank on April 1, 2010, that paid 8% per annum. Assume that you withdrew the money on September 30, 2010. (Further assume, hypothetically, that there were no phoney charges or hidden withdrawal fees.)

(A) Calculate the future value of the account as of September 30, 2010 (i.e., after 6 months), using your simple interest formula.

(B) Perform the same calculation using the compound interest formula, under the assumption that your interest, at the same rate, is compounded quarterly.

(C) Which method yields greater interest? Why?

Solution (A): $P = \$10{,}000$
$r = .08$
$t = 6 \text{ months} / 12 \text{ months} = .50$
$V_F = P\ (1 + rt)$
$V_F = (\$10{,}000) \times (1 + (.08)(.50)) =$
$V_F = (\$10{,}000) \times (1 + .04) =$
$V_F = \$10{,}400$
Total Interest Earned: $400

Solution (B): $V_F = P\ (1 + i)^n$
$P = \$10{,}000$
$r = 8\%$ per annum
$i = .08 \times .25$ (quarterly interest = 8% per annum / 4) = .02
$n = 6$ months / 2 periods (i.e., 3 month quarters) = 2
$V_F = \$10{,}000\ (1 + .02)^2$
$V_F = \$10{,}000\ (1.02)^2$
$V_F = \$1{,}000\ (1.02)^2 = \$10{,}404$
Total Interest Earned: $404

Solution (C): Compounding yields greater interest because with respect to the first quarter of the loan, you are collecting interest on the interest, which adds an extra $4.

Over 200 years ago, while George Washington and his soldiers were freezing their rear ends off at Valley Forge, a rich Pennsylvania merchant named Jacob De Haven allegedly loaned $450,000 to the Continental Congress in order to rescue the troops. Apparently no part of the loan was ever repaid. In the case that follows, De Haven's descendants sued the United States in Claims Court in an attempt to rescue the cold principal and interest. The United States opposed the claim (Is that ingratitude or what?).

Lunaas v. United States

United States Court of Appeals for the Federal Circuit,
936 F.2d 1277 (1991)

Pauline Newman, Circuit Judge.

Thelma Weasenforth Lunaas et al. appeal the decision of the United States Claims Court, dismissing their claim for repayment of loans made to the Continental Congress by their ancestor, Jacob De Haven. We affirm.

Background

For the purpose of reviewing the Claims Court's dismissal of the complaint, the plaintiffs' statement of the facts is taken as true.

During the winter of 1777–78 Jacob De Haven made a loan to the revolutionary government, in the form of specie and supplies, with an estimated value of $450,000. This loan contributed to the survival of General Washington's army at Valley Forge. The loan was contracted, pursuant to authorization of the Continental Congress, on the following terms and conditions: interest was to be paid at the rate of six percent; the faith of the United States was pledged to the lender for the repayment of the sum borrowed, and for payment of the interest; interest was to be paid annually; and the principal was to be repaid at the end of three years. To date, despite "numerous and tenacious pleas by Mr. De Haven, his heirs and their descendants," no part of the loan or interest has been repaid.

Ms. Lunaas states that she is a descendant of an heir at law of Jacob De Haven, and that she is entitled to a proportionate share of the repayment. On March 17, 1989 suit was filed in the Claims Court to recover the proceeds due. The court granted the government's motion to suspend discovery; and ruled that the claim is barred in accordance with 28 U.S.C. § 2501: Every claim of which the United States Claims Court has jurisdiction shall be barred unless the petition thereon is filed within six years after such claim first accrues. This appeal followed.

Discussion

A.

The principal of the De Haven loan became due and payable in 1780–81. The exact dates are not provided by and are apparently unknown to the plaintiffs. On March 1, 1781 the Articles of Confederation were adopted, providing:

> ARTICLE XII. All bills of credit emitted, monies borrowed and debts contracted by, or under the authority of Congress, before the assembling of the United States, in pursuance of the present confederation, shall be deemed and considered as a charge against the United States, for payment and satisfaction whereof the said United States, and the public faith are hereby solemnly pledged.

This protection of creditors was extended in the subsequently adopted Constitution:

> ARTICLE VI, cl.1. All Debts contracted and Engagements entered into, before the Adoption of this Constitution, shall be as valid against the United States under this Constitution, as under the Confederation.

Thus the nation undertook to assure creditors that the adoption of the Constitution would not erase existing obligations recognized under the Articles of Confederation. Due to the precarious financial circumstances of the new government, the first Congress adopted, on August 4, 1790, a plan of Secretary of the Treasury Alexander Hamilton for settling debts incurred by the governments of the Confederation and the individual states. 1 Stat. 138. The plan required, inter alia, that creditors consent to an enlargement of the term when payment was due. Subsequent congressional enactments renewed this settlement plan, until it finally expired on March 3, 1837. See, e.g., Act of July 14, 1832, ch. 245, 4 Stat. 602; J. Res. 5, 5 Stat. 200 (1837).

The United States Court of Claims was formed in 1855 to hear claims against the government and make recommendations to Congress; the court could not render final judgments against the United States. Act of February 24, 1855, ch. 122, 10 Stat. 612. The absence of a statute of limitations was a recognized deficiency. Thus in 1863 Congress assigned to the Court of Claims authority to render final judgments against the government. Act of March 3, 1863, ch. 92, § 3, 12 Stat. 765. A six-year limitations clause was enacted. See United States v. Wardwell, 172 U.S. 48, 52, 19 S. Ct. 86, 43 L. Ed. 360 (1898) (the statute of limitations is a jurisdictional limitation on the ability of the Court of Claims to hear the case); Finn v. United States, 123 U.S. 227, 232, 8 S. Ct. 82, 31 L. Ed. 128 (1887). A savings clause in the 1863 Act provided that claims which accrued six years before enactment would not be barred if the petition were filed within three years after enactment, i.e., by 1866. Act of March 3, 1863, ch. 92, § 10, 12 Stat. 767:

> Sec. 10. And be it further enacted, That every claim against the United States, cognizable by the Court of Claims, shall be forever barred unless the petition setting forth a statement of the claim be filed in the court or transmitted to it under the provisions of this act within six years after the claim first accrues; Provided, That claims which have accrued six years before the passage of this act shall not

> be barred if the petition be filed in the court or transmitted as aforesaid within three years after the passage of this act.

In 1874 the savings clause was dropped, for it "was then needless, the time of the saving thereby created with respect to the claims to which it related having before expired." Clark v. United States, 99 U.S. 493, 495, 25 L. Ed. 481 (1878). A six year limitation period has continued to apply, for the Court of Claims and its successor tribunals.

B.

According to the plaintiffs, Mr. De Haven made several attempts to recover the proceeds during his lifetime. After his death in 1812 various De Haven descendants and their representatives petitioned the Congress for relief during the 1850s, 1870s, 1890s, and the early 1900s. It is apparent from the complaint that Mr. De Haven and his descendants knew or should have known of the claim after the government's liability for prior debts was reaffirmed in the Constitution. See Hopland Band of Pomo Indians v. United States, 855 F.2d 1573, 1577 (Fed. Cir. 1988) (a cause of action against the government has first accrued when all the events that fix the government's alleged liability have occurred, and the plaintiff was or should have been aware of their existence). Even if the claim were deemed not to have accrued until actual demand for payment was made, see United States v. Taylor, 104 U.S. 216, 222, 26 L. Ed. 721 (1881) (right of landowner to recover money that the government held for him as his trustee did not accrue until demand had been made at the treasury), or until Congress in 1863 created a court with authority to adjudicate contract claims against the government, it is not disputed that demands were made before and after 1863, and that no action was filed in the Court of Claims or its successor court until this suit, in 1989.

Congress has the power to impose reasonable limits upon a cause of action. Koshkonong v. Burton, 104 U.S. 668, 675, 26 L. Ed. 886 (1881) (statute of limitations applied against municipal bond-holders not unconstitutional impairment of obligations of contracts). Courts must give effect to statutes of limitations in accordance with congressional intent. United States v. Kubrick, 444 U.S. 111, 125, 100 S. Ct. 352, 62 L. Ed. 2d 259 (1979). The legislative history shows the careful attention given in 1863 to the limitation assigned to actions brought in the Court of Claims. It was intended to deal fairly with the large number of stale claims that were periodically reasserted to the Congress. As stated by Congressman A. G. Porter of Indiana:

> This is a very important section. It will soon set at rest the claims which for a generation have been persistently pressed upon successive Congresses, because repeated rejections did not affect the right still further to press them, and which seemed to grow stronger as time deprived the Government of the means of refuting the fictitious pretenses on which they were based. It will require also that claims shall be prosecuted while the facts on which they rest are recent and the evidence is conveniently attainable. Cong. Globe, 37th Cong., 2d Sess., App. at 124 (1862).

The validity of the plaintiffs' claim is not here at issue. Although the plaintiffs argue that the adoption of the Constitution created a new and separate cause of action, distinct from an action on the debt itself, this distinction does not affect the conclusion. The statutory period "is the term of time the statute specifies for the continuance of such liability, and on which it makes the payment of the debt and the loss of its evidence a presumption juris et de jure." Battelle v. United States, 7 Ct. Cl. 297, 301 (1871).

Plaintiffs argue that regardless of when their claim accrued it can not be barred by limitations enacted by Congress, because the Constitution specifically recognized such claims as an obligation of the United States. However, such a claim if filed after 1893, six years after Congress extended the jurisdiction of the Court of Claims to constitutional claims, Act of March 3, 1887, 24 Stat. 505, would also be barred by the statute of limitations. See Block v. North Dakota, 461 U.S. 273, 292, 103 S. Ct. 1811, 75 L. Ed. 2d 840 (1983) ("A constitutional claim can become time-barred just as any other claim can. Nothing in the Constitution requires otherwise") (citations omitted).

We conclude that the statutory period for this claim has expired, and that the Claims Court correctly so held. No costs.

Affirmed.

Notes and Problems

1. If Jacob De Haven had it do all over again, do you think that he should have left George Washington to freeze instead of loaning him the money? What kind of message does this case send to future revolutionaries?

2. At around 1777, the financial situation of our young break-away republic was seriously deteriorating. The individual states, at this juncture, were not making any voluntary payments to Congress. At the same time the Continental money was required to compete with a variety of other currencies; the states, for example, issued their own individual currencies. The British, meanwhile, in an effort to destroy the funding system of the Continental Congress, began a covert program of counterfeiting Continental dollars. These dollars were printed and then distributed throughout the former colonies by the British army and agents who were loyal to Britain. This led to a rapid inflation and corresponding depreciation of the Continental dollar. By the end of 1777 inflation had reduced the value of the "Continental" to about 20% of its value when originally issued. This rapid decline in value became a serious problem for Congress, and represents one of our earliest experiences with inflation.

3. Be suspicious of language in contracts and advertisements that happens to be less than precise when stating interest rates. Terms such as "five per hundred," "5 per 100," "5 on the 100" or "5 on the dollar" might lead one to believe that the borrower is paying 5% simple interest. But if the loan is being paid off periodically (monthly, weekly, etc.) over a one year period, the borrower does not actually have the money for an entire year. Therefore, the interest payment of a "flat" five dollars would be more akin to 10% interest. And if the payment period were less than 1 year, let's say 6 months, then the actual interest rate might be closer to 20%. It is rumored that this deception was created by used car sellers. What would be the actual interest rate on a January 1 "5 on the dollar" loan where all principal and interest is due and payable on June 30? On March 31?

4. Assume that on December 31, 1777, De Haven loaned the United States the amount alleged, $450,000, at 6% compounded monthly up through 1990 (De Haven's heirs filed suit in 1989). How much would his heirs be owed as of 1990? Solution:

$$V_F = P\ (1 + i)^n$$

P = $450,000
i = .06 x 1/12 = .005 (6% per annum /12)
t = 1990 - 1777 = 213
n = 213 x 12 (months/year) = 2556
$V_F = \$450{,}000\ (1 + .005)^{2556}$
$V_F = \$154{,}762{,}723{,}400$

5. What if the De Haven interest had been compounded annually instead of monthly? What if George never got the money?

White v. Sweeney

138 Cal.App.2d 199, 291 P.2d 77 (1955)

PETERS, Presiding Justice.

Plaintiffs appeal from an adverse judgment based on the sustaining without leave to amend of a demurrer to their first amended complaint.

The challenged pleading seeks a declaration that a certain promissory note is usurious, and also requests that the note be reformed to show the true amount borrowed. The pleading alleges that on March 22, 1950, plaintiffs borrowed $3,000 from defendant, executing and delivering to defendant their promissory note in the sum of $3,500, plus interest at 6 per cent on the unpaid balance, payable in 69.13 monthly installments of $60 each, commencing April 15, 1950; that plaintiffs have paid defendant $2,100; that the interest charged in this obligation is usurious, being 11.94 per cent; that the law fixes a maximum rate of 10 per cent.

The trial court apparently believed that, as a matter of law, the transaction was not usurious, and so sustained the demurrer without leave to amend.

The question presented is whether the interest here involved is in excess of the 10 per cent per annum limitation imposed by Art. XX, Sec. 22, of the state Constitution. The parties agree that the $500 differential between the amount loaned and the amount of the note must be treated as interest, that such interest should be spread over the entire period of the loan, and must be added to the 6 per cent per annum interest provided in the note. The cases so hold. Grall v. San Diego Bldg. & Loan Ass'n, 127 Cal.App. 250, 15 P.2d 797; Haines v. Commercial Mortgage Co., 200 Cal. 609, 254 P. 956, 255 P. 805, 53 A.L.R. 725; Connor v. Minier, 109 Cal.App.Supp. 770, 288 P. 23; Otis v. I. Eisner Co., 7 Cal.App.2d 496, 46 P.2d 235. The parties differ, however, as to how the $500 differential should be spread over the entire period of the loan.

The trial court amortized over the entire 69.13 month period the $500 bonus assuming that the full $3,000 was owed for all of that period. By this method of computation the interest rate chargeable to the $500 item is 2.8 per cent. When this is added to the 6 per cent provided in the note the result is 8.8 per cent, which, of course, is well under the 10 per cent constitutional limitation.

Respondent cites French v. Mortgage Guarantee Co., 16 Cal.2d 26, 104 P.2d 655, 130 A.L.R. 67, and Grall v. San Diego Bldg. & Loan Ass'n, 127 Cal.App. 250, 15 P.2d 797, in support of this method of computation. These cases, although containing some general language which, when separated from the facts to which the language was intended to apply, supports respondent, actually stand for no such general proposition. In those cases the borrower had the power to prepay the loan. It was properly held that in determining whether the loan was usurious the full term of the loan should be considered rather than a shorter period created when the loan was prepaid. The basis of this common sense rule is that the borrower cannot be permitted to convert an otherwise legal loan into a usurious one by his own act of prepayment.

The two cases cited are of no help in determining the problem here involved. There is obviously a basic fallacy in the contention that the $500 bonus should be treated as interest on $3,000 over 69.13 months, when actually each month the principal is proportionately reduced by the $60 monthly payment. Where interest and principal payments are made periodically over a fixed period as set forth in the note, as in the instant case, it is a simple mathematical fact that each month the principal decreases. In the instant case the principal was $3,000 payable at $60 per month over 69.13 months. The $60 monthly payment substantially exceeded the monthly interest, so that each month the principal was reduced.

One does not have to be an accountant or mathematician to know that the rate of interest produced by $500 amortized on an ever-decreasing principal over 69.13 months is greater than the rate of interest determined by amortizing the $500 on a full $3,000 over the entire 69.13 months. Appellants submitted to the trial court and have included in their briefs what they call a 'standard formula' by which the rate of interest on a decreasing principal is computed. By this formula appellants claim that as a mathematical certainty the total interest rate here charged exceeds 11 per cent, and certainly exceeds the 10 per cent permitted by law. Appellants ask us to hold, as a matter of law, based on this formula, that the interest rate is excessive.

We are not inclined to take judicial notice of this formula and to rule on this complicated mathematical problem as a matter of law. In some cases the court, with even its limited mathematical abilities, can easily determine whether a certain interest charge does or does not exceed the lawful rate. In such cases it is proper to decide the issue as a matter of law on demurrer. But where the determination of the rate requires complicated calculations it would appear that the issue should be decided as a question of fact based on expert evidence. The courts generally, in such cases, have held or suggested that the issue is one of fact and not of law. Middlekauf v. Vinson, 106 Cal.App.2d 204, 234 P.2d 742; Traders' Credit Corp. v. Thyle, 116 Cal.App. 252, 2 P.2d 568; Milana v. Credit Discount Co., 27 Cal.2d 335, 163 P.2d 869, 165 A.L.R. 621.

Respondent also seeks to sustain the judgment by the claim that the appeal was not filed in time. The judgment was entered December 28, 1953. The notice of appeal, which appears in the record, was filed February 16, 1954. Respondent contends that this appeal was abandoned on February 23, 1954; that thereafter appellants moved under section 473 of the Code of Civil Procedure to vacate the judgment; and that after denial of this motion, appellants filed a second notice of appeal on March 9, 1954. This was, of course, more than 60 days after entry of the judgment.

Respondent is in no position to raise the point. The only record now before us shows the filing of the notice of appeal on February 16, 1954, well within the time prescribed by law. Moreover, the other Division of this court, in May of 1954, denied a motion to dismiss this appeal, made on the very grounds now urged. This alone is sufficient basis for refusing to consider the point for a second time. Tyrrell v. Baldwin, 78 Cal. 470, 21 P. 116. In any event, under Rule 3(b) of the Rules on Appeal, the so-called second appeal, if one was taken, was filed within the time permitted by law.

The judgment appealed from is reversed.

BRAY and FRED B. WOOD, JJ., concur.

Notes and Problems

1. With respect to standard loan amortization methodology, the court in *White* declared "*We are not inclined to take judicial notice of this formula and to rule on this complicated mathematical problem as a matter of law.*" Was the problem presented to the court truly complicated?

2. Perform your own *White* calculations by allowing the principal (P) to equal the amount that the lender actually paid to the borrower, then determine whether or not the interest actually charged to the borrower exceeded 10%.

3. Why do you think usury laws existed? Would it surprise you to learn that some of the largest credit card companies have charged interest that exceeded well over 50% per annum?

§ 9.13 *Annuities*

An annuity is a future obligation to make periodic payments over a period of time. It may also be defined as a series of payments required to be made or received over time at regular intervals. The most common payment intervals are yearly, semi-annually, quarterly, and monthly. Examples of annuities include: Mortgages, rent, pension fund payments, insurance premiums. The annuity also provides a common device for settling large personal injury actions.

A. Ordinary Annuity

An **ordinary annuity** is one wherein each periodic payment is due at the end of the particular periodic interval. For ordinary annuities it is always presumed that the periodic payments will be equal. The future value (V_F) of an ordinary annuity with a periodic payment amount (A) for n periods at interest rate i, is calculated as follows:

$$V_F = (A) \frac{(1 + i)^n - 1}{i}$$

Example 9-4: To pay for college, Vito decides to put aside \$50 on the last day of each month for his godson Guido, a masculine child, who was born today. Assuming Guido lives to be 18, how much will be available for his college education on his eighteenth birthday? Where, as here, the money is to be used for a specific purpose, it is sometimes referred to as a "Sinking Fund." Assume an interest rate of 5% compounded monthly.

Solution: A = \$50

$r = .05$

$i = 0.05/12$ or 0.004166

$n = 18$ (years) $\times$ 12 (months) = 216 periods

$V_F = (A)\ ((1 + i)^n - 1) / i$

$V_F = (\$50)\ ((1 + .004166)^{216} - 1) / .004166$

$V_F = \$17{,}460.10$

Example 9-5: In the preceding example, what lump sum of money would Vito have to make (at the same interest rate) to have the same amount ready for Guido at age 18?

Solution: Here, we are required to calculate the Present Value (V_P) of the annuity. Using the same values for i, n, and V_F we now have a comound interest problem that can be resolved by solving:

$V_F = V_P\ (1 + i)^n$

$17{,}460.10 = V_P\ (1 + .04166)^{216}$

$V_P = 17{,}460.10 / (1 + .04166)^{216}$

$V_P = \$7{,}112.03$

Example 9-6: Kenny's woman Peaches wants to put away \$200 for the next 35 years so that she and Kenny will have some vacation money when Kenny gets out of

Leavenworth; Kenny is serving 35 years for money laundering. If Peaches earns 8.75% per annum, how much would be in the fund when the Worm is released?

Solution:

$$V_F = (A)\frac{(1 + i)^n - 1}{i}$$

$$i = 1/12 \times 0.0875 = 0.0072916$$

$$n = 35 \text{ years} \times 12 \text{ months} = 420 \text{ months}$$

$$V_F = (200)\frac{((1 + 0.0072916\ i)^{420} - 1)}{0.0072916} = 552{,}539.96$$

$$V_F = \$552{,}539.96$$

B. Annuity Due

The confusing term **annuity due** is not designed to imply that the ordinary annuity will *not* become due. On the contrary, as opposed to ordinary annuities, the "annuity due" annuity simply means that each periodic payment is due at the beginning of the payment cycle instead of the end. This means that an additional period of interest will be paid. The future value (V_F) of an annuity due with a periodic payment amount (A) is equal to the future value of an ordinary annuity times (1 + i), or:

$$V_{F(\text{annuity due})} = V_{F(\text{ordinary annuity})} \times (1 + i)$$

$$V_{F(\text{annuity due})} = \frac{A\ (1 + i)^n - 1}{i} \times (1 + i)$$

Because the obligation to pay comes at the beginning of the cycle, it is best analogized to rental income. Consider a situation where the annual rent for a parcel of land is paid in a lump sum at the beginning of each lease period (i.e., year). A similar arrangement for a three-year period, in the form of an annuity due would require four payments at: (1) the beginning of year 1, (2) at the end of year 1, (3) the end of year 2, and (4) the end of year 3. If you were to ignore completely the first payment, the result would be identical to an ordinary annuity for three years.

§ 9.14 Present Value of Future Payments

The right to receive $100 today is obviously worth more than the right to receive the same amount in a year, or two years, or any future date. Theoretically, any sum of money received today (as compared with the same sum to be received in the future) would immediately start to draw interest—not so for money that has yet to come into your hands. Therefore, if I were to pay you *today* for a debt that does not become due until some time in the future, I should be entitled to a discount that is based upon the amount of interest that you will earn from today through the date that the money is actually owed. We would refer to the interest rate as the "discount rate." This concept is particularly important for calculating damages in the form of future lost wages (See § 9.61).

Most calculations involving future lost wages will have a point at some future time; this is because people tend to die. The end point would often be determined by mortality tables for permanent disabilities. Consider, however, the present value (V) of a promise to pay a fixed amount per year (P) forever (Ignore the rule against perpetuities), assuming an interest rate of "I" :

$$V = \frac{P}{I}$$

Example 9.7 What is the present value of the right to receive $100,000 per year in perpetuity (i.e., forever), assuming that the average interest rate over time (a long time) will be 4% ?

Solution:

$$R = \frac{\$100{,}000}{.04} = \$2{,}500{,}000$$

§ 9.20 Basic Accounting

There are many different types of statements and accounting documents,[5] but there are two that are particularly fundamental and ubiquitous: The *Balance Sheet* and the *Income Statement*. The balance sheet (a.k.a. "Position Statement") speaks as of a specific point in time (i.e., a particular date), whereas the income statement (a.k.a. "Profit and Loss Statement") speaks as to a particular period of time.

§ 9.21 *The Balance Sheet*

The balance sheet is one of the two most important portraits of a business, its liquidity and its general health. Often referred to as a "snapshot," the balance sheet summarizes what a business owns (assets) and the amount that the business owes (liabilities). It is constructed according to the following fundamental equation (Owner's Equity is most often referred to as "Net Worth"):

Assets = Liabilities + Owner's Equity

From a pure mathematical perspective, the equation displays some interesting properties. Most notable is the manner in which the value of a business' assets is directly proportional to its liabilities. In other words, the greater your liabilities, the greater your assets. And yet, in the mind of the public, large assets seem to imply wealth not debt. It makes you wonder how many depositors were told by defunct Indy Bank and Washington Mutual: "*Don't worry, we have billions in assets.*" Berny Madoff could make the same claim from his prison cell.

5. Others include "The Cash Flow Summary," and "Statement of Changes in Shareholder Equity."

A favorite classroom exercise involves a simple application of the equation by asking the students how long it would take any given person to accumulate $1,000,000.00 in assets. The answer, of course, suggests that it would only take a matter of minutes. Two students, Willy and Cupcake, could merely exchange promissory notes for a million dollars each. The note to Willy could bear a slightly different due date from the note issued to Cupcake; that way, the two obligations would be something less than totally cancellable. If the transaction seems illusory, it is; but it points out one manner in which consumers are bilked on a regular basis. Moreover, it is far less knavish than many corporate shenanigans. Nevertheless, the real wealth and health of the entity is measured by its Net Worth:

Net Worth = Assets - Liabilities

Because of its universal recognition, the balance sheet carries a number of structural conventions. A typical balance sheet may contain a single snapshot for a single point in time, or it may contain two snapshots, one at the beginning of a period (such as January 1) and another at the end (possibly December 31). They normally contain three major headings (Assets, Liabilities and Owners' Equity). There is also a common method for arranging the order in which items appear on the balance sheet. Here is a typical example:

Example 9-8:

ENDRUN INDUSTRIES, INC.

Balance Sheet

As of December 31, 2011 and January 1, 2011

	January 1	December 31
ASSETS		
Current Assets:		
Cash	$2,714	$1,327
Receivables	21, 630	16,295
Inventories	10,861	10,058
Prepaid Expenses	879	718
Total Current Assets	$36,084	$28,398
Fixed Assets:		
Land & Buildings	$349	$345
Machinery & Equipment	6,037	4,965
Leasehold Improvements	2,190	2,149
Accumulated Depreciation & Amortization	(4,397)	(4,028)
Total Fixed Assets	$4,179	$3,431
Other Assets:	895	863
Total Assets	**$41,158**	**$32,692**
LIABILITIES & SHAREHOLDERS' EQUITY		
Current liabilities:		
Notes Payable	$00	$417
Accounts Payable	9,342	5,712
Income Taxes	1,378	746
Accrued Expenses	7,166	5,367
Total Current Liabilities	$17,886	$12,242
Allowance (Retirement Benefits)	770	770
Deferred Royalties	432	541

Shareholders' Equity:		
Common shares	$2,182	$2,181
Capital in Excess of Par Value	5,605	5,603
Retained Earnings	14,603	11,406
	$22,390	$19,190
Less Treasury Stock	320	51
Total Shareholder Equity	22,070	19,139
Total Liability & Shareholders' Equity	**$41,158**	**$32,692**

Notes and Problems

1. Why are the entries for Accumulated Depreciation & Amortization presented as negative amounts?

2. Why are "retirement benefits" singled out from the other categories?

3. As you may recall from *Property Law*, leasehold improvements that become fixtures of the underlying leased land become the property of the landlord, not the lessee. Why then, are leasehold improvements treated as assets, as in Example 9-8?

4. Treasury stock (sometimes referred to as "reacquired shares") is stock that is repurchased by the issuing corporation thereby reducing the number of outstanding shares on the open market. Stock repurchases are oftentimes used as a tax-savings device to put cash into shareholders' hands without paying dividends (which are taxable to the recipient). Often, companies do this when the stock is undervalued on the open market. Other times, companies may wish to provide a bonus to incentive compensation plans for employees. Rather than receive cash, recipients receive an asset that might appreciate in value faster than cash in a bank account. Another motive for stock repurchase is to protect the company against a hostile takeover threat.

On the balance sheet, treasury stock is normally listed under shareholder equity as a negative number (Notice that in Example 9-8, it is treated as a subtracted positive number; the result is the same). On the books, the accounts are sometimes referenced as "equity reduction" or "contra-equity." Note that any transaction involving treasury stock cannot increase the amount of retained earnings.

If treasury stock is sold for more than cost, the paid-in capital treasury stock account is increased, not retained earnings. In auditing financial statements, it is a common practice to check for this switch in order to detect possible attempts to "cook the books." Under what circumstances would treasury stock *not* be considered as an asset?

5. The so-called "underwater mortgage" occurs when the value of a home diminishes below the amount of the outstanding mortgages. Would a large dip in value affect a homeowner's overall balance sheet?

§ 9.22 The Income Statement

Whereas the Balance Sheet speaks as of a point in time, the Income Statement covers an interval or period of time. Often referred to as a "Profit and Loss Statement," the Income Statement is designed to illustrate precisely what the profits or losses happen to be for a particular period of time. The structure of the statement can vary significantly by industry but certain components have important common meanings because of their significance with respect to tax accounting. Here is one structure for calculating the income statement:

Gross Receipts
– Cost of Goods Sold

Gross Profits
– Administrative Expenses
– Depreciation
– Amortization
– Payroll & Fringe Benefits
– Other Expenses

Earnings Before Taxes
– Taxes

Net Income (Net Profits)

CPA's who conduct audits will no doubt be concerned about the integrity of all of the above components; so too should lawyers. But because lawyers often have an opportunity to construct and define the terms and clauses that put the money into motion, particular attention should be directed toward the term: "Net Profits." Because it is subject to creative definition and, at the same time possesses an accepted meaning in the world of accounting, it is particularly useful as an instrument of fraud, overreaching and unconscionability. Many are the lawyers who have been snookered by the term.

The greatest abuses and strategic re-definitions of the term net profits can be found in the entertainment industry. Film producers and the major studios can work magic when it comes to making tens of millions in profits disappear. It's a risky business, they say. And they are correct; one in ten products is profitable. Less are tolerable. This, they claim, justifies contracts of adhesion that uniformly deny artists, writers and actors a reasonable return on their percentage-based agreements.

In Phase 1 of his lawsuit against Paramount, Art Buchwald established that the studio based its Eddie Murphy 1988 smash hit *Coming to America* on a short story that he had previously written ("King for a Day"). The court found that Paramount breached a contract it had signed with Buchwald and Alan Bernheim (his producer) by failing to pay them an initial $265,000 together with a minimum of 19% of the film's "net profits." Phase 2 involved computing the profits that Buchwald was entitled to as compensation. The spotlight shifted from the issue of liability to the so-called "*standard studio contract.*" Even though the film grossed $350 million in worldwide box office receipts, Paramount alleged that the movie suffered a shortfall of $18 million. We now proudly present Phase 2:

Art Buchwald v. Paramount Pictures Corporation

Superior Court of the State of California,
County of Los Angeles, No. C706083 (1990)

Schneider, Judge.

I. Preliminary Statement

In the first phase of this case, this court ruled that Paramount's film *Coming to America* was "based upon" the screen treatment written by plaintiff Art Buchwald. In the second phase of the case the court has been presented with numerous issues, including whether: (i) The contract between plaintiff Bernheim (a producer) and Paramount is a contract adhesion; (ii) the contract, or any provision thereof, is unconscionable; (iii) the relationship between Bernheim and Paramount was that of co-venturers; (iv) Paramount owed a fiduciary duty to Bernheim and (iv) conduct on the part of Paramount breached the implied covenant of good faith and fair dealing. The court has also been presented with the task of interpreting other contract provisions, including the so-called "consultation" clause; the "turnaround" provision; and paragraph D.2.b. of the Bernheim Deal Memo.

II. The Contract

In order to understand the issues presented to the court in this phase of the proceeding, it is important to identify the components of the contract that present those issues. These components are:

1. The February 14, 1983, Deal Memo (consisting of six pages) entered into between Alma Productions, Inc. (Alain Bernheim's loan-out company) and Paramount;

2. The so-called "turnaround" agreement (consisting of three pages);

3. Additional Terms and Conditions (consisting of six pages); and

4. Paramount's standard net profit participation agreement (consisting of 23 pages), with two attachments relating to royalties.

III. Discussion

A. Contract of Adhesion

A "contract of adhesion" signifies a standardized contract, which, imposed and drafted by the party of superior bargaining strength, relegates to the subscribing party only the opportunity to adhere to the contract or reject it. Graham v. Scissor-Tail Inc., 28 Cal. 3d 807, 817 (1981). As the Court in *Graham* stated:

> Such contracts are, of course, a familiar part of the modern legal landscape, in which the classical model of "free" contracting by parties of equal or near-equal bargaining strength is often found to be unresponsive to the realities brought about by increasing concentrations of economic and other power. They are also an inevitable fact of life for all citizens—businessman and consumer alike.
>
> While not lacking in social advantages, they bear within them the clear danger of oppression and overreaching. It is in the context of this tension—between social advantage in the light of modern conditions on the one hand, and the danger of oppression on the other—that courts and legislatures have sometimes acted to prevent perceived abuses. 28 Cal.3d. at 817–818.

In the present case, the court finds that Bernheim's compensation package, as set forth in the Deal Memo, was negotiated by Bernheim's agent and Paramount's representative, as were other provisions of the Deal Memo not relevant to this case. The court finds, however, that the "boilerplate" language of the Deal Memo was not negotiated. The court further finds that the "turnaround" provision, the Additional Terms and Conditions, and the net profit participation agreement were not negotiated. With respect to the latter three parts of the Bernheim—Paramount contract, there is not the slightest doubt that they were presented to Bernheim on a "take it or leave it" basis. Indeed, the evidence reveals that Bernheim did not have the "clout" to make a better deal.

It is true Paramount has submitted evidence that it freely negotiates its net profit formula with the talent with which it deals. The court is not impressed with Paramount's evidence. To the contrary, the court concludes plaintiffs have proved by a preponderance of the evidence that Paramount negotiates its net profit formula with only a relatively small number of persons who possess the necessary "clout," and even these negotiations result in changes that are cosmetic, rather than substantive. Indeed, if, as Paramount contends, it freely negotiates with respect to its net profit formula, the court presumes it would have been inundated with examples of contracts where this was done. Succinctly stated, this has not occurred.

The evidence also discloses that the entire contract was drafted by Paramount and that the "turnaround" and net profit participation provisions were standard form provisions. Indeed, there is evidence in the record that Paramount's net profit formula is standard in the film industry. Further, there is evidence in the record to support the conclusion that essentially the same negotiations are conducted at all studios, and that when one studio revises a provision of its net profit formula, that revision is adopted by the other studios.

The above factors lead to the inescapable conclusion that the Bernheim—Paramount contract is a contract of adhesion. The fact that a portion of the contract was negotiated, i.e., Bernheim's compensation package in the Deal Memo, does not require a different conclusion. In Graham, supra, the Court held that the contract before it was a contract of adhesion, even though some of the terms were negotiated between the parties.

B. Unconscionability

In Graham, supra, the court stated: To describe a contract as adhesive in character is not to indicate its legal effect. It is, rather, "the beginning and not the end of the analysis in so far as foreeabil ity of its terms is concerned." Thus, a contract of adhesion is fully enforceable according to its terms unless certain other factors are present which, under established legal rules—legislative or judicial—operate to render it otherwise.

> Generally speaking, there are two judicially imposed limitations on the enforcement of adhesion contracts or provisions thereof. The first is that such a contract or provision which does not fall within the reasonable expectations of the weaker or "adhering" party will not be enforced against him. The second—a principle of equity applicable to all contracts generally—is that a contract or provision, even if consistent with the reasonable expectation of the parties, will be denied enforcement if, considered in its context, it is unduly oppressive or "unconscionable." 28 Cal.3d at 8 19–20.

Unconscionability—All or Any Provision of the Contract

There is no question that the law relating to the doctrine of unconscionability permits a court to strike down an entire contract or any provision thereof. Indeed, Civil Code section 1670.5 so provides.

Paramount, while apparently recognizing the above quoted law, argues that it would be impermissible to apply the unconscionability doctrine to this case. As the court understands it, Paramount's argument has two prongs. First, Paramount argues that a court may strike an unconscionable clause of a contract only where that clause is "divisible." Paramount contends that in the present case, plaintiffs are impermissibly attacking "financially interrelated provisions" and demanding "an individual defense of each." Second, relying on a number of so-called "price" cases, Paramount argues that "profitability is not relevant to unconscionability."

Addressing the last argument first, it is apparent that the events that occurred at the November 8, 1990, hearing in this case have rendered Paramount's second argument moot. A little discussion of the history of this case is required in order to validate this conclusion.

In many documents filed with the court prior to November 8, 1990, Paramount argued that its net profit formula was justified, and indeed required, in order to permit it to remain in business. For example, in the Response of Defendant Paramount Pictures Corporation to Plaintiffs' Preliminary Statement of Contentions, filed on May 21, 1990 (hereinafter referred to as the "5/21/90 Memo") Paramount argued:

> In agreeing to underwrite what it could thus anticipate to be a $66.50 million investment or, if produced, would not commercially succeed and that its investment would be lost. In contrast, Bernheim and Buchwald risked nothing.
>
> Not surprisingly, Paramount obtained from Buchwald and Bernheim, as it does in varying degrees of all net participants, the right to attain gross receipts in excess of its direct out-of-pocket costs before it began sharing those receipts with participants. This simply reflects an attempt by the studio to balance the enormous economic risks attendant to motion picture production by insuring that the studio will reap a fair portion of the rewards resulting from a commercial success. As a means for compensating for an allocation of risks in the motion picture industry that places all of the uncertainties on the studio, Paramount's contracts with Bernheim and Buchwald are not unconscionable.

Similarly, in its 7/24/90 Memo Paramount stated:

> "As forty years of studio—talent bargaining has established, a studio is entitled to a return commensurate with the risks of movie-making. Otherwise, it could not remain a viable business." There is nothing unfair or unreasonable about how the "Net Profits" formula strikes this balance.
>
> The level of return allowed to Paramount under its "Net Profits" formula is more than offset by the risks that the studio alone takes. As plaintiffs' experts readily conceded, "Net Profits" participants bear no risk; if a film flops, participants have no obligation to take up the shortfall and their up-front fee is guaranteed.
>
> In contrast, the studio's risks are enormous. When it signed the Buchwald and Bernheim contracts, Paramount assumed the risk that, despite substantial script development costs (nearly $500,000), the picture might never be made and that, even if made, the picture would not make money. Paramount spent $40 million to produce Coming to America and committed another $35 million to an advertising and a promotional campaign with no assurance that a single theater admission would be sold.
>
> The risk of failure in the motion picture business is ever present, immense, and unmitigable.

The Court interpreted the above quoted statements of Paramount, and many others like them, to mean that Paramount was attempting to justify its net profit formula on the ground that this formula was necessary for Paramount's survival. Indeed, when Paramount's counsel stated, "[o]therwise it could not remain a viable business," the Court understood Paramount to mean what its counsel had stated.

It was because Paramount argued that its net profit definition was justified by the exigencies of the film industry that the court decided to appoint its own accounting expert, pursuant to Evidence Code section 730. Indeed, the November 8, 1990, hearing was scheduled for the specific purpose of defining the tasks to be performed by the court's expert. This would have included, of course, an examination of Paramount's books and records to determine the accuracy of Paramount's representation with respect to its profitability, the number of films that make and lose money, and whether it was necessary for successful films to subsidize unsuccessful films. Remarkably, it was at this same hearing that counsel for Paramount abandoned the argument that Paramount's net profit formula was required by the nature of the motion picture business.

Paramount's abandonment of its "justification" argument rendered inquiry into Paramount's profitability moot and the appointment of the court's expert unnecessary. This abandonment also renders inapplicable the so-called "price" cases relied upon by Paramount. These "price" cases were submitted to the Court, according to Paramount, to establish the point that "profitability is not relevant to unconscionability." Since Paramount no longer seeks to defend its net profit formula on the ground it is justified by the nature of its business, it is clear Paramount's profitability is irrelevant to the determination of whether the contract involved in this case is unconscionable.

As indicated above, Paramount also argues that the court may not strike down all or any portion of the net profit definition because that definition is part of the entire compensation package between Paramount and Bernheim. Paramount further argues that it would not have paid Bernheim as much "up-front" money if it had known many of the components of the net profit formula would be invalidated, and that Bernheim will reap a windfall if the court finds unconscionable portions of the net profit formula.

Paramount's argument is based on the proposition that the dispute between the parties is one over price. The court is not convinced that this is the case. However, even if Paramount is correct, it is "clear that the price term, like any other term in a contract, may be unconscionable." Perdue, supra, at 926. In fact, in Perdue the Court stated:

> The courts look to the basis and justification for the price, including "the price actually being paid by ... other similarly situated consumers in a similar transaction." The cases, however, do not support defendant's contention that a price equal to the market price cannot be held unconscionable. While it is unlikely that a court would find a price set by a freely competitive market to be unconscionable, the market price set by an oligopoly should not be immune from scrutiny. Thus courts consider not only the market price, but also the cost of the goods or services to the seller, the inconvenience imposed on the seller, and the true value of the product or service.

In the present case, the court has already found the Bernheim-Paramount contract to be adhesive. Moreover, it is clear, as the court has already found, that contractual relations between certain talent and studios, at least talent such as Bernheim who lack the "clout" of major stars, do not take place in a freely competitive market. Rather, it is clear that if a talent such as Bernheim wishes to work in the film industry, he must do so on terms substantially dictated by the studio.

This is particularly true with respect to the net profit formula contained in the contract involved in this case. As previously indicated, Paramount simply does not negotiate with respect to its net profit formula with talent such as Bernheim.

Moreover, Paramount's argument that net profits represented a relatively insignificant part of Bernheim's total compensation package flies in the face of other evidence in the record. For example, in his Supplemental Declaration, Carmen Desiderio, Paramount's Vice-President of Contract Accounting, testified that Paramount had paid more than $150 million in net profits over the past 15 years, using the net profit formula contained in Bernheim's contact, or one similar to it. Additionally, Paramount itself admitted that "'Net Profits' are a valuable form of contingent compensation, not the 'cruel hoax' that plaintiffs insinuate." Indeed, Paramount's "turnaround" provision provides for Paramount to receive net profits in the event Bernheim was successful in convincing another studio to make a film based on Buchwald's treatment. Further, the doctrine of unconscionability would be rendered nugatory if a contracting party could escape its application by negotiating some monetary provisions, while at the same time imposing unjustifiably onerous provisions with respect to other contract provisions. Yet, that is precisely what Paramount agrees is permissible.

Paramount has referred the court to four cases which, it is contended, supports Paramount's position that the court may not strike down certain provisions of its net profit formula while enforcing the remainder of the contract with Bemheim. Paramount's argument is totally refuted by the provisions of Civil Code section 1670.5, which specifically permits the Court to "enforce the remainder of the contract without the unconscionability clause" or to "limit the application of any unconscionable clause as to avoid any unconscionable result." Moreover, none of the four cases relied upon supports Paramount's argument, and at least one refutes it. (New York v. Georgia-Pacific Corp., 58 F. Supp. 1265 [N.D. Miss. 1984]; Sykes v. Perry, 162 Kan. 365 [1947]; IMO Development Corp. v. Dow Coming Corp., 135 Cal. App. 3d 451 [19821; and Chow v. Levi Strauss, 49 Cal. App. 3d 315 [1975].)

The Court in IMO Development stated:

> What IMO does allege is that its consent was obtained by economic duress. Business or economic duress exists when threats to business or property interests by way of coercion and/or wrongful compulsion are present. *That, however, is not tantamount to a showing of unconscionability. In other words, the presence of a supposed unconscionable contract provision, such as would admit to differential enforcement, does not logically provide for differential rescission.* (Emphasis in original opinion.)

In sum, the court concludes that there is nothing about the contract involved in this case, or the circumstances surrounding its execution, which precludes the court from addressing the issue of whether certain component parts of the net profit definition are unconscionable. The next issue is the appropriate manner of applying the doctrine of unconscionability to the contract involved in this case.

Unconscionability—The Doctrine Applied

Plaintiffs have challenged as unconscionable a number of provisions of Paramount's net profit formula. The challenged provisions include: 15 percent overhead on Murphy and Landis [profit] participation; 15 percent overhead on Eddie Murphy Productions operational allowance; 10 percent advertising over-head; 15 percent overhead; interest on negative cost balance without credit for distribution fees; interest on overhead; interest on profit participation payments; the interest rate not being in proportion to actual cost of funds; exclusion of 80 percent of video cassette receipts from gross receipts; distribution fee on video royalties; charging as distribution costs residuals on 20 percent video royalties; charges for services and facilities in excess of actual costs; no credit to production cost for reusable items retained or sold; charging taxes offset by income tax credit; charging interest in addition to distribution fees; 15 percent overhead in addition to distribution fees; and 10 percent advertising overhead in addition to distribution fees.

Paramount has never argued that any of these provisions are individually fair and reasonable. Rather, as has been indicated, Paramount has argued that the Bernheim-Paramount contract must be considered as a whole, that the contract is fair and reasonable and, therefore, the court is not permitted to focus on individual provisions of the net profit formula to determine if such provisions are unconscionable. As discussed above, the court rejects the argument that it is impermissible for it to focus on individual provisions of the net profit formula. Plaintiffs, by contrast, have presented evidence which they argue supports their position that each of the challenged provisions are unconscionable.

The court is not persuaded that plaintiffs have sustained their burden of proof with respect to each challenged item. In fact, with respect to a number of challenged items it appears plaintiffs would like the court to make a finding unconscionability based upon the mere description of the item and without supporting evidence. This the court is not prepared to do. However, with respect to a number of provisions plaintiffs have sustained their burden of proving such provisions are "overly harsh" and "one-sided." Indeed, in light of Paramount's "all or nothing" approach to unconscionability, plaintiffs' evidence stands unrefuted. The court finds the following provisions of Paramount's net formula unconscionable for the reasons indicated:

1. Fifteen Percent Overhead on Eddie Murphy Productions Operational Allowance. The court finds this provision unconscionable because an additional 15 percent charge is made for overhead "on top of" this item. In effect, this results in charging overhead on overhead. The court is able to perceive no justification for this obviously one-sided double charge and Paramount has offered none.

2. Ten Percent Advertising Overhead Not in Proportion to Actual Costs. This flat overhead charge, which has no relation to actual costs, adds significantly to the amount that must be recouped by Paramount before the picture will realize net profits. Again, the court is able to discern no justification for this flat charge and Paramount has offered none.

3. Fifteen Percent Overhead Not in Proportion to Actual Costs. Paramount's charge of a flat 15 percent for overhead yields huge profits, even though the overhead charges do not even remotely correspond to the actual costs incurred by Paramount. In this connection it should be observed that although Paramount originally contended that this charge was justified because "winners must pay for losers" this justification was abandoned by Paramount during the November 8, 1990 hearing held in this case.

4. Charging Interest on Negative Cost Balance Without Credit for Distribution Fees. Paramount accounts for income on a cash basis, while simultaneously accounting for cost on an accrual basis. This slows down the recoupment of negative costs and inflates the amount of interest charged. The court finds this practice to be "one-sided" in the absence of a justification for the practice.

5. Charging Interest on Overhead. Paramount receives revenues in the form of distribution fees and overhead charges, neither of which are taken into account in determining whether costs have been recouped. This results in "interest" becoming an additional source of unjustified profit. The court finds this practice to be "overly harsh" and "one sided," and thus unconscionable.

6. Charging Interest on Profit Participation Payments. Paramount charges the payments made to gross participants to negative costs. In fact, these payments are not paid until the film has derived receipts. Accordingly, Paramount has not in any real sense advanced this money. Nevertheless, Paramount charges interest on gross participation shares. This is unconscionable.

7. Charging an Interest Rate Not in Proportion to the Actual Cost of Funds. Paramount charges an interest rate which can be as much as 20 to 30 percent, even when no funds have been laid out by Paramount. This is a one-sided, and thus unconscionable, provision.

In sum, the court concludes that the foregoing provisions of Paramount's net profit formula are unconscionable. In sum, the court concludes that the foregoing provisions of Paramount's net profit formula are unconscionable. The conclusion that these provisions are unconscionable is by no means the end of the analytic trail. While this conclusion does actuate the court's powers under Civil Code section 1670.5, it remains to be decided how those powers should be invoked.

As noted in A & M Produce Co., supra, "unconscionability is a flexible doctrine designed to allow courts to directly consider numerous factors which may adulterate the contractual process." Similarly, in Frostifresh Corporation v. Reynoso, 274 N.Y.S.2d 757, 759 (1966) the court stated that paragraph 2-302 of the Uniform Commercial Code, upon which Civil Code section 1670.5 is based, gives "the courts power 'to police explicitly against the contracts or clauses which they find to be unconscionable.'"

This court interprets the cases dealing with the doctrine of unconscionability as authorizing the court to use its powers under Civil Code section 1670.5 to produce an equitable result. Indeed, "equitable" would appear to be the antithesis of "unconscionable." In *Graham v. Scissor-Tail, Inc.*, supra, the court specifically recognized that the doctrine of unconscionability involves "a principle of equity applicable to all contracts generally.... that a contract or provision, even if consistent with the reasonable expectation of the parties, will be denied enforcement if, considered in its context, it is unduly oppressive or 'unconscionable.'" See also Slaughter v. Jefferson Federal Savings and Loan Association, 361 F. Supp. 590, 602 (D.C.D.C. 1973) in which the court, after concluding the provisions of a contract were unconscionable, stated that in such "circumstances the court has broad discretion to fashion relief appropriate to the situation presented."

Since it is the task of the court to achieve an equitable result, the question before the court is: What decision is necessary in order to produce such a result? Plaintiffs answer this question by arguing that Bernheim is entitled to receive compensation provided for in paragraph D.2.b of the Bernheim Deal Memo, after all of the unconscionable provisions are stricken and after permitting Paramount to recoup its actual costs plus a reasonable rate of return on its investment.

Counsel for Paramount, although specifically asked by the court during oral argument on December 6, 1990, stated he had no position with respect to this issue in light of his view that the court could not determine that individual provisions of the net profit formula were unconscionable.

After careful consideration, the court has concluded plaintiffs' approach must be rejected because it does not produce an equitable result. There are a number of reason's for the court's conclusion.

If the court were to strike all of the challenged provisions of the net profit formula that it has found to be unconscionable and permit Paramount only to recover its costs, plus a reasonable rate of return, the result would be an inequitable windfall to Bernheim. Stated another way, accepting plaintiffs' argument would result in Bernheim receiving a profit far beyond the contemplation of the parties at the time the contract was entered into and, apparently, far beyond the profit a producer with Bernheim's experience and track record would reasonably have been expected to earn.

The court believes it does not have sufficient facts to fix the amount that Paramount should be required to pay Bernheim in this case. The court intends, therefore, to defer to the third phase of this trial the amount of damages to which Bernheim is entitled and the manner in which such damages could be calculated. The court anticipates that expert testimony may be required. Further, the court desires to hear argument from counsel concerning these issues, particularly with respect to the factors that the court should consider in arriving at an equitable award. Although counsel for Paramount has heretofore declined to take a position with respect to these issues, the court assumes that, in light of the views expressed by the court herein, counsel will now proffer Paramount's position.

The court also desires to emphasize that its focus in the third phase will be on awarding damages to Bernheim which are fair and reasonable, but which will not result in Bernheim receiving a windfall, i.e., an award far beyond the reasonable expectations of the parties when the contract was executed. The court also intends to defer ruling on the amount to which Buchwald is entitled until after the amount due Bernheim is fixed. The court observes, however, that under the contracts as written, Buchwald was to receive only a fraction of the net profits to which Bernheim would have been entitled (1½ percent for Buchwald; 17½ to 40 percent for Bernheim). The court will in all likelihood be influenced by this fact in setting the amount due Buchwald.

C. The Juxtaposition of Unconscionability and the Consultation Clause

Paragraph D.2.b of the Bernheim Deal Memo contains the so-called "consultation clause." That clause provides that Bernheim "will be consulted on gross and net-profit participations granted by PPC to third parties, but PPC's decision shall be final."

Bernheim contends that Paramount breached the consultation clause by not consulting with him. Paramount argues that the consultation clause is not significant since Paramount retained the right to make the final decision with respect to granting gross and net—profit participations. The court finds it unnecessary to resolve this dispute. If Bernheim is correct, the result would be that he is entitled to receive 33.5 percent of the net profits on Coming to America under the net profit formula contained in the contract as written. This conclusion follows from Bernheim's position that, by reason of Paramount's breach of the "consultation" clause, he is entitled to the highest percentage of net profit permissible under paragraph D.2.b of the Deal Memo and Bernheim's concession that that highest percentage is 33.5 percent.

If Paramount is correct, the result would be that [because of the shares of third parties] Bernheim is entitled to receive only 17½ percent of net profits (the floor established in Section D.2.b of the Bernheim Deal Memo) under the net profit formula contained in the contract as written.

In the preceding section of this Tentative Decision, the court has concluded that a number of provisions of the net profit formula as written are unconscionable. The court has also determined that it will follow a different path in arriving at equitable compensation for Bernheim and Buchwald in light of such unconscionability. Since, pursuant to the court's ruling, the net profit formula as written—no longer exists, it makes no difference whether Bernheim or Paramount is correct with respect to the percentage of net profits to which Bernheim is entitled. This factor also makes Paramount's alleged breach of the consultation clause irrelevant.

D. The "Turnaround" Provision

As indicated above, one of the component parts of the contract between Paramount and Bernheim is the so—called "turnaround" provisions. The purpose of the "turnaround" provision is to permit a producer to take his project to another studio if the first studio is no longer interested in pursuing it, while at the same time permitting the first studio to recoup its development costs if the project is undertaken by the second studio. Insofar as is pertinent to the present case, the "turnaround" agreement provides:

> If, prior to the expiration of the turnaround period, the project is not placed elsewhere and/or if Lender has not complied with the conditions above, including, without limitation, complete reimbursement to Paramount, then at the end of the turnaround period, Lender's rights with respect to the project shall cease and Paramount's ownership thereof and all properties and rights encompassed therein shall be absolute.

The facts with respect to the application of the "turnaround" agreement to the present case are these: In March 1985 Paramount purported to give notice that it was abandoning the project that had been inspired by Buchwald's treatment. In May 1985 Paramount permitted its option with respect to the Buchwald material to expire. Paramount contends that since Bernheim failed to set up the project at another studio within the 12—month period ending in March 1986, the "turn-around" agreement extinguished any obligations Paramount had with respect to Bernheim.

It is true, as Paramount argues, that if the "turnaround" provision is considered in isolation, it would appear Bernheim's rights to compensation ended in March 1986. The vice of Paramount's argument is that the "turnaround" provision cannot be considered in isolation. Paragraph D. 1 of the Bernheim Deal Memo provides, in pertinent part that "[i]f the Picture is produced, Lender will furnish the services of Artist, who shall be employed by PPC to personally render all customary services as producer."

The court has already concluded that the picture was made, i.e., that *Coming to America* was "based upon" Buchwald's treatment entitled "King for a Day." In light of this conclusion, it is clear Paramount was required to employ Bernheim as producer on *Coming to America* and that Paramount breached its contract with Bernheim by failing to do so. It would make no sense to conclude that Paramount breached the agreement by failing to employ Bernheim, while at the same time concluding Bernheim's right to compensation was terminated by application of the "turnaround" provisions.

In reality, and the court so finds, it was never contemplated that the "turnaround" provision would apply in a situation such as is presented by the facts of this case. Moreover, to the extent that there exists an ambiguity by reason of the existence of paragraph D.1 and the "turnaround" provision, it is clear that such ambiguity must be resolved against Paramount as drafter of the agreement. Civil Code section 1654; Jacobs v. Freeman, 104 CaL.App.3d 177, 189 (1980).

Finally, the court observes that one of the important purposes, perhaps the most important purpose of the "turnaround" provision, from Paramount's perspective, was to permit it to recoup its costs in the event Bernheim placed the project at another studio. In the present case that purpose has been satisfied since it is too clear to doubt Paramount has recovered all of its costs on *Coming to America.*

E. The Co-Venturer and Fiduciary Duty Issues

Bernheim contends that he and Paramount were co-venturers and that Paramount owed a fiduciary duty to him. With one exception to be discussed below, the court is unable to agree with either of these contentions.

Whether or not the relationship between parties is that of co-venturer is essentially a question of fact. Few, if any, of the features that usually characterize a joint venture are present in this case. Bernheim did not have a right at all times to inspect and copy the purported venture's books and records and Paramount had pervasive control over the purported venture. Moreover, while there was an agreement between Bernheim and Paramount with respect to the sharing of profits (but not losses), Paramount retained the virtually unlimited power to determine whether Bernheim ever received any profit. The factors present in this case do not point to the existence of a joint venture between Bernheim and Paramount.

The court is also unable to find the existence of a fiduciary relationship between Paramount and Bernheim, except with respect to Paramount's duty to render accounting, Waverly Productions v. RKO General, Inc., 217 Cal.App.2d 721(1963). In fact, the court disposed of Bernheim's fiduciary duty claim in the Statement of Decision that was issued in the first phase of this case. In its Statement of Decision the court stated:

> In addition to their contract claims, plaintiffs have advanced several tort theories of recovery, namely, bad faith denial of existence of contracts, bad faith denial of liability on their contracts, tortious breach of the implied 'covenant of good faith and fair dealing,' breach of fiduciary duty, fraudulent concealment by a fiduciary and constructive trust. The obvious reason plaintiffs have asserted tort causes of action is to recover punitive damages since, absent such damages, the court is able to discern no difference between any tort damages plaintiffs might recover and their contract damages.

The court has concluded, as indicated, that Coming to America was based upon Buchwald's treatment. The court is unable to find, however, any tortious conduct on the part of Paramount or any of its representatives.

In order to award punitive damages to plaintiffs, the court would be required to find by clear and convincing evidence that defendant was guilty of fraud, oppression or malice, as those terms are defined in Civil Code Section 3294. While the court rejects Paramount's contention that Coming to America is not "based upon" "King for a Day," the court is unable to conclude that Paramount's conduct was in bad faith, let alone fraudulent, oppressive or malicious. Accordingly, while plaintiffs are entitled to recover on their breach of contract claims, the court finds the defendant is entitled to judgment on plaintiff' tort claims.

In light of the court's finding that Paramount's conduct was not tortious, the issue of whether a fiduciary duty existed between Bernheim and Paramount and, if so, whether Paramount breached that duty has been rendered moot. As indicated, however, the court does find that a fiduciary duty exists with respect to Paramount's duty to render an accounting.

F. The Covenant of Good Faith and Fair Dealing

Plaintiffs argue that Paramount breached the implied covenant of good faith and fair dealing by improperly or excessively charging a number of different items as costs on Coming to America. Paramount has countered by arguing that plaintiffs will be given the opportunity to challenge these costs in the third (damage) phase of this trial.

In a preceding section of this Tentative Decision, the court has ruled that a number of provisions of Paramount's net profit formula are unconscionable. The court also indicated that it intends to fashion relief that will produce an equitable result in this case. In light of the court's ruling, it appears to the court that application of the doctrine of unconscionability will produce damages at least equal to damages that could be awarded for a breach of the covenant. The court finds it unnecessary, therefore, to determine whether a breach of covenant has in fact occurred.

Batfilm Productions, Inc. v. Warner Bros., Inc.

California Superior Court, Los Angeles County
BC-051653 and BC 051654 (1994)

Hon. David P. Yaffe, Judge of the Superior Court

Phase I Statement of Decision

The Court divided the trial of this case into two phases. Phase I consisted of a bench trial of plaintiffs' non-jury claims. Those claims primarily concern Plaintiffs' "Net Profits" participation in the Batman motion pictures.

The plaintiffs are two individuals, Benjamin Melniker and Michael Uslan, and the two corporations that furnish their services, Batfilm Productions, Inc., and Franklin Enterprises, Ltd. The defendants are Warner Bros. and Polygram Pictures, Inc. (The Court previously granted the summary judgment motion of defendants Peter Guber, Jon Peters, and the Guber-Peters Entertainment Co.) Polygram Pictures did not participate in the bench trial.

In 1979, Mr. Melniker and Mr. Uslan obtained an option on the motion picture rights to the Batman comic book characters. In November 1979, they made a deal with Casablanca Productions (Polygram's predecessor) for the development and production of a motion picture to be based on those characters (the "Casablanca Agreement"). Under the Casablanca Agreement, Mr. Melniker and Mr. Uslan were entitled to receive certain fixed and contingent compensation if a Batman motion picture was produced.

In 1981, Polygram assigned to Warner Bros. its rights and obligations under the Casablanca Agreement. In 1988, Mr. Melniker and Mr. Uslan and Warner Bros. signed a written amendment to the Casablanca Agreement (the "Warner Agreement"). Under the Warner Agreement, Mr. Melniker and Mr. Uslan were entitled to receive $300,000 in fixed compensation for Batman, plus $100,000 "deferment" once the film generated a certain level of receipts, plus 13% of the so-called "Net Profits" as defined in an attachment to the Warner Agreement.

Warner Bros. has paid Messrs. Melniker and Uslan the $300,000 fixed fee and $100,000 deferment. Under the Warner Agreement, Warner Bros. has also paid Melniker and Uslan an additional $700,000 in fixed fees on two additional motion pictures (Batman Returns and Batman: Mask of the Phantasm). Warner Bros. will have similar financial obligations to plaintiffs on each additional Batman motion picture. Although Batman has generated more revenue than any other Warner Bros. film, it has not generated any "Net

Profits" under plaintiffs' contract. Melniker and Uslan filed suit in 1992 claiming, inter alia, they were denied their fair "Net Profits" compensation.

The primary claims originally to be tried to the Court were the Tenth Cause of Action for an accounting of the revenues and expenses of Batman and the Eleventh cause of action for a declaration that the plaintiffs' "Net Profits" definition is unconscionable and, thus, unenforceable. On the first day of trial, however, plaintiffs dismissed their accounting claims. Warner Bros. is therefore entitled to prevail on that cause of action.

At the close of plaintiffs' case, Warner Bros. moved for judgment pursuant to section 631.8 of the Code of Civil Procedure. In reviewing the evidence, the Court believed that Mr. Melniker and Mr. Uslan had offered evidence to prove that the Warner Agreement was a contract of adhesion that should be strictly interpreted against Warner Bros. and should not be interpreted in a way that would be contrary to plaintiffs' reasonable expectations.

But a contract of adhesion is a contract, and a contract of adhesion is not the same as an unconscionable contract, which is no contract at all. "Unconscionability" requires a far different level of proof. The plaintiffs did not prove that they are to be relieved of their contract with Warner Bros. on the ground of unconscionability.

Mr. Melniker negotiated the Warner Agreement on his and Mr. Uslan's behalf. No one is less likely to have been coerced against his will into signing a contract like the Warner Agreement than Mr. Melniker. This former general counsel and senior executive of a major motion picture studio (Metro-Goldwyn-Mayer) knew all the tricks of the trade; he knew inside and out how these contracts work, what they mean, and how they are negotiated.

Even with Mr. Melniker's knowledge and experience, plaintiffs complain that Warner Bros. knew when the parties signed the Warner Agreement in 1988 that Batman would not generate "Net Profits." Plaintiffs did not explain the relevance of this to the issue of whether their contract is unconscionable. Even if they had, however, they failed to prove that Warner Bros. knew in 1988 that Batman would not generate any "Net Profits."

At the core of plaintiffs' case is their argument that the contract was not fair to them because Warner Bros. and others earned millions of dollars on Batman and plaintiffs did not. The answer to that argument is that ever since the King's Bench decided Slade's case in 1602, right down to today, courts do not refuse to enforce contracts or remake contracts for the parties because the court or the jury thinks that the contract is not fair. That principle is not some medieval anachronism. This society, this country, this culture operates on the basis of billions of bargains struck willingly every day by people all across the country in all walks of life. And if any one of those people could have their bargain reexamined after the fact on the ground that it was not fair or on an assertion that it was not fair, we would have a far different type of society than we have now; we would have one that none of the parties to this case would like very much.

When one talks about a motion picture and the claims of this type that are made, they all have one thing in common: the plaintiff comes in and says "Without me, they would have had nothing, and look how they treated me." But the process of making a motion picture consists of the process of bargaining with many talented people on many different and inconsistent bases, and making bargains with them that cannot rationally be compared to one another. It would not be good for the motion picture business or for the parties to this case if any one of those people on any motion picture could come back and ask a court to remake the bargain that he made on the ground that he now asserts, after the fact and in light of the success of the picture, that he was not fairly treated in comparison with others. Whether a contract is fair is not the issue. A contract is not unconscionable simply because it is not fair. Plaintiffs claim that the Warner Agreement is unconscionable within the meaning of Civil Code section 1670.5. To be unconscionable, a contract must "shock the conscience" or, as plaintiffs alleged in paragraph 139(b) of their complaint must be "harsh, oppressive, and unduly one-sided."

After considering all the evidence, the Court finds that the plaintiffs have failed to prove that the Warner Agreement, taken as a whole, is unconscionable. That, however, is not the end of the inquiry that the Court must make. Under Civil Code section 1670.5, if the evidence shows that any part of a contract is unconscionable, the Court may refuse to enforce that part of the contract.

During the trial, plaintiffs claimed that eight elements of the Warner Agreement's "Net Profits" definition were unconscionable: (1) the 10% advertising overhead charge; (2) Warner Bros.' retention of any economic value of United States tax credits created by the payment of taxes in the foreign territories where Batman was distributed; (3) application of the 15% production overhead charge on participation payments to third parties; (4) application of the 15% production overhead charge on the $100,000 deferment; (5) all of the interest charges; (6) the costs charged by Pinewood Studios in England for holding sets and stages after completion of photography; (7) application of the 15% production overhead charge to the costs incurred at the Pinewood Studio lot; and (8) the inclusion in "gross receipts" of only 20% of the revenue from videocassettes, less a distribution fee. (These items, and the dollar amounts associated with them, are listed on Exhibit B9.)

In considering Warner Bros.' motion for judgment under Code of Civil Procedure section 631.8, the Court had little difficulty in rejecting seven of plaintiffs' claims.

As to all of the items relating to overhead charges (Items One, Three, Four, and Seven), the Court granted Warner Bros.' motion for judgment because the plaintiffs failed to prove that historically Warner Bros.' indirect general administrative expenses for motion picture, production and advertising—"overhead"—do not equal or exceed the amount charged under the "Net Profits" definition, namely, 15 percent of production costs and 10 percent of advertising expenditures. As a matter of fact, plaintiffs conceded that they could not show that the overhead charges under the "Net Profits" definition exceeded Warner Bros.' actual overhead costs, taken as a whole.

Plaintiffs argued that charging overhead on certain production costs, advertising expenses, gross participations, deferred payments, and payments paid to foreign studios was unconscionable because the administrative cost of providing those goods or services was less than the contractual 10 or 15 percent overhead surcharge. Plaintiffs did not prove that allegation. And, more important, the test is not whether Warner Bros.' overhead charge on a particular direct cost item exceeded the "actual" administrative or other indirect expenses associated with providing that one item or service to the production or advertising of a movie. As the accounting experts for both sides testified, overhead cannot be assessed with such precision. Under the circumstances, the test must be whether the production and advertising overheads charged by using the percentage allocations are, in total, unconscionably higher than Warner Bros.' actual production and advertising overhead costs on a motion picture. Plaintiffs offered no evidence to support such a finding.

Plaintiffs also failed to show that the advertising costs, gross participations, deferred payments, and payments paid to foreign studios were not historically included in the pool of costs that were compared to Warner Bros.' general and administrative expenses to estimate its rate of overhead. In sum, plaintiffs simply failed to prove that any of the overhead charges are unconscionable.

The Court also granted Warner Bros.' motion for judgment as to Item Two, the foreign tax credit. According to plaintiffs, when a motion picture is distributed overseas, many countries impose a tax on the receipts generated. That tax payment gives rise to a credit that can be used under certain circumstances to offset United States income tax obligations. Plaintiffs claimed that, in calculating their "Net Profits," it is unconscionable for Warner Bros. to deduct foreign taxes as a distribution expense without adding something for the value of the foreign tax credits. The plaintiffs failed to prove, however, that Warner Bros. received any foreign tax credits on Batman, or the amounts thereof, or that Warner Bros. received any actual financial benefit from those tax credits when calculating and paying its United States tax obligations. Even if such a credit had been received, the plaintiffs failed to prove that they ever asked Warner Bros. to agree that, in computing "Net Profits," Warner Bros. would augment the gross receipts of the picture by the amount of the tax credits. No such provision is contained in plaintiffs' contract and there was no evidence that they ever expected such treatment of the tax credits.

The Court also granted the motion for judgment as to Item Six, the Pinewood Studios sound stage holdover costs, because there was no evidence that the holdover charge is not properly a cost of the first Batman movie.

The Court granted the motion for judgment as to Item Eight, videocassette distribution, on the ground that Mr. Melniker knew that a 20 percent royalty was standard in the industry. He never questioned it. He never asked that it be changed. The plaintiffs did not prove that the 20 percent royalty unconscionably exceeded the actual revenues, less expenses, from videocassette distribution. They also offered no evidence that a "distribution fee" on the distribution of videocassettes was unconscionable. Nor did they prove that they could have negotiated a better deal elsewhere at the time this deal was made, in which a higher percentage of video revenue, without deduction of a distribution fee, would be credited to the picture in calculating "Net Profits."

Item Five concerned the "interest" charge on production costs. Under paragraph 2A of plaintiffs' contract, "Net Profits" became payable once the picture generates enough gross receipts to cover the specified distribution fees, distribution expenses, and production costs. Until then, under paragraphs 2A and 9 of plaintiffs' "Net Profits" definition, the production costs bear an interest charge. Under the contract, Warner Bros. reduces the interest—bearing balance of production costs with only those gross receipts that remain after deducting the distribution fees and expenses. Plaintiffs claim that it is unconscionable for Warner Bros. to not credit the interest—bearing production cost balance with all of the gross receipt of the picture. They also claim that because the distribution fee represents a source of "profit" for Warner Bros., this method of calculating interest is unconscionable because it allows Warner Bros. to charge interest on the cost of production after the picture has generated revenues in excess of that amount.

Plaintiffs did present sufficient evidence to require Warner Bros. to defend its method of computing interest under the contract. After listening to the evidence presented by Warner Bros. and the arguments of counsel, however, the Court finds that Warner Bros. met its burden of showing that the method of calculating interest provided in their contract is not unconscionable. Warner Bros. met its burden in a number of ways.

Warner Bros. showed that the interest provision in the Warner Agreement is really the same provision found in the 1979 Casablanca Agreement that Warner Bros. did not have anything to do with. Plaintiffs were bound by that contract before they ever dealt with Warner Bros. They cannot complain that they were harmed by being required to abide by a similar provision with the same effect.

Warner Bros. also showed that plaintiffs would not have gotten any better deal on the calculation of interest if they had borrowed the production costs from a third party lender, had produced Batman themselves as independent producers, and had hired Warner Bros. (or presumably anybody else) just to distribute it for them. In that case, plaintiffs would not have been able to use all of the gross receipts generated by the film to repay their lender. Just as in their contract with Warner Bros., they would have been unable to repay the production financier only with the gross receipts left over after the distributor retained enough to cover the distribution fees and expenses.

And, if there is a "profit" embedded within Warner Bros.' distribution fee, plaintiffs did not prove the amount of it or that it prevented the picture from showing a net profit. All of the evidence is sufficient to overcome the plaintiffs' evidence as to the unconscionability of the method of calculating interest under their "Net Profits" contract.

Separately, plaintiffs argued that the language of their "Net Profits" contract did not permit Warner Bros. to continue charging interest once the gross receipts of the picture—prior to the deduction of distribution fees and expenses—exceed the total production costs. The duty of the Court is to find out what the parties meant by the language of their contract. If the contract is one of adhesion, the Court interprets it so that it does not defeat the reasonable expectations of the party who was forced to adhere to it. But the Court will not substitute its own interpretation of the contract if that is not what the evidence shows that the parties intended.

The Court rejects plaintiffs' argument because there was no evidence that plaintiffs ever interpreted the language of the interest provisions in the manner claimed at trial. Mr. Melniker was an old hand at motion picture agreements of this type and had negotiated other "Net Profits" contracts like this himself. He had experience with similar provisions yet he never mentioned the interest issue with anyone at Warner Bros. Plaintiffs offered no evidence that they expected Warner Bros. to compute interest in any other manner. They have thus failed to prove that the contract defeated their reasonable expectations.

Given the Court's decision in favor of Warner Bros. on plaintiffs' unconscionability claim, Warner Bros. is entitled to prevail on plaintiffs' Thirteenth Cause of Action for "unfair competition" because that claim was dependent on a finding that their "Net Profits" contract was unconscionable.

Finally, Warner Bros. is entitled to prevail on plaintiffs' Fourteenth Cause of Action arising from the exhibition of the animated Batman television series. Plaintiffs presented no evidence on this cause of action at trial.

Notes and Problems

1. How would you reconcile *Buchwald* with *Batfilm*? Was Art Buchwald any less sophisticated with respect to the film industry or any less familiar with "the tricks of the trade" than Mr. Melniker?

2. If you placed each of the *Buchwald* deductions on a scale from 1 to 10 (1 being completely legitimate; 10 being ridiculous), what number would you assign to each of them?

3. With some attention to simple math, the problems presented by the percentage-of-profits disputes could be avoided by focusing on gross receipts instead of net profits. In particular, if the contract provided the obligee with a percentage of "gross receipts," there would be less room for dispute because that term is normally easy to determine and typ-

ically defined as "revenue and receipts from all sources, foreign and domestic." The definition leaves no wiggle room for deducting inflated charges, salaries or expenses. The percentage would, of course be less (i.e., 10% of net profits might equal only 2% of gross). This alternative method would work best where the ratio of profits to gross receipts is somewhat predictable.

§ 9.30 Tracing

Tracing the ownership of a financial asset (along with its accrued interest, profits and debts) and thereafter determining an equitable allocation for that asset often challenges the court's accounting and mathematical skills.

Tracing problems can arise in several contexts such as: The imposition of a constructive trust upon misappropriated property, the division of marital property in conjunction with a divorce, and reallocating money recovered from a Ponzi or pyramiding scheme.[6] Our attention turns to issues involving community property.

Charles Ponzi
[1911 Police Mugshot]

In a community property jurisdiction (Alaska, Arizona, California, Idaho, Louisiana, Nevada, New Mexico, Texas, Washington, Wisconsin and Puerto Rico), property acquired during the marriage (excepting gifts or inheritances) is owned jointly by both spouses and is normally divided upon divorce, annulment or death.

Property acquired prior to marriage and the interest and benefits earned thereon are deemed to be the separate property; gifts and inheritances received during mar-

---

6. The *Ponzi Scheme*, made famous by its namesake Charles Ponzi, involves three key elements: [1] The promise of a large return of money for a relatively small investment, [2] paying off prior investors with money from new incoming suckers, and [3] skipping town at the right moment.

There is a misconception that the earlier investors (who are fully paid both principal and interest) are basically "home free." This is not the case. At best, the earlier victims might be able to lawfully keep their returned principal, but they are not entitled to retain the interest, which essentially consists of funds stolen from later victims.

riage are also considered as separate property. Where there is doubt, joint ownership is presumed by law in the absence of specific evidence that would point to a contrary conclusion. Division of community property may take place by item, by splitting all items or by value.

In some states, such as California, a 50/50 division of community property is strictly mandated by statute. In other states, a divorce court may decree an "equitable distribution" of community property, which may result in an unequal division of such. Division of community debts may not be the same as the division of community property; in California, community property is required to be divided "equally" while community debt is required to be divided "equitably."

In re Marriage of Marsden

130 Cal.App.3d 426, 181 Cal.Rptr. 910, Cal.App.,1982

BARRY-DEAL, Associate Justice.

In this opinion we will consider the issues raised in two appeals (1 Civ.No. 47922 and 1 Civ.No. 51438) which were consolidated by this court "for the purposes of oral argument and decision" by an order dated May 22, 1981. Both appeals involve the same proceeding commenced by respondent (wife) to dissolve her marriage to appellant (husband) in the Superior Court of Santa Clara County on July 17, 1978. At trial the issue of the dissolution of the marriage was bifurcated from the remaining issues presented in the proceeding. An interlocutory judgment of dissolution of marriage was entered on February 27, 1979. An interlocutory judgment pertaining to the disposition of the community property was entered on April 5, 1979. This latter interlocutory judgment provides in part: "The value, disposition and division of husband's pensions/annuities with CREP (# P-38492) and TIAA (# A-115014) was [sic] previously bifurcated by stipulation of counsel. In the event the parties are unable to agree, the court specifically retains jurisdiction to determine the value, disposition, and division of these two pensions/annuities."

The appeal in 1 Civil No. 47922 is from the interlocutory judgment entered on April 5, 1979. In this appeal husband contends (1) the trial court incorrectly determined the community interest in the leasehold and house he acquired prior to the marriage on which payments were made after marriage from community funds; (2) the trial court incorrectly found that certain shares of stock were community property on the ground that these shares of stock were purchased with funds from accounts in which separate and community property or funds had been commingled; (3) the trial court incorrectly concluded that there had been no periods of legal separation from the date of marriage (Feb. 1, 1971) until July 10, 1978; and (4) attorney's fees were improperly awarded to wife. An interlocutory judgment pertaining to husband's pensions and/or annuities was filed on April 8, 1980. The appeal in 1 Civil No. 51438 is from the interlocutory judgment filed on April 8, 1980. Husband also contends on his appeal in 1 Civil No. 51438 that the trial court incorrectly determined that there had been no periods of legal separation from the date of marriage until July 10, 1978. Husband further contends in this second appeal that the interlocutory judgment, filed on April 8, 1980, is ambiguous in that it fails to specify the date of separation as the appropriate valuation date for the retirement benefits. The facts necessary to resolve each issue will be set out in the discussion of the issue.

Husband contends that there were two periods of legal separation between the parties prior to July 10, 1978. Husband alleges that the parties were separated between (1) April of 1975 and September of 1976; and (2) June of 1977 and July of 1978. The resolution of this issue is important to husband because Civil Code section 5118 provides that the earnings and accumulations of a spouse while living "separate and apart from the other spouse, are the separate property of the spouse." We have chosen to consider this issue first as the resolution of the issue may affect the determination of the community interest in the leasehold and house as well as certain shares of stock and the retirement benefits. Husband states that the interlocutory judgment entered on April 5, 1979, "with respect to the allocation of real property values and stock can only be explained by a finding that the parties did not live separate and apart."

The interlocutory judgment filed on April 8, 1980, provides in part: "The parties were married February 11, 1971 and separated July 10, 1978. During the period February 11, 1971 to July 10, 1978, the parties did not live separate and apart." Husband contends that the finding that parties did not live separate and apart "was erroneous in view of the overwhelming evidence of [wife's] intent to terminate the marriage to her husband during these periods of separation."

The parties were married on February 11, 1971. On April 12, 1975, wife moved out of the house on Lathrop Drive, Stanford, in which she had lived with husband and moved into an apartment on Kingsley Street in Palo Alto. On April 25, 1975, wife filed a petition for dissolution of marriage. The April 25, 1975, action was not dismissed by wife until after she filed the present action for dissolution in July of 1978.

Wife testified although she filed the petition for dissolution in April of 1975, she did not want a divorce. Wife stated she simply wanted to work out the parties' differences. Wife rented the apartment in Palo Alto because it was close to the house on Lathrop and she wanted to remain near her husband. The parties' sexual relationship continued after wife moved out of the house on Lathrop. The parties also saw a marriage counselor during this period.

Wife went to Puerto Vallarta, Mexico, in June of 1975. Husband went to Bogota, Colombia, in the summer of 1975 to "give some lectures." On his way to Bogota and on his return trip, husband spent a few days with wife in Puerto Vallarta. Husband testified that he and wife agreed that upon her return from Puerto Vallarta in August or September of 1975 they would dissolve the marriage.

Husband was invited to lecture in Iran and asked wife if she would like to accompany him. Husband and wife left for the Middle East in September of 1975 and were gone for five or six weeks. Upon their return from the Middle East, wife continued to live in the apartment on Kingsley. In December of 1975 husband and wife went to Puerto Vallarta for Christmas.

Wife returned to Puerto Vallarta in June of 1976 and remained there for the summer. In September of 1976 husband met wife in Los Angeles and drove with her to the Palo Alto area. Shortly thereafter husband went on a three-week trip, and when he returned in the first week of October, "I found that completely to my surprise [wife] and her two daughters had moved back into the house."

Husband was on sabbatical leave from Stanford University from June of 1977 until July of 1978. Since husband planned to travel around the world during his leave, the parties decided to rent the house on Lathrop and that wife would live in Puerto Vallarta where husband would visit her. During this year, husband spent approximately 14 weeks in Puerto Vallarta. Husband testified that he did not believe he and wife were separated during his sabbatical because of any marriage problems. During this period he received warm and loving letters from wife. Husband subsequently learned that during this year, wife was writing letters to her mother indicating that there were problems with the marriage. Wife also testified that during this year, she wrote to her mother and stated that she was considering terminating the marriage. However, wife testified that she was merely thinking about the possibility of ending the marriage, and "it depended on how he acted toward me." Wife returned from Mexico in July of 1978 and removed her belongings from the house on Lathrop.

What little law defines separation under Civil Code section 5118 holds that 'living separate and apart refers to "that condition when spouses have come to a parting of the ways with no present intention of resuming marital relations." That husband and wife may live in separate residences is not determinative. The question is whether the parties' conduct evidences a complete and final break in the marital relationship." (In re Marriage of Baragry (1977) 73 Cal.App.3d 444, 448, 140 Cal.Rptr. 779.)

In Baragry, husband moved out of the family residence and took an apartment with his girl friend. Husband ate frequently at the family residence and took his family on trips. Husband went with wife to Sun Valley for a week without the children. Husband attended social functions with wife and sent gifts and cards to her on holidays. The parties continued to file joint income tax returns and husband maintained his voting registration at the family residence. This arrangement continued for four years, although remaining nonsexual. (Id., at p. 447, 140 Cal.Rptr. 779.)

The court in Baragry found that the parties' conduct did not demonstrate "a complete and final break in the marital relationship. Here the only evidence of such a break is the absence of an active sexual relationship between the parties and husband's cohabitation elsewhere with a girlfriend. In our view such evidence is not tantamount to legal separation." (Id., at p. 448, 140 Cal.Rptr. 779.)

The finding of the trial court that during "the period February 11, 1971 to July 10, 1978, the parties did not live separate and apart" is supported by substantial evidence. We will first consider the time period between April of 1975 and September of 1976. Admittedly wife moved out of the family residence in April of 1975 and filed a petition for dissolution. However, she took absolutely no further legal action for over three years. The parties continued their sexual relationship and attempted to resolve their marital differences with the aid of a marriage counselor. Husband joined wife in Mexico in the summer of 1975 and traveled with her to the Middle East in September and October of 1975. The parties spent Christmas together in 1975 in Mexico. They again traveled together in September of 1976. When husband returned from a trip in the first week of October of 1976, wife had moved back into the family residence.

We do not feel that such evidence establishes as a matter of law that the parties had come to a parting of the ways with no present intention of resuming marital relations. Rather, the parties' conduct would appear to be an attempt to effect a reconciliation on an international scale and certainly does not reflect a complete and final break in the marital relationship.

The only evidence that would support a finding that the parties were living separate and apart during the time period between June of 1977 and July of 1978 was the testimony regarding the letters wife wrote to her mother. However, wife testified she was merely considering the possibility of terminating the marriage during this period. The evidence clearly shows that the parties were living apart during this period because husband was on a sabbatical leave and was required to do a great deal of traveling. "Living separate and apart ... does not apply to a case where a man and wife are residing temporarily in different places due to economic or social reasons, ...” (Makeig v. United Security Bk. & T. Co. (1931) 112 Cal.App. 138, 143, 296 P. 673.)

Valuation of Leasehold and House

Husband concedes that there is a community interest in the leasehold and house. He contends, however, that the trial court incorrectly calculated the community interest. We agree.

The pertinent facts are as follows: in 1962, husband paid to Stanford University $6,300 for an 80-year lease of a parcel of property owned by Stanford. That same year, he had a house constructed on the property which he financed by a $2,000 cash payment and the proceeds from a $30,000 loan from Stanford. Thus, the cost of the property in 1962 was $38,300. By the time the parties were married in February 1971, husband had reduced the loan by $7,000, so the outstanding balance was $23,000. During the marriage, payments from community funds further reduced the principal due on the loan by $9,200. Between the time of separation in July 1978 and trial in February 1979, husband paid $655 on the principal.

The trial court found the fair market value (FMV) of the house and leasehold interest was $65,000 at the time of the marriage in February 1971 and $182,500 at the time of trial. It computed the appreciation as $136,700, the difference between the equity at the time of trial ($182,500 less a loan balance of $13,800 equals $168,700) and the equity in 1971 ($65,000 less the $23,000 loan balance equals $42,000). It then allocated $77,632 as the community interest based on the ratio of separate property loan payments ($7,000 or 42.21 percent) to the community property loan payments ($9,200 or 56.79 percent). This was clearly in error.

“Where community funds are used to make payments on property purchased by one of the spouses before marriage ‘the rule developed through decisions in California gives to the community a pro tanto community property interest in such property in the ratio that the payments on the purchase price with community funds bear to the payments made with separate funds.’” (In re Marriage of Moore (1980) 28 Cal.3d 366, 371–372, 168 Cal.Rptr. 662, 618 P.2d 208.) In clarifying the application of the rule, the Supreme Court has recently reaffirmed a formula articulated in In re Marriage of Aufmuth (1979) 89 Cal.App.3d 446, 454–457, 152 Cal.Rptr. 668, and approved in In re Marriage of Lucas (1980) 27 Cal.3d 808, 816–817, 166 Cal.Rptr. 853, 614 P.2d 285, to determine the respective community and separate interests in the property. (In re Marriage of Moore, supra, 28 Cal.3d 366, 373–374, 168 Cal.Rptr. 662, 618 P.2d 208.) The formula gives recognition to the economic value of any loan proceeds contributed toward the purchase of the property. Where, as in Moore and in the case before us, the loan was extended before marriage and was based on separate assets, it is a separate property contribution. The Moore court also negated the inclusion of such expenses as loan interest and taxes in the computation, thereby clearing up any ambiguity created by the holding in Vieux v. Vieux (1926) 80 Cal.App. 222, 251 P. 640. (In re Marriage of Moore, supra, 28 Cal.3d at p. 371, 168 Cal.Rptr. 662, 618 P.2d 208.)

Under the Moore/Lucas/Aufmuth formula, husband’s separate property percentage interest is determined by crediting the separate property with the down payments and the full amount of the loan, less the amount by which the community property payments reduced the principal balance of the loan ($8,300 plus ($30,000 less $9,200) equals $29,100). This sum is divided by the purchase price for the separate property percentage share ($29,100 divided by $38,300 equals 75.98 percent). The community property percentage interest is found by dividing the community property payments on the loan principal by the purchase price ($9,200 divided by $38,300 equals 24.02 percent).

Husband proposes a modification of the above formula. He contends, and we agree, that he should have the benefit of approximately nine years of appreciation in the value of the property before the marriage in 1971. Husband argues that his separate property percentage interest should be 85.85 percent, based on the fair market value (FMV) of the property at the time of marriage ($65,000), rather than the purchase price in 1962 ($38,300).

Husband does not cite any cases directly on point, and we can find none. The Moore/Lucas/Aufmuth formula makes no provision for prenuptial appreciation on property and therefore offers no guidance. In Moore, however, the court stated: "This rule [giving the community a pro tanto community property interest for community contributions to separate property] has been commonly understood as excluding payments for interest and taxes. For example in Bare v. Bare [(1967) 256 Cal.App.2d 684, 690, 64 Cal.Rptr. 335], the Court of Appeal directed the trial court to determine the increase in equity in the house during marriage and the fair market value of it before and after the marriage, stating: 'the community is entitled to a minimum interest in the property represented by the ratio of the community investment to the total separate and community investment in the property. In the event the fair market value has increased disproportionately to the increase in equity the wife is entitled to participate in that increment in a similar proportion.' [Citations.]" (In re Marriage of Moore, supra, 28 Cal.3d at p. 372, 168 Cal.Rptr. 662, 618 P.2d 208, emphasis added.) The court further explained that "[a]mounts paid for interest, taxes and insurance do not contribute to the capital investment and are not considered part of it. A variety of expenses may be incurred in the maintenance of investment property, but such expenses are not considered in the valuation of the property except to the extent they may be relevant in determining its market value from which in turn the owners' equity is derived by subtracting the outstanding obligation." (Id., at p. 372, 168 Cal.Rptr. 662, 618 P.2d 208, emphasis added.)

It is clear that part of our confusion results from the use of the word "equity." It is defined as "5. [t]he amount or value of a property or properties above the total of liens or charges." (Webster's New Internat. Dict. (2d ed. 1935) p. 865.) The fair market value at or near the time of purchase is usually equivalent to the purchase price. In short, the appraised value equals the cost value. Such was the case in Moore, where the respondent wife had purchased the property eight months before the marriage, and thus no question of the fair market value of the home at the time of marriage was posed. (Id., at p. 370, 168 Cal.Rptr. 662, 618 P.2d 208.)

Where the separate property is owned for a considerable period before marriage, the increase in value in an inflationary market, such as we have had for the past several decades, is substantial. The fair market value at the time of marriage would usually be significantly greater than the purchase price, and this is true in the case before us. We think it is equitable to credit the separate property interest with this prenuptial appreciation. Although we are bound by the Moore/Lucas/Aufmuth formula, and adjustments must fit within that formula we can infer from the language in Moore and Bare that recognition of prenuptial appreciation in the separate property estate is appropriate. (We note that the trial court is given greater flexibility in determining the community and separate interests in property where one spouse has contributed "industry" rather than community capital to separate property.) As previously stated, the Moore court in explaining the pro tanto allocation rule noted the direction in Bare to the trial court "to determine the increase in equity in the house during marriage and the fair market value of it before and after the marriage.

We therefore compute the pro tanto community and separate property interests in the house and leasehold interest as follows:

Purchase price in 1962		$38,300.00	
Less community payments		9,200.00	(24.02%)
Separate property interest		$29,100.00	(75.98%)
FMV at time of trial		$182,500.00	
Less purchase price	$38,300.00		
Appreciation before marriage	26,700.00	65,000.00	
Appreciation during marriage		$117,500.00	
Separate property interest:			
Down payments	$8,300.00		
Loan payments			
Before marriage	7,000.00		
After separation	655.00		
		$15,955.00	
Appreciation before marriage		26,700.00	
75.98% of appreciation after marriage		89,276.50	
			$131,931.50
Community property interest:			
Loan payments	$9,200.00		
24.02% of appreciation after marriage	28,223.50		
		37,423.50	
Balance on loan at time of trial		13,145.00	
		$182,500.00	

Based upon the above computations, the trial court should have awarded husband a separate property interest of $131,931.50, plus one-half of the community interest, or $18,711.75, for a total of $150,643.25. The wife's one-half share of the community interest is $18,711.75, and husband, of course, is responsible for the balance due on the loan.

Stocks

Husband contends that the trial court "incorrectly allocated the shares of stock owned by the husband at the termination of the marriage between separate and community property." Prior to marriage husband owned a substantial amount of shares of stock. During the marriage husband had a checking account at Union Bank in his name. Husband's pay check was automatically deposited into his account at Union Bank. Husband also had an account in his name at Great Western Savings and Loan. During the time the parties were married, husband sold some of the shares of stock he owned and purchased others. Husband admits that some of the proceeds from the sale of certain shares of stock were deposited in his account at Union Bank. Husband concedes "that the Union Bank account was commingled to the extent that proceeds from stock were sometimes deposited therein along with his salary checks, the salary checks being community property." However, husband states "there were numerous stock sales and purchases which never went into the Union Bank Account." Husband asserts that the "[p]roceeds from the sale of stocks, dividends and interest were deposited generally in the Great

Western Savings and Loan account which he carried in his name and which he had owned prior to marriage" and the trial court incorrectly determined that the shares of stock that were purchased from funds from the Great Western Savings and Loan account were community property. Husband claims he adequately proved that certain shares of stock were purchased from funds in his Great Western Savings and Loan account or other separate property funds through the testimony of Kevin Loney, a certified public accountant. Husband states Loney "logged each stock transaction on a multipaged worksheet showing each sale of stock owned prior to the marriage, the disposition of the proceeds of the sale, and the shares of stock purchased with the disposition of the proceeds of the sale. He traced each transaction by brokerage statement and amount into an account either with the Great Western Savings and Loan, Union Bank, California Federal Savings and Loan, or a stockbrokerage account.... Mr. Loney's worksheet showed exactly where the money came from, where it went and how it was disposed of." Accordingly, husband contends since he adequately demonstrated that certain shares of stock were purchased from his separate property account at Great Western, these shares of stock remained his separate property although purchased during marriage.

Husband made the same contention in the trial court. The trial court in its memorandum of decision gave the following explanation for its rejection of husband's contention: "Husband also argues that since the funds from the Great Western account were his separate property that stock purchased from his fund are still his separate property and that the only question of identity centers around stocks purchased from funds in the [Union] Bank. Husband's arguments along this line would be persuasive if the Court were satisfied as to the truth of husband's contentions, but it is not. The Court cannot find that husband has carried his burden of proof in establishing what stocks were purchased from funds in the Great Western account. Husband could have avoided this difficulty by some rudimentary record keeping, but he kept virtually no records in this respect and it is far too late to reconstruct them with any degree of persuasive proof."

"Property acquired by purchase during a marriage is presumed to be community property, and the burden is on the spouse asserting its separate character to overcome the presumption. The presumption applies when a husband purchases property during the marriage with funds from an undisclosed or disputed source, such as an account or fund in which he has commingled his separate funds with community funds." (See v. See (1966) 64 Cal.2d 778, 783, 51 Cal.Rptr. 888, 415 P.2d 776.) "'The mere commingling of separate with community funds in a bank account does not destroy the character of the former if the amount thereof can be ascertained.' [Citations.] As the court in Patterson [Patterson v. Patterson (1966) 242 Cal.App.2d 333, 341, 51 Cal.Rptr. 339] stated: 'If the property, or the source of funds with which it is acquired, can be traced, its separate property character remains unchanged. But if separate and community property or funds are commingled in such a manner that it is impossible to trace the source of the property or funds, the whole will be treated as community property....'" (In re Marriage of Mix (1975) 14 Cal.3d 604, 611, 122 Cal.Rptr. 79, 536 P.2d 479.)

The presumption that all property acquired by either spouse during the marriage is community property may be overcome. (Id., at p. 611, 122 Cal.Rptr. 79, 536 P.2d 479.) "Generally speaking there are two methods of carrying the burden of showing property purchased during the marriage to be separate: (1) direct tracing to a separate property source or (2) proof that at the time of purchase all community income was exhausted by family expenses." (Estate of Murphy (1976) 15 Cal.3d 907, 918, 126 Cal.Rptr. 820, 544 P.2d 956.)

The issue for this court to determine is whether husband, as a matter of law, introduced sufficient evidence to directly trace the source of the funds used to acquire each item of disputed property to his separate property in accordance with the standards set out by the courts of this state. "[T]he burden of establishing a spouse's separate interest in presumptive community property is not simply that of presenting proof at the time of litigation but also one of keeping adequate records. 'The husband may protect his separate property by not commingling community and separate assets and income. Once he commingles, he assumes the burden of keeping records adequate to establish the balance of community income and expenditures at the time an asset is acquired with commingled property.'" (Estate of Murphy, supra, 15 Cal.3d at p. 919, 126 Cal.Rptr. 820, 544 P.2d 956.) "Evidence which merely establishes the availability of separate funds on particular dates without also showing any disposition of the funds is not sufficient proof of tracing to overcome the presumption in favor of community property." (Id., at p. 918, 126 Cal.Rptr. 820, 544 P.2d 956, see also In re Marriage of Mix, supra, 14 Cal.3d 604, 613–614, 122 Cal.Rptr. 79, 536 P.2d 479.)

In Mix, wife "introduced into evidence a schedule compiled by herself and her accountant from her records which itemized chronologically each source of separate funds, each expenditure for separate property purposes, and the balance of separate property funds remaining after each such expenditure." (Id., at p. 613, 122 Cal.Rptr. 79, 536 P.2d 479.) In Mix, wife conceded "that she was unable to support the schedule by correlating each itemized deposit and withdrawal on the schedule with an entry in a particular bank account due to the unavailability of various bank records as well as to the lack of such records of her own." (Id., at p. 614, 122 Cal.Rptr. 79, 536 P.2d 479.)

The California Supreme Court in Mix held "that the schedule by itself is wholly inadequate to meet the test prescribed by Hicks v. Hicks [1962] 211 Cal.App.2d 144, 27 Cal.Rptr. 307, and to support the trial court's finding that [wife] 'identified and traced' the separate property. However, the schedule was not the only evidence introduced by [wife] to effect the tracing. She personally testified that the schedule was a true and accurate record, that it accurately reflected the receipts and expenditures as accomplished through various bank accounts, although she could not in all instances correlate the items of the schedule with a particular bank account, ..." (Ibid.) The California Supreme Court in Mix noted that the "trial court evidently believed" wife and was "warranted in inferring from this evidence that the bank records if introduced would fully verify the schedule as supported by [wife's] testimony to the effect that 'separate funds ... continue[d] to be on deposit when a withdrawal [was] made ... for the purpose of purchasing specific property, and ... [that] the intention of the drawer ... [was] to withdraw such funds therefrom....'" (Ibid.)

There is substantial evidence in the instant case to support the trial court's finding that husband did not introduce sufficient evidence to directly trace the source of the funds used to acquire each item of disputed property to his separate property to overcome the presumption in favor of community property. Husband merely established that at most times under consideration he had sufficient separate funds available to make the purchases of the shares of stock in question. Husband has failed to show that he actually expended said funds for the purchases of the disputed shares of stock. Contrary to the assertion made by husband, the work sheet and testimony of Loney did not show "exactly where the money came from, where it went and how it was disposed of." In preparing the work sheet, Loney looked at husband's position at the end of the year. Loney looked to the stocks husband owned at the time of marriage, and then for each year the parties were married, calculated the difference between the amount of money needed to purchase the stock and the amount of money received from the sale of various stocks in a particular year. Loney assumed that the difference between these two figures was separate property of husband regardless of where the funds for the purchase were obtained or where the proceeds from the sale were deposited. For example, Loney stated that husband in 1973 purchased certain shares of stock for $11,991 and the funds to purchase the stocks came from Great Western and Union Bank. When husband sold stock the proceeds from the sale "would be spread among [husband's] accounts in some way." Loney admitted that if husband purchased stock from the account at Union Bank and then sold this particular stock and placed the funds in his account at Great Western, his work sheet would not show that the funds originally were from the account at Union Bank. In other words, husband did not attempt to deposit the proceeds from the sale of stock into the same account from which the funds were obtained to purchase the stock. In 1975, the amount of money expended to purchase stock exceeded the amount of money received from the sale of stock or dividends in the amount of $1,334. In 1976, husband purchased stock in an amount which exceeded the proceeds realized from the sale of stock in that year in the amount of $1,500. These sums did not come from separate property funds.

The work sheet was inadequate by itself to overcome the presumption in favor of community property. Husband was unable to support the schedule by correlating each purchase and sale on the schedule with an entry in a particular bank account. Husband apparently concedes that the funds in the Union Bank and the stock purchased from this account are community property because the funds were so commingled that they could not be traced. Husband did not keep adequate records to establish which stocks were purchased from his account at Great Western.

Loney looked at husband's position at the end of the year and was not always able to ascertain from which bank account husband obtained the funds to make his stock purchases or in which bank account he deposited the proceeds of a sale. Husband used some community property funds in his stock transactions, but the record does not establish in which transactions such funds were used. Husband indiscriminately deposited and withdrew funds from his bank accounts for his stock transactions. As noted above, Loney admitted that if husband purchased stock from the account at Union Bank and then sold this particular stock and placed the funds in his account at Great Western, his work sheet would not show that the funds originally were from the account at Union Bank. The proceeds from the sale of stock were not necessarily deposited in the account from which the funds were obtained to purchase that particular stock.

Since the records introduced by husband were prepared after the stock transactions took place and the work sheet did not correlate each stock transaction with an entry in a particular bank account and it is close to impossible to trace the source of most of the shares of stock because husband indiscriminately deposited and withdrew funds from his bank accounts for his stock transactions, the trial court did not err in determining that all of the stock purchased by husband during the marriage was community property on the ground that it could not determine which stocks were purchased from funds in the Great Western account. This determination was a question of fact for the trial court and its finding on the issue is supported by substantial evidence.

Husband also states that certain shares of stock were acquired prior to marriage and certain other shares of stock were inherited by him "and were not traded." Husband contends that these shares of stock were his separate property. In regard to this contention husband states: "the evidence is clear that of the shares he owned prior to marriage, he never dealt with the following: 200 shares of American Telephone and Telegraph, 125 shares of Northern Illinois Gas Company, now known as Nicor, 240 shares of J. P. Morgan, 225 shares of Suburban Propane, and 100 shares of Stanford Bank, now Union Bank Corp. In addition, he had a small amount of American Telephone which was converted by that company into preferred American Telephone and was never traded by him. 33 shares of Union Electric Corporation, 10 shares of First Union Bank Corporation, 370 shares of Dividend Shares, 37 shares of Anchor Income Fund, 35 shares of Wellington Fund and an $8,000 American Financial Debenture were inherited by him and were not traded. All of these shares should have been held to be separate property. The Trial Court, admittedly faced with a large stock portfolio, only found that the American Telephone and Telegraph, Morgan Guaranty Trust, Northern Illinois Gas Company and Suburban Propane Gas were the husband's separate property."

Husband testified at trial that four of the stocks he had at the time of trial, he had acquired prior to marriage and had not engaged in any transactions involving these stocks during the marriage. Husband testified that these four stocks were American Telephone and Telegraph, J. P. Morgan, Northern Illinois Gas (now known as Nicor) and Suburban Propane Gas. Husband also contends that he owned shares of Union Bank Corporation prior to marriage and purchased additional shares after marriage. Appellant asserts that the shares of Union Bank Corporation that he acquired prior to the marriage are his separate property. The trial court did not err in determining that the shares of Union Bank Corporation were community property. Husband did not produce any adequate documentation which showed the number of shares he acquired prior to the marriage. In regard to the shares of stock husband contends that he received from his parents during the marriage, husband failed to introduce any evidence that in fact he received these stocks from his parents either as an inheritance or a gift.

Husband contends that a significant portion of the shares of stock that the trial court determined to be community property "was bought on a margin account on which, at the beginning of January, 1979, $38,220 was owed...." In regard to the margin account, husband states the "trial court demonstrated its lack of knowledge of what constitutes a margin account because it did not take into allowance the brokerage firm's debit of $38,000 for the account. This corresponded to the market value of a substantial amount of the stock held at the time of the trial. Later on, these were sold to wipe out the margin account and to avoid the currently very high interest rates. That $38,000 actually should have been a debit to the appellant." Husband did testify that at the time of trial (Jan. 10, 1979) his "current" obligation of his margin account was about $38,000. However, husband failed to introduce any evidence that his obligation on the margin account at the time of trial was on account of any of the stocks that the trial court determined to be community property of the parties. Husband had the opportunity to produce evidence on this issue at trial, and since he failed to produce evidence regarding the transactions that resulted in a $38,000 obligation on his margin account, he cannot now contend that the trial court should have considered the amount owing on his margin account in dividing the shares of stock.

Attorney's Fees

We now turn to husband's argument that attorney's fees were improperly awarded to wife. Husband does not challenge the jurisdiction of the court in a dissolution proceeding to award attorney's fees. (See Civ.Code, § 4370; In re Marriage of Holmgren (1976) 60 Cal.App.3d 869, 873, 130 Cal.Rptr. 440.) But he does claim that the award of $2,500 for attorney's fees to wife was excessive. He further argues that there is no evidence in the record that wife was in need of attorney's fees given the fact that wife was awarded "community property assets in the approximate amount of $200,000." Husband in making this latter argument states that there is evidence in the record that wife had sufficient funds with which to engage an attorney separate and apart from her share of the community property assets.

Civil Code section 4370, subdivision (a), authorizes the court in a dissolution proceeding to order the husband or wife "to pay such amount as may be reasonably necessary for the cost of maintaining or defending the proceeding and for attorneys' fees; ..." The need of a spouse for an award of attorney's fees and the amount of that award are matters addressed to the sound discretion of the trial court. (6 Witkin, Summary of Cal.Law (8th ed. 1974) Husband and Wife, § 146, pp. 5013–5014.) The exercise of this discretion will not be disturbed on appeal "without a clear showing of abuse." (In re Marriage of Holmgren, supra, at p. 873, 130 Cal.Rptr. 440.)

The fact that wife was awarded a substantial amount of community property assets in the instant case does not require a finding that the trial court abused its discretion in awarding her attorney's fees. A similar contention was considered in In re Marriage of Borson (1974) 37 Cal.App.3d 632, 639, 112 Cal.Rptr. 432, and rejected. (See 6 Witkin, Summary of Cal.Law (8th ed. 1974) Husband and Wife, § 141, pp. 5008–5009.)

Nor does it appear that the award of attorney's fees in the instant case was excessive. "'The question of the reasonableness of the order for attorney's fees is addressed to the sound discretion of the trial court [citations], and in the absence of a clear showing of abuse its determination will not be disturbed on appeal.'" (In re Marriage of Lopez (1974) 38 Cal.App.3d 93, 113, 113 Cal.Rptr. 58.)

Valuation Date for the Retirement Benefits

Husband contends that the date of separation rather than the date of trial is the appropriate valuation date for his retirement benefits. Husband reasons "[w]hile there is case law and code authority [Civil Code § 4800(a)] indicating that the assets shall be valued as nearly as practical to the time of trial, in the case of retirement assets which continued to increase in value because of the employment of the spouse during the period of separation, utilizing the date of trial as the valuation date would give an inaccurate allocation." The interlocutory judgment filed on April 8, 1980, provides in part: "[Husband's] TIAA/CREF pensions/annuities are to be divided in accordance with the court's prior interlocutory decree dividing the parties' community property. All contributions during the marriage and the accumulations thereon are the community property of both parties and are to be divided equally between the parties." Husband states "the judgment does not specify the date on which the retirement benefits are to be valued. To that extent the judgment is uncertain and ambiguous."

Contrary to husband's contention retirement benefits are not valued at the time of separation. The California Supreme Court has indicated that there are two basic solutions to the problem of the proper method for the division of pension or retirement benefits: "first, a determination by the trial court of the present value of the rights or benefits adjudged to be marital property and an equal division or adjustment of the same [citations]; and second, 'if the court concludes that because of uncertainties affecting the vesting or maturation of [such] rights ... it should not attempt to divide the present value ... it can instead award each spouse an appropriate portion of each ... payment as it is paid.'" (In re Marriage of Skaden (1977) 19 Cal.3d 679, 688, 139 Cal.Rptr. 615, 566 P.2d 249, emphasis added.) If the nonemployee spouse chooses to wait to receive the retirement benefits until the employee spouse actually retires, the "nonemployee may thereby ensure some protection for the future and may be able to share in the increased value of the pension plan." (In re Marriage of Gillmore (1981) 29 Cal.3d 418, 428.) Accordingly, the appropriate date of valuation of retirement or pension benefits is the date of trial or the date of payment of benefits. The date of valuation of retirement or pension benefits depends on when the community interest in such benefits is divided. The community interest in retirement benefits is that portion of the retirement benefits or pension attributable to employment from the date of marriage until separation.

The judgment in 1 Civil No. 47922 is reversed insofar as it determines the respective interest of the parties in the house and leasehold and divides the community property. The judgment in 1 Civil No. 51438 is affirmed. Each party shall pay his or her own attorney's fees and costs on appeal.

WHITE, P. J., and FEINBERG, J., concur.

Hearing denied; MOSK, J., dissenting.

Notes and Problems

1. Do you agree with the *Marsden* holding with respect to the house and leasehold? The valuation of the stocks? The valuation of the retirement benefits?

2. How, if at all, should a declining housing market and interest rates affect the outcome of the *Marsden* valuations?

§ 9.40 Financial Indicators

Several ratios are commonly used to better describe the economic health or financial soundness of a business entity with respect to its liquidity, activity, leverageability and profitability. No rocket science here; merely long division.

§ *9.41 Liquidity*

Liquidity measures the ability of a company to stay solvent by meeting its short term obligations. The most common components of liquidity are represented by the following ratios:

A. The Current Ratio

The Current Ratio, often referred to as "the working capital ratio," is obtained by simply dividing Current Assets by Current Liabilities. It is a common measure of a company's liquidity and ability to meet the demands of its creditors:

$$CR = \frac{\text{Current Assets}}{\text{Current Liabilities}}$$

A "current asset" on the balance sheet is one that can be sold or otherwise used up in a relatively short period of time, normally within one year, or one operating cycle, whichever is longer. Typically, current assets include cash, cash equivalents, various liquid bank accounts, accounts receivable, prepaid liabilities, short term investments and inventory.

A "Current Liability" is basically a company obligation that is due within one year. Current liabilities normally include accounts payable, bills, short term debts and various other accrued liabilities. The higher the Current Ratio, the better the company's financial stability and liquidity.

For example, if Bilco, Inc. Has current assets in the sum of \$50,000 and if its current liabilities happen to equal \$25,000 then its Current Ratio would be:

$$CR = \frac{\$50{,}000}{\$25{,}000} = 2.0$$

which constitutes a ratio of 2:1. This particular ratio happens to equal the ratio that many look to as a general rule of thumb with respect to short term financial health. Simply stated, it means that you have two readily available dollars in the bush for every dollar that you may shortly be expected to pay. The ratio indicating health or sickness for any particular business will vary from industry to industry. But you might think twice before investing in a business with a Current Ratio of less than 1:1.

B. Quick Ratio

Sometimes termed the "Acid Test Ratio" or "Liquid Ratio," the so-called Quick Ratio measures the ability of a company to quickly, if not immediately discharge all of its short term debts. Notice that the equation basically equals the Current Ratio but excludes inventory from the numerator, thereby guaranteeing that $QR \leq CR$. The general rule of thumb suggests that a company should have a Quick Ratio of at least 1:1; but, as is the case with the Current Ratio, the recommended Quick Ratio will vary among industries.

$$QR = \frac{\text{Current Assets - Inventory}}{\text{Current Liabilities}}$$

C. Net Working Capital to Total Assets Ratio

This particular ratio is claimed by many to provide a good indicator of whether or not a company is on the road to bankruptcy. It takes a different perspective from the Quick Ratio by focusing on current assets as compared to total assets. This ratio will normally be greater than 0 for most firms, but there are circumstances where it might be negative. This, of course, would occur where the company's Current Liabilities exceed its Current Assets.

$$\text{NWC to TAR} = \frac{\text{Current Assets - Current Liabilities}}{\text{Total Assets}}$$

Example 9-9: Examine the Endrun Inc. Balance Sheet (See Example 9-8); as of January 1, 2011, calculate the: (A) Current Ratio, (B) Quick Ratio and (C) Net Working Capital to Total Assets Ratio.

Solution: (A) Current Ratio (CR)

$$CR = \frac{\text{Current Assets}}{\text{Current Liabilities}}$$

$$CR = \frac{\$36{,}084}{\$17{,}886}$$

$$CR = 2.0174$$

(B) Quick Ratio (QR)

$$QR = \frac{\text{Current Assets - Inventory}}{\text{Current Liabilities}}$$

$$QR = \frac{\$36{,}084 - \$10{,}861}{\$17{,}886} = \frac{\$25{,}223}{\$17{,}886}$$

$$QR = 1.4102$$

(C) Net Working Capital to Total Assets Ratio (NWC to TAR)

$$\text{NWC to TAR} = \frac{\text{Current Assets - Current Liabilities}}{\text{Total Assets}}$$

$$\text{NWC to TAR} = \frac{\$36{,}084 - \$17{,}886}{\$41{,}158}$$

$$\text{NWC to TAR} = \frac{\$18{,}198}{\$41{,}158}$$

$$\text{NWC to TAR} = 0.4421$$

§ *9.42 Activity*

Activity measures the efficiency and ability of a company to generate sales as compared to its assets. In a roundabout way, it might be compared to the yield of a fixed asset.

A. Inventory Turnover Ratio

The Inventory Turnover Ratio provides a general view of the inventories of a company. In order to calculate it you should divide the cost of goods sold by its average inventory. The "Cost of Goods Sold" is an accounting term of art that encompasses all of the costs required to bring the "Goods" to their final form so they are ready for sale.

The Inventory Turnover Ratio essentially measures the number of times, on average that the inventory is sold during the subject period. Its purpose is to measure the liquidity of the inventory. A popular variation of the this ratio involves computing the average number of days required to sell off a given amount of inventory.

$$ITR = \frac{\text{Cost of Goods Sold}}{\text{Average Inventory}}$$

B. Fixed Asset Turnover Ratio

The Fixed Asset Turnover is the ratio of sales (From the Income Statement) to the value of the fixed assets (From the Balance Sheet). It indicates the efficiency with which the company is using its fixed assets to generate sales.

Generally, the higher the ratio, the better it is for the financial health of the company. Moreover, a high ratio indicates that the business uses less assets to generate more revenue.

$$\text{FATR} = \frac{\text{Sales}}{\text{Net Fixed Assets}}$$

C. Total Assets Turnover Ratio

The total assets turnover ratio measures the effective use of all assets in terms of sales, by comparing sales with net total assets. This ratio will always be equal to, or lower than, the Fixed Asset Turnover Ratio.

$$\text{TATR} = \frac{\text{Sales}}{\text{Total Assets}}$$

§ 9.43 *Leverageability*

The degree of the company's indebtedness affects its ability to meet both short and long term obligations. Its ability to borrow additional money is important and depends upon the degree to which it has already made debt a part of its assets.

A. Debt Ratio

The Debt Ratio is a simple basic measure that shows the proportion of company assets that is financed through debt. If the Ratio is less than 0.5, it may be said that most of the company's assets are financed through equity. If, however, the Ratio is greater than 0.5, most of its assets are financed through debt. Companies that have high Debt Ratios are said to be "highly leveraged." By the same token, such companies are not "highly liquid."

A business with a high debt ratio could be on shaky ground if creditors start to demand repayment of the debt. The higher the ratio, the greater risk that will be associated with the firm's operation. In addition, the high ratio may indicate a low borrowing capacity of a firm, which in turn will lower the firm's financial flexibility. On the other side of the coin, however, high liquidity often makes the company a takeover target. As is the case with all financial ratios, a company's debt ratio should be compared with its industry averages and other competing firms.

$$\text{Debt Ratio} = \frac{\text{Total Liabilities}}{\text{Total Assets}}$$

B. Debt to Equity Ratio

The debt-to-equity ratio, sometimes referred to simply as the "D/E," is a very commonly used financial ratio that provides the relative proportion of the shareholders' equity and debt. Because it is a prime tool for leveraging measurement, it is also known as "Risk" or just "Leverage." The two factors are often pulled from the company's balance sheet or statement of financial position. The ratio may, however, also be calculated by using market values for both, particularly if the corporation's debt and equity are publicly traded.

The D/E Ratio is also employed by the Internal Revenue Service when dealing with attempts to avoid federal corporate income taxes. When forming a new corporation, a high ratio might signal an attempt by the equity shareholders to convert their equity interest (which consists of non-depreciable assets) into a debt, which can then be repaid in the form of non-taxable payments as opposed to taxable dividends. If the shareholders get too cute (i.e., the D/E is too high), the Internal Revenue Service may treat the loan repayments as dividends and tax them accordingly.

$$\text{DTER} = \frac{\text{Long Term Debt}}{\text{Shareholders' Equity}}$$

C. Equity Multiplier Ratio

The equity multiplier ratio is used to measure a particular company's total assets as compared to its total stockholder's equity. This ratio provides a method by which investors can examine the level to which a company uses debt to finance its assets and operations. A high equity multiplier indicates a more highly-leveraged company. The equity multiplier, like other leverage ratios, can help investors determine whether a company is heading for financial problems due to an excessive debt load.

$$\text{EMR} = \frac{\text{Total Assets}}{\text{Total Equity}}$$

Example 9-10: Compare Endrun Inc.'s EMR at the start of 2011 to its EMR at the end of 21011 (See Example 9-8).

Solution: As of January 1, 2011:

$$\text{EMR} = \frac{\$41{,}158}{\$22{,}070} = 1.865$$

As of December 31, 2011:

$$\text{EMR} = \frac{\$32{,}692}{\$19{,}139} = 1.708$$

For the year 2011, Endrun's EMR deceased 0.157.

§ 9.44 *Profitability*

In the last analysis, the measure of returns and profits tells us how successful or unsuccessful a business happens to be. After all is said and done, profitability is what it is all about.

A. The Gross Profit Margin

Gross Margin, Gross Profit Margin and Gross Profit Rate are terms used to describe the same basic concept. It is the difference between the company's sales and its production costs, excluding overhead, payroll, taxation, and interest payments. The Gross Margin can be defined as the amount of contribution or increase to a business enterprise, after allowing for direct-fixed and direct-variable unit costs required to cover overheads (fixed commitments). It expresses a relationship between Gross Profits and Sales Revenue. It is in essence a measure of how well each dollar of a company's revenue is utilized to cover the costs of goods sold. It can be expressed in more specific terms by the following equation:

Gross Margin = Net Sales - Cost of goods sold + annual sales return

or it can also be expressed as the ratio of gross profit to sales revenue, usually in the form of a percentage:

$$\text{Gross Margin Percentage} = \frac{\text{Revenue - Cost of Goods Sold}}{\text{Revenue}}$$

The Cost of Sales (Cost of Goods Sold) includes variable costs and fixed costs that are directly linked to the sale of the subject goods, such as material costs, labor, supplier profit, shipping costs, etc. It does not include indirect fixed costs such as office expenses, rent or administrative costs. Generally, higher gross margins for a manufacturer reflect greater efficiency in converting raw materials into money.

For a retailer, the markup over wholesale will be the key factor in determining profits and efficiency. Larger gross margins are generally "good" for companies, with the exception of discount retailers; often, financing allows them to operate with tiny margins.

$$\text{Gross Profit Margin} = \frac{\text{Gross Profit}}{\text{Sales}}$$

B. The Net Profit Margin

The Net Profit Margin (sometimes referred to as "Profit Margin" or "Net Profit Ratio") is employed to measure a company's basic profitability. It is frequently confused with "Margin."

$$\text{NPM} = \frac{\text{Net Income (after taxes)}}{\text{Total Revenue}} \times 100$$

C. Return on Assets

The Return on Assets (ROA) percentage shows how profitable a company's assets are with respect to its ability to generate revenue. The ROA suggests what the company can do with the assets that it has at its disposal (i.e., how many dollars of earnings it can derive from each dollar of assets it controls). It is a useful number for comparing competing companies within a particular industry. The number will vary widely across different industries. The ROA also provides an indication of the capital intensity of the company.

$$\text{ROA} = \frac{\text{Net Income}}{\text{Total Assets}}$$

D. Return on Equity

The Return on Equity (ROE) measures the rate of return as compared to the ownership interest (i.e., shareholders' equity) of the common stock owners. It measures the company's efficiency with respect to generating profits. The ROE also demonstrates how well a company uses its investment funds to generate earnings growth. It is also used to compare companies in the same industry.

$$\text{ROE} = \frac{\text{Net Income}}{\text{Equity}}$$

E. Price/Earnings (P/E) Ratio

The Price to Earnings Ratio (also referred to as the "P/E" or simply the "Multiple") is a measure of the price paid for a share relative to its annual net income, or profit per share. It is one of the most important indicators used in deciding whether or not to purchase stock in a particular company.

$$\text{P/ER} = \frac{\text{Common Stock Price/Share}}{\text{Earning Per Share}}$$

F. Earnings Per Share Ratio

The Earnings Per Share Ratio is another tool to help determine the value of a given stock and whether or not it might be a prudent investment.

$$\text{EPS} = \frac{\text{Earnings for Common Shareholders}}{\text{Number of Common Shares Outstanding}}$$

G. Market to Book Ratio

The Market-to-Book Ratio, is the ratio of the current share price to the book value per share. It measures how much a particular company is worth at the present time as compared to the amount of capital invested by current and past shareholders.

$$\text{MTBR} = \frac{\text{Common Stock Price Per Share}}{\text{Common Stock Value Per Share}}$$

Notes and Problems

1. A fellow with a fine, blank look is aptly named Lloyd Blankfein, CEO of Goldman Sachs, who apparently could not foresee the 2008 collapse of the mortgage/housing industry. Which, if any, of the above ratios should have triggered suspicion that the home mortgage industry was headed for collapse? What is the probability that, as claimed, Mr. Blanklook could not predict that mortgages nationally advertised as follows would fail to remain solvent?

Need money?
No job?
No credit?
No assets?
No life?
No problem!

2. Rumor has it that there exists an extremely important ratio that only CEOs consider and know about. Call it the "Stashed Cash Per Year Ratio." It is also known as the Skilling-Maddoff Ratio. It is calculated as follows:

$$\text{SCPYR} = \frac{\text{Money Stashed in Swiss Bank Account}}{\text{Length of Prison Term (in Years)}}$$

3. In less litigious times, accountants and financial professionals enjoyed limited immunity for their bad advice and misdeeds. See *Ultramares v. Touche Niven & Co.*, 255 N.Y. 170 (1931), wherein Judge Cardozo famously said no to: "Liability in an indeterminate amount for an indeterminate time to an indeterminate class." Today, of course, those professionals face both civil and criminal liability for their misbehavior and unfounded prognostications.

§ 9.45 *Valuing The Business*

All of the previous financial indicators contribute to determining the ultimate value of a business. After all is said and done, it is the *value* that represents the summarized end product that we lawyers (both transactional and litigating) are most concerned with. Buy-sell agreements, fraudulent misrepresentation, SEC matters, taxation wars, divorce battles, and even child support fights may all turn debates over the value of a business enterprise.

While the preceding equations and ratios may pretend to provide mathematical precision, in truth, no such formula exists. For publically held companies, stock prices on an open exchange may provide a measure, but even that yardstick is subject to variation and speculation. Therefore, unless a willing buyer is ready and willing to pay a fixed price to a willing seller, the value of the enterprise will always be subject to debate. No one can predict a business' future with certainty.

Arguably, it is even more difficult in our current unpredictable global economy to determine and set a company's value than it was a few decades ago. Not so long ago, business assets were more tangible; there were bricks, buildings, plants, offices, and inventory. Today, there is much more in the way of smoke, mirrors, impressive web sites and *face-looks*. Despite the variations and problems, however, there do exist some recognized, persuasive methods for attempting to estimate a business' value. The starting point and simplest measure is the "Book" value, which is derived from the company's balance sheet (See § 9.21). Recall the equation:

Net Worth = Assets - Liabilities

A. Balance Sheet Based Valuations

The term "Net Worth" can be substituted for "Owner's Equity," or for purposes of our present discussion, "Book Value." It is a down-and-dirty quick measure and one that is often modified.

Book Value = Assets - Liabilities

For a corporation, a common measure involves the "Book Value Per Share," which is employed to roughly determine the value of a single share of a corporation's stock:

Book Value / Share = (Assets - Liabilities) / Number of Shares Outstanding

The concept seems to be predicated upon the mathematical notion that "the whole equals the sum of its parts." This in turn implies that the value of a company is basically equal to sum of the values of the individual assets (less the liabilities). Think of this from a raider's perspective. How much money can you get if you were to sell off the assets, one by one, Gordon Gekko style? In truth, a company is equal to more than just sum of its parts. For a successful company, the assets working together (not separately) produce a product or service capable of generating revenue.

A distinction might be drawn between "Book Value" and "Net Tangible Book Value." The latter term is designed to measure the book value less intangible assets such as goodwill and intellectual property. Normally intangible assets are very difficult to value and they are often worth little or nothing when separated from the remainder of the company.

Net Tangible Book Value = Assets - Intangible Assets - Liabilities

A different modification designed to keep things simple but less restrained by the balance sheet is the "Adjusted Book Value." Instead of valuing the assets at their book value, this method individually measures the current market value of the assets. It also tends to ignore the fact that the value of the business may very well be greater than the value of the sum of its assets.

The "Net Asset Value," unlike the preceding valuation methods, treats a business as a living breathing enterprise, a "going concern," whose goodwill and history have meaning. Under this method of accounting, goodwill is included as an intangible and measurable asset of the business. The "Market To Book Ratio" (MTBR) is designed to measure the difference between the Book Value and the Net Asset Value (See § 9.44) :

MTBR = Market Price Per Share / Book Value Per Share

For privately owned companies, there is no market price. But that doesn't deter attempts to calculate a Net Asset Value. But by using the MTBR of a "comparable" publically traded company, and applying it to the privately owned enterprise, the private enterprise can achieve the same result.

Example 9-11: Macebook Inc., a privately owned, closely held corporation has a per share Book Value of $20.00; there are 10,000 shares of Macebook outstanding. A comparable publically traded corporation, Tastebook, Inc., has a Net Asset Value of $10,000,000, with one million shares of stock outstanding. The Book Value of Tastebook is $,8,000,000. What is the Net Asset Value of Macebook?

Solution: Tastebook's Market Price Per Share is $10.00; its Book Value Per Share would be $8.00 price per When calculating our M/B Ration, it makes no difference whether we deal with the Net Asset Value or the "Per Share" value. For Tastebook, the ratio is going to be 1.25 regardless.

MTBR = 10 / 8 = 1.25

MTBR = 10,000,000 / 8,000,000 = 1.25

Macebook's Book Value = $20 x 10,000 = $200,000

Next, apply the MTBR to Macebook's Book Value:

MTBR (Tastebook) = Macebook's MPPS / Macebook's BVPS

1.25 = Macebook's Market Price Per Share / 20.00

1.25 x 20.00 = Macebook's Market Price Per Share = $25.00

Macebook's Net Asset Value = 10,000 x $25.00 = $250,000.00

B. Income Statement Based Valuations

The income statement, which traces the movement of dollars in and out of the company, tends to present a more dynamic picture of the business than the frozen point-in-time balance sheet. Using an income statement as a basis for determining the value of the company can be a tricky affair, requiring some rather subjective elements. It is predicated upon the notion that a company's power to generate revenue is more important than the sum total of its individual assets. The most common and widely recognized method is referred to the "Capitalization of Earnings" calculation. If V equals the value that we are seeking, and E represents the earnings per share (averaged or "normalized" over a period of several years), with R equaling the company's "Capitalization Rate," then:

$$V = E / R$$

The normalized period, often 5 years, can be subject to debate. The capitalization rate can also be somewhat subjective even though the following formula is often employed (See § 9.44) :

$$R = \frac{1}{PE}$$

The "P/E," the price to earnings ratio is a very common device used to compare the price of a share of stock to the amount of revenue that it earns within a given period of time. For publicly traded corporations, the ratio can be easily determined from widely published sources including the Wall Street Journal.

Despite its common use, there are a number of problems and concerns with this oversimplified methodology. Many of the problems relate to the definition of "earnings," which can assume many forms depending upon the layout and structure of the income statement.

C. Cash-Flow Based Valuations

From an investor's perspective, dividends and cash flow are often the most important elements in the valuation process. Predicting future dividends and cash flow thus become paramount. The "Dividend Discount Method" of evaluation attempts to combine information about the company's dividend history with attempts to predict its future growth pattern. The problem with this type of method resides in the simple fact that many very successful companies pay little or no dividends. For a small company, corporate and income taxation rarely favor the issuance of dividends.

A far more popular flow-based evaluation is the "Discounted Cash Flow Method." This methodology presents a multi-step process that commences with an attempt to forecast the company's cash flow for the period that the investor plans to own the stock. Next comes an attempt to evaluate the company at the end of the targeted holding period. Discount rates are then factored into the process to arrive at a present value of the company.

In re Watt and Shand

452 Pa. 287, 304 A.2d 694 (1973)

EAGEN, Justice.

Appellant, Laura Watt O'Connor, is the owner of 813 shares of common stock of Watt & Shand, a Pennsylvania corporation, which operates department stores in Lancaster. At the corporation's annual meeting on April 24, 1969, the majority shareholders amended the articles of incorporation to eliminate cumulative voting. Mrs. O'Connor dissented from the plan and in accordance with the Act of May 5, 1933, P.L. 364, s 515, as amended, 15 P.S. s 1515 demanded the fair value payment of her shares of stock.

Subsection A of s 515 gives the following summary of the purposes of the statute: 'If any shareholder of a business corporation objects to any proposed plan of action of such corporation authorized under any section of this act, and such section provides that such shareholders shall be entitled to the rights and remedies of dissenting shareholders, such shareholders shall be entitled to the following rights and remedies.' Dissenter's rights were instantly provided by the Act of July 20, 1968, P.L. 459 which provides in part: 'If any amendment to the articles shall ... eliminate cumulative voting for directors ... the holder of any outstanding shares the ... cumulative voting rights of which are so cancelled ... shall be entitled to the rights and remedies of dissenting shareholders....' Subsection B of s 515 provides in part: 'If any shareholder of a business corporation shall file with such corporation, prior to the commencement of the voting by shareholders upon the plan at the meeting of shareholders at which a plan is submitted to a vote, a written objection to such plan, and shall not vote in favor thereof, and such shareholder, within twenty days after the date on which the vote approving the plan was taken, shall also make written demand on the corporation ... for the payment of the fair value of his shares, such corporation shall pay to such shareholder the fair value of his shares as of the day prior to the date on which the vote was taken without regard to any depreciation or appreciation thereof in consequence of the plan....' Thus, what is presently at issue is the value of these shares on April 23, 1969. It is unquestioned that Mrs. O'Connor pursued her claim with procedural precision.

Watt & Shand offered to pay appellant $93.46 per share which was refused. The corporation then petitioned the Court of Common Pleas of Lancaster County to determine the fair value of appellant's stock. Exercising its prerogative under the statute, the trial court appointed an appraiser to receive evidence and make a recommendation concerning the fair value of the stock. At the close of hearings, the appraiser recommended appellant be paid $84.56 per share. The trial court dismissed appellant's exceptions and adopted the findings of the appraiser. On appeal we vacated the judgment and remanded for a new determination of fair value because the court's acceptance of the appraiser's recommendation was not supported by competent and substantial evidence, which, incidentally, is the compass of our review.

On remand, counsel for both parties agreed the new determination of fair value should be decided upon the record previously made. On May 23, 1972, the trial court, without appointing an appraiser, found the fair value to be $102.15 per share and entered judgment for appellant in the amount of $83,047.95, without interest, but with costs (other than all fees and expenses of counsel and expert witnesses) to be paid equally by the parties.

Two contentions are advanced by this appeal: (1) the trial court again failed to make a proper determination of fair value, and (2) error was committed in refusing to award interest and in failing to impose all costs on the corporation. For the reasons hereinafter discussed, we remand the case to the trial court for further consideration.

Preliminarily it should be observed that while the term 'fair value' is hardly self-executing in its clarity, the object of an appraisal proceeding is to determine the value of the dissenter's shares on a going concern basis. Lowry v. General Waterworks Corp., 26 Pa.D. & C.2d 154, 159 (1961); Austin v. City Stores Co. (No. 1), 89 Pa.D. & C. 57 (1953). In a Note, Shareholders: Appraisal Rights: Compensation to Shareholders Dissenting From Mergers and Consolidations, 40 Calif.L.Rev. 140, 142—43 (1952) it was observed:'Although courts have long been faced with the problem in many differing situations they have been unable to set out clearly what value means, practically or philosophically.'Some difficulty may be due to a conceptual failure to understand the nature of the share to be valued. A shareholder does not own a pro rata share in the corporation's assets except upon dissolution; other than his right to participate in management, his only right of value is his right to share in earnings as distributed. Essentially what is being valued is a right to share the future returns of a going concern. In determining what figure represents this true or intrinsic value, consideration must be given to all factors and elements which reasonably might enter into the fixing of value. As the late Judge Allessandroni wrote in Austin v. City Stores Co. :

> 'Some of the factors that must be considered in rendering an intelligent decision are: Asset value; market value; market prices of comparable companies; market price and earnings ratio; management and its policies; earnings; dividends; valuation of assets; reserves for various contingencies; tax liabilities; future earnings; predictions of future business events and etc. The list seems interminable and yet all factors must be considered and given their proper weight in order that a just result might be attained.'

In an attempt to render the unwieldy, wieldable, courts have distilled all of these factors into three principal methods of valuation which have been variably used, commonly in combination, in the actual judicial determination of intrinsic value: (1) net asset value; (2) actual market value; and (3) investment value. This case presents an instance where a certain method of valuation has no applicability whatever. Watt & Shand is a closely-held family corporation having unlisted stock and therefore, no public market. Shares are sold too infrequently for market value to play any part in these proceedings, a fact correctly realized by the lower court.

Net Asset Value is the share which the stock represents in the value of the net assets of the corporation. Such assets include every kind of property and value, whether realty or personalty, tangible or intangible, including goodwill and the corporation's value as a going concern. Investment Value is an estimate of present worth in light of past, present and prospective financial records of the company and is obtained by capitalizing earnings. There are two basic steps in the capitalization process: calculation of a representative annual earnings figure, and choice of a capitalization ratio which reflects the stability and predictability of earnings of the particular corporation. Market Value refers to the price at which the stock was selling on the market prior to the action which is objected to, disregarding any change in price due to the action.

Upon remand the trial judge undertook to conduct the appraisal as was his right. His opinion indicates that while he studied closely the valuations of the respective experts who testified, he did not choose to adopt either conclusion, opting instead to work out his own valuation. By taking the annual earnings of $11.35, multiplied by 10, or $113.50 less a 10% Discount, the trial judge reached a fair value figure of $102.15. Appellant's expert, Fitzpatrick, found the fair value of each share to be $145.00. He arrived at this figure by multiplying the diluted earnings per share for the year ended January, 1969 of $11.35 by 15 which obtained a figure of $170.35. Then, because the stock was not freely traded, he discounted it to $145 per share (a discount of approximately 14.7%).Watt & Shand's expert, Castle, multiplied $11.35 by 10 and from the resulting $113.50 applied a 25% Discount, thus reaching a final valuation of $85.13 per share.

Appellant asserts the court's price earnings ratio is incorrect, claiming that the earnings per share for the year ended January 1969 were $12.93 and not $11.35. Appellant apparently does not realize that use of $11.35 inures to her ultimate benefit, else she would not press this claim. The experts for both sides used the $11.35 earnings figure in their computations. This diluted earnings figure was obtained by subtracting $.55 (a debenture reserve) from $11.90, the stated earnings figure. The exhibit entitled Watt & Shand and Hager & Bro., Inc. Pro Forma Combined Statement of Income for the Year Ended January 25, 1969, does indeed list earnings per share as $12.93. It should be pointed out that on May 31, 1968, Watt & Shand acquired all outstanding stock of Hager & Bro., a rival Lancaster department store.

The lower court used but a single year [the year ended January 1969] to obtain the average earnings of Watt & Shand. In our judgment this was error, but since neither party complains of the action, we will not review it. However, we do point out that in this kind of proceeding '[i]t is best to average earnings over several years to avoid undue emphasis on one exceptionally good or bad recent year.' Note, Valuation of Dissenter's Stock under Appraisal Statutes, 79 Harv.L.Rev. 1453, 1464 (1966). The purpose of any average is to balance off extraordinary profits against extraordinary losses, in order that a hypothetical figure of what might be considered the ordinary profit or inherent earning capacity of the business may become discernible.

In his treatise, Financial Policy of Corporations, Prof. Dewing notes that a period of five years is frequently used, although he goes on to explain at p. 376 that: "There is no mystical significance in taking a ten-, or a five-, or a three-year average of earnings to get a more accurate picture of the fundamental earning capacity of the business than is given by the earnings of a single year; and there is no reason why one period should be used for all businesses. It must, however, to use the words of a legal decision, be broad enough to '... cover a sufficient period to show the settled condition of things.'" See also Bonbright, Valuation of Property, p. 253, where the author states: 'There is still an agreement among writers that a capitalization of average earnings over a period of 3 to 5 years (occasionally 10 to 15 years) is preferable, in most cases, to a capitalization of the earnings of any single year.' While our research disclosed no cases on point in this jurisdiction, the Delaware Chancery Court has held it error to use only the two years prior to appraisal. Adams v. R. C. Williams & Co., 39 Del.Ch. 61 (1960), preferring to average at least the five previous years. In Sporborg v. City Specialty Stores, 35 Del.Ch. 560 (1957), the appraiser considered only the fiscal year immediately preceding the effective date of the merger in arriving at the earnings value to be capitalized. The court found this to be error, saying: 'True, the upward trend of this Corporation in the last fiscal year may suggest the likelihood that future earnings may be even greater than those of preceding years. But this does not justify the use of only a single year's earnings.' Again the five-year period was insisted upon.

If the corporation's earnings are computed over five years using $12.93 as the Year I figure, the average obtained is $10.79. If the same operation is performed using $11.90, the five year average is $10.58. The parties and the lower court are wrong in thinking a single year constitutes the average and while appellant may be correct that $12.93 represents the true earnings for the year ended January, 1969, she has nonetheless gained $.56 per share by not disturbing the $11.35 figure. For example, the earnings per share for the year ended January, 1968, were $17.13. However, more than half this amount represented the gain Watt & Shand made on the sale of its parking facilities to the city of Lancaster. The following table lists Watt & Shand's earnings per share for the other years necessary to our computation:

Year Ended January:	1968	1967	1966	1965
	$17.13	$8.32	$10.75	$4.81

The capitalization ratio used instantly was ten which we feel is within the range of reason. Many courts still look to Dewing's treatise, The Financial Policy of Corporations, as a guide to the selection of a multiplier. Therein are described seven common types of businesses and suggested capitalization rates ranging from one to ten times earnings. Dewing assigns a value of ten times the net earnings to 'old established businesses with large capital assets and excellent goodwill' and adds 'very few industrial enterprises would come within this category. A value of eight times net earnings is assigned to '(b)usinesses, well established, but requiring considerable managerial care. To this category would belong the great number of old, successful industrial businesses—large and small.' Dewing's work was last revised in 1953 and in certain situations may be dated, although we feel the ratio used instantly was appropriate and are fortified in this belief by the record testimony concerning the price earnings ratio of Strawbridge and Clothier, a somewhat comparable department store in Philadelphia.

Appellant also contends that the determination of fair value was incorrect because the lower court placed little or no significance on the net asset or book value of this stock. As the lower court's opinion noted, the book value on January 25, 1969, was $126 per share and when adjusted for an up-to-date appraisal of the real estate became $141.91 per share.

While net asset value must be given significance, especially in situations where there is no reliable market value for the stock, see, Felder v. Anderson, Clayton & Co., 39 Del.Ch. 76, 159 A.2d 278 (1960), courts must be extremely judicious with respect to the amount of weight assigned this factor. As one authority has pointed out '(t)here is nearly complete agreement that book value does not accurately represent the fair value of the corporate assets.' Note, 79 Harv.L.Rev., supra, at 1457. This is so for the reason that the balance sheet usually lists assets at original cost, which may differ greatly from present useable value or the present cost of equivalent assets. Normally, the total assets are worth more than the sum of their parts, because they include qualities useful in the context of a particular business that they would lose if put to different uses. In Borg v. International Silver Co., 11 F.2d 147, 152 (2 Cir. 1926), the court cautioned, 'The suggestion that the book value of the shares is any measure of their actual value is clearly fallacious. It presupposes, first, that book values can be realized on liquidation, which is practically never the case; and second, that liquidation values are a measure of present values. Everyone knows that the value of shares in a commercial or manufacturing company depends chiefly on what they will earn, on which balance sheets throw little light.'

In its opinion the lower court expressed reservation about factoring in discount in this particular case, a reservation which we share because of the opinion's conclusion that 'book value is not inflated but very conservatively stated.' We feel a more appropriate weight can be given to net assets by not employing a discount in this situation. Hence, the fair market value of the stock is $113.50.

We also conclude the lower court erred in declining to award appellant interest and costs. At one time Pennsylvania law made no provision for the award of interest and our courts uniformly refused to allow it. See, e.g., Austin v. City Stores Co. (No. 2), 89 Pa.D. & C. 75 (1954). In 1959 certain amendments were added to the Corporation Code which rectified this situation.

Subsection G of s 515 provides: 'The judgment shall make due allowance for any distribution to the shareholders between the day before the date of the vote on the plan and the date of their demand for the fair value of their shares and for such interest as the court may find to be fair and equitable in all the circumstances.' Subsection H of s 515 provides as follows: 'The costs and expenses of any such proceeding shall be determined by the court and shall be assessed against the corporation, but all or any part of such costs and expenses may be apportioned and assessed as the court may deem equitable against any or all of the dissenting shareholders who are parties to the proceeding to whom the corporation shall have made an offer to pay for the shares if the court shall find that the action of such shareholders in failing to accept such offer was arbitrary or vexatious or not in good faith. Such expenses shall include reasonable compensation for and reasonable expenses of the appraisers but shall exclude the fees and expenses of counsel for and experts employed by any party, but if the fair value of the shares as determined materially exceeds the amount which the corporation offered to pay therefor, or if no offer was made, the court in its discretion may award to any shareholder who is a party to the proceeding such sum as the court may determine to be reasonable compensation to any expert or experts employed by the shareholder in the proceeding.'

We believe the better view, as well as the intent of the statute, is to allow interest as a matter of course in all cases absent the existence of some inequitable situation. 'Since the corporation has had the use of the dissenter's money without his consent from the time he demanded appraisal, it seems that interest is justified.' Note, 79 Harv.L.Rev., supra, at 1472.

Despite the fact that the lower court declined to make these awards, there was no explicit finding that Mrs. O'Connor's refusal to accept Watt & Shand's offer was arbitrary, vexatious or not in good faith, nor is there a shred of evidence in the record to support such a finding. The difference between the corporation's offer and the trial judge's undiscounted appraisal is $16,292.52. This figure materially exceeds the amount the corporation offered to pay and, therefore, appellant should also receive reasonable expenses for the employment of experts. Appellant further requests reimbursement for attorney's fees but the statute does not permit it.

The wording of 15 P.S. s 1515 H indicates the lower court upon a hearing will determine the proper rate of interest, as well as the costs and expenses of the proceeding and the compensation for experts employed. We do not wish to indicate any view as to how interests should be computed; undoubtedly, there are several techniques, any one of which could be employed. The problem was considered at length in the case of Felder v. Anderson, Clayton & Co., supra, which may be of some assistance. On the question of costs, guidance can be obtained from Lowry v. General Waterworks Corp., 26 Pa.D. & C.2d 154, appeal dismissed, 406 Pa. 152, 177 A.2d 82 (1962), where the court allowed record costs, fees of expert witnesses and costs of service, but disallowed the cost of copies of the notes of testimony, of multilithing briefs or of photostats, saying that these latter expenses were incurred for the convenience of the plaintiff and were not contemplated by the Corporation Code.

We also recommend the Form of the Lowry opinion be followed by the several courts of common pleas in future stock appraisal proceedings. By this we mean that the opinion set forth the appraiser's tables stating the percentage weight assigned to each value factor. One of the more obvious benefits is to simplify the task of review.

Judgment vacated, and the record is remanded for any further proceeding necessary for the entry of a judgment conforming to this opinion.

ROBERTS and NIX, JJ., concur in the result.

Notes and Problems

1. Of all of the financial indicators in the preceding sections, which would be most subject to manipulation or outright fraud?

2. How might the valuation of intangible assets interrelate with one another? For example, might goodwill be dependent upon the strength of a trademark?

3. According to *In Re Watt*, when determining the fair value of dissenter's shares, consideration should be given to *all* factors and elements which reasonably might enter into the fixing of value, including: Asset value, market value, market prices of comparable companies, market price and earnings ratio, management and policies, earnings, dividends, valuation of assets, reserves for various contingencies, tax liabilities, future earnings, and predictions of future business events. Is this multi-factor approach judicially economical? How do the "three principal methods of valuation" suggested by *In Re Watt* compare to the valuation methods discussed immediately prior to that case?

4. Is it true, or even possible, that the whole can mathematically exceed the sum of its parts? What is the logical point that this mathematical inequality seeks to highlight?

5. A primary function of the Securities and Exchange Commission is to ensure a fair, level playing field with respect to information affecting the value of companies traded on public stock exchanges. Theoretically, the SEC purports to be able to regulate a seemingly infinite amount of information about companies and neatly pigeonhole the data into two slots: *insider information* and *public knowledge*. This fiction permits the Justice Department to spend all of its time chasing the benign Martha Stewarts while ignoring the toxic Bernie Madoffs. And then there are those cynics among us who believe that a common investor in the stock market is really akin to a stranger who sits down at a draw poker game, only to be handed cards by a dealer who already knows what they are. No one can deny that buying and selling stocks is truly a zero sum game.

6. The so-called "Efficient Capital Market Hypothesis" presumes that the price of a stock is accurately reflected by the dissemination of information into the market place. Whatever happened in the past is, according to this theory, already priced into the stock. The theory, widely recognized, gained some lip service in *Basic Inc. v. Levinson*, 485 U.S. 224, 108 S.Ct. 978 (1988), a stock fraud case involving Rule 10(b)—5. The class action lawsuit considered the impact of a company's undisclosed and disclosed information in terms of "fraud on the market."

7. Regardless of the amount of regulation by the SEC or the honesty of the individuals involved, won't the corporate insiders (those who live with a company on a daily basis) always have information that is superior to the street investor? The only thing that you can truly rely upon: If the CEO of a large company suddenly sells all of his shares and heads for Bolivia to escape indictment, perhaps it's time for you to sell.

8. For those who are lazy or too cheap to retain accountants or other schooled professionals, there are some very crude rules of thumb that have been used to value businesses over the years. They are neither accurate, nor scientific:

> **Double Annual Gross Revenues:** This method of valuation, often employed for small businesses, requires the prospective purchaser do nothing more than merely look at the gross revenues for the preceding year and double it (or triple it, depending upon the naivete of the prospective buyer). The method, of course, presumes that the business will, in the future, continue to earn as it has in the past.

Five Times Earnings: With a greater pretense of accuracy than the preceding method, this methodology at least takes into account the expense end of the equation in addition to gross revenues.

Asset Evaluation: Assets can be evaluated from several different perspectives. The straight *book value* might reflect the original cost of the assets, adjusted by depreciation and amortization. A *fair market value* might provide a more accurate prognostication. And a *liquidation value* approach provides yet another alternative. This particular methodology, regardless of the selected alternative seems to discount the company's goodwill and its value as a going concern. It certainly ignores any suggestion that a company might be worth more than sum of its parts.

Historical Evaluation: This approach recognizes the fact that solid profitable businesses can sustain losses. Examining the company's track record, say some, can provide the best indicator of future value.

None of the preceding methods should be employed as an exclusive measure of the actual value of a company. Rather, they might better be used a discussion-opener and potential basis for future discussions and negotiations.

9. The "*Z-Score Formula,*" a popular, widely accepted predictor-of-bankruptcy, was developed by NYU Professor of Finance Edward I. Altman in 1968. Theoretically, the formula is used to predict the likelihood that a firm will become bankrupt within two years. Z-scores have been employed frequently to predict corporate defaults, and at the same time measure financial distress. The formula has been modified for various factors; the original equation states that:

$$\mathbf{Z = 0.012\ T1 + 0.014\ T2 + 0.033\ T3 + 0.006\ T4 + 0.009\ T5}$$

T_1 = Working Capital / Total Assets.

T_2 = Retained Earnings / Total Assets.

T_3 = Earnings Before Interest and Taxes / Total Assets.

T_4 = Market Value of Equity / Book Value of Total Liabilities.

T_5 = Sales/ Total Assets.

Based upon the above equation and the listed factors, a company's likelihood of bankruptcy, according to Altman, would be reflected by the following chart:

Z-SCORE	STATUS
Z < 1.81	Bankruptcy Likely
1.81 < Z < 2.99	Future Uncertain
Z > 2.99	Safe Zone

10. Be particularly wary of "new" devices that are not thoroughly understood or explained. Often times accounting and tax devices that are permissible for specific, well defined functions and situations are misapplied. For example, the term "*Mark-to-Market*"

(sometimes termed "Fair Value Accounting") permits businesses to depart from classic, historical cost accounting methodologies in favor of a looser method of evaluation. For commodities whose prices are not objectively determined, inflated valuations could be used to enhance the balance sheet. For Enron, this was an excellent way to commit accounting fraud, especially since the market price of electricity could not be objectively determined. There is no real day-to-day market available or the asset value to be derived from some traded commodities such as crude oil futures. Following the Enron scandal, changes were, nevertheless, made to mark-to-market accounting through the *Sarbanes–Oxley* Act in 2002, which required companies to implement stricter accounting standards.

Another red flag term is "Special Purpose Entity," which Enron also misused to its advantage. Such entities, often corporations, Limited Partnerships and LLC's are constructed for a specific purpose as opposed to perpetual entrepreneurship. That doesn't necessarily mean that there is fraud afoot; on the contrary, there are many legitimate reasons for creating such special limited liability entities. All of the major motion picture studios, for example, create a separate corporation for each new film produced, even the bombs. The entities provide efficient mechanisms with respect to production, accounting and taxation. But again, in the wrong hands, the special purpose entity can become a tool of fraud for a variety improper purposes, which include hiding company losses. See Rapaport and Dharan, *Enron Corporate Fiascos And Their Implications* (Foundation Press, 2004).

§ 9.50 Tax Accounting

The subject of taxation fills libraries not just books. In order to assist those buried in the complexities of the subject, regulations are promulgated by the various taxing agencies that purport to describe and simplify statutory language. Often, however, the regulations themselves cause confusion. We will briefly examine some very basic concepts, touch upon the topic's complexity, then leave.

Mistakes in this arena can send you to jail. Civil penalties can also cause grief. Historically, the concept of taxation dates back to the biblical Big Bang. In the Book of Genesis, the first ever tax statute was relatively simple: "When the crop comes in, give a fifth of it to Pharaoh (Genesis 47:24)." From then on, the history of taxation has been a downhill journey on the world stage, at least with respect to simplicity.

On a positive note, The United States came into existence because of tax dissatisfaction. It then proceeded to create a mathematical monstrosity. The Internal Revenue Code of 1954, as amended, has been called the most complex law in the history of the human race.[7] According to Albert Einstein: "The hardest thing in the world to understand is the income tax." But income tax is but one type; there are many.[8]

---

7. The Internal Revenue Code of 1954, 26 U.S.C. et seq. was replaced by the so-called Tax Reform Act of 1986. A few loopholes were covered up, others were created; little if anything, however, was simplified.

8. To name a few: Ad Valorem Tax, Capital Gains Tax, Sales Tax, Property Tax, Excise Tax, Inheritance Tax, Social Security Tax, Transfer Tax, Value Added Tax, on and on.

§ 9.51 *The Mathematics of Taxation*

Structurally, both direct taxes (such as income taxes) and indirect taxes (such as sales taxes) are predicated upon the same mathematical notions; no rocket science here.

$$\text{Tax} = \text{Rate} \times \text{Base}$$

All taxes imply the existence of a tax rate, which may be fixed or scaled. The federal income tax is a graduated or scaled tax. A state sales tax, however, normally applies a fixed rate. Our more complex taxes introduce the notion of deductions and credits. While the regulations may be ambiguous or complex, the basic mathematics are not. For an income tax, for example, we sum-up all of the applicable deductions, subtract that sum from the tax base, then subtract whatever exclusion (minimum cut-off) might apply (For the 2010 Federal Income Tax, for example, the exclusion was $5.00). Lastly, tax credits are deducted. Notice that credits are more valuable than deductions because they (unlike deductions) are worth 100% of their face value.

$$\text{Tax} = \text{Rate} \times (\text{Base} - \Sigma d_i - \text{Exclusion}) - \Sigma c_j$$

The above equation is a generic structure that does not take into account the complex alternative calculations that are common to many taxes (most notably, the 1040 Standard Deduction and the infamous Alternative Minimum Tax).

Notwithstanding the general incomprehensibility of the Internal Revenue Code ("The Code"), its consistency and methodology regarding the descriptions of its required mathematical operations are relatively stable though not entirely unambiguous:

A. Addition

The word "plus" and the term "The sum of..." both denote the mathematical operation of addition. There is no mystery here. Do not, however, assume that the word "and" refers to addition. On the contrary, "and" is a boolean, as opposed to a mathematical operator (see supra, §3.12).

B. Subtraction

Your Form 1040 will tell you what and when to "subtract." The word "difference" is, likewise, an indication that subtraction is warranted. These common terms can cause ambiguity where negative numbers are added and subtracted from one another. For that reason, the Code prefers the term "the excess of... over...." Consider the most common example, as embodied in Section 1001(a) of the Code:

> The gain from the sale or other disposition of property shall be **the excess** of the amount realized therefrom **over** the adjusted basis provided in section 1011 for determining gain, and the loss shall be the excess of the adjusted basis provided in such section for determining loss over the amount realized.

It is understood that if the required subtraction happens to provide a negative number, the use of the term "excess... over" will yield a zero. Often, the phrase "if any" is inserted to modify the word "excess" to make that point clear.

The Code may also employ the language: "to the extent that... exceeds..." or "to the extent of..." In either event, subtraction is indicated. Consider Code Section 1211(a):

> In the case of a corporation, losses from sales or exchanges of capital assets shall be allowed only to the extent of gains from such sales or exchanges.

C. Multiplication

Multiplication is relatively straightforward since the Code will use the term "multiplied" or "multiplied by." The Code may also direct multiplication by requiring that a "percentage" of a given number be taken.

D. Division

Division, like multiplication is relatively unambiguous. The term "divided by" is universally understood. Always remember, however, that the single most illegal operation in the entire realm of mathematics is division by zero (i.e., a zero may never appear in the denominator).

E. Max and Min

"Max," which is a mathematical symbol meaning "maximum," and "min," a symbol meaning "minimum," are two simple functions that are somewhat self- explanatory. For example:

$$\text{Max}\ (2, 6, 4, 0) = 6$$

$$\text{Min}\ (1, 5) = 1$$

These two functions, which appear frequently in Code are signaled by the terms: "The lesser of" or "the greater of." The words "higher" and "lower" may also be used to denote the two functions. Consider Section 1211(b) of the Code:

> In the case of a taxpayer other than a corporation, losses from sales or exchanges of capital assets shall be allowed only to the extent of the gains from such sales or exchanges, plus (if such losses exceed such gains) **the lower of**—(1) $3,000 ($1,500 in the case of a married individual filing a separate return), or (2) the excess of such losses over such gains.

F. Ratios

Up to this point in the discussion, the Code's mathematical functions are fairly straightforward, though not crystal clear. The Code's treatment of ratios, by comparison, leaves much to be desired. The phrase signaling the infamous Code Ratio is: "An amount which bears the same ratio to... as... bears to..." A mathematician, of course, would have provided a simple algebraic equation in place this inherently ambiguous verbiage involving four variables. To make life our lives easier let's, therefore, insert letters into the blank spaces:

> An amount (x) which bears the same ratio to p (annual payment) as y (cost of annuity) bears to z (total return).

Assume that the "amount" we are seeking = x. The fixed ratio governing that amount is y / z. Therefore, the amount can simply be represented by the equation:

$$x / p = y / z \text{ which implies that } x = py / z$$

Consider Section 72(b) of the Code:

> Gross income does not include that part of any **amount** received as an annuity under an annuity, endowment, or life insurance contract **which bears the same ratio to** such amount **as** the investment in the contract (as of the annuity starting date) **bears to** the expected return under the contract (as of such date).

Example:

> Assume that you paid $1,000 for an annuity contract that promised to pay you $150 per year for the rest of your life. If we further assume that your life expectancy is 10 years, you would expect to receive a total of $1,500 for your $1,000 investment. It makes sense that 2/3 of the money received is nothing more than

a return of your own capital (which was already taxed), and should, therefore, be excluded from your taxable income. (y / z = 1,000/1500 = 2/3.) Therefore, for each payment of $150, the amount of $100 should be excluded.

$$x = (150)\ (1000)\ /\ 1500 = (150)\ (2/3) = 100$$

§ 9.52 The Complexity of Taxation

The Internal Revenue Code of 1954, as amended, is not a product of mathematics, but rather one of politics and special interests. It is a good thing that the Code is incomprehensible. Were that not the case, the average low-to-middle income wage earner would be able to read it. If the average citizen could understand the Code, the resulting uprising might rival 1776. We all know that the Code is long and complex, but it may surprise you to learn that its creators can't even count the number of pages in the book:[9]

> "The income tax code and its associated regulations contain almost 5.6 million words—seven times as many words as the Bible. Taxpayers now spend about 5.4 billion hours a year trying to comply with 2,500 pages of tax laws."
>
> —U.S. Rep. Rob Portman (OH)
>
> "At 3,458 pages, twice the length of the Bible, it's impossible for the average taxpayer to know, understand, and accurately apply its provisions. The length is twice that of the Bible! Even tax experts cannot do so reliably."
>
> —U.S. Rep. J.C. Watts, Jr. (OK)
>
> "The Internal Revenue Code and regulations now come in at one million words and 9,000 pages."
>
> —U.S. Rep. Bill Archer (TX)
>
> "The tax code runs 17,000 pages and contains a mind-boggling 5.5 million words. By way of comparison, War and Peace is only 1,444 pages and the Bible checks in at 1,291 pages."
>
> —U.S. Rep. Vito Fossella (NY)
>
> "The federal tax code with its 44,000 pages, 5.5 million words, and 721 different forms is a patchwork maze of complexity and a testament to confusion over common sense."
>
> —U.S. Rep. Jim DeMint (SC)
>
> "The current tax code is almost 60,000 pages, longer than the Bible"
>
> —U.S. Rep. Bobby Jindal (LA)
>
> "The tax code is a complicated mess. You realize, it's a million pages long."
>
> —President George W. Bush

Notes and Problems

1. For a brief taste of IRC (Internal Revenue Code) confusion, refer back to § 1.31.

9. See official press releases on www.house.gov.

2. A basic premise of corporate accounting teaches that there are three methods of transferring money from a corporation to an individual:

> **First:** A dividend, from a taxation perspective is the worst way to move money from the corporation to an individual. It is both taxable to the recipient and non-deductible to the corporation. This is a common invitation to double taxation.
>
> **Second:** A salary or bonus is taxable to the recipient but deductible to the corporation; however, special "payroll" taxes apply to this form of distribution.
>
> **Third:** Expense reimbursement and fringe benefits are non-taxable to the individual but nevertheless deductible to the corporation. From a tax perspective, therefore, these are the best ways to transfer money from a corporation to an individual.

The quirks, blunders, errors and inequities of our tax systems could fill entire libraries let alone volumes. The discussion of the case that follows, is but one example among countless that illustrate the manipulability of the nation's tax laws.

The Crane Doctrine

A notorious blunder by the United States Supreme Court once resulted in a great gift to tax lawyers, tax dodgers and inventive accountants; it consisted of a single infamous footnote in the case of *Crane v. Commissioner*, 331 U.S. 1 (1947). Appreciating *Crane* requires an understanding of some elementary arithmetical tax rules concerning the purchase and sale of capital assets.

Capital Gain (Loss) = Amount Realized - Adjusted Basis

First, to determine your taxable gain from the sale of property, you subtract your *basis* from the amount that you realize upon sale. In other words, subtract what you paid from what you received upon sale; simple enough. Second, mortgages and debts don't count; because every loan requires a repayment, debts, loans and mortgages are non-taxable transactions. Thus, money that you borrowed to buy property is irrelevant in determining your gain. A discharge of a debt, however, is a different matter; it results in taxable income. Finally, prior to sale, your basis is reduced by the amount of any depreciation that you may have used as a deduction to reduce your prior taxes.[10]

Crane inherited an apartment building worth $300,000 that was subject to outstanding mortgages and debts in the sum of $300,000, which arguably implied that he inherited nothing.[11] For the next 7 years, Crane operated the apartment complex, collecting and reporting income in the form of rents, and deducting expenses. He also claimed annual depreciation deductions at the rate of $4,000 per year (for a total of $28,000). Facing foreclosure, in the 7th year Crane sold the property subject to the outstanding mortgages and debts. He received $3,000 which he reported as a long term capital gain. Crane reasoned that $3,000 - 0 = $3,000. The IRS Commissioner arrived at a different conclusion, claiming that the $28,000 should be deducted from Crane's adjusted basis, resulting in a gain of $28,000 + $3,000 = $31,000.

The High court ruled in favor of the Commissioner finding that it was "irrelevant" whether or not the mortgagor was a debtor or liable on the mortgage. The Court held

---

10. Gain (or loss) = Amount Realized - Adjusted Basis.
The Amount Realized = Acquisition Price + Discharged Liabilities.
The Adjusted Basis = Acquisition Basis - Depreciation.

11. Numbers are rounded for simplification.

that: "[W]e are no more concerned with whether the mortgagor is, strictly speaking, a debtor on the mortgage, than we are with whether the benefit to him is, strictly speaking, a receipt of money or property. We are rather concerned with the reality that an owner of property, mortgaged at a figure less than that which the property will sell, must and will treat, the conditions of the mortgage exactly if they were his personal obligations." This holding, known as the Crane Doctrine was designed to permit the government to recapture the tax deductions enjoyed by the taxpayer prior to sale.

Ignoring the laws of simple arithmetic, the Supreme Court found it necessary to footnote the above statement as follows:

> "Obviously, if the value of the property is less than the amount of the mortgage, a mortgager who is not personally liable cannot realize a benefit equal to the mortgage. Consequently, a different problem might be encountered where a mortgagor abandoned the property or transferred it subject to the mortgage without receiving boot. This is not the case."

The implications that flowed from the above obiter dictum served an underlying predicate for one of the greatest tax loopholes in history. Common sense and a knowledge of 3rd Grade Arithmetic leads to the conclusion that receipt of a single penny could trigger a depreciation recapture of thousands or even millions depending upon the circumstances.

Notes and Problems

1. Benny Billion wants a tax write-off. A no-talent screenwriter named David Linched wants to sell Benny Billion all of the rights to his latest screenplay titled: *Godzilla v. Billy The Kid* for $1,000,000. The terms of the sale are as follows: Benny is to pay David $50,000 down and execute a non-recourse promissory note for the balance. Prior to plugging the Crane loophole, why would this have been a beneficial deal wherein Benny would profit?

2. In the above set of facts, explain how both parties would benefit? After the screenplay is rejected by every sane studio in the free world, should the parties merely abandon the asset or should David repossess it for non-payment of the note?

3. Forensic Accountants: While all accountants are "experts" of sorts, if you intend to have one testify as such, you might want to ensure that his or her qualifications are sufficient enough to overcome the hurdles of Rule 702 of the Federal Rules of Evidence and the challenges of a *Daubert* Hearing (See § 1.22). In particular, you may wish to hire someone from a modern-day litter of forensic accountants, often thought of as the pit bulls of accountancy (an otherwise mild-mannered occupation). Unlike earlier forensic accountants, whose primary concern was unearthing *his* assets in bitter divorce cases, today's experts handle a never-ending list of complex financial, economic and operational matters. Their duties often include the dissection of tricky Enron-like business dealings, not to mention antitrust, fraud, regulatory, tax, and political matters. In addition, they must be able to provide the type of simplification and testimony that jurors can understand. Forensic Accountants are becoming more useful and necessary as complex regulatory measures (such as The *Sarbanes-Oxley Act of 2002* and The *Dodd-Frank Wall Street Reform and Consumer Protection Act of 2009*) continue to dot the economic landscape.

4. Calculate the "*official*" number of pages in the Internal Revenue Code by taking the mean of all of the numbers stated by the babbling politicians (none of whom apparently had the sense to merely grab a code and glance at the last page number).

§ 9.60 Damages

Damages take many shapes and forms. Perhaps the only thing that binds all of the non-equitable remedies together is the requirement that they be reduced to some dollar amount. There are many ways to classify the subject matter of damages. Perhaps the broadest scheme would carve the topic into four types: Compensatory, Punitive, Declaratory and Equitable. We will focus on the first two measures which are quantitative in nature.

§ *9.61 Employment Related Damages*

In personal injury actions future lost wages often form a significant part of the damages that the plaintiff is entitled to recover. Measuring the value of those wages is somewhat akin to measuring the future value of an annuity. In particular, periodic future payments are similar to the periodic payment of wages. The so-called "structured settlement" of a personal injury claim might provide the plaintiff with an annuity that would compensate him or her for the lost wages. More often than not, however, a single lump sum results in a final disposition of the case. The defendant, during settlement negotiations will want to raise the following ubiquitous points and arguments in an effort to reduce the lump sum:

a. Take home pay would be significantly less than gross wages due to deductions for payroll taxes.

b. A fixed settlement eliminates the uncertainty of discharge or layoff.

c. A fixed settlement eliminates the uncertainty of an economic downturn that might result in the bankruptcy or insolvency of the employer.

d. A fixed settlement eliminates the uncertainty that the plaintiff might be injured on the job and thereby unable to work or otherwise earn a liveable wage.

By the same token, the plaintiff would argue that there are countervailing positive factors that should not be ruled out and must, therefore, also be taken into account when arriving at a fair settlement:

a. A fixed settlement rules out the possibility that the plaintiff might have been promoted or switched to a higher paying job.

b. A fixed settlement does not account for periodic raises and bonuses.

c. A fixed settlement does not account for fringe benefits.

d. A fixed settlement does not account for inflation, subsequent cost of living increases and other positive adjustments.

All of the above factors should be raised by the respective parties during the course of settlement negotiations. Some are off-setting. The most likely offset occurs with respect to the totally unpredictable factors of inflation and raises in pay. These two "unpredictables" are often offset by pointing out to the plaintiff that he or she is getting a present lump sum as opposed to a periodic payout of wages. Money *now* is, of course, worth more than the same amount of money at some future date.

In the event that there is no settlement, the personal action proceeds to trial, in which case, the so-called "Single Judgment Rule" will apply. In such an instance, the closing argument of counsel for the plaintiff should include the following pitch:

> Ladies and gentlemen of the jury, as you deliberate and decide how much compensation my client should receive, keep in mind that your verdict should compensate him for all related future damages, not just those that he has already sustained and suffered. By law, we are only entitled to a single judgment, regardless of how much pain he may suffer in the future. Once you have rendered your verdict, we cannot come seeking any more. Your verdict should be increased to reflect this. Assuming that he will never work again, your verdict should also include lost wages for the remainder of his life.

By the same token, counsel for the defendant might be expected to advance the following arguments:

> Ladies and gentlemen of the jury, the Single Judgment Rule mentioned by the plaintiff also includes the proposition that, like the plaintiff, the defendant cannot, after you render your verdict, seek any adjustments in the event that the plaintiff experiences a moderate or miraculous recovery.
>
> In addition, keep in mind that the lost wages that he seeks would never have been delivered presently in a lump sum. Wages, as we all know, are paid after they are earned, by working month after month, year after year. Any verdict that you render now will, therefore, be worth more than Plaintiff would have otherwise received.

During summation, both sides should be armed with their charts and equations in order to provide visual support to their respective arguments. One of the most important equations involves the present value of future payments.

The lump sum (P) representing the present of value of future payments can be calculated using the following formula where i is the interest rate, n is the number of periods and A is amount of each periodic payment:

$$P(1 + i)^{n} = \frac{A\ (1 + i)^{n} - 1}{i}$$

Example 9-12: Benny became permanently disabled in an automobile mishap. At the time of the accident he was earning $2,000 per month. Benny's life expectancy is 25 years. Assume that had it not been for the accident, Benny would have continued earning $2,000 per month for remainder of his life. Ignore the factors discussed above (taxes, inflation, etc.) and calculate the present value of the amount that Benny would have received for the remainder of life, assuming an interest rate of 10.5%.

Solution:

Our goal is to determine the value of P.

$i = 1/12 \times 10.5\% = 0.00875$

$n = 25 \text{ years} \times 12 \text{ months} = 300 \text{ months}$

$A = \$2,000$

$$P(1 + i)^{n} = \frac{2,000\ (1 + 0.00875)^{300} - 1}{0.00875}$$

$$P(1 + 0.00875)^{300} = \frac{2,000\ (1 + 0.00875)^{300} - 1}{0.00875}$$

Calculate the value of $(1 + 0.00875)^{300}$ using a computer or calculator.

Then perform the required arithmetic to obtain the result:

P = $211,823.60.

[Note: Total amount of earnings over the 25 year period would have amounted to 25 × 12 × $2,000 = $600,000.]

It is notable that the same basic equation for determining the present value of future payments may also be used to calculate the amortization of a simple interest loan.

A much simpler method involves using the sum total of all of the future lost earnings (T) and applying a discount rate, based upon a fixed interest rate (I). The lump sum (P) would then be calculated as follows:

$$P = \frac{T}{(1 + i)^n}$$

The true value of the preceding equations can be found in the framework that they provide for intelligent attempts to arrive at settlements. There is no way to accurately predict what a injured person's future employment would have yielded. It is, however, convenient that two of the most important factors (the salary growth potential and the interest factor) happen to be offsetting. For settlement purposes, it is very convenient to engage in the fiction that they are perfectly offsetting, and therefore, require no calculation. This is sometimes referred to as the "Alaska Rule," due to the official recognition that this fiction garnered from *Beaulieu v. Elliot*, 434 P.2d 665 (Alaska, 1967). More recently and authoritatively, in *Jones s & Laughlin Steel Corp. V. Pfeifer*, 462 U.S. 523 (1983), a federal suit under the Longshoreman's And Harbor Worker's Compensation Act, the Supreme Court weighed in by holding that the use of a setoff between prospective wage increases and discount rates (based upon inflation) may be simple and expedient, but it is not mandatory.

§ 9.62 *Punitive Damages*

Sometimes referred to as exemplary damages, punitive damages are designed to punish the defendant as opposed to compensating the plaintiff. The purpose is to deter the action or inaction that led to the harm as opposed to making an injured party whole.

In a products liability setting, when weighing the "calculus of negligence" (See § 2.70) corporate boards may be deterred from spending money to recall or modify dangerous products when it is cheaper to defend the resulting lawsuits by merely letting the consumers die. A similar matter arose in *Grimshaw v. Ford Motor Company*, 119 Cal.App.3d 757 (1981), wherein the appellate court resolved the issue as to whether or not a company's decision predicated up on a liability calculus could equate to a level of "malice" required for the imposition of punitive damages. In *Grimshaw*, California's high court: "[O]bserved that the availability of punitive damages has not been limited to cases in which there is an actual intent to harm plaintiff or others … conscious disregard of the safety of others is an appropriate description of the animus malus required." In that case, an exploding Pinto caused the plaintiff to receive severe disfiguring burns from a relatively simple rear end collision. The jury awarded the Plaintiff $2,000,000, and the heirs $550,000 in compensatory damages, but then awarded $125,000,000 in punitive damages for refusing to correct a long-standing deadly gas tank defect. The award was affirmed on appeal. The amount of punitive damages was 49 times greater that the compensatory damages awarded.

$2,000,000 + $550,000 = $2,550,000 [Total Compensatory Damages]

$125,000,000 / $2,550,000 = 49.0196 : 1 [Ratio of Punitive to Compensatory]

Is there a limit? Or any guideline with respect to the amount that an angry jury can award? In *Philip Morris USA v. Williams*, 549 U.S. 346 (2007), the Supreme Court set aside $79,500,000 in punitive damages awarded to the widow of a heavy cigarette smoker, eventually reducing it to $32,000,000. In an earlier case, *State Farm Mutual Automobile Insurance Company v. Campbell*, 538 U.S. 408 (2003), the Supreme Court let it be known that there were limits; a jury's punitive damage award of $145 million for an automobile liability insurer's bad-faith failure to settle for the policy limits (where the compensatory damages were $1 million) was held to be neither reasonable nor proportionate to the wrong committed and, therefore, violated the due process clause. The High Court mentioned factors such as the wealth of the defendant, the severity of the defendant's conduct and the need for deterrence. But it provided no bright line or ceiling. The Supreme Court did, however, indicate its dissatisfaction with ratios exceeding 9-1:

> "[T]he Court has been reluctant to identify concrete constitutional limits on the ratio between harm, or potential harm, to the plaintiff and the punitive damages award; but, in practice, few awards exceeding a single-digit ratio between punitive and compensatory damages will satisfy due process. Single-digit multipliers are more likely to comport with due process, while still achieving the State's deterrence and retribution goals, than are awards with 145-to-1 ratios, as in this case. Because there are no rigid benchmarks, ratios greater than those that this Court has previously upheld may comport with due process where a particularly egregious act has resulted in only a small amount of economic damages, but when compensatory damages are substantial, then an even lesser ratio can reach the outermost limit of the due process guarantee."

§ 9.63 *Liquidated Damages*

Liquidated damages, provided in private contracts and agreements are not without their own limiting ratios. As is the case with punitive damages, they often exceed the amount of actual compensatory damages suffered by the plaintiff and, therefore, require scrutiny. The reasonableness of a liquidated damages clause depends upon whether or not it is so excessive as to be deemed a "penalty" in which case, it may be deemed unenforceable.

A ratio worth considering would compare the actual injury or damages to the amount specified as liquidated damages. More leeway and, therefore, a larger amount would be deemed acceptable where damages are difficult to determine or prove. See Uniform Commercial Code § 2-718(1) and Restatement of Contracts § 356(1). Nor is it necessarily required that the calculation be simple or non-creative. See *GFI Brokers, LLC v. Santana*, 2009 WL 2482130 (S.D.N.Y., 2009).

§ 9.64 *Pain And Suffering*

Theoretically, plain and suffering is subjective and unquantifiable. Experienced personal injury lawyers know otherwise due to what is commonly referred to as a "per diem argument." A small trick with small numbers can sometimes prompt large verdicts. Consider this closing argument by plaintiff:

> Look at Ralph over there, who must spend most of waking hours in pain (depending upon how drugged-up he is). Ladies and gentlemen of the jury, it would seem as if his pain should be worth at least $10 per hour. That is all that we are asking for.

Ten dollars is a modest amount. But if you multiply it by a 16-hour day, times a 365-day year, and if mortality tables provide Ralph with a modest 20-year life expectancy, the jury might actually be awarding a not-so-modest amount for future pain and suffering, which is often speculative at best:

$10 x 16 Hours x 365 Day x 20 Years = $1,168,000.00

§ 9.65 *Hedonic Damages*

Hedonic damages are strange animals. As is the case with pain and suffering, they purport to measure the unmeasurable. Predicated upon the notion that we all have a basic constitutional right to fun, hedonic damages are designed to provide money when we are deprived of that fun. An injury that prevents a plaintiff from playing tennis, for example might arguably warrant damages that are calculated in a manner similar to those awarded for pain and suffering.

Defense attorneys should, perhaps, consider an even newer untested concept—Anti-Hedonic Damages. Incapacitation could arguably be a valuable offsetting asset for avoiding family get-togethers, dinners with in-laws and weddings.

Notes and Problems

1. Assume that Plaintiff, who earned $48,000 per year, was injured on February 1, 2012 and was totally incapacitated. His case came to trial on February 1, 2013. Defendant was totally at fault. Plaintiff has a life expectancy of 10 years. What can you deduce about his entitlement to: (a) Back wages, (b) Future Lost Income, (c) Prejudgment Interest, and (d) Post Judgment Interest, if any?

2. How would you quantify hedonic damages?

3. Lastly: Have you ever personally been misled or suffered any losses from any of the schemes or misrepresentations discussed in this treatise?

About the Author

Ashley S. Lipson received his B.A. in Telecommunications from Michigan State University, and his J.D. from St. Johns University Law School. He also obtained a Masters (M.A.) in Mathematics from Wayne State University, an LL.M in Taxation from Wayne State University Law School, an As.D. in Computer Programming and a Second As.D. in Computer Information Systems. He also studied Physics at the University of Michigan.

He is currently a full Professor of Law at the University of La Verne College of Law, a computer programmer and author of several published books along with more than 70 articles.

The former columnist for *Law Office Computing Magazine* was the developer and creator of the *Lawyers Comprehensive Document System* (See Lipson, *Law Office Automation*, Prentice Hall, Inc., 1986), but is better known as the creator and programmer of the *Objection!* computer game series, the first computer games to be accredited by state bar associations for purposes of *Mandatory Continuing Legal* education.

As a law professor, Lipson constructed and taught *"The Law of The Computer Game Industry"* in 2002, the first law school course devoted entirely to statutes, laws and cases pertaining to video/computer games. He also developed and taught the first law school course on the topics of Mathematics, Physics and Finance (the subject matter of this text) designed entirely for future lawyers.

His experience as an adjunct professor of law includes courses in Entertainment Law, Multimedia Law and The Law of Cyberspace at The Detroit College of Law, Michigan State University and Pepperdine along with practice related courses at UCLA.

Lipson is still is a member of several bar associations, including the California Bar Association, Michigan Bar Association (Emeritus), New York Bar Association (Certified), and the Beverly Hills Bar Association (Order of Distinguished Attorneys).

Index